CIRCUIT ANALYSIS ESSENTIALS:
A SIGNAL PROCESSING APPROACH

James S. Bryant
DeVry University
Columbus, OH

THOMSON

DELMAR LEARNING

Australia Canada Mexico Singapore Spain United Kingdom United States

Circuit Analysis Essentials: A Signal Processing Approach

James S. Bryant

Vice President, Technology and Trades ABU:

David Garza

Director of Learning Solutions:

Sandy Clark

Senior Acquisitions Editor:

Stephen Helba

Senior Product Manager:

Michelle Ruelos Cannistraci

Senior Channel Manager:

Dennis Williams

Marketing Coordinator:

Stacey Wiktorek

Production Director:

Mary Ellen Black

Senior Production Manager:

Larry Main

Senior Project Editor:

Christopher Chien

Art/Design Coordinator:

Francis Hogan

Technology Project Specialist:

Linda Verde

Senior Editorial Assistant:

Dawn Daugherty

Library of Congress Cataloging-in-Publication Data:
Bryant, James S.
  Circuit analysis essentials:
a signal processing approach /
James S. Bryant.
     p. cm.
  ISBN 1-4018-5041-3
1. Electric circuit analysis.
2. Electronic circuit design—
   Data processing.
3. Signal processing.
I. Title.
  TK454.B766 2005
  621.319'2—dc22

2005025290

NOTICE TO THE READER

DEDICATION

I dedicate this text to the many students and colleagues who over the years have helped me to discover the joy of teaching. It is my hope that I have in return helped my students to discover the joy of learning. It is not possible to have one without the other.

TABLE OF CONTENTS

PREFACE

PURPOSE OF THE TEXT

Circuit Analysis Essentials: A Signal Processing Approach is designed to support a twelve- to fifteen-week course in the first year of an electrical engineering or electronics engineering technology curriculum. The text should be considered as an introduction to analog circuit analysis and design with an emphasis on signal processing applications.

STUDENT PREPARATION

This textbook is intended for students who have either completed or are concurrently enrolled in a college-level algebra course. It occasionally introduces references to calculus, such as when expressing the voltage across an inductor as $v_L = L \, di/dt$. Much of the mathematical computation can be accomplished through the use of an engineering calculator. The calculator instructions presented in the text are specific to the Texas instruments TI-89; however, comparable calculators from other manufacturers may also be used. Although MATLAB is not required, it is used to create several of the figures. MATLAB commands for selected figures are contained in an appendix for instructor or student use. The text also uses MultiSIM to simulate many of the circuits and signals, and computer simulation is used in several of the examples and student exercises.

RATIONALE FOR THE TEXT

Introductory circuit analysis has long been a standard required first course in electrical and electronic engineering technology curricula. While the content and emphasis of introductory circuit analysis texts has not changed dramatically over the years, the nature of electronic circuits and systems has been reborn many times over. The one constant throughout the evolution of electronic circuits and systems has been that electrical and electronic engineering technology remain in large part the science of creating, processing, and utilizing electronic signals.

Historically, electrical engineering technology texts were developed from the study of electrical power distribution circuits, machinery, and systems. Most circuit analysis texts therefore favored power distribution topics rather than electronic signal processing topics. Over time, the study of electronic circuits and systems began to emerge as a science closely related to, yet distinct from, electrical power engineering. Recently, many excellent introductory circuit analysis texts include circuit elements such as diodes, transistors, and operational amplifiers. This broadens the scope and usefulness of the circuits presented and gives the texts more of an electronics orientation and less of a power distribution influence.

The text emphasizes the relationship between an input signal and an output signal. Concepts such as gain, decibel gain, and the transfer function are presented to establish a clear signal processing orientation. My objective in writing this text is to preserve the important fundamentals of traditional analog circuit analysis and design, while at the same time orienting the student to the signal processing aspects of electronic circuits.

ORGANIZATION OF THE TEXT

Chapter 1 An Introduction to Electronic Signals: Chapter 1 introduces the physical nature of electronic signals. It also introduces the concepts of current and voltage, emphasizing that electronic signals exist either as a current or voltage waveform. The chapter continues with a discussion of the instruments used to measure electronic signals and concludes with an introduction to notations, symbols, and prefixes commonly found in the study of electronic circuits.

Chapter 2 Summary Essentials of DC Circuit Analysis: Chapter 2 presents the essential concepts of the DC series resistive and the DC parallel resistive circuit. The student learns to determine total equivalent resistance for resistors connected in either series or parallel. Important circuit laws such as Ohm's law, Kirchhoff's laws, the current divider rule, and the voltage divider rule are introduced and illustrated through several examples.

Chapter 3 Summary Analysis of Combination DC Circuits: Chapter 3 presents the laws and techniques that are commonly used to analyze more complex series-parallel combination DC electronic circuits. Although only DC circuits are analyzed in this chapter, the principles and analysis techniques learned are also utilized in the analysis of circuits whose currents and voltages vary over time. The analysis of more complex series-parallel circuits is accomplished through the careful application of Ohm's law and Kirchhoff's laws.

Chapter 4 Summary The Sine Wave and Linear Response: Chapter 4 is partly devoted to the study of the sine wave. The sinusoidal waveform is the most important electronic signal to be considered in this text; it is represented mathematically as a function of time in the time domain and as a phasor in the phasor domain. The chapter also introduces the linear components—such as resistors, inductors, and capacitors—that are utilized by engineers and engineering technologists to design the electronic circuits that create, process, and transmit electronic signals; it considers the physical nature of each component, derives its mathematical model, and determines how each of the components responds to sinusoidal excitation.

Chapter 5 Essentials of AC Circuit Analysis: Chapter 5 demonstrates how the circuit analysis laws and techniques learned in previous chapters are used to analyze circuits that contain resistors, capacitors, and inductors whose input signal is a sine wave. Since the sine wave voltage changes polarity every half cycle and the sine wave current changes direction every half cycle, the circuits are referred to as alternating current, or AC, circuits. This chapter considers only circuits whose currents and voltages are sinusoidal and shows how to determine the magnitudes and phase relationships of the current and voltage in a variety of analog circuits by applying Ohm's law, Kirchhoff's laws, and the voltage and current divider rules.

Chapter 6 Signal Processing Circuit Analysis and Design: Chapter 6 demonstrates how the circuit analysis laws and techniques learned in previous chapters are used to analyze and design several practical analog electronic signal processing circuits, and shows how to determine the characteristics of an output signal from a circuit, given a specified input signal, by applying Ohm's law, Kirchhoff's laws, and the voltage and current divider rules.

Chapter 7 Transient Circuit Analysis: Chapters 5 and 6 are dedicated to the analysis of circuits whose current and voltage signals are sinusoidal waveforms; many other waveforms are common in electronic circuits, although the sine wave is the most important signal to be considered. Chapter 7 considers the transient and steady-state response of the R-C and R-L circuit to DC excitation. Many practical electronic circuits such as amplifiers contain both DC and AC waveforms, so that an appreciation of DC response is necessary to understand the behavior of such circuits.

Chapter 8 Network Analysis and Selected Theorems: Chapter 8 introduces methods that are typically used to analyze networks that contain large numbers of components or branches. Although the networks presented in the examples in this chapter are not overly complex, the methods demonstrated are most suitable for analysis of larger networks.

Chapter 9 Power Considerations in Electronic Circuits: Chapter 9 discusses power considerations for electronic circuits, which can be divided into three related but separate considerations. The first section of Chapter 9 discusses power in an AC circuit from the perspective of the power source requirements; the source must supply power to the circuit and so it is important to calculate how much power is required by the resistors, capacitors, and inductors that comprise the circuit. The chapter then considers power from a signal processing perspective, which is to ensure that maximum power from a signal source is delivered to a specified load. The maximum power transfer theorem defines the conditions necessary for maximum power transfer, and techniques to achieve maximum power transfer are presented. The final two sections of the chapter consider the power ratings required for the components that comprise the electronic circuit. Components must be selected so that their maximum power ratings are not exceeded. The maximum power ratings of components vary with ambient temperature and must be adjusted or derated to compensate for temperature effects.

Chapter 10 Analysis of Nonlinear Analog Circuits: The circuits presented in the previous chapters of this text consisted only of linear bilateral components, including resistors, capacitors, and inductors. Many other useful electronic circuits contain components such as the diode and transistor that are neither linear nor bilateral. Chapter 10 explores the fundamental characteristics of the p-n junction diode, junction transistor, and operational amplifier, and demonstrates how to analyze several signal processing circuits that contain these devices.

Chapter 11 An Introduction to Nonsinusoidal Signals: The previous chapters in this text are dedicated to the analysis and design of discrete analog circuits whose voltage and current waveforms are either DC or sinusoidal. Other waveforms are common in electronic circuits, although the sine wave remains the most important time-varying signal. Chapter 11 introduces circuit analysis techniques that can be used to analyze circuits whose input signals are nonsinusoidal. The material presented in this chapter is not intended to provide a rigorous mathematical analysis of the Fourier series, nor is it intended to investigate overly complex nonsinusoidal waveforms. The chapter takes advantage of computer-assisted analysis to illustrate how circuits respond to nonsinusoidal signal input waveforms where appropriate.

Student Exercises: The end-of-chapter student exercises are separated into two categories based upon degree of difficulty. The Essential Learning Exercises are the least difficult, and closely follow the solutions that are demonstrated in the examples within the body of the text. The answers to odd-numbered Essential Learning Exercises are given in Appendix A of the text. The Essential Learning Exercises are typically arranged in pairs such that an odd-numbered problem with its answer given in the appendix is immediately followed by a similar problem for which the answer is not given. The exercises for which answers are not given may be used for graded homework assignments or as examination problems.

The end-of-chapter Challenging Learning Exercises are more advanced than the Essential Learning Exercises. Several of the Challenging Learning Exercises require

the student to extend the material presented in the chapter beyond the level of the examples illustrated within the text. Answers to selected Challenging Learning Exercises are given in Appendix A of the text.

In addition to the many examples whose solutions are demonstrated within the body of the text, several Practice Exercises are strategically placed to give students the opportunity to evaluate their understanding of the material. The answers to the Practice Exercises are given immediately following the exercise, and complete solutions to the Practice Exercises are given in Appendix A of the text.

Back-of-Book CD: The text contains figures of circuit schematics created using MultiSIM (version 7) and figures created using MATLAB. All of the respective figures are available for instructor or student access on the accompanying CD. The CD permits the instructor to quickly recreate the figures for in-class instructional use. The CD also provides a valuable student learning resource because it enables students to access the figures and interact by taking measurements on the existing circuits, or modifying the circuits to observe the effects of changing component values or connections.

SUPPLEMENTS

Lab Manual: (ISBN: 1-4018-5042-1) The lab manual includes 20 experiments that cover material in the analysis and design of DC and AC circuits. Each lab requires calculations, construction/measurement, and simulation. Four challenging design projects emphasize circuit design of a practical application.

e.resource CD. (ISBN: 1-4018-5043-X) A CD for instructors is available upon adoption of the text. It includes the following teaching aids:

- Textbook Solutions
- Lab Manual Solutions
- Computerized Testbank
- PowerPoint Presentation Slides
- Lab.Builder template
- Image Library with all images from the textbook to customize handouts and tests.

ACKNOWLEDGMENTS

The Author and Thomson Delmar Learning would like to thank the following reviewers:

John Ashcraft, Blue Mountain College, Pendleton, OR

Don Barrett, DeVry University, Irving, TX

Seddik Benhamida, DeVry University, Arlington, VA

John Blankenship, DeVry University, Decatur, GA

Lance Crimm, Southern Polytechnic State University, Marietta, GA

Yolanda Guran, Oregon Institute of Technology, Portland, OR

Richard Jones, Old Dominion University, Norfolk, VA

Anthony Leovaris, DeVry University, Tinley Park, IL

Palmerino Mazzucco, Mesa Community College, Mesa, AZ

Ali Rahbar, DeVry University, Fremont, CA

Daniel Lee Saine, DeVry University, Phoenix, AZ

Jeffrey Schwartz, DeVry University, Long Island City, NY

John Sebeson, DeVry University, Addison, IL

Hesham Shaalan, Texas A & U University, Corpus Christi, TX

William Shepherd, Owens Community College, Perrysburg, OH

Dr. Ehsan Sheybani, U. of Southern Mississippi, Hattiesburg, MS

Rajeswari Sundararajan, Arizona State University, Mesa, AZ

Mike Wilson, Kansas State University, Salina, KS

Quihou Zhou, Miami University, Oxford, OH

AN INTRODUCTION TO ELECTRONIC SIGNALS

"IMAGINATION IS MUCH MORE IMPORTANT THAN INTELLIGENCE."

—Albert Einstein

LEARNING OBJECTIVES

Upon successful completion of this chapter you will be able to:

- Describe several important uses of electronic signals in our society.
- Describe the nature of current and voltage.
- Describe what is meant by an analog signal.
- Describe how physical quantities are converted to electronic signals.
- Describe the instruments used to measure electronic signals.
- Define the units and notations used in the science of electronics.
- Select the appropriate prefix symbol for a current or voltage.
- Describe software typically utilized in the analysis and design of electronic circuits.

CHAPTER ONE

INTRODUCTION

This chapter discusses the abundance and importance of electronic signals in a modern advanced society. The chapter introduces the physical nature of current and voltage and discusses how physical quantities such as temperature, pressure, or sound can be converted into electronic signals. The chapter continues with a discussion of the symbols and notation that are used to represent current and voltage, and their proper use is illustrated. The chapter describes several important instruments that are used to measure current and voltage. The chapter concludes with a discussion of the software typically used by electrical and electronic engineers to analyze and design electronic circuits.

1.1   THE WORLD OF ELECTRONIC SIGNALS

Can you imagine our ancestors crouched along an ancient river patiently awaiting the sound of a distant drum to signal the arrival of a friendly, or perhaps not so friendly, neighboring tribe or clan? Can you imagine driving through the streets of New York City following a power outage that suddenly disabled all traffic signals? Can you imagine the panic and confusion at O'Hare airport if the communication signals between pilots and air traffic controllers were suddenly disrupted? Signals have played an essential role in human activities since the dawn of civilization and arguably define modern culture. Indeed, our physical senses are themselves marvelous signal-processing systems that allow us to respond to signals that we hear, see, smell, taste, and feel. Like it or not, we are organisms whose lives are in large part comprised of a series of responses to a variety of stimuli or signals.

Of the many signals that inundate our society, none are more contemporary, more useful, or more abundant than the signals that are electrical or electronic in nature. Electronic signals control the robotic arm that applies paint to a new Harley Davidson, enable us to view a live soccer match in Brazil, allow us to call anywhere in the world from the cell phone at our fingertips, facilitate the processing and storage of data in our computers, and in the case of the artificial pacemaker, provide the steady signals to keep hearts beating in proper rhythm.

It is difficult to envision the billions of electronic signals that travel invisibly through our atmosphere, race at the speed of light through wire bundles, or fly through fiber-optic cables carrying information and data of unfathomable volume and variety. From lifelike video games to the delicate microscopic procedures of nanosurgery, from the awesome computer-generated images that entertain us at movie theatres to the stunning images that scientists have retrieved from the planet Mars, electronic signals impact every facet of our lives.

The illustrations in Figure 1–1 indicate some of the innumerable ways in which electronic signals shape modern advanced societies.

1.2   AN INTRODUCTION TO CURRENT AND VOLTAGE

A **signal** may be defined as any scheme by which information is transferred from one location to another. **Electronic signals** exist in the form of either a voltage or a current.

FIGURE 1–1

Electronic Signals Define a Modern Society

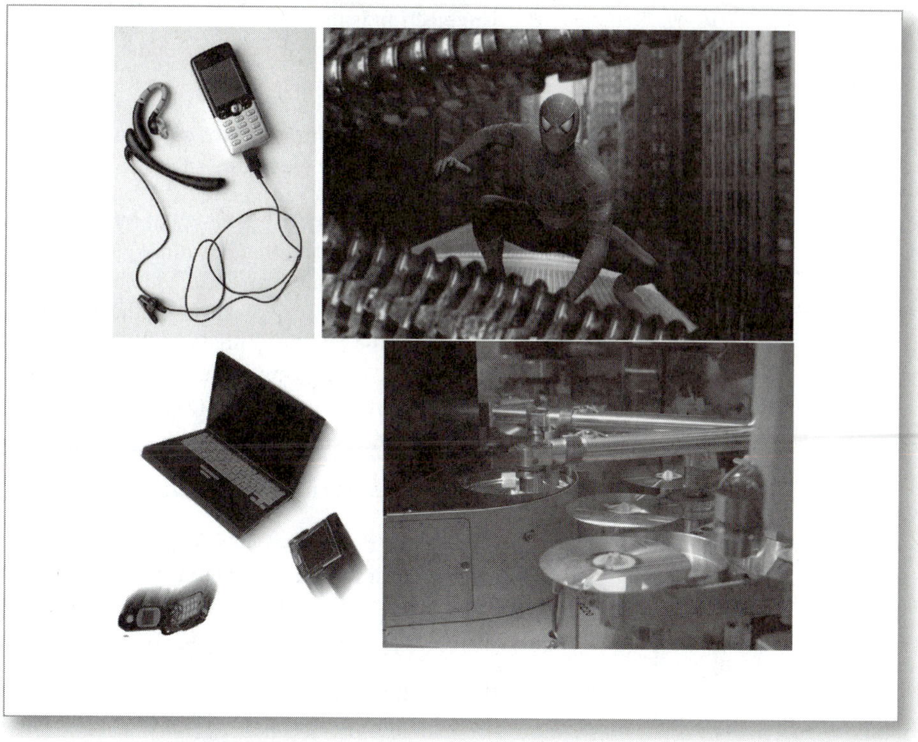

Current and voltage can be defined by relating them to the electrically charged particles that are a part of all atoms. The *electron* is an orbiting atomic particle that carries a small negative electrical charge as illustrated in Figure 1–2.

FIGURE 1–2

Illustration of the Particles Contained in All Atoms

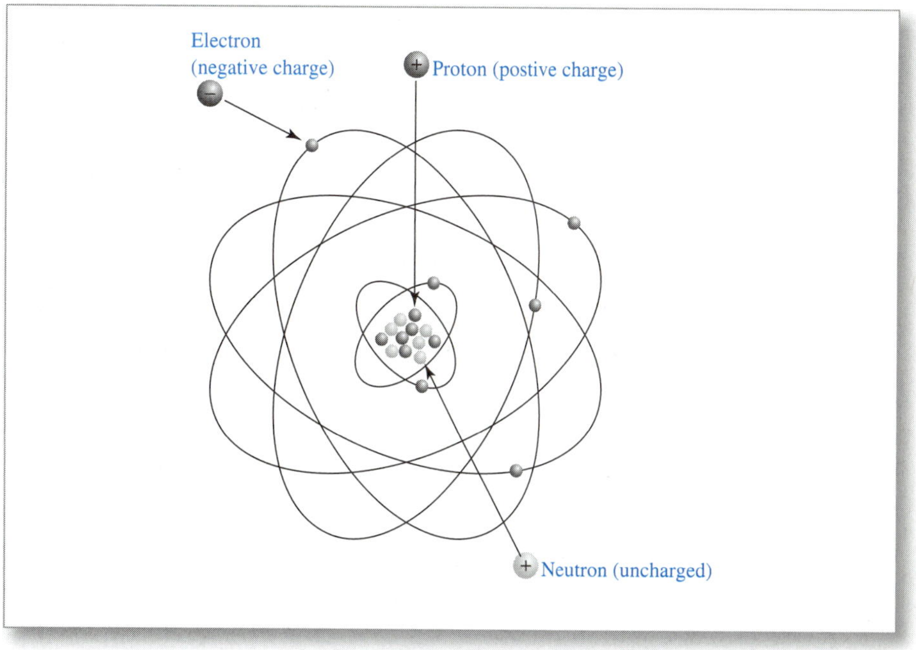

The proton and neutron shown in Figure 1–2 are also particles that are contained in every atom; however, they are not important to the study of electronics. The neutron possesses no electrical charge, so it is not relative to the study of electronics.

The proton shown in Figure 1–2 does possess an electrical charge; however, the proton is tightly bound in the center or nucleus of the atom. The study of electronics considers only charged particles that are relatively easy to move from one atom to another.

The electron is the only atomic particle that meets both conditions of possessing charge and being relatively free to move about.

The electrons in the atoms of most metals, such as copper and gold, are relatively free to move about within a wire or printed circuit board track. Materials that contain an abundance of **free electrons** are called **electrical conductors**.

An electrical force may be created by causing an imbalance of charge to exist between two points. The point that has more electrons is considered to be negative, and the point with fewer electrons is considered to be positive. If such a force is placed across the ends of a conductor, the free electrons in the conductor will migrate or flow in an orderly fashion toward the point of imbalance that is more positive, and away from the point that is more negative. Keeping these concepts in mind, we can define two important quantities that exist in the study of electronic circuits and signals.

Current is defined as the controlled flow of electrical charge. The unit of measure of current is the ampere, and the symbol traditionally chosen to represent current is the letter "I."

Voltage is defined as the force that sustains the flow of current. Voltage is created by an imbalance of charge between two points and is measured in units of volts. The symbol traditionally chosen to represent voltage is the letter "V." The voltage that exists between two points is often referred to as the **voltage difference** between the points.

If a voltage exists between two points, then one of the points must be more positive than the other and is indicated with a (+) sign. The more negative point is indicated with a (−) sign. The (+) and (−) signs together indicate the **polarity** of the voltage.

A voltage is identified by indicating its magnitude in volts (V) and its polarity (+, −). A current is identified by indicating its magnitude in amperes (A) and its direction of flow.

While the direction of electron flow is from negative to positive, electrical and electronic engineers traditionally assume that current in an electronic circuit flows from a more positive point (+) to a more negative point (−). The current is referred to as **conventional current** and will be assumed throughout this text. The apparent disagreement between the direction of electron flow and conventional current flow does not create contradiction when analyzing or designing electronic circuits. The circuit analyzer must be aware of which current flow is being used and obey the conventions for that particular current flow. The essential results of circuit analysis, whether using electron flow or conventional current flow, are identical. As noted, this text uses conventional current flow throughout.

Figure 1–3 illustrates a voltage source connected across the ends of a conductor causing conventional current to flow in the direction indicated.

FIGURE **1–3**

A Voltage Source Causing
Conventional Current Flow
through a Conductor

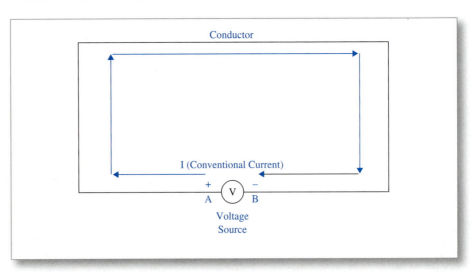

The voltage source causes a voltage difference between points A and B. Point A as indicated is more positive than point B. Note that conventional current flows from the more positive terminal of the voltage source through the conductor to the negative terminal of the voltage source. So long as the voltage source maintains a voltage difference between its terminals, current will continue to flow.

1.3   THE ELECTRONIC SIGNAL AS A FUNCTION OF TIME

A voltage or current may either be constant over time or vary over time.

Voltage and current that are constant over time are referred to respectively as direct voltage and direct current and are designated by the capital letters **DC**.

A voltage or current that varies with time is referred to respectively as alternating voltage or alternating current and is designated by the letters **AC**.

Although DC voltage and current are important electronic circuit quantities, most electronic signals are AC or waveforms that vary with time.

An **analog signal** is a voltage or current waveform that has a unique value at every instant of time and is often referred to as a continuous signal or continuous waveform.

The sounds that we hear result from continuous sound pressure waves that travel through air from the sound source to our ears. The images that our eyes receive are continuous variations of color and light intensity. Since both sound and visual images are continuous waveforms or signals, they are classified as analog signals. In general, most signals that originate in the physical world are continuous or analog signals.

Figure 1–4 illustrates an analog voltage signal waveform as a function of time.

FIGURE **1–4**

An Analog Voltage Waveform
as a Function of Time

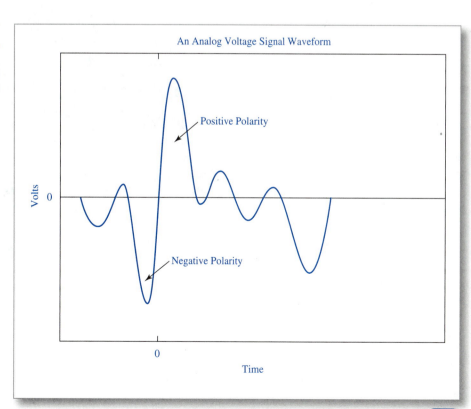

A unique value of voltage exists at every point in time, so the signal in Figure 1–4 is a continuous or analog voltage waveform.

1.4 CREATING AND TRANSMITTING ELECTRONIC SIGNALS

Signals that exist in the physical world are typically analog; however, they may not be electronic (voltage or current) waveforms and therefore cannot be processed electronically. For example, we may wish to store temperature variations in a computer memory or display the temperature variations on a video screen, or we may want to send a visual image of the temperature variations as an e-mail attachment. Many ingenious devices have been invented to convert various physical quantities such as sound, temperature, pressure, velocity, and light intensity into voltage or current signals that can be processed electronically. Devices that convert physical quantities into electronic signals are referred to as **transducers.**

The familiar microphone, for example, converts sound pressure variations into an electronic voltage waveform. The tachometer is a transducer that converts the revolutions per minute (rpm) of a motor into an electronic voltage signal that may be displayed or used to automatically control the speed of the motor.

A digital camera is a transducer that scans an image and creates voltage signals whose value at an instant of time is proportional to the intensity of light at the corresponding point on the image. The resulting electronic signals can be stored in electronic memory in the camera or downloaded to a computer for processing, storage, or printing, as shown in Figure 1–5.

FIGURE **1–5**

Electronic Signals from a Digital Camera can be Printed or Stored

Electronic signals may also be created directly by electronic circuits. A *signal generator* is a common laboratory instrument whose electronic circuits can create a variety of voltage waveforms that are typically used to test or troubleshoot electronic circuits. An *oscillator*

is a circuit that creates the electronic signals that carry information from commercial broadcast radio and television stations.

Electronic signals may be transmitted through electrical conductors such as telephone land lines or coaxial cables such as those used to transmit and receive cable television signals. Cellular telephone signals are transmitted through the atmosphere and are referred to as **wireless transmission.** Still other electronic signals are transmitted to and from satellite relay stations positioned miles above the earth, while some electronic signals are converted into light waveforms and transmitted through fiber-optic cables buried beneath the earth's surface.

1.5   MEASURING ELECTRONIC SIGNALS

An **oscilloscope** is an important instrument that converts electronic voltage signals into visual patterns that can be observed on its display screen. An oscilloscope produces visual images in much the same way that a computer video monitor or conventional television produces visual images. These devices produce a stream of electrons that bombard the display screen. The display screen is coated with a phosphorescent material that emits light as it is struck by the electrons. The position of the electron beam is controlled to coincide with the voltage pattern of the input signal.

The image produced by an oscilloscope is a visual reproduction of the input signal voltage waveform as a function of time. Many of the figures illustrated in this text are plots of electronic signals as functions of time and can be considered to simulate images as they would appear on the screen of an oscilloscope. A digital oscilloscope displaying an electronic voltage waveform is shown in Figure 1–6.

FIGURE **1–6**

A Voltage Waveform
Displayed on an Oscilloscope

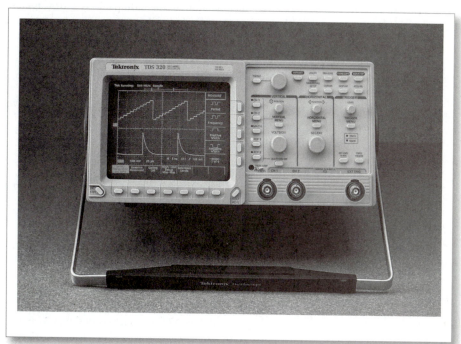

The **digital multimeter (DMM)** shown in Figure 1–7 is used to measure the value of electronic signals in units of volts or amperes.

Although the DMM does not create a visual display of a signal, the magnitude of a voltage or current signal is an important measurement for analyzing and troubleshooting electronic circuits.

FIGURE **1-7**

A Digital Multimeter (DMM)

1.6 UNITS AND NOTATION

Current signals are represented by the letter I. A current represented by an uppercase I designates a current that is constant or that does not vary over time and is referred to as a direct current or **DC current**. Similarly, a **DC voltage** is one that is constant over time and is represented by an uppercase V.

Current and voltage signals that vary over time are represented by lowercase letters (i, v, respectively) and are referred to as alternating or **AC current** or **AC voltage**.

Because it is often necessary to designate more than one voltage or current in a circuit, an alphanumeric subscript is used to distinguish one voltage or current from another. As an example, three DC voltages may be designated as:

$$V_1, V_2, \text{ and } V_3 \text{ or alternatively as } V_A, V_B, \text{ and } V_C$$

Three time-varying currents may be designated as:

$$i_1, i_2, \text{ and } i_3 \text{ or alternatively as } i_a, i_b, \text{ and } i_c$$

The voltage difference between two points may be designated as:

$$V_{12} \text{ or } V_{AB} \text{ for a DC voltage or as } v_{12} \text{ or } v_{AB} \text{ for an AC voltage}$$

Note that the subscript for voltage notation consists of two numbers or letters. The notation is referred to as double subscript notation. The two numbers or letters indicate the two points between which the voltage is measured.

The voltage "V_{12}" indicates the voltage between the points designated as 1 and 2. The notation is stated as V (one, two) not V (twelve). The voltage "V_{34}" indicates the voltage between the points designated as 3 and 4. The notation is stated as V (three, four) not V (thirty-four). The voltage V_{AB} indicates the voltage between points designated as A and B. The notation is stated as V (AB). The notations may seem a bit confusing at this point but they will become second nature as we progress through the text.

The magnitudes of electronic signals can sometimes be quite small. Engineers have adopted the prefix symbols used by other scientist to avoid the use of excessively small numbers. Table 1–1 indicates the prefix symbols commonly used by engineers along with their respective numerical equivalents.

Additional prefix symbols are used to indicate quantities that are exceptionally large and will be introduced as needed in later chapters.

TABLE **1–1**	Symbol	Name	Numerical Equivalent
Table of Common Engineering Prefixes	m	milli	10^{-3}
	μ	micro	10^{-6}
	n	nano	10^{-9}
	p	pico	10^{-12}

EXAMPLE 1–1   Selecting Appropriate Prefix Symbols

a. Given a DC current of 0.003 A, write an expression for the current using the appropriate prefix symbol from Table 1–1.

✔ SOLUTION

The current may be expressed as 3×10^{-3}. Selecting the appropriate prefix symbol, we obtain:

$$I = 3 \text{ mA}$$

Note that an uppercase letter is used to indicate the current because the given current is a DC current.

b. Given an AC current of 0.0000065 A, write an expression for the current using the appropriate prefix symbol from Table 1–1.

✔ SOLUTION

The current may be expressed as 6.5×10^{-6}. Selecting the appropriate prefix symbol, we obtain:

$$i = 65 \ \mu\text{A}$$

Note that a lowercase letter is used to indicate the current because the given current is an AC current.

c. Given an AC voltage of 0.028 V, write an expression for the voltage using the appropriate prefix symbol from Table 1–1.

✔ SOLUTION

The voltage may be expressed as 28×10^{-3}. Selecting the appropriate prefix symbol, we obtain:

$$v = 28 \text{ mV}$$

The solutions to Example 1–1 can be obtained quickly and accurately using an appropriate engineering calculator. Place the calculator in "engineering mode."

For Example a, enter 0.003 and press ENTER. The calculator display should read 3 E-3 or 3 mA.

For Example b, enter 0.0000065 and press ENTER. The calculator display should read 6.5 E-6 or 6.5 μA.

For Example c, enter 0.028 and press ENTER. The calculator display should read 28 E-3 or 28 mV.

CALCULATOR NOTE: By setting the calculator to "engineering mode," the only powers of ten that will be displayed by the calculator correspond directly to the prefix symbols in Table 1–1.

PRACTICE EXERCISE 1–1

Use an engineering calculator set to "engineering mode" to select the appropriate prefix symbol from Table 1–1. Write an appropriate expression for each of the given quantities.

   a. An AC current of 0.0000021 A

   b. A DC current of 0.000043 A

   c. A DC voltage of 0.129 V

✓ ANSWERS

   a. $i = 2.1\ \mu\text{A}$

   b. $I = 43\ \mu\text{A}$

   c. $V = 129\ \text{mV}$

(Complete solutions to the Practice Exercises can be found in Appendix A.)

1.7 SOFTWARE UTILIZED IN CIRCUIT ANALYSIS

The engineering calculator is an essential computational tool for the analysis and design of electronic circuits. The mathematical operations that can be performed on a modern engineering calculator are too numerous to list. Calculator operations will be introduced at appropriate points in the text. In addition, the calculator keystrokes will be indicated for some of the more complex calculator operations.

Circuit simulation software such as MultiSIM provides a nearly indispensable aid to the study of circuit analysis. Every circuit presented in this text can be simulated to produce visual evidence of the accuracy of the analysis. This visual check of a solution provides immediate feedback and is an invaluable reinforcement for student homework exercises. Many of the examples in this text are accompanied by a printout of a MultiSIM figure to verify analytical results. It is suggested that you verify many of your homework exercises using software such as MultiSIM.

Figure 1–8 shows a MultiSIM simulation of an electronic signal displayed on an oscilloscope.

Mathematics-based software such as MATLAB is utilized by many electronic engineers to obtain solutions to complex equations and to plot graphs of a wide variety of waveforms. Although MATLAB is not required to use this text, many MATLAB figures are included in a variety of applications throughout the text. If MATLAB is available, it is suggested that you reproduce several of the MATLAB figures as a first step in learning how to use this valuable engineering software tool.

FIGURE **1-8**

An Electronic Signal Waveform Simulated using MultiSIM

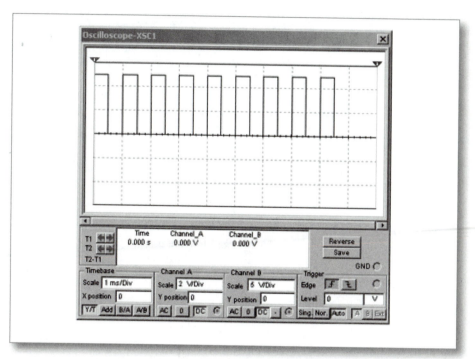

A MATLAB plot of an electronic signal voltage waveform is shown in Figure 1–9.

FIGURE **1-9**

An Electronic Signal Waveform Created in MATLAB

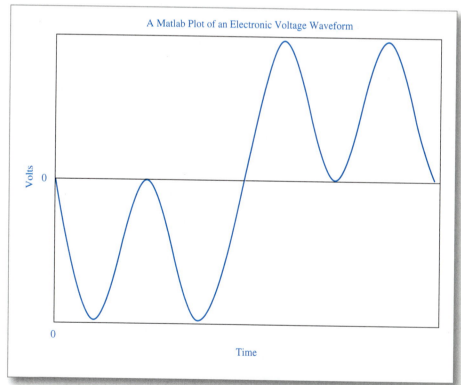

SUMMARY

The abundance and variety of uses of electronic signals in a modern society are beyond measure. Electronic signals that travel through wires and cables, through fiber-optic links, and through space impact every facet of our working and leisure activities.

Electronic signals exist either as a voltage waveform or as a current waveform that is a function of time. Transducers convert quantities such as sound, light, velocity, and temperature into electronic signals. The microphone, for example, converts sound pressure waves into electronic voltage or current waveforms.

A digital multimeter may be used to measure the magnitude of electronic signals. In addition, the DMM can indicate the polarity of a voltage signal or the direction of flow of a current signal. Electronic voltage signals can be visually displayed using an oscilloscope to reveal important characteristics of the signal.

Electronic signals are processed by a nearly infinite variety of ingenious electronic circuits to send or retrieve information or data.

Engineers typically utilize a variety of sophisticated test instruments and dedicated software to assist in the analysis and design of electronic circuits.

ESSENTIAL LEARNING EXERCISES

(ANSWERS TO ODD-NUMBERED ESSENTIAL LEARNING EXERCISES CAN BE FOUND IN APPENDIX A.)

1. Define the following:

   a. Signal

   b. Electronic signal

   c. Current

   d. Voltage

2. List five activities that you perform daily that utilize electronic signals.

3. Substances that contain an abundance of free electrons are referred to as _____

4. What is meant by voltage polarity?

5. A voltage is identified by indicating its _____ and its _____

6. A current is identified by indicating its _____ and its _____

7. What is meant by "conventional current flow"?

8. What is the function of a transducer?

9. What transducer converts sound into an electronic signal?

10. What transducer converts rotational speed into an electronic signal?

11. What device is used to convert voltage signal waveforms into a visual display?

12. What is meant by a DC voltage or DC current?

13. What is meant by an AC voltage or AC current?

14. What symbol is used to represent the following quantities?

    a. DC voltage

    b. DC current

    c. AC voltage

    d. AC current

15. What symbol is used to represent the voltage difference between two points, A and B?

16. Why are engineering prefixes commonly used in the science of electronics?

17. Indicate the equivalent power of ten for each of the following symbols:

    a. m

    b. μ

    c. n

    d. p

18. Write the complete prefix for each prefix symbol in Exercise 17.

19. Convert the following quantities using an engineering calculator to determine an appropriate engineering prefix. (Ensure the calculator is set to "engineering" mode.)

    a. 0.0085 A

    b. 0.026 V

20. Convert the following quantities using an engineering calculator to determine an appropriate engineering prefix. (Ensure the calculator is set to "engineering" mode.)

    a. 0.00003 V

    b. 0.000000032 A

CHALLENGING LEARNING EXERCISES

(ANSWERS TO SELECTED CHALLENGING LEARNING EXERCISES CAN BE FOUND IN APPENDIX A.)

1. Explain in your own words how a digital camera converts an image into electronic signals.

2. A monitoring system measures the temperature of a refrigerated storage room at fifteen-minute intervals. The temperatures are converted to a voltage that is proportional to the temperature. Will the resulting electronic signal be an analog signal? Explain.

3. Explain why voltage must be measured between two points.

4. A water distribution system can be compared to an electronic system. What quantity in the water system would be analogous to current flow? What quantity in the water system would be analogous to voltage? What part of the water system would be analogous to a conductor?

ESSENTIALS OF DC CIRCUIT ANALYSIS

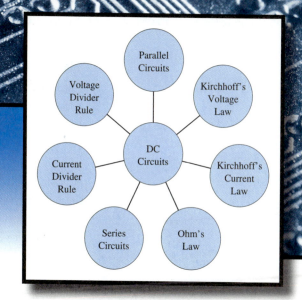

"AN IDEA IS THE MOST EXCITING
THING THERE IS."

—John Russell

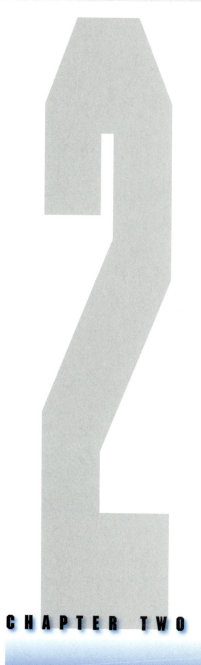

CHAPTER OUTLINE

LEARNING OBJECTIVES

Upon successful completion of this chapter you will be able to:

- Identify a DC series circuit.
- Calculate total equivalent resistance for series resistors.
- Apply Ohm's law to calculate current and voltage in a DC series circuit.
- Apply Kirchhoff's voltage law.
- Apply the voltage divider rule to determine the voltage in a series DC circuit.
- Identify a DC parallel circuit.
- Calculate total equivalent resistance for parallel resistors.

C H A P T E R   T W O

- Calculate total equivalent conductance.
- Apply Kirchhoff's current law at a circuit node.
- Apply the current divider rule.

INTRODUCTION

This chapter introduces the series and parallel DC circuit. Ohm's law, Kirchhoff's voltage law, and Kirchhoff's current law are used to calculate the value of current and voltage in a DC circuit. The voltage divider rule is used to calculate voltage in a series circuit, and the current divider rule is used to calculate branch currents in a parallel circuit. The concept of circuit ground is described, and the single line circuit diagram is illustrated.

2.1   THE DC CIRCUIT

An **electronic circuit** is a collection of electronic components connected together to create, process, or transmit electronic signals.

The only components typically considered in the study of introductory DC circuit analysis are resistors. While the DC circuit is the most fundamental electronic circuit, the circuit analysis laws and procedures learned in the study of DC circuits will be applied to more complex circuits in later chapters.

2.2   THE RESISTOR

Electronic circuits are designed to perform a seemingly endless variety of functions to create electronic communications systems, industrial controls systems, and computer data processing systems to name a few. All electronic circuits, no matter how complex, accomplish their function by controlling or processing voltage and current signals in some way.

Resistors are utilized in virtually every electronic circuit either to limit the amount of current flow or to develop a voltage in a circuit.

As current flows through a resistor, the atoms of the resistor material collide with the charged particles that comprise the current. The resistor therefore offers opposition to current flow in much the same way that friction impedes motion in a mechanical system. To carry the mechanical analogy a step farther, resistors convert electrical energy into heat much in the same way that friction creates heat in mechanical systems.

The opposition to current flow from a resistor is called **resistance** and is measured in units of **ohms** (Ω).

Resistors are manufactured to rigid specifications so that circuit designers can precisely control the amount of current that flows through various branches of a circuit. Figure 2–1a is a photograph of common commercially available resistors of the type that are found in nearly all electronic circuits. The schematic symbol used to represent a resistor is shown in Figure 2–1b.

FIGURE **2–1**

a. Common Commercially Manufactured Resistors
b. Schematic Symbol used to Represent a Resistor

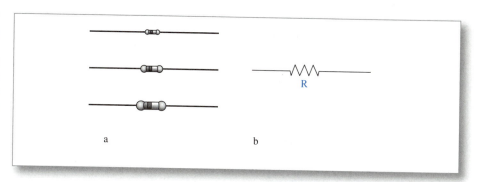

a   b

The resistance value of a resistor may be determined from a series of colored bands printed on the body of the resistor. The resistance value may also be printed directly on the resistor. The procedure for determining the value of a resistor from the color code bands is demonstrated in the laboratory manual that accompanies this text. An easy-to-use resistor color code calculator is available on the Internet at http://www.electrician.com (look under "Calculators" and select "Resistor Color Code Calculator").

Figure 2–2 is a photograph of a *resistor array* that contains several resistors packaged in a sixteen dual inline pin (DIP) configuration. Resistor arrays are economical and are easily mounted on printed circuit boards.

FIGURE **2–2**

A Commercially Manufactured Resistor Array for Mounting on Printed Circuit Board

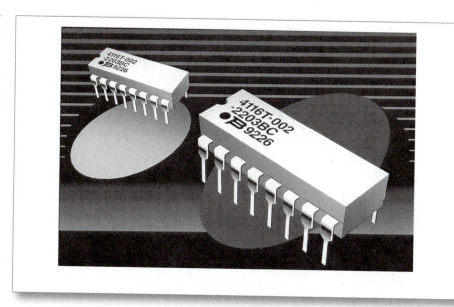

Typically, resistor arrays such as the one in Figure 2–2 contain only resistors of equal value. Electronic circuits are often designed using components of equal value when possible to minimize the number of different components that must be utilized.

2.3 THE DC VOLTAGE SOURCE

DC voltage and current are constant over time. A familiar DC voltage source is the miniature nickel-cadmium (Nicad) or lithium batteries found in portable electronic systems such as a digital camera or an electronic wristwatch. The lead-acid battery that provides the energy to start an automobile engine is a 12 V DC voltage source.

Another important DC voltage source is created by converting commercial (AC) voltage to DC voltage. Electronic equipment that is connected to a commercial outlet contains electronic circuits designed to convert commercial (AC) voltage to any required value of DC voltage. A variable DC voltage source is an indispensable laboratory instrument that can be adjusted to provide DC voltage, typically ranging from 0 V to 25 V.

The schematic symbol used by electrical and electronic engineers to represent a 12 V DC voltage source and a 24 V DC source respectively are shown in Figure 2–3.

FIGURE **2–3**

Schematic Symbol Used
to Represent a
DC Voltage Source

12 V   24 V

As noted in Chapter 1, a voltage is specified by indicating the magnitude of the voltage measured in volts and the polarity of the voltage. The long bar of the DC voltage symbol shown in Figure 2–3 is always considered to be the positive (+) terminal of the source and the short bar is the negative terminal.

The plot of a 12 V DC voltage as a function of time is shown in Figure 2–4.

FIGURE **2–4**

A Time Plot of a DC
Voltage Indicating That
DC Voltage is Constant
Over Time

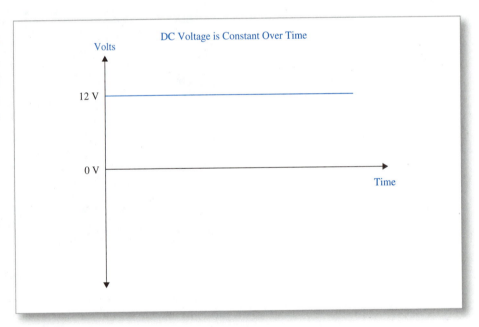

Note that the voltage is constant at 12 V at every point in time. The current that flows from a DC voltage source will be DC current, which is also constant over time.

2.4 THE RESISTOR AND OHM'S LAW

Intuitive reasoning would conclude that the amount of current that flows through a resistor is inversely proportional to resistance of the resistor. If the DC voltage placed across a resistor is held constant, the amount of DC current that will flow will decrease if the resistance is increased.

Nearly 160 years ago, a scientist named Georg Simon Ohm (1787–1854) determined the mathematical relationship that exists between voltage, current, and resistance.

His discovery, known as **Ohm's law,** can be stated mathematically for DC voltage and current as:

$$I = \frac{V}{R} \tag{2.1}$$

Ohm's law remains one of the most important relationships to be considered when analyzing or designing electronic circuits. Through algebraic manipulation the alternate forms of Ohm's law may be derived as:

$$V = IR \tag{2.2}$$

and

$$R = \frac{V}{I} \tag{2.3}$$

The form of Ohm's law in Equation (2.1) suggests that a resistor may be used to limit the amount of current flow. Increasing the amount of resistance will decrease the amount of current that flows. The form of Ohm's law in Equation (2.2) suggests that current flowing through a resistor will produce a voltage across the terminals of the resistor that is directly proportional to the resistance measured in ohms. Both of these important resistor applications are illustrated in Figure 2–5.

FIGURE **2–5**

a. A Resistor Used to Limit Current
b. A Resistor Used to Develop an Output Voltage

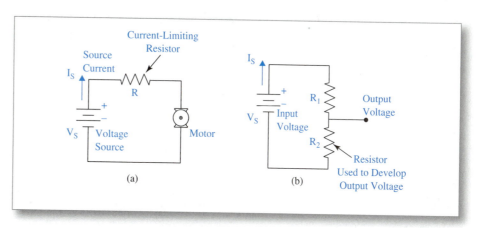

(a)   (b)

The resistor R shown in Figure 2–5a is used to limit the amount of current flowing to the industrial motor. If the resistance is increased, the amount of current flowing to the motor will decrease, which will reduce the speed of the motor. The resistor in Figure 2–5b is used to produce an output voltage across resistor R_2. The value of the output voltage can be varied by changing the value of R_2. The output voltage may be connected to other electronic circuits as a DC voltage source.

The ohm (Ω) is the unit of measure for resistance. The resistance of resistors typically used in the design of electronic circuits is often several thousand or even several million ohms. Just as prefixes have been adopted to indicate very small quantities as introduced in Chapter 1, prefixes have also been adopted to indicate very large quantities, as illustrated in Table 2–1.

TABLE **2–1**

Table of Common Engineering Prefixes

Symbol	Name	Numerical Equivalent
T	tera	10^{12}
G	giga	10^{9}
M	mega	10^{6}
k	kilo	10^{3}

A resistor whose resistance value is 1200 Ω is expressed as 1.2 kΩ, and a resistor whose resistance value is 1,600,000 Ω is expressed as 1.6 MΩ.

EXAMPLE 2-1   Using Ohm's Law

a. A DC voltage source of 36 V is connected across a 4 kΩ resistor. Calculate the magnitude of the current that flows through the resistor.

✓ **SOLUTION**

Ohm's law may be used to calculate the current through the resistor as:

$$I = \frac{V}{R} = \frac{36 \text{ V}}{4 \text{ k}\Omega} = 9 \text{ mA}$$

Recall that the 4 kΩ is entered into the calculator as 4 EE 3.

b. A current of 16 mA is flowing through 2.2 kΩ resistor. Calculate the magnitude of the voltage across the resistor.

✓ **SOLUTION**

Ohm's law may be used to calculate the voltage across the resistor as:

$$V = IR = (16 \text{ mA})(2.2 \text{ k}\Omega) = 35.2 \text{ V}$$

Recall that the 2.2 kΩ is entered into the calculator as 2.2 EE 3 and the 16 mA is entered as 16 EE −3.

PRACTICE EXERCISE 2-1

a. A voltage source of 18 V is connected across a 5 kΩ resistor. Calculate the magnitude of the current that flows through the resistor.

b. A current of 2.8 mA is flowing through 4.7 kΩ resistor. Calculate the magnitude of the voltage across the resistor.

✓ **ANSWERS**

a. $I = 3.6$ mA;

b. $V = 13.16$ V

(Complete solutions to the Practice Exercises can be found in Appendix A.)

2.5 KIRCHHOFF'S VOLTAGE LAW (KVL)

Gustav Kirchhoff (1824–1887) discovered that the sum of voltages around any closed loop in an electronic circuit must equal zero. His discovery is referred to as **Kirchhoff's voltage law (KVL)** and may be stated mathematically as:

$$\sum V_{closed \ loop} = 0 \qquad (2.4)$$

If the sum of voltages around a closed loop is to be zero, then some of the voltages must be positive and some must be negative. An agreement or convention has been established among engineers to determine which voltages are positive and which ones are negative.

When applying KVL, a voltage that transitions from positive to negative (+ to −) in the direction of travel around the loop is considered to be a negative voltage, a voltage that transitions from negative to positive (− to +) in the direction of travel is considered to be a positive voltage.

The diagram in Figure 2–6 illustrates a closed circuit loop indicating four voltage changes with their respective polarities. KVL may be applied to obtain:

$$V_1 - V_2 - V_3 - V_4 = 0 \qquad (2.5)$$

FIGURE **2-6**

An Illustration of Kirchhoff's
Voltage Law

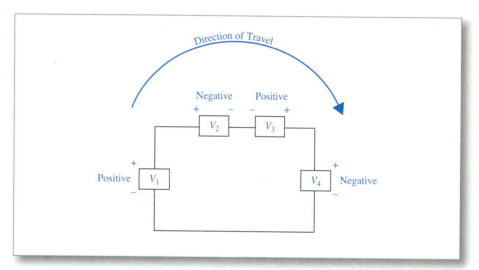

Note that as we travel clockwise around the loop in Figure 2–6, V_1 and V_3 transition from (−) to (+) and are therefore entered into Equation (2.5) as a *positive voltage*, while V_2 and V_4 transition from (+) to (−) in the direction of travel and are therefore entered as a *negative voltage*.

When applying KVL, the starting point in the loop is irrelevant because the law requires the summation of every voltage in a complete loop. When applying KVL, voltages that transition from positive to negative (+ to −) are commonly called *voltage drops*, and voltages that transition from (−) to (+) are commonly called *voltage rises*.

KVL may be stated as: The sum of voltage rises around a closed loop must equal the sum of voltage drops.

EXAMPLE 2-2   Using KVL to Determine an Unknown Voltage

Determine the voltage V_X in the diagram in Figure 2–7.

✓ SOLUTION A

Observe that neither the magnitude nor the polarity of the voltage V_X is indicated in the circuit diagram in Figure 2–7. The polarity of V_X in Figure 2–7 is arbitrarily assigned as indicated.

Observing that the diagram is a closed loop, KVL may be expressed moving clockwise around the loop as:

$$-9\,\text{V} + 3\,\text{V} + 4\,\text{V} - V_X = 0 \qquad (2.6)$$

FIGURE **2–7**

Arbitrarily
Assigned Polarity
for Example 2–2
Solution A

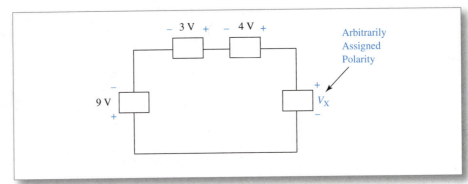

Note that the arbitrarily assigned polarity for V_X has been observed when writing KVL so that V_X is entered into KVL as a negative quantity. Solving for V_X in Equation (2.6), we obtain:

$$V_X = -2 \text{ V} \tag{2.7}$$

The magnitude of the voltage V_X is 2 V as indicated in Equation (2.7).

The negative sign in Equation (2.7) indicates that the arbitrarily assigned polarity was not correct and so the polarity must be reversed to be correct.

✓ SOLUTION B

Solve for V_X again, but this time arbitrarily assign the polarity as indicated in Figure 2–8.

Note that the arbitrarily assigned polarity in Figure 2–8 is the reverse of the polarity that was arbitrarily assigned in Figure 2–7 in Solution A.

FIGURE **2–8**

Arbitrarily
Assigned Polarity
for Example 2–2
Solution B

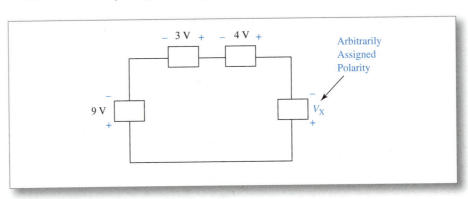

KVL can be expressed moving clockwise around the closed loop as:

$$-9 \text{ V} + 3 \text{ V} + 4 \text{ V} + V_X = 0 \tag{2.8}$$

Note that the arbitrarily assigned polarity for V_X has been observed when writing KVL so that V_X is entered into KVL as a positive voltage.

Solving for V_X in Equation (2.8), we obtain:

$$V_X = +2 \text{ V} \tag{2.9}$$

The magnitude of the voltage V_X is 2 V as indicated in Equation (2.9). *The positive sign in Equation (2.9) indicates that the arbitrarily assigned polarity is correct.*

It is important to observe that the ultimate results of Solutions A and B are identical. Both solutions indicate that the magnitude of V_X is 2 V. Both solutions also indicate that the upper terminal of V_X is negative with respect to the bottom terminal.

When applying KVL, the assignment of polarity to an unknown voltage is arbitrary. If the mathematical solution for the unknown voltage is positive, the arbitrarily assigned polarity is correct. If the mathematical solution for the unknown voltage is negative, the arbitrarily assigned polarity must be reversed.

EXAMPLE 2–3   Using KVL to Determine an Unknown Voltage

Use KVL to determine the magnitude and polarity of V_X in the circuit shown in Figure 2–9.

FIGURE **2–9**

Example 2–3: Applying Kirchhoff's Voltage Law

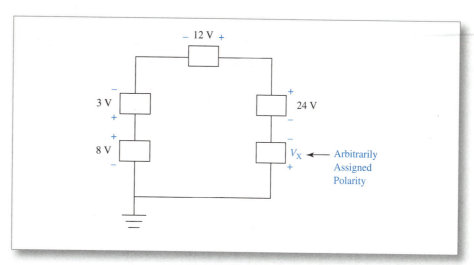

✓ SOLUTION

The polarity of V_X may be arbitrarily assigned to be $(-)$ to $(+)$ traveling clockwise so that KVL may be expressed beginning at circuit ground and traveling clockwise as:

$$8\ V - 3\ V + 12\ V - 24\ V + V_X = 0$$

and solving for V_X to obtain:

$$V_X = +7\ V$$

Since the value of V_X is a positive quantity, the arbitrarily assigned polarity is correct.

Note that the same magnitude and polarity for V_X is obtained by expressing KVL traveling counterclockwise as:

$$-V_X + 24\ V - 12\ V + 3\ V - 8\ V = 0$$

and solving for V_X to obtain:

$$V_X = +7\ V$$

PRACTICE EXERCISE 2–2

Use KVL to determine the magnitude and polarity of V_X in the circuit shown in Figure 2–10.

FIGURE **2–10**

Circuit Diagram for
Practice Exercise 2–2

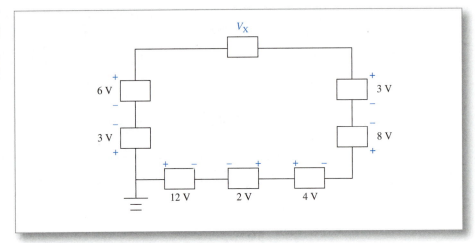

✓**ANSWER**

$V_X = +22$ V(+) to (−) traveling clockwise.

(Complete solutions to the Practice Exercises can be found in Appendix A.)

2.6 THE SERIES DC CIRCUIT

The most fundamental way in which circuit components can be connected together is so that the same current will flow consecutively through each component.

Resistors are connected in **series** in an electronic circuit so that the same current flows through each resistor. The voltage developed across each series resistor is directly proportional to the resistance of each respective resistor.

The circuit diagram in Figure 2–11 shows an electronic circuit consisting of three resistors and a DC voltage source connected in series.

FIGURE **2–11**

A DC Series Circuit

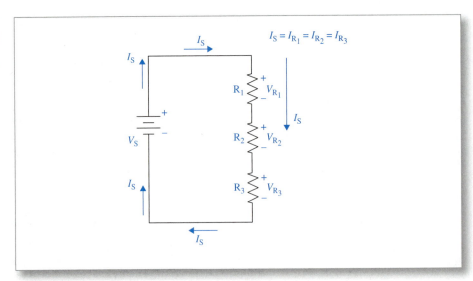

It is important to observe that the source current I_S in Figure 2–11 flows from the positive terminal of the DC voltage source V_S, through each successive resistor, and returns to the negative side of the voltage source. Since the same current flows throughout the entire circuit, by definition, the resistors and the voltage source are connected in series.

Recall that current flowing through a resistor will develop a voltage across the resistor as indicated by Ohm's law ($V = IR$). The voltage developed across each resistor in Figure 2–11 is indicated on the circuit diagram respectively as V_{R1}, V_{R2}, and V_{R3}.

The **polarity** of the voltage developed across each resistor is $(+)$ to $(-)$ in the direction of current flow through the resistor as indicated in Figure 2–11.

Recall that a voltage is specified by magnitude and polarity and that current is specified by magnitude and direction of flow, as indicated on the circuit schematic shown in Figure 2–11.

We now consider how Ohm's law may be applied to a circuit comprised of resistors connected in series.

Ohm's law may be used to calculate the source current I_S that flows from a voltage source V_S as:

$$I_S = \frac{V_S}{R_T}$$

(2.10)

where I_S is the source current, V_S is the source voltage, and R_T is the total equivalent resistance connected between the terminals of the source.

The **total equivalent resistance** of resistors connected in series is defined as the sum of the individual resistances such that:

$$R_T = R_1 + R_2 + R_3 + \cdots + R_N$$

(2.11)

where R_N represents the last resistor in the series circuit.

EXAMPLE 2–4   Analysis of a Series DC Circuit

Given the series DC circuit in Figure 2–12, determine the total equivalent resistance R_T, source current I_S, and voltage across each resistor.

FIGURE **2–12**

Circuit Diagram for Example 2–4

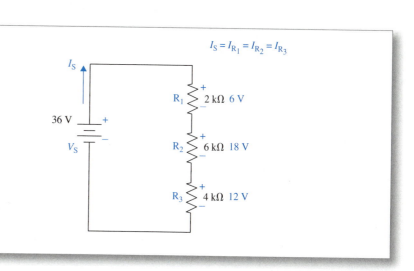

✓ SOLUTION

Observe that the same current I_S flows consecutively through each of the resistors, so the resistors by definition are connected in series.

The total equivalent resistance may be determined using Equation (2.11) as:

$$R_T = R_1 + R_2 + R_3 = 2 \text{ k}\Omega + 6 \text{ k}\Omega + 4 \text{ k}\Omega = 12 \text{ k}\Omega$$

Ohm's law may be applied to determine source current using Equation (2.10) to obtain:

$$I_S = \frac{V_S}{R_T} = \frac{36 \text{ V}}{12 \text{ k}\Omega} = 3 \text{ mA} \tag{2.12}$$

Since the circuit in Figure 2–12 is a series configuration, 3 mA will flow through each resistor.

Ohm's law may be used to calculate the voltage developed across each respective resistor as:

$$V_{R1} = I_S R_1 = (3 \text{ mA})(2 \text{ k}\Omega) = 6 \text{ V} \tag{2.13}$$

$$V_{R2} = I_S R_2 = (3 \text{ mA})(6 \text{ k}\Omega) = 18 \text{ V}$$

$$V_{R3} = I_S R_2 = (3 \text{ mA})(4 \text{ k}\Omega) = 12 \text{ V}$$

KVL may be expressed for the circuit to obtain:

$$V_S - V_{R1} - V_{R2} - V_{R3} = 0$$

so that: $36 \text{ V} - 6 \text{ V} - 18 \text{ V} - 12 \text{ V} = 0$

The MultiSIM simulation of Figure 2–12 is shown in Figure 2–13.

MULTISIM

FIGURE **2–13**

MultiSIM
Simulation for
Example 2–4

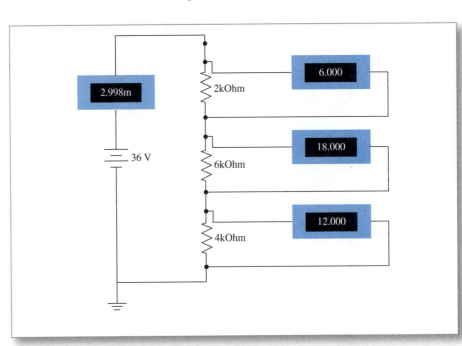

The source current in Figure 2–12 was calculated by dividing the source voltage by the total equivalent resistance (12 kΩ) of the circuit. Many other circuits can be created whose total equivalent resistance is 12 kΩ as well. In all of these cases, the source current will be 3 mA because the source current depends only upon the source voltage and total equivalent resistance.

Note that prefixes have been used throughout the calculations in the preceding examples. The calculator can greatly simplify calculations using prefixes as well as indicate which prefix is most appropriate.

The calculator should be set to the "engineering" exponential format for all calculations.

The following calculator keystrokes are specific to the TI–89 calculator manufactured by the Texas Instruments Corporation.

Calculating the source current I_S in Equation (2.12):

Keystrokes: ③⑥ ÷ ①② EE ③ ENTER to obtain: 3E − 3. Selecting the appropriate prefix and units, we obtain: $I_S = 3$ mA.

Calculating the voltage across R_1 in Equation (2.13):

Keystrokes: ③ EE − ③ x ② EE ③ ENTER to obtain: 6E0. Selecting the appropriate prefix and units, we obtain: $V_{R1} = 6$ V.

EXAMPLE 2–5   Analysis of a DC Series Circuit

Given the series DC circuit in Figure 2–14, determine the total equivalent resistance R_T, source current I_S, and voltage across each resistor. Write the expression of KVL for the circuit.

FIGURE **2–14**

Circuit Diagram for
Example 2–5

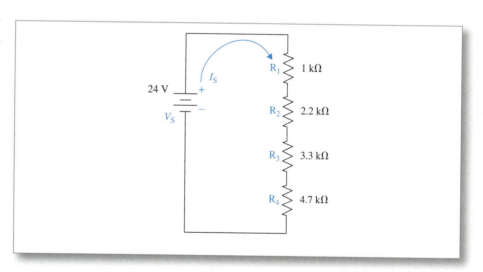

✓ SOLUTION

The total equivalent resistance of the circuit is calculated as:

$$R_T = R_1 + R_2 + R_3 + R_T + R_4 + 1 \text{ k}\Omega + 2.2 \text{ k}\Omega + 3.3 \text{ k}\Omega + 4.7 \text{ k}\Omega = 11.2 \text{ k}\Omega$$

Ohm's law may be used to calculate the source current as:

$$I_S = \frac{V_S}{R_T} = \frac{24 \text{ V}}{11.2 \text{ k}\Omega} = 2.14 \text{ mA (rounded)}$$

Ohm's law may be used to calculate the voltage across each respective resistor as:

$$V_{R1} = I_S R_1 = (2.14 \text{ mA})(1 \text{ k}\Omega) = 2.14 \text{ V}$$

$$V_{R2} = I_S R_2 = (2.14 \text{ mA})(2.2 \text{ k}\Omega) = 4.71 \text{ V}$$

$$V_{R3} = I_S R_2 = (2.14 \text{ mA})(3.3 \text{ k}\Omega) = 7.06 \text{ V}$$

$$V_{R4} = I_S R_2 = (2.14 \text{ mA})(4.7 \text{ k}\Omega) = 10.06 \text{ V}$$

KVL is expressed as:

$$V_S - V_{R1} - V_{R2} - V_{R3} - V_{R4} = 0 \text{ V}$$

so that: $24 \text{ V} - 2.14 \text{ V} - 4.71 \text{ V} - 7.06 \text{ V} - 10.06 \text{ V} = 0.03 \text{ V}$

The sum of voltages is not exactly zero due to rounding. The slight difference from zero (0.03 V) is not considered to be significant error.

The rather strange resistor values in this example are in fact standard values that are produced by commercial manufacturers.

The MultiSIM simulation of the circuit in Figure 2–14 is shown in Figure 2–15.

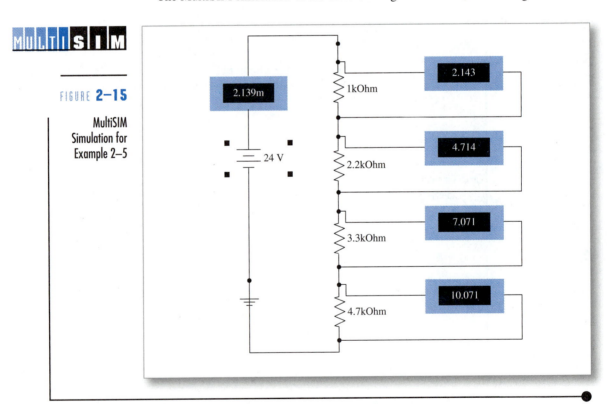

FIGURE **2–15**

MultiSIM
Simulation for
Example 2–5

Determine the total equivalent resistance R_T, source current I_S, and voltage across each resistor in Figure 2–16. Indicate the polarity of the voltage across each resistor. Write the expression of KVL for the circuit.

FIGURE **2-16**

Circuit Diagram for
Practice Exercise 2–3

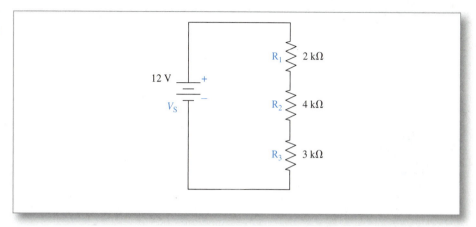

✓ANSWER

$R_T = 9$ kΩ; $I_S = 1.33$ mA; $V_{R1} = 2.67$ V; $V_{R2} = 5.33$ V; $V_{R3} = 4$ V;

KVL: 12 V $- 2.67$ V $- 5.33$ V $- 4$ V $= 0$

(Complete solutions to the Practice Exercises can be found in Appendix A.)

2.7 THE VOLTAGE DIVIDER RULE

The circuit in Figure 2–17 consists of three resistors connected in series with a voltage source V_S. The voltage across each resistor is indicated respectively as V_{R1}, V_{R2}, and V_{R3}.

FIGURE **2-17**

Deriving the Voltage
Divider Rule

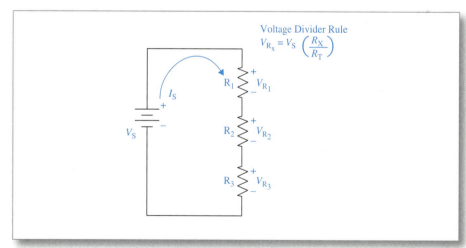

Ohm's law may be used to calculate the source current as:

$$I_S = \frac{V_S}{R_T}$$

(2.14)

Ohm's law may also be used to calculate the current through R_1 as:

$$I_{R1} = \frac{V_{R1}}{R_1}$$

(2.15)

However, the resistors and the voltage source in Figure 2–17 are connected in series, so that I_S and I_{R1} must be the same current.

Setting Equation (2.14) equal to Equation (2.15), we obtain:

$$\frac{V_S}{R_T} = \frac{V_{R1}}{R_1} \tag{2.16}$$

Rearranging Equation (2.16), we obtain:

$$V_{R1} = V_S \frac{R_1}{R_T} \tag{2.17}$$

Equation (2.17) can be generalized so that the voltage V_{Rx} across any resistor (R_x) in a series circuit may be calculated as:

$$V_{RX} = V_S \frac{R_X}{R_T} \tag{2.18}$$

Equation (2.18) is referred to as the voltage divider rule and may be used to calculate the voltage across any resistor in a series circuit. It is important to remember that the voltage divider rule can be applied only to components that are connected in series.

EXAMPLE 2-6   Applying the Voltage Divider Rule

Use the voltage divider rule to determine the voltage across each resistor in Figure 2–18. Calculate the voltage across each resistor using Ohm's law and compare to the voltage calculated using the voltage divider rule. Verify KVL. Verify the calculations using computer simulation.

FIGURE **2-18**

Example 2–6:
Applying the
Voltage Divider
Rule

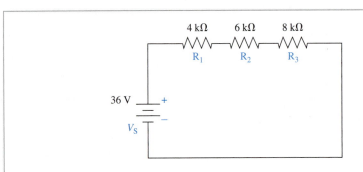

✓ SOLUTION

Applying the voltage divider rule defined by Equation (2.18), we obtain:

$$V_{R1} = V_S \frac{R_1}{R_T} = \frac{(36 \text{ V})(4 \text{ k}\Omega)}{(18 \text{ k}\Omega)} = 8 \text{ V}$$

$$V_{R2} = V_S \frac{R_2}{R_T} = \frac{(36 \text{ V})(6 \text{ k}\Omega)}{(18 \text{ k}\Omega)} = 12 \text{ V}$$

$$V_{R3} = V_S \frac{R_3}{R_T} = \frac{(36 \text{ V})(8 \text{ k}\Omega)}{(18 \text{ k}\Omega)} = 16 \text{ V}$$

where: $R_T = 4 \text{ k}\Omega + 6 \text{ k}\Omega + 8 \text{ k}\Omega = 18 \text{ k}\Omega$

The sum of the voltages obtained applying the voltage divider rule satisfies KVL because the sum of voltage drops across the resistors (36 V) is equal to the source voltage (36 V, voltage rise).

Ohm's law may be used to calculate the source current as:

$$I_S = \frac{V_S}{R_T} = \frac{36 \text{ V}}{18 \text{ k}\Omega} = 2 \text{ mA}$$

Ohm's law may be used to calculate the voltage across each respective resistor as:

$$V_{R1} = I_S R_1 = (3 \text{ mA})(4 \text{ k}\Omega) = 8 \text{ V}$$

$$V_{R2} = I_S R_2 = (2 \text{ mA})(6 \text{ k}\Omega) = 12 \text{ V}$$

$$V_{R3} = I_S R_2 = (2 \text{ mA})(8 \text{ k}\Omega) = 16 \text{ V}$$

Note that the voltage across each resistor calculated by Ohm's law is identical to the respective voltages obtained by applying the voltage divider rule.

Figure 2–19 shows the MultiSIM computer simulation indicating the voltage across each resistor.

MULTISIM

FIGURE 2–19

MultiSIM
Simulation for
Example 2–6

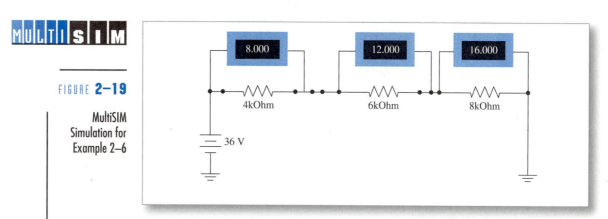

PRACTICE EXERCISE 2–4

Use the voltage divider rule to calculate the voltage across each resistor for the circuit shown in Figure 2–16. Compare the results to those obtained in Practice Exercise 2.3.

(Complete solutions to the Practice Exercises can be found in Appendix A.)

2.8 MEASURING VOLTAGE WITH RESPECT TO CIRCUIT GROUND

Voltage at any point in an electronic circuit is often measured as the voltage difference between the specified point and circuit ground.

Circuit ground in an electronic circuit is an arbitrarily selected point in the circuit and is identified by the symbol ⏚. The voltage at circuit ground in most electronic circuits is typically assumed to be 0 V, and this convention is assumed throughout this text.

Single subscript notation is used to represent a voltage at a point in an electronic circuit measured with respect to circuit ground. The symbol used to represent the voltage at point A measured with respect to ground is V_A. The symbol used to represent the voltage at point B measured with respect to ground is V_B.

Voltages V_A and V_B indicated in the circuit diagram in Figure 2–20 are understood to be measured with respect to circuit ground.

FIGURE **2–20**

Measuring Voltage with Respect to Circuit Ground

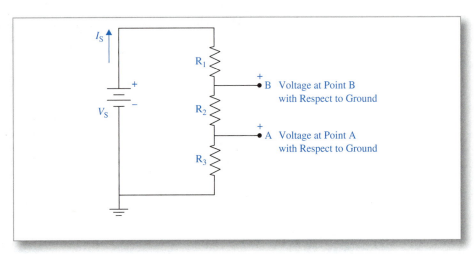

Care must be exercised to avoid confusing the voltage measured across a resistor with a voltage that is measured at a point with respect to circuit ground.

EXAMPLE 2–7   Determining a Voltage with Respect to Circuit Ground

Determine V_A and V_B in the circuit diagram in Figure 2–21.

FIGURE **2–21**

Example 2–7: Calculating Voltage with Respect to Circuit Ground

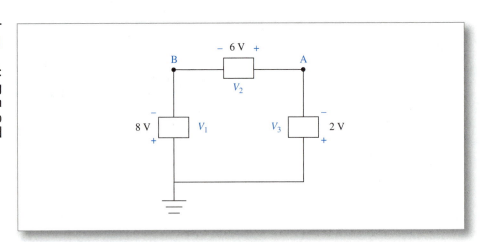

✓ **SOLUTION**

The voltage V_A is determined as the algebraic sum of voltages (observing polarities) from circuit ground to point A. The voltage V_A in Figure 2–21 may be determined traveling counterclockwise from circuit ground as:

$$V_A = V_3 = -2\,\text{V} \tag{2.19}$$

The voltage V_A may also be determined by traveling clockwise from circuit ground as:

$$V_A = V_1 + V_2 = -8\,V + 6\,V = -2\,V \qquad (2.20)$$

The negative signs in Equations (2.19) and (2.20) indicate that the voltage at point A is 2 V below ground where ground is assumed to be 0 V.

V_B may be determined traveling clockwise from circuit ground as:

$$V_B = V_1 = -8\,V \qquad (2.21)$$

The voltage V_B may also be determined traveling counterclockwise from circuit ground as:

$$V_B = V_3 - V_2 = -2\,V - 6\,V = -8\,V \qquad (2.22)$$

The negative signs in Equations (2.21) and (2.22) indicate that the voltage at point B is 8 V below circuit ground.

PRACTICE EXERCISE 2–5

Determine V_A and V_B in the circuit diagram in Figure 2–22.

FIGURE **2–22**

Circuit Diagram for Practice Exercise 2–5

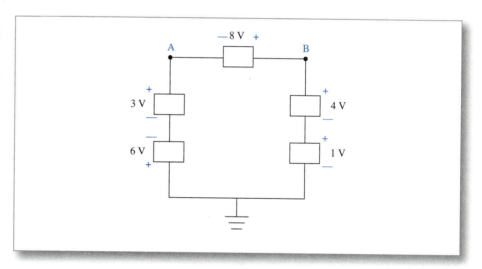

✓ANSWER

$V_A = -3\,V; \quad V_B = 5\,V$

(Complete solutions to the Practice Exercises can be found in Appendix A.)

2.9 THE SINGLE LINE CIRCUIT DIAGRAM

Single line circuit diagrams are often created to simplify or clarify circuit connections. Consider the series DC circuit in Figure 2–23.

FIGURE **2–23**

Circuit Diagram of a
Series DC Circuit

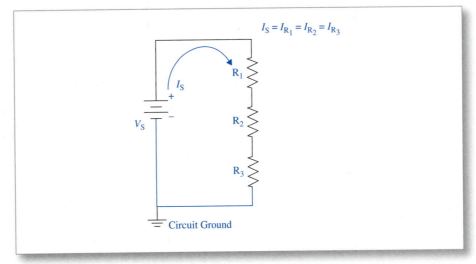

Note that all of the points along the conductor that are shaded blue in Figure 2–23 are at circuit ground. Rather than drawing the circuit to show every circuit ground point, the circuit may be drawn indicating only the most important circuit ground connections, as shown in Figure 2–24.

FIGURE **2–24**

Single Line Circuit
Diagram of a Series
DC Circuit

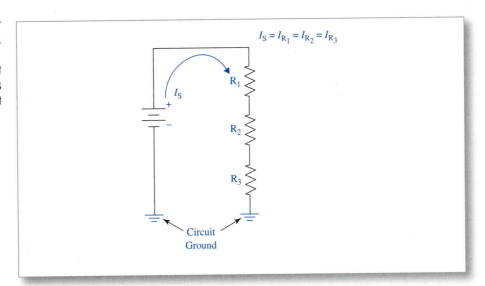

The circuit ground symbol in Figure 2–24 is used to indicate that the negative terminal of the voltage source V_S and the lower terminal of resistor R_3 are connected to circuit ground.

All circuit ground points in an electronic circuit diagram indicated by the symbol ⏚ are understood to be connected together even though the connection may be not shown in the circuit diagram.

The circuits shown in Figure 2–23 and Figure 2–24 are equivalent in every way. The circuit ground points in Figure 2–24 are understood to be connected so that current can return to the negative terminal of the voltage source through the circuit ground connection.

Most electrical and electronics engineers prefer single line circuit drawings as shown in Figure 2–24, and the single line drawing will frequently appear throughout the remainder of this text.

2.10 THE PARALLEL DC CIRCUIT

We have seen that resistors connected in series have the same current flowing through them and that the voltage developed across each resistor is directly proportional to the resistance of each respective resistor.

Resistors are connected in **parallel** in an electronic circuit so that the same voltage exists across each resistor. The current through resistors connected in parallel is inversely proportional to the resistance of each respective resistor.

The voltage source V_S in Figure 2–25 is connected directly across each of the resistors. The current that flows through each resistor is indicated respectively as I_1, I_2, and I_3.

FIGURE **2–25**

A DC Parallel Circuit

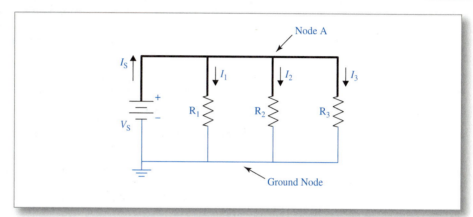

A **circuit node** is defined as a collection of points in a circuit that are at the same voltage with respect to circuit ground.

The boundary of node A is shaded black and the boundary of the circuit ground node is shaded blue in Figure 2–25. All points within the boundary of a given node are at the same voltage level because there is no resistor or voltage source within either node boundary that would cause a voltage change. Specifically, all of the points within the boundary designated as node A in Figure 2–25 are at the voltage $+V_S$ with respect to circuit ground, and all of the points within the boundary of the circuit ground node are at ground voltage typically assumed to be zero.

Resistors that are connected in parallel in an electronic circuit must be connected between the same two circuit nodes.

Note that each resistor and the voltage source V_S are connected between the same two nodes in Figure 2–25 and are therefore connected in parallel.

2.11 KIRCHHOFF'S CURRENT LAW [KCL]

Kirchhoff's current law (KCL) states that the sum of the currents entering a circuit node must equal the sum of the currents that leave the node. The statement may be expressed mathematically as:

$$\sum I_{\text{entering node}} = \sum I_{\text{leaving node}}$$

(2.23)

The illustration in Figure 2–26 illustrates a circuit node that has currents entering and leaving the node as indicated. KCL may be applied at the node to obtain:

$$I_1 + I_4 = I_2 + I_3$$

FIGURE **2–26**

An Illustration of Kirchhoff's Current Law

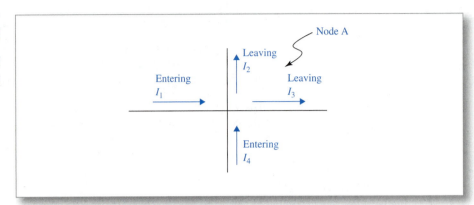

Note that currents I_1 and I_4 are entering the node and currents I_2 and I_3 are leaving the node. KCL requires that the sum of the currents entering a node must equal the sum of the currents leaving the node.

KCL may be applied at node A of the circuit in Figure 2–25 to obtain:

$$I_S = I_1 + I_2 + I_3 \tag{2.24}$$

where I_S is the current *entering* node A and $(I_1 + I_2 + I_3)$ is the sum of currents *leaving* node A.

KCL may also be applied at the ground node in Figure 2–25 to obtain:

$$I_1 + I_2 + I_3 = I_S$$

where $(I_1 + I_2 + I_3)$ is the sum of currents entering the ground node and I_S is the current leaving the circuit ground node and returning to the negative terminal of the voltage source V_S.

2.12 TOTAL EQUIVALENT RESISTANCE AND CONDUCTANCE

Consider the circuit in Figure 2–27.

FIGURE **2–27**

Equivalent Resistance of Parallel Resistors

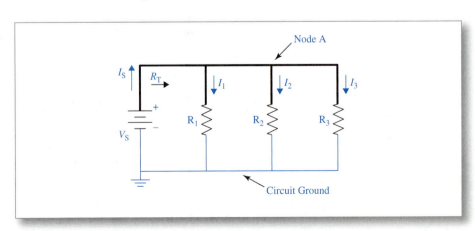

The current paths in a parallel circuit are often referred to as *branches* so that the currents I_1, I_2, and I_3 may be referred to as *branch currents*.

Ohm's law may be used to determine the respective branch currents in Figure 2–27 as:

$$I_1 = \frac{V_S}{R_1}; \quad I_2 = \frac{V_S}{R_2}; \quad \text{and} \quad I_3 = \frac{V_S}{R_3} \qquad (2.25)$$

Ohm's law may be used to calculate the source current as:

$$I_S = \frac{V_S}{R_T}$$

where R_T represents the total equivalent resistance of the parallel resistors.

Applying KCL at node A, we obtain:

$$I_S = I_1 + I_2 + I_3$$

and substituting equivalent expressions from Equation (2.25), we obtain:

$$\frac{V_S}{R_T} = \frac{V_S}{R_1} + \frac{V_S}{R_2} + \frac{V_S}{R_3} \qquad (2.26)$$

Dividing each term of Equation (2.26) by V_S, we obtain:

$$\frac{1}{R_T} = \frac{1}{R_1} + \frac{1}{R_2} + \frac{1}{R_3} \qquad (2.27)$$

The expression in Equation (2.27) may be extended to include any number of resistors connected in parallel so that the total equivalent resistance of N resistors connected in parallel may be determined as:

$$\frac{1}{R_T} = \frac{1}{R_1} + \frac{1}{R_2} + \frac{1}{R_3} + \cdots + \frac{1}{R_N} \qquad (2.28)$$

The product-over-sum method is an alternative method frequently used to calculate equivalent resistance of resistors connected in parallel.

Considering only two parallel resistors, Equation (2.28) can be expressed as:

$$\frac{1}{R_T} = \frac{1}{R_1} + \frac{1}{R_2} \qquad (2.29)$$

A common denominator for the terms on the right-hand side of Equation (2.29) is $R_1 R_2$ so that Equation (2.29) can be expressed as:

$$\frac{1}{R_T} = \frac{R_2}{R_1 R_2} + \frac{R_1}{R_1 R_2} = \frac{R_1 + R_2}{R_1 R_2} \qquad (2.30)$$

Taking the reciprocal of both sides of Equation (2.30), we obtain:

$$R_T = \frac{R_1 R_2}{R_1 + R_2} \qquad (2.31)$$

The equivalent resistance of two resistors connected in parallel may be calculated as the product of the two resistor values divided by the sum of the two resistor values. Although the product-over-sum method for calculating equivalent resistance is limited to only two resistors, the method can be reiterated for any number of parallel resistors.

Conductance (G) is defined as $\dfrac{1}{R}$ and is measured in units of siemens (S).

Equation (2.28) may be expressed in terms of conductance as:

$$G_T = G_1 + G_2 + G_3 + \cdots + G_N \tag{2.32}$$

where G_T represents the **total equivalent conductance** of the circuit.

Total equivalent resistance may be calculated as the reciprocal of total equivalent conductance so that:

$$R_T = \frac{1}{G_T} \tag{2.33}$$

EXAMPLE 2-8 Calculating Total Equivalent Resistance

Calculate the total equivalent resistance of the parallel resistors in Figure 2–28.

FIGURE **2-28**

Example 2–8:
Calculating
Total Equivalent
Resistance

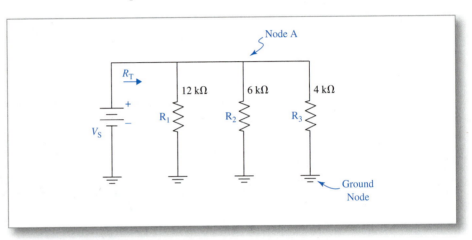

✓ **SOLUTION**

The resistors and the voltage source are connected between node A and circuit ground and are therefore by definition connected in parallel.

Equation (2.28) may be used to calculate total equivalent resistance as:

$$\frac{1}{R_T} = \frac{1}{R_1} + \frac{1}{R_2} + \frac{1}{R_3} = \frac{1}{12 \text{ k}\Omega} + \frac{1}{6 \text{ k}\Omega} + \frac{1}{4 \text{ k}\Omega}$$

so that: $\qquad\qquad R_T = 2 \text{ k}\Omega$

The calculation of total equivalent resistance can be keyed as:

(①②⌈EE⌉③∧−①+⑥⌈EE⌉③−①+④⌈EE⌉③∧−①) ⌈ENTER⌉ to obtain 0.5E-3 ; keying ∧−① ⌈ENTER⌉ gives the reciprocal required by Equation (2.28) to obtain $R_T = 2 \text{ k}\Omega$.

The total equivalent resistance may also be calculated as the reciprocal of the total equivalent conductance as:

$$G_1 = \frac{1}{R_1} = \frac{1}{12 \text{ k}\Omega} = 83.33 \text{ }\mu\text{S}$$

$$G_2 = \frac{1}{R_2} = \frac{1}{6 \text{ k}\Omega} = 166.67 \text{ }\mu\text{S}$$

$$G_3 = \frac{1}{R_3} = \frac{1}{4\text{ k}\Omega} = 250.00\ \mu\text{S}$$

so that:

$$G_T = G_1 + G_2 + G_3 = 500\ \mu\text{S}$$

The total equivalent resistance is calculated as the reciprocal of total equivalent conductance to obtain:

$$R_T = \frac{1}{G_T} = \frac{1}{500\ \mu\text{S}} = 2\text{ k}\Omega \qquad (2.34)$$

The total equivalent resistance may also be determined using the product-over-sum method to obtain:

$$R_{1//2} = \frac{R_1 R_2}{R_1 + R_2} = \frac{(12\text{ k}\Omega)(6\text{ k}\Omega)}{(12\text{ k}\Omega + 6\text{ k}\Omega)} = 4\text{ k}\Omega$$

$$R_T = \frac{R_{1//2} R_3}{R_{1//2} + R_3} = \frac{(4\text{ k}\Omega)(4\text{ k}\Omega)}{(4\text{ k}\Omega) + 4\text{ k}\Omega)} = 2\text{ k}\Omega$$

The symbol $R_{1//2}$ represents the equivalent resistance of R_1 and R_2 connected in parallel. Note how the product-over-sum method can be reiterated to obtain the total equivalent resistance of any number of resistors connected in parallel.

PRACTICE EXERCISE 2–6

A 2 kΩ, 8 kΩ, 16 kΩ, and 24 kΩ resistor is connected in parallel; calculate the total equivalent resistance using Equation (2.28). Calculate the total equivalent resistance using the product-over-sum method. Also calculate the total equivalent resistance as the reciprocal of total equivalent conductance.

✓ANSWER

$R_T = 1.37\text{ k}\Omega$

(Complete solutions to the Practice Exercises can be found in Appendix A.)

2.13 ANALYSIS OF A PARALLEL DC CIRCUIT

EXAMPLE 2–9   Analysis of a Parallel DC Circuit

Calculate the branch currents (I_1, I_2, and I_3) and the source current (I_S) for the circuit in Figure 2–29. Verify the values of the currents using computer simulation.

FIGURE **2-29**

Circuit Diagram
for Example 2–9

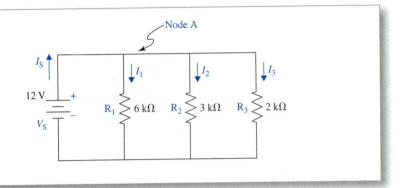

✔ **SOLUTION**

Ohm's law may be used to determine the current that flows through each respective (branch) resistor. The voltage across each (branch) resistor is the source voltage V_S so that:

$$I_1 = \frac{V_S}{R_1} = \frac{12 \text{ V}}{6 \text{ k}\Omega} = 2 \text{ mA}$$

$$I_2 = \frac{V_S}{R_2} = \frac{12 \text{ V}}{3 \text{ k}\Omega} = 4 \text{ mA}$$

$$I_3 = \frac{V_S}{R_3} = \frac{12 \text{ V}}{2 \text{ k}\Omega} = 6 \text{ mA}$$

KCL may be applied at node A to obtain the source current as:

$$I_S = I_1 + I_2 + I_3 = 2 \text{ mA} + 4 \text{ mA} + 6 \text{ mA} = 12 \text{ mA}$$

The source current may also be determined by calculating total equivalent resistance as:

$$\frac{1}{R_T} = \frac{1}{R_1} + \frac{1}{R_2} + \frac{1}{R_3} = \frac{1}{6 \text{ k}\Omega} + \frac{1}{3 \text{ k}\Omega} + \frac{1}{2 \text{ k}\Omega}$$

so that:
$$R_T = 2 \text{ k}\Omega$$

and applying Ohm's law to obtain:

$$I_S = \frac{V_S}{R_T} = \frac{12 \text{ V}}{1 \text{ k}\Omega} = 12 \text{ mA}$$

The definition of conductance may be used to determine the total equivalent conductance of the circuit as:

$$G_1 = \frac{1}{R_1} = \frac{1}{6 \text{ k}\Omega} = 166.67 \text{ μS}$$

$$G_2 = \frac{1}{R_2} = \frac{1}{3 \text{ k}\Omega} = 333.33 \text{ μS}$$

$$G_3 = \frac{1}{R_3} = \frac{1}{2 \text{ k}\Omega} = 500.00 \text{ μS}$$

so that:
$$G_T = G_1 + G_2 + G_3 = 1000 \text{ μS} = 1 \text{ mS}$$

The total equivalent resistance is calculated as the reciprocal of total equivalent conductance to obtain:

$$R_T = \frac{1}{G_T} = \frac{1}{1 \text{ mS}} = 1 \text{ k}\Omega$$

The product-over-sum method may also be used to calculate the total equivalent resistance of the parallel resistors by calculating the equivalent resistance of the first two resistors as:

$$R_{1//2} = \frac{R_1 R_2}{R_1 + R_2} = \frac{(6 \text{ k}\Omega)(3 \text{ k}\Omega)}{(6 \text{ k}\Omega + 3 \text{ k}\Omega)} = 2 \text{ k}\Omega$$

$$R_T = \frac{R_{1//2} R_3}{R_{1//2} + R_3} = \frac{(2 \text{ k}\Omega)(2 \text{ k}\Omega)}{(2 \text{ k}\Omega + 2 \text{ k}\Omega)} = 1 \text{ k}\Omega$$

The MultiSIM simulation of the circuit is shown in Figure 2–30.

FIGURE **2–30**

MultiSIM Simulation for Example 2–9

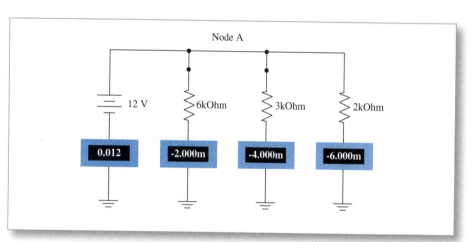

Note that the branch currents indicated by the ammeters in the MultiSIM simulation are negative, indicating that the currents are flowing away from node A. The source current ammeter is a positive value, indicating that the source current is flowing into node A.

PRACTICE EXERCISE 2–7

A 3 kΩ, 6 kΩ, and 12 kΩ resistor is connected in parallel with a 36 V voltage source. Determine the current through each resistor and source current. Calculate total equivalent resistance and total equivalent conductance.

✓ANSWERS

$I_{3 \text{ k}\Omega} = 12$ mA, $I_{6 \text{ k}\Omega} = 6$ mA, $I_{12 \text{ k}\Omega} = 3$ mA; $R_T = 1.714$ kΩ; $G_T = 583.43$ μS

(Complete solutions to the Practice Exercises can be found in Appendix A.)

2.14 THE CURRENT DIVIDER RULE

The current divider rule can be used to determine branch currents in a parallel circuit.

The voltage across each resistor of the parallel circuit in Figure 2–31 is the source voltage V_S.

FIGURE **2–31**

Deriving the
Current Divider Rule

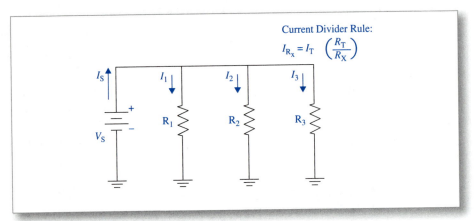

Ohm's law may be used to express the source voltage as the voltage across R_1 so that:

$$V_S = V_{R1} = I_1 R_1 \tag{2.35}$$

Ohm's law may also be used to express the source voltage as the voltage using total equivalent resistance as:

$$V_S = I_S R_T \tag{2.36}$$

The expressions in Equations (2.35) and (2.36) may be set equal to each other to obtain:

$$I_1 R_1 = I_S R_T \tag{2.37}$$

Equation (2.37) can be rearranged as:

$$I_1 = I_S \frac{R_T}{R_1} \tag{2.38}$$

Equation (2.38) is referred to as the current divider rule and may be used to calculate the branch currents for resistors connected in parallel.

The **general current divider rule** may be expressed as:

$$I_{Rx} = I_T \frac{R_T}{R_X} \tag{2.39}$$

where I_{Rx} represents the branch current through resistor R_X, I_T is the total current flowing into the parallel combination of resistors, and R_T is the total equivalent resistance of the parallel resistors.

EXAMPLE 2–10   Applying the Current Divider Rule

Use the current divider rule to determine each branch current for the circuit in Figure 2–32. Verify KCL at node A.

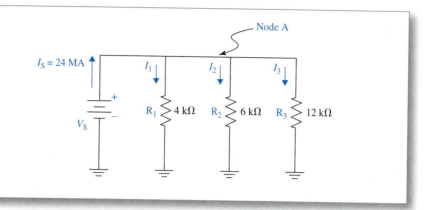

FIGURE **2–32**

Circuit Diagram for Example 2–10

✓ SOLUTION

The total equivalent resistance of the circuit may be calculated as:

$$\frac{1}{R_T} = \frac{1}{4\ k\Omega} + \frac{1}{6\ k\Omega} + \frac{1}{12\ k\Omega}$$

so that: $R_T = 2\ k\Omega$

The total current flowing into the parallel circuit in Figure 2–32 is the source current I_S, which is given as 24 mA. The current divider rule may be used to determine respective branch currents as:

$$I_1 = I_T \frac{R_T}{R_1} = \frac{(24\ mA)(2\ k\Omega)}{4\ k\Omega} = 12\ mA$$

$$I_2 = I_T \frac{R_T}{R_2} = \frac{(24\ mA)(2\ k\Omega)}{6\ k\Omega} = 8\ mA$$

$$I_3 = I_T \frac{R_T}{R_3} = \frac{(24\ mA)(2\ k\Omega)}{12\ k\Omega} = 4\ mA$$

Applying KCL at node A, we obtain:

$$\sum I_{\text{enterning node}} = \sum I_{\text{leaving node}}$$

so that: $I_S = I_1 + I_2 + I_3 = 12\ mA + 8\ mA + 4\ mA = 24\ mA$

EXAMPLE 2–11   Analysis of a Parallel Resistive Circuit

Given the parallel resistive circuit in Figure 2–33, calculate the total equivalent resistance. Calculate the total equivalent resistance using the product-over-sum method. Calculate the source current and branch currents using Ohm's law. Calculate the branch currents using the current divider rule. Verify KCL at node A. Verify the value of the currents using computer simulation.

✓ SOLUTION

The total equivalent resistance may be calculated as:

$$\frac{1}{R_T} = \frac{1}{12\ k\Omega} + \frac{1}{4\ k\Omega} + \frac{1}{3\ k\Omega}$$

so that: $R_T = 1.5\ k\Omega$

FIGURE **2–33**

Circuit Diagram
for Example 2–11

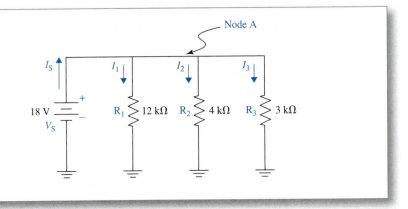

Total equivalent resistance may be calculated using the product-over-sum method to obtain:

$$R_{1//2} = \frac{R_1 R_2}{R_1 + R_2} = \frac{(12\ \text{k}\Omega)(4\ \text{k}\Omega)}{(12\ \text{k}\Omega + 4\ \text{k}\Omega)} = 3\ \text{k}\Omega$$

$$R_T = \frac{R_{1//2} R_3}{R_{1//2} + R_3} = \frac{(3\ \text{k}\Omega)(3\ \text{k}\Omega)}{(3\ \text{k}\Omega + 3\ \text{k}\Omega)} = 1.5\ \text{k}\Omega$$

Ohm's law may be used to calculate the source current to obtain:

$$I_S = \frac{V_S}{R_1} = \frac{18\ \text{V}}{1.5\ \text{k}\Omega} = 12\ \text{mA}$$

Ohm's law may be used to calculate the branch current through each respective resistor to obtain:

$$I_1 = \frac{V_S}{R_1} = \frac{18\ \text{V}}{12\ \text{k}\Omega} = 1.5\ \text{mA}$$

$$I_2 = \frac{V_S}{R_2} = \frac{18\ \text{V}}{4\ \text{k}\Omega} = 4.5\ \text{mA}$$

$$I_3 = \frac{V_S}{R_3} = \frac{18\ \text{V}}{3\ \text{k}\Omega} = 6\ \text{mA}$$

Using the current divider rule to calculate the branch current through each respective resistor, we obtain:

$$I_1 = I_T \frac{R_T}{R_1} = \frac{(12\ \text{mA})(1.5\ \text{k}\Omega)}{12\ \text{k}\Omega} = 1.5\ \text{mA}$$

$$I_2 = I_T \frac{R_T}{R_2} = \frac{(12\ \text{mA})(1.5\ \text{k}\Omega)}{4\ \text{k}\Omega} = 4.5\ \text{mA}$$

$$I_3 = I_T \frac{R_T}{R_3} = \frac{(12\ \text{mA})(1.5\ \text{k}\Omega)}{3\ \text{k}\Omega} = 6\ \text{mA}$$

where: $I_S = I_T$

Applying KCL at node A, we obtain:

$$I_S = I_1 + I_2 + I_3 = 1.5\ \text{mA} + 4.5\ \text{mA} + 6\ \text{mA} = 12\text{mA}$$

The MultiSIM simulation is shown in Figure 2–34.

FIGURE **2–34**

MultiSIM
Simulation for
Example 2–11

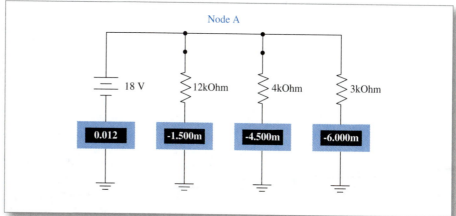

Note that the branch currents indicated by the ammeters in the MultiSIM simulation are negative, indicating that the currents are flowing away from node A. The source current ammeter is a positive value, indicating that the source current is flowing into node A.

PRACTICE EXERCISE 2–8

Given the parallel DC circuit in Figure 2–35, determine the total equivalent resistance, the total equivalent conductance, the source current, and the branch currents. Use the current divider rule to calculate the branch currents. Verify KCL at the ground node.

FIGURE **2–35**

Circuit Diagram
for Practice Exercise 2–8

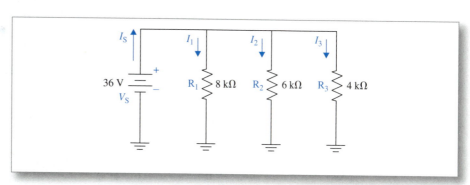

✓**ANSWERS**

$G_T = 541.67$ μS, $R_T = 1.85$ kΩ, $I_S = 19.46$ mA

$I_1 = 4.5$ mA, $I_2 = 6.0$ mA, $I_3 = 9.0$ mA

KCL: 19.5 mA $= 4.5$ mA $+ 6.0$ mA $+ 9.0$ mA

(Complete solutions to the Practice Exercise can be found in Appendix A.)

SUMMARY

An electronic circuit is a collection of components connected together to perform a specific signal-processing function.

Resistors are used in an electronic circuit either to limit the amount of current flow or to develop a voltage across the resistor.

Resistors in an electronic circuit are connected in series so that the same current flows through each resistor. The voltage across each series resistor is proportional to the resistance of each respective resistor.

The total equivalent resistance of resistors connected in series is calculated as the sum of the individual resistances:

$$R_T = R_1 + R_2 + R_3 + \cdots + R_N$$

Ohm's law states that the current that flows through a resistor is directly proportional to the voltage across the resistor and inversely proportional to the resistance value measured in ohms so that:

$$I_R = \frac{V_R}{R}$$

Ohm's law may be used to calculate the current that flows from a voltage source as:

$$I_S = \frac{V_S}{R_T}$$

where I_S is the source current, V_S is the source voltage, and R_T is the total equivalent resistance of the circuit.

Kirchhoff's voltage law states that the sum of voltages around a closed loop in a circuit is equal to 0. KVL may be stated mathematically as:

$$\sum V_{closed\ loop} = 0$$

Resistors in an electronic circuit are connected in parallel so that the current through each resistor, or branch current, is independent of any other branch current. Resistors are connected in parallel when they are connected between the same two circuit nodes. The current through each parallel resistor is inversely proportional to the resistance of each respective resistor.

The total equivalent resistance of resistors connected in parallel is calculated as:

$$\frac{1}{R_T} = \frac{1}{R_1} + \frac{1}{R_2} + \frac{1}{R_3} + \cdots + \frac{1}{R_N}$$

The total equivalent conductance of resistors connected in parallel is calculated as:

$$G_T = G_1 + G_2 + G_3 + \cdots + G_N$$

where G is defined as $\frac{1}{R}$.

Kirchhoff's current law states that the sum of the currents entering a circuit node must equal the sum of the currents leaving the node and may be stated mathematically as:

$$\sum I_{\text{entering node}} = \sum I_{\text{leaving node}}$$

The voltage divider rule states that the voltage across a resistor in a DC series circuit is equal to the source voltage multiplied by the ratio of the resistor value to the total equivalent resistance. The voltage divider rule may be stated mathematically as:

$$V_R = V_S \frac{R}{R_T}$$

The current divider rule states that the current that flows through a resistor in a DC parallel circuit is equal to the total current flowing into the parallel circuit multiplied by the ratio of the total equivalent résistance to the resistor value and may be stated mathematically as:

$$I_R = I_S \frac{R_T}{R}$$

Analysis of electronic circuits is accomplished through the careful application of Ohm's law, KVL, KCL, and the voltage and current divider rules.

Computer simulation provides an excellent verification of analytical calculations.

ESSENTIAL LEARNING EXERCISES

(ANSWERS TO ODD-NUMBERED ESSENTIAL LEARNING EXERCISES CAN BE FOUND IN APPENDIX A.)

1. Determine the nominal value of a resistor whose color bands are yellow, violet, and red.

2. Determine the nominal value of a resistor whose color bands are red, red, and brown.

3. Calculate the total equivalent resistance for the circuit in Figure 2–36. Calculate the source current and the voltage across each resistor. Verify KVL for the circuit.

FIGURE **2–36**

Circuit Diagram
for Essential Learning
Exercise 3

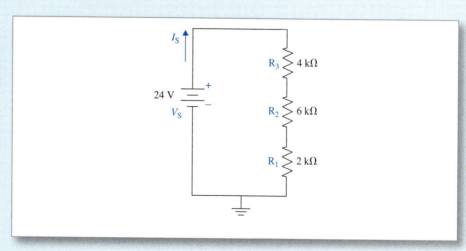

4. Calculate the total equivalent resistance for the circuit in Figure 2–37. Calculate the source current and the voltage across each resistor. Verify KVL.

FIGURE **2–37**

Circuit Diagram
for Essential Learning
Exercise 4

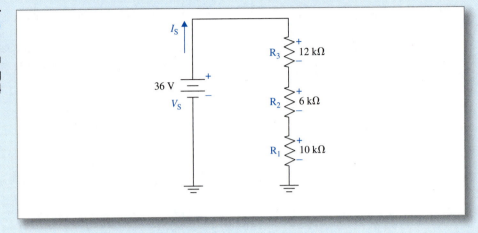

5. Calculate the voltage at point A with respect to circuit ground in Figure 2–38.

FIGURE **2–38**

Circuit Diagram
for Essential Learning
Exercise 5

6. Calculate the voltage at point A with respect to circuit ground in Figure 2–39.

FIGURE **2–39**

Circuit Diagram
for Essential Learning
Exercise 6

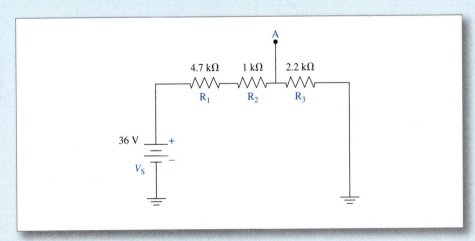

7. Use the voltage divider rule to determine the voltage across each resistor in Figure 2–40. Verify KVL for the circuit.

FIGURE **2–40**

Circuit Diagram
for Essential Learning
Exercise 7

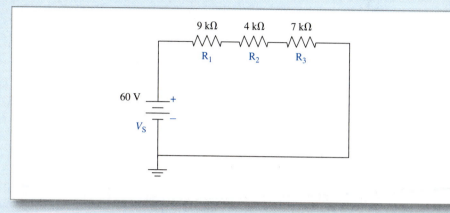

8. Use the voltage divider rule to determine the voltage across each resistor in Figure 2–41. Verify KVL for the circuit.

FIGURE **2–41**

Circuit Diagram
for Essential Learning
Exercise 8

V_S +28 V

R_1 2 kΩ

R_2 3.3 kΩ

R_3 4.7 kΩ

R_4 1 kΩ

9. Use the voltage divider rule to determine the voltage at point A in Figure 2–42. Determine the voltage at point A using Ohm's law and compare results.

FIGURE **2–42**

Circuit Diagram
for Essential Learning
Exercise 9

V_S +36 V

R_1 12 kΩ

R_2 6 kΩ

A

R_3 4 kΩ

R_4 2 kΩ

10. Use the voltage divider rule to determine the voltage at point A and point B in Figure 2–43. Determine the voltage at point A and point B using Ohm's law and compare results.

FIGURE **2–43**

Circuit Diagram for Essential Learning Exercise 10

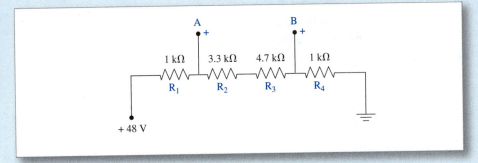

11. Verify KVL for each of the diagrams in Figure 2–44.

FIGURE **2–44**

Circuit Diagram for Essential Learning Exercise 11

12. Use KVL to determine the magnitude and polarity of V_X for each of the diagrams in Figure 2–45.

FIGURE **2–45**

Circuit Diagram
for Essential Learning
Exercise 12

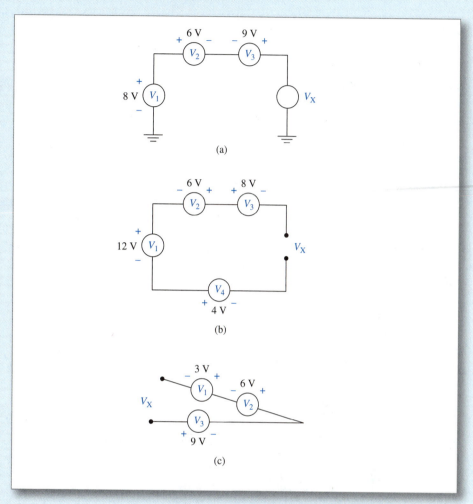

13. Determine the voltage across each resistor in Figure 2–46. Verify KVL for the circuit.

FIGURE **2–46**

Circuit Diagram
for Essential Learning
Exercise 13

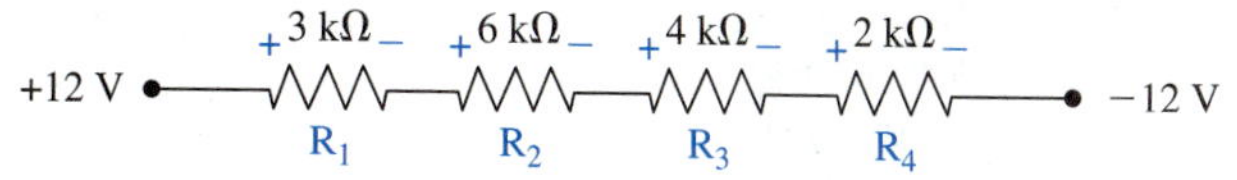

14. Determine the value of R_2 in Figure 2–47.

FIGURE **2–47**

Circuit Diagram
for Essential Learning
Exercise 14

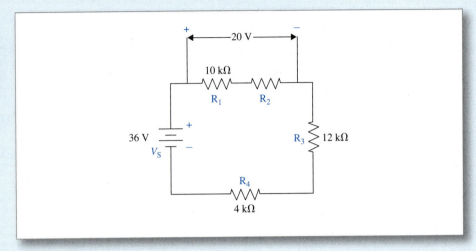

15. Determine the value of R_1 and R_3 in Figure 2–48.

FIGURE **2–48**

Circuit Diagram
for Essential Learning
Exercise 15

16. The voltages indicated in Figure 2–49 are measured with respect to circuit ground. Determine the voltage across each resistor and the voltage at point A with respect to circuit ground.

FIGURE **2–49**

Circuit Diagram
for Essential Learning
Exercise 16

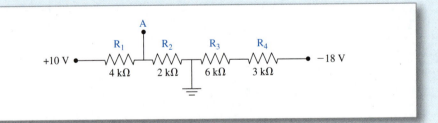

17. Determine the voltage across each resistor and the voltage at point A with respect to circuit ground in Figure 2–50.

FIGURE **2–50**

Circuit Diagram
for Essential Learning
Exercise 17

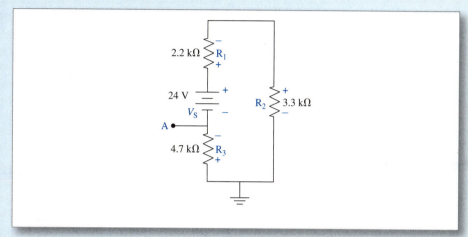

18. Determine the total equivalent resistance and total equivalent conductance for the circuit in Figure 2–51.

FIGURE **2–51**

Circuit Diagram
for Essential Learning
Exercise 18

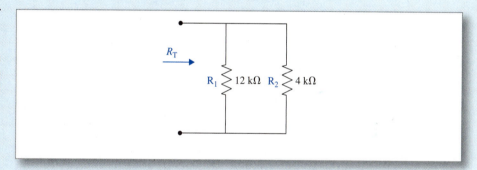

19. Determine the total equivalent resistance and total equivalent conductance for the circuit in Figure 2–52.

FIGURE **2–52**

Circuit Diagram
for Essential Learning
Exercise 19

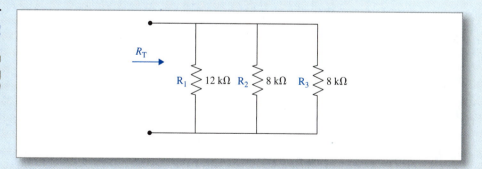

20. Determine the source current and the current through each resistor (branch) in Figure 2–53. Verify KCL at Node A.

FIGURE **2–53**

Circuit Diagram
for Essential Learning
Exercise 20

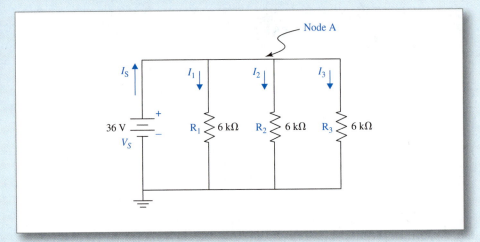

21. Determine the source current and the current through each resistor (branch) in Figure 2–54.

FIGURE **2–54**

Circuit Diagram
for Essential Learning
Exercise 21

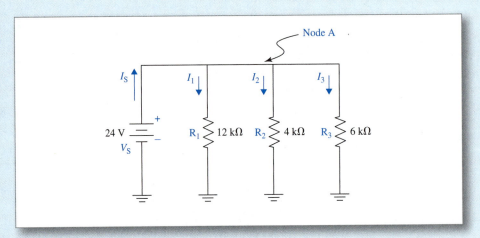

22. Determine the source current and the current through each resistor (branch) in Figure 2–55. Verify KCL at Node A.

FIGURE **2–55**

Circuit Diagram
for Essential Learning
Exercise 22

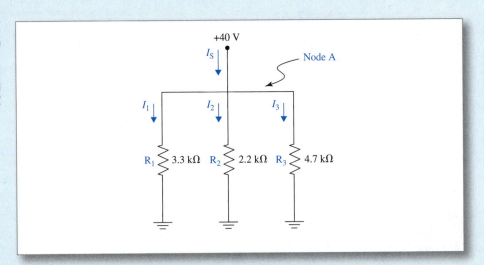

23. Determine the magnitude and direction of *I* in Figure 2–56.

FIGURE **2-56**

Circuit Diagram
for Essential Learning
Exercise 23

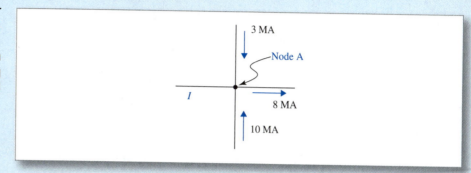

24. Determine the magnitude and direction of I_1 and I_2 in Figure 2–57.

FIGURE **2-57**

Circuit Diagram
for Essential Learning
Exercise 24

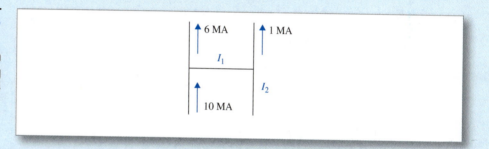

25. Determine the source current and current through each resistor in Figure 2–58. Verify KCL at node A.

FIGURE **2-58**

Circuit Diagram
for Essential Learning
Exercise 25

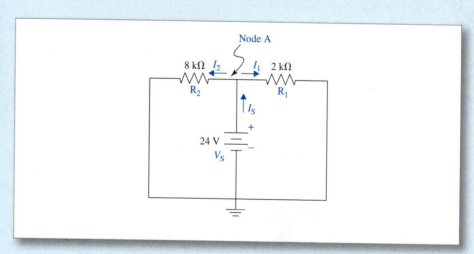

26. Use the current divider rule to determine the current through each resistor in Figure 2–59. Verify KCL at node A.

FIGURE **2–59**

Circuit Diagram for Essential Learning Exercise 26

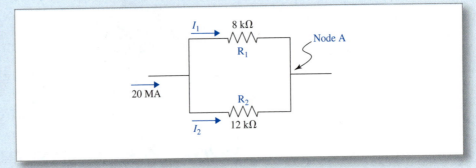

27. Use the current divider rule to determine the current through each resistor in Figure 2–60. Verify KCL at node A.

FIGURE **2–60**

Circuit Diagram for Essential Learning Exercise 27

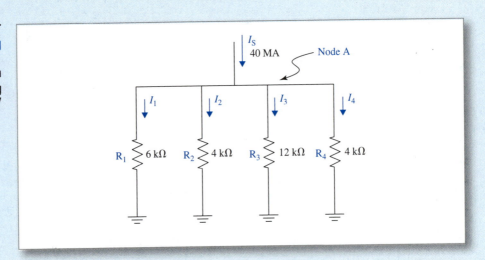

28. Determine the total equivalent resistance of the circuit in Figure 2–61. Determine the source current and current through each resistor.

FIGURE **2–61**

Circuit Diagram for Essential Learning Exercise 28

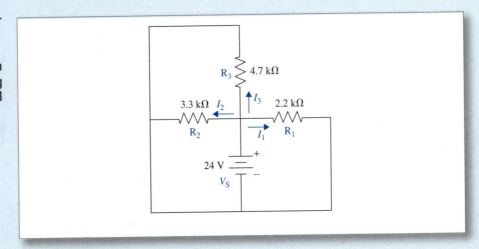

29. The voltage at point A (V_A) in Figure 2–62 is 36 V is given as 36 V. Determine the current through each resistor.

FIGURE **2–62**

Circuit Diagram
for Essential Learning
Exercise 29

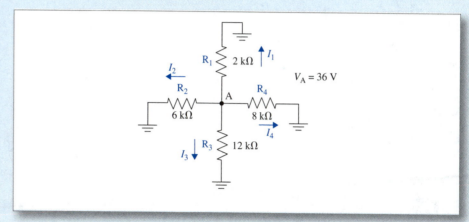

30. Determine the value of the source current in Figure 2–63.

FIGURE **2–63**

Circuit Diagram
for Essential Learning
Exercise 30

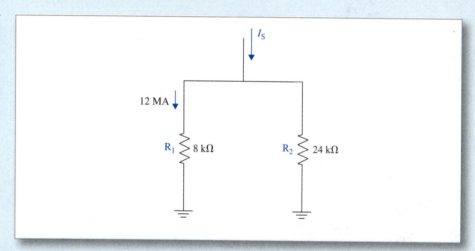

CHALLENGING LEARNING EXERCISES

(ANSWERS TO SELECTED CHALLENGING LEARNING EXERCISES CAN BE FOUND IN APPENDIX A.)

1. Use Ohm's law and KVL to determine the output voltage (V_{out}) with respect to circuit ground in Figure 2–64. Verify your solution using computer simulation.

FIGURE **2–64**

Circuit Diagram
for Challenging Learning
Exercise 1

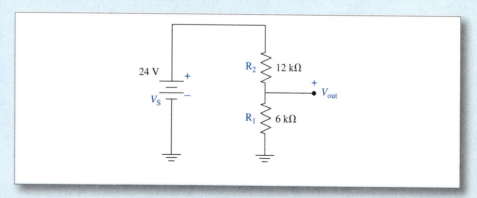

2. Use Ohm's law and KVL to determine the voltage at point A (V_A) with respect to circuit ground in Figure 2–65. Verify your solution using computer simulation.

FIGURE **2–65**

Circuit Diagram for Challenging Learning Exercise 2

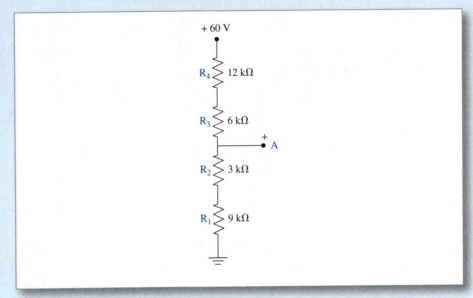

3. Use the voltage divider rule to determine the output voltage (V_{out}) with respect to circuit ground in Figure 2–64. Compare your results to those obtained in Challenging Learning Exercise 1.

4. Use the voltage divider rule to determine the voltage at point A (V_A) with respect to circuit ground in Figure 2–65. Compare your results to those obtained in Challenging Learning Exercise 2.

5. The voltage at point A in Figure 2–66 is given as 16 V. Determine the value of the unknown resistor R_4.

FIGURE **2–66**

Circuit Diagram for Challenging Learning Exercise 5

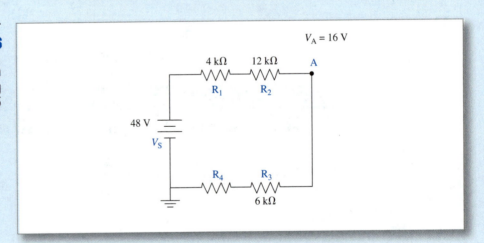

6. Determine the voltage at point A and point B with respect to circuit ground in Figure 2–67.

FIGURE **2–67**

Circuit Diagram
for Challenging Learning
Exercise 6

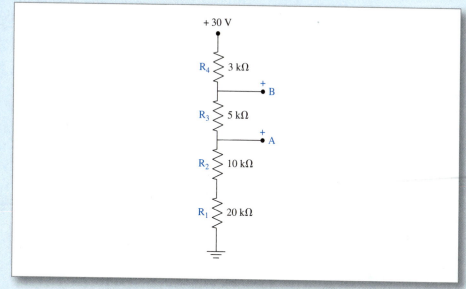

7. Determine the voltage at point A and point B with respect to circuit ground in Figure 2–68. Verify your solution using computer simulation.

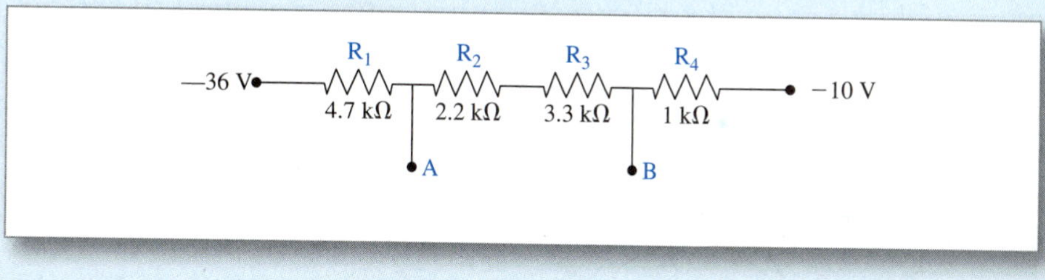

FIGURE **2–68**

Circuit Diagram
for Challenging Learning
Exercise 7

8. Use KVL to determine the voltage V_X for the diagrams in Figure 2–69.

FIGURE **2–69**

Circuit Diagrams
for Challenging Learning
Exercise 8

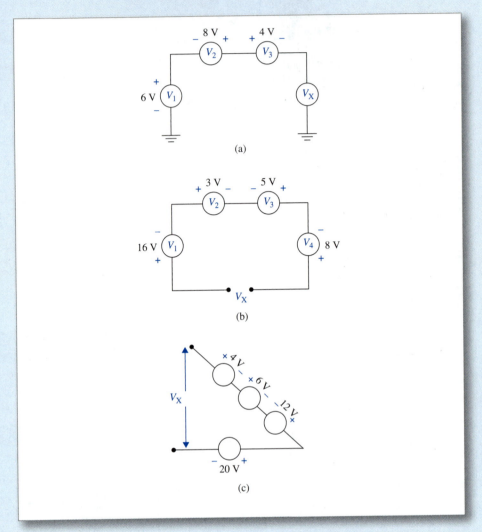

9. Use KVL to determine the voltage V_X and V_Y for the circuit diagram in Figure 2–70.

FIGURE **2–70**

Circuit Diagram
for Challenging Learning
Exercise 9

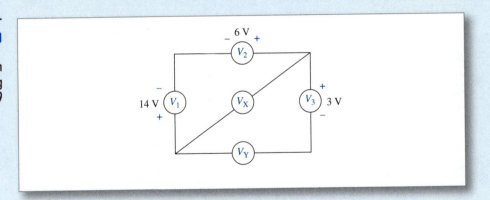

10. Determine the current through each resistor in Figure 2–71.

FIGURE **2–71**

Circuit Diagram
for Challenging Learning
Exercise 10

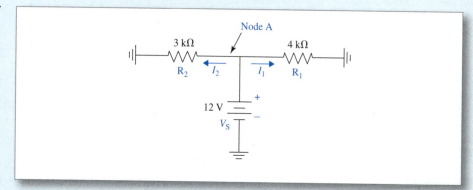

11. Use the current divider rule to determine the current through each resistor in Figure 2–72.

FIGURE **2–72**

Circuit Diagram
for Challenging Learning
Exercise 11

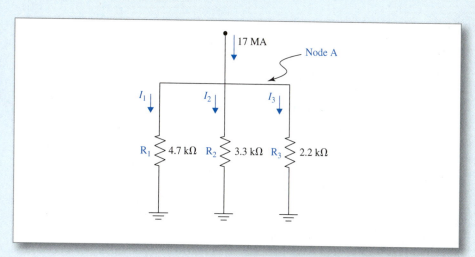

12. Determine the current I_X in Figure 2–73.

FIGURE **2–73**

Circuit Diagram
for Challenging Learning
Exercise 12

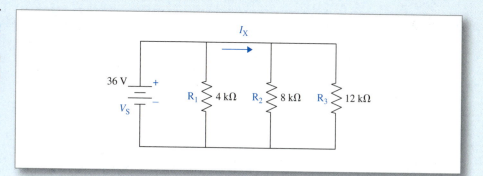

13. Determine the current I_X, I_Y, and I_Z in Figure 2–74.

FIGURE **2–74**

Circuit Diagram
for Challenging Learning
Exercise 13

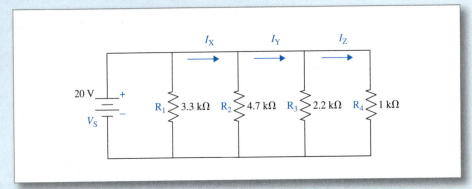

14. Determine the source current and the current through each resistor in Figure 2–75. Verify your solution using computer simulation.

FIGURE **2–75**

Circuit Diagram
for Challenging Learning
Exercise 14

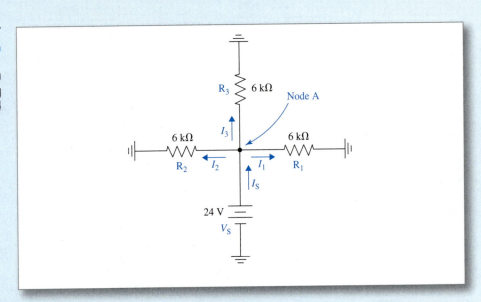

15. Design a series circuit consisting of two resistors and a 24 V DC voltage source. The circuit is to produce an output voltage of 14 V. Determine appropriate resistor values and sketch a diagram of the circuit. Verify your solution using computer simulation.

The circuit shown in Figure 2–76 is a DC voltage divider network. The function of the circuit is to produce several voltages from a single DC source.

FIGURE 2–76

Circuit Diagram for Team Activity

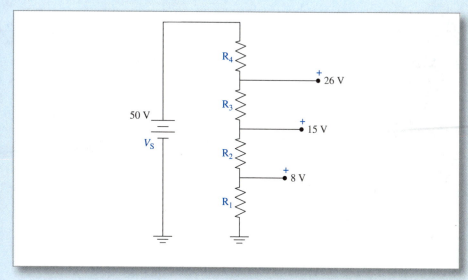

The members of your study group are asked to determine the values of R_1, R_2, R_3, and R_4 to produce the voltages indicated in Figure 2–76. Verify your design using computer simulation.

ANALYSIS OF DC COMBINATION CIRCUITS

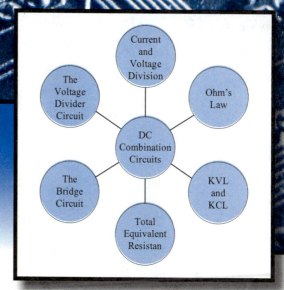

"THE ONLY DIFFERANCE BETWEEN
SUCCESS AND FAILURE IS
THE ABILITY TO TAKE ACTION."

—Alexander Graham Bell

CHAPTER OUTLINE

LEARNING OBJECTIVES

Upon successful completion of this chapter you will be able to:

- Calculate total equivalent resistance for a DC combination circuit.
- Determine the current through any resistor in a DC combination circuit.
- Determine the voltage across any resistor in a DC combination circuit.
- Determine voltage at a point with respect to circuit ground.
- Determine the voltage between two points in a DC combination circuit.
- Apply Kirchhoff's voltage law in a DC combination circuit.
- Apply Kirchhoff's current law in a DC combination circuit.
- Determine the output voltages for a DC voltage divider circuit.

CHAPTER THREE

- Determine open circuit voltage.
- Determine short circuit current.
- Calculate the voltage across a bridge circuit.

INTRODUCTION

This chapter introduces the laws and techniques that are commonly used to analyze complex electronic circuits. Although only DC circuits are analyzed in this chapter, the principles and analysis techniques learned are also utilized in the analysis of circuits whose currents and voltages vary over time. The vast majority of circuit analysis is accomplished through the careful application of Ohm's law and Kirchhoff's laws as demonstrated in this chapter.

3.1 THE COMBINATION DC CIRCUIT

It is often necessary to design electronic circuits that are more complex than the series or parallel configurations presented in previous chapters.

Combination circuits combine both series and parallel connections within the same circuit and are often referred to as series-parallel circuits.

The analysis of combination circuits is accomplished through the use of Ohm's law, Kirchhoff's laws, and the voltage and current divider rules, although the correct application of the laws is somewhat more challenging when analyzing a combination circuit.

3.2 DETERMINING TOTAL EQUIVALENT RESISTANCE

A DC circuit often consists of several resistors to form a combination circuit. Regardless of how many resistors may be present in a DC circuit or how they are connected, the combination can always be reduced to a single equivalent resistance.

The total equivalent resistance of a DC combination circuit is determined by sequentially reducing the individual series and parallel combinations to obtain a single equivalent resistance value.

3.3 OHM'S LAW AND THE COMBINATION DC CIRCUIT

Ohm's law can be used to determine the current that flows from a voltage source in a combination DC circuit by dividing the voltage of the source by the total equivalent resistance that exists between the terminals of the source.

3.4 KIRCHHOFF'S LAWS AND THE COMBINATION DC CIRCUIT

Kirchhoff's laws require that the sum of voltages around any closed loop in a circuit must be zero and that the sum of currents entering a node must equal the sum of the currents leaving the node. Kirchhoff's laws are frequently used to determine voltage or current in a combination DC circuit.

3.5 CURRENT AND VOLTAGE DIVISION IN A COMBINATION DC CIRCUIT

The voltage divider rule and current divider rule are useful tools when analyzing combination DC circuits. The voltage divider rule can be applied to resistors that are connected in series. The current divider rule can be applied to resistors or circuit branches that are connected in parallel.

3.6 VOLTAGE WITH RESPECT TO CIRCUIT GROUND

Voltage can exist between any two points in a combination DC circuit. One of the points is called the reference, and the voltage at the other point is said to be measured with respect to the reference point. The reference point for most voltage measurements is designated as circuit ground.

All numerical voltage values indicated on a circuit diagram are measured with respect to circuit ground. All voltages indicated by a single subscript such as V_A are measured between point A and circuit ground.

3.7 VOLTAGE DIFFERENCE BETWEEN TWO POINTS

Voltage difference can also be measured between two points, neither one of which is circuit ground. The two points are specified by using double subscript notation such as V_{AB}.

The second subscript of double subscript notation is always considered to be the reference point for the voltage measurement. The voltage V_{AB} indicates the voltage at point A in the circuit with respect to reference point B.

Voltage may also be measured across and individual resistor and is indicated by a subscript that identifies the resistor such as V_{R1}. The voltage V_{R1} indicates the voltage that exists across resistor R_1.

3.8 ANALYSIS OF COMBINATION DC CIRCUITS

EXAMPLE 3–1   Analysis of a Combination DC Circuit

Consider the combination circuit in Figure 3–1. Determine the total equivalent resistance, the source current, the current through each resistor, and the voltage across each resistor. Verify KVL and KCL. Determine the voltage at node A with respect to ground. Verify the solutions using computer simulation.

FIGURE **3–1**

Circuit for
Example 3–1

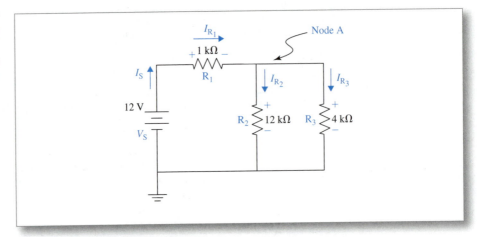

Observe that resistors R_2 and R_3 are both connected between the same two nodes (node A and circuit ground) and are therefore, by definition, connected in parallel. Observe also that the source current I_S flows through R_1 and *through the parallel combination* of R_2 and R_3. We may therefore conclude that R_1 is in series with the parallel combination of R_2 and R_3.

It is important to understand that R_1 is not in series with either R_2 or R_3, but is in series with the parallel combination of R_2 and R_3.

The total equivalent resistance of the circuit may be expressed as:

$$R_T = R_1 + (R_2 /\!/ R_3) \tag{3.1}$$

where the symbol $/\!/$ is used to indicate a parallel connection and $+$ is used to designate a series connection.

Note that the parentheses in Equation (3.1) clarify how the resistors are connected in the circuit in much the same way that parentheses are used to clarify groupings of variables in algebraic notation. The expression in Equation (3.1) indicates the resistors R_2 and R_3 are connected in parallel and the parallel combination is connected in series with R_1.

The product-over-sum method can be used to calculate the equivalent resistance of the parallel combination of $R_2/\!/R_3$ to obtain:

$$(R_2 /\!/ R_3) = \frac{(R_2)(R_3)}{(R_2 + R_3)} = \frac{(12 \text{ k}\Omega)(4 \text{ k}\Omega)}{(12 \text{ k}\Omega + 4 \text{ k}\Omega)} = 3 \text{ k}\Omega \tag{3.2}$$

The total equivalent resistance can be calculated as:

$$R_T = R_1 + (R_2 /\!/ R_3) = 1 \text{ k}\Omega + 3 \text{ k}\Omega = 4 \text{ k}\Omega$$

The total equivalent resistance of an electronic circuit is the resistance that is "sensed" by the voltage source (V_S). Neither the number of circuit components nor their particular configuration is relevant to the voltage source. The current that flows from a voltage source depends only upon the voltage of the source and the total equivalent resistance between the terminals of the source.

Ohm's law can be used to calculate source current in Figure 3–1 to obtain:

$$I_S = \frac{V_S}{R_T} = \frac{12\ V}{4\ k\Omega} = 3\ mA$$

Note that the source current flows directly through R_1 so that:

$$I_{R1} = I_S = 3\ mA$$

The *current* divider rule can be applied at node A to obtain:

$$I_{Rx} = I_T \frac{R_T}{R_X} \tag{3.3}$$

so that:

$$I_{R2} = I_T \frac{R_{(1//2)}}{R_2} = \frac{(3\ mA)(3\ k\Omega)}{12\ k\Omega} = 0.75\ mA$$

and:

$$I_{R3} = I_T \frac{R_{(1//2)}}{R_3} = \frac{(3\ mA)(3\ k\Omega)}{4\ k\Omega} = 2.25\ mA$$

where:

$$I_S = I_T$$

A very important point regarding the use of the current divider rule should be clarified. Note that $R_{(1//2)}$ is substituted for R_T in the current divider rule expressed in Equation (3.3). The current divider rule applies only to the parallel combination of R_2 and R_3 because the source current divides between these two resistors at node A. Only the equivalent resistance of the parallel resistors is placed in the numerator of the current divider rule.

KCL can be applied at node A to obtain:

$$I_{R1} = I_{R2} + I_{R3} = 0.75\ mA + 2.25\ mA = 3\ mA$$

Ohm's law can be used to obtain the voltage across each respective resistor as:

$$V_{R1} = I_{R1}R_1 = (3\ mA)(1\ k\Omega) = 3\ V$$

$$V_{R2} = I_{R2}R_2 = (0.75\ mA)(12\ k\Omega) = 9\ V$$

$$V_{R3} = I_{R3}R_2 = (2.25\ mA)(4\ k\Omega) = 9\ V$$

Note that the current used to calculate the voltage across each resistor is the respective current for each resistor.

KVL can be expressed as:

$$V_S = V_{R1} + V_{R2} = 3\ V + 9\ V = 12\ V$$

$$V_S = V_{R1} + V_{R3} = 3\ V + 9\ V = 12\ V$$

Node voltages are often used as test points when troubleshooting electronic circuits, so it is important to determine these voltages. Node voltages are measured with respect to circuit ground and are designated by single subscript notation. The voltage at node A is designated as V_A and is understood to be measured with respect to circuit ground.

The voltage at node A is determined as the algebraic sum of the voltages from any circuit ground point to node A. Beginning at source ground and traveling clockwise, V_A is determined as:

$$V_A = +12\,V - 3\,V = +9\ V$$

The voltage at node A can also be determined by traveling counterclockwise from circuit ground as either V_{R2} or V_{R3} to obtain:

$$V_A = +9\ V$$

The + sign is not required when expressing V_A; however, the + sign does emphasize that node A is 9 V above ground voltage.

Complex circuits may often be solved using alternate methods to obtain solutions. The method selected is in large part individual preference. Obtaining solutions using more than one method is an excellent way to verify that the solutions are correct. All solutions should agree regardless of the method used.

The equivalent resistance of the parallel combination of R_2 and R_3 was calculated using the product-over-sum method. The equivalent resistance could also have been calculated as:

$$\frac{1}{R_A} = \frac{1}{R_2} + \frac{1}{R_3}$$

so that:

$$R_A = 3\ k\Omega$$

which agrees with the solution obtained in Equation (3.2).

The MultiSIM simulation is shown in Figure 3–2.

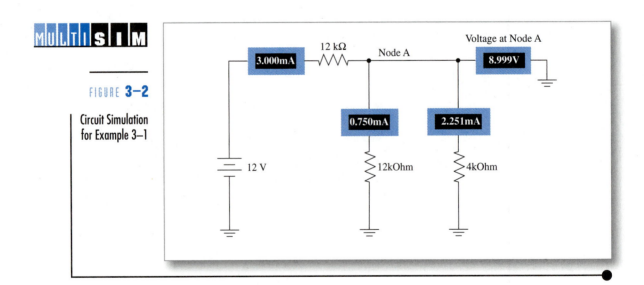

MULTISIM

FIGURE **3–2**

Circuit Simulation for Example 3–1

EXAMPLE 3–2   Analysis of a Combination DC Circuit

Given the combination circuit in Figure 3–3, determine the current through each resistor and the voltage across each resistor. Verify KCL at node A and verify KVL around each closed loop. Determine the voltage at node A. Verify the solutions using computer simulation.

FIGURE **3—3**

Circuit for
Example 3—2

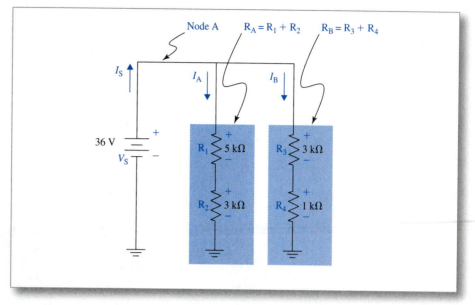

✓ **SOLUTION**

It is often helpful to view a combination circuit as a collection of branches rather than
as a collection of individual components.

The circuit in Figure 3–3 consists of two branches designated as R_A and R_B. R_A
is the equivalent resistance of the series combination of R_1 and R_2, and R_B is the
equivalent resistance of the series combination of R_3 and R_4.

The equivalent branch resistance may be calculated as:

$$R_A = R_1 + R_2 = 5 \text{ k}\Omega + 3 \text{ k}\Omega = 8 \text{ k}\Omega$$

$$R_B = R_3 + R_4 = 3 \text{ k}\Omega + 1 \text{ k}\Omega = 4 \text{ k}\Omega$$

Observe that the two branches R_A and R_B are connected between node A and circuit
ground, so the two branches are connected in parallel.

The total equivalent resistance of the entire circuit may be calculated as:

$$R_T = R_A \, /\!/ \, R_B = 8 \text{ k}\Omega \, /\!/ \, 4 \text{ k}\Omega = 2.67 \text{ k}\Omega$$

Ohm's law can be used to calculate the source current as:

$$I_S = \frac{V_S}{R_T} = \frac{36 \text{ V}}{2.67 \text{ k}\Omega} = 13.48 \text{ mA}$$

The current divider rule can be used to determine branch currents as:

$$I_A = I_S \frac{R_A \, /\!/ \, R_B}{R_A} = \frac{(13.48 \text{ mA}) \, (2.67 \text{ k}\Omega)}{8 \text{ k}\Omega} = 4.5 \text{ mA}$$

$$I_B = I_S \frac{R_A \, /\!/ \, R_B}{R_B} = \frac{(13.48 \text{ mA}) \, (2.67 \text{ k}\Omega)}{4 \text{ k}\Omega} = 9 \text{ mA}$$

The current that flows through each resistor may now be determined as:

$$I_A = I_{R1} = I_{R2} = 4.5 \text{ mA}$$

$$I_B = I_{R3} = I_{R4} = 9 \text{ mA}$$

Ohm's law can be used to calculate the voltage across each respective resistor as:

$$V_{R1} = I_{R1}R_1 = (4.5 \text{ mA})\,(5 \text{ k}\Omega) = 22.5 \text{ V}$$

$$V_{R2} = I_{R2}R_2 = (4.5 \text{ mA})\,(3 \text{ k}\Omega) = 13.5 \text{ V}$$

$$V_{R3} = I_{R3}R_3 = (9 \text{ mA})\,(3 \text{ k}\Omega) = 27 \text{ V}$$

$$V_{R4} = I_{R4}R_4 = (9 \text{ mA})\,(1 \text{ k}\Omega) = 9 \text{ V}$$

Two closed loops exist in the circuit. One closed loop is the path from source ground, across V_S, across R_1 and R_2, to circuit ground, so that KVL may be expressed as:

$$36 \text{ V} - 22.5 \text{ V} - 13.5 \text{ V} = 0$$

A second closed loop is the path from circuit ground, across V_S, across R_3 and R_4, to circuit ground, so that KVL may be expressed as:

$$36 \text{ V} - 27 \text{ V} - 9 \text{ V} = 0$$

KCL can be applied at node A to obtain:

$$I_S = I_A + I_B = 4.5 \text{ mA} + 9 \text{ mA} = 13.5 \text{ mA}$$

The voltage at node A is determined as:

$$V_A = V_S = 36 \text{ V}$$

The MultiSIM simulation is shown in Figure 3–4.

MULTISIM

FIGURE 3–4

Circuit Simulation for Example 3–2

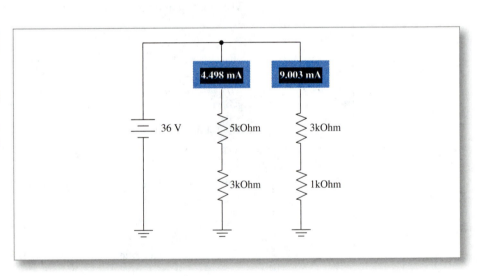

Note that the branch currents in this example could also have been determined using Ohm's law as:

$$I_A = \frac{V_S}{R_A} = \frac{36 \text{ V}}{8 \text{ k}\Omega} = 4.5 \text{ mA}$$

and:

$$I_B = \frac{V_S}{R_B} = \frac{36 \text{ V}}{4 \text{ k}\Omega} = 9 \text{ mA}$$

EXAMPLE 3-3   Analysis of a Combination DC Circuit

Given the combination circuit in Figure 3–5, determine the total equivalent resistance, the current through each resistor, and the voltage across each resistor. Verify KCL at node A. Verify KVL for each closed loop. Determine the voltage at node A. Verify the solutions using computer simulation.

FIGURE **3–5**

Circuit for
Example 3–3

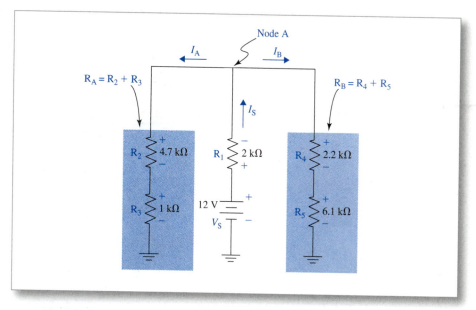

✓ SOLUTION

It is helpful to consider the series combinations of R_2 and R_3, and R_4 and R_5, to be branches of the circuit rather than as individual components. R_A represents the equivalent resistance of the series combination of R_2 and R_3, and R_B represents the equivalent resistance of the series combination of R_4 and R_5.

The branches R_A and R_B are both connected between node A and circuit ground and are therefore connected in parallel. The current that flows through R_1 flows through the parallel combination of R_A and R_B so that R_1 is in series with the parallel combination of R_A and R_B.

The total equivalent resistance of the circuit may be expressed as:

$$R_T = (R_A // R_B) + R_1$$

and may be calculated as:

$$R_A = 4.7 \text{ k}\Omega + 1 \text{ k}\Omega = 5.7 \text{ k}\Omega$$

$$R_B = 2.2 \text{ k}\Omega + 6.1 \text{ k}\Omega = 8.3 \text{ k}\Omega$$

so that: $$R_A // R_B = 5.7 \text{ k}\Omega // 8.3 \text{ k}\Omega = 3.38 \text{ k}\Omega$$

$$R_T = (R_A // R_B) + R_1 = 3.38 \text{ k}\Omega + 2 \text{ k}\Omega = 5.38 \text{ k}\Omega$$

Ohm's law can be used to calculate the source current as:

$$I_S = \frac{V_S}{R_T} = \frac{12 \text{ V}}{5.38 \text{ k}\Omega} = 2.23 \text{ mA}$$

The current divider rule may be used to calculate the branch currents I_A and I_B as:

$$I_A = I_S \frac{R_A \parallel R_B}{R_A} = \frac{(2.23 \text{ mA})(3.38 \text{ k}\Omega)}{5.7 \text{ k}\Omega} = 1.32 \text{ mA}$$

(3.4)

$$I_B = I_S \frac{R_A \parallel R_B}{R_B} = \frac{(2.23 \text{ mA})(3.38 \text{ k}\Omega)}{8.3 \text{ k}\Omega} = 0.91 \text{ mA}$$

Care must be taken when applying the current divider rule. The source current divides between the branch resistances R_A and R_B in Figure 3–5, as indicated by the expression in Equation (3.4).

The current through each respective resistor may now be determined as:

$$I_A = I_{R2} = I_{R3} = 1.32 \text{ mA}$$

$$I_B = I_{R4} = I_{R5} = 0.91 \text{ mA}$$

$$I_{R1} = I_S = 2.23 \text{ mA}$$

Ohm's law can be used to calculate the voltage across each respective resistor as:

$$V_{R1} = I_{R1} R_1 = (2.23 \text{ mA})(2 \text{ k}\Omega) = 4.46 \text{ V}$$

$$V_{R2} = I_{R2} R_2 = (1.32 \text{ mA})(4.7 \text{ k}\Omega) = 6.21 \text{ V}$$

$$V_{R3} = I_{R3} R_3 = (1.32 \text{ mA})(1 \text{ k}\Omega) = 1.32 \text{ V}$$

$$V_{R4} = I_{R4} R_4 = (0.91 \text{ mA})(2.2 \text{ k}\Omega) = 2 \text{ V}$$

$$V_{R5} = I_{R5} R_5 = (0.91 \text{ mA})(6.1 \text{ k}\Omega) = 5.55 \text{ V}$$

KCL can be applied at node A to obtain:

$$I_S = I_A + I_B = 1.32 \text{ mA} + 0.91 \text{ mA} = 2.23 \text{ mA}$$

KVL can be expressed for the three closed loops *traveling clockwise* to obtain:

$$V_{R3} + V_{R2} + V_{R1} - V_S = 0$$

so that: $$1.32 \text{ V} + 6.21 \text{ V} + 4.46 \text{ V} - 12 \text{ V} = 0$$

$$V_S - V_{R1} - V_{R4} - V_{R5} = 0$$

so that: $$12 \text{ V} - 4.46 \text{ V} - 2 \text{ V} - 5.55 \text{ V} = 0$$

$$V_{R3} + V_{R2} - V_{R4} - V_{R5} = 0$$

so that: $$1.32 \text{ V} + 6.21 \text{ V} - 2 \text{ V} - 5.55 \text{ V} = 0$$

The slight errors in the KVL expressions are due to rounding.

The voltage at node A may be determined as:

$$V_A = V_S - V_{R1} = 12 \text{ V} - 4.46 \text{ V} = 7.54 \text{ V}$$

or

$$V_A = V_{R3} + V_{R2} = 1.32 \text{ V} + 6.21 \text{ V} = 7.53 \text{ V}$$

or

$$V_A = V_{R5} + V_{R4} = 5.54 \text{ V} + 2 \text{ V} = 7.54 \text{ V}$$

The MultiSIM simulation is shown on Figure 3–6.

FIGURE **3–6**

Circuit Simulation
for Example 3–3

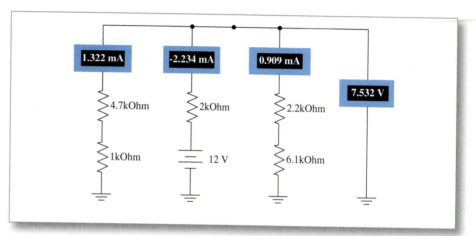

The example solutions have shown that the currents and voltages in a complex combination circuit may be obtained in a variety of ways. Although only one solution is required, solutions obtained from alternate approaches verify that calculations were performed correctly. Computer simulation also provides excellent verification.

PRACTICE EXERCISE 3–1

Given the combination DC circuit in Figure 3–7, determine the total equivalent resistance, the current through each resistor, and the voltage across each resistor. Verify KCL at node A and verify KVL around the source loop. Determine the voltage at node A.

FIGURE **3–7**

Circuit for
Practice Exercise 3–1

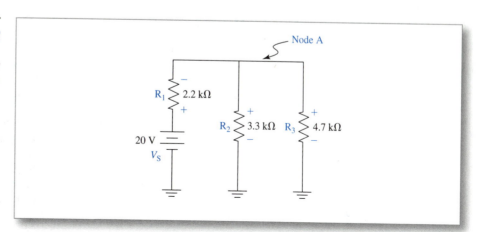

$R_T = 4.14$ kΩ; $I_{R1} = 4.83$ mA; $I_{R2} = 2.4$ mA; $I_{R3} = 1.99$ mA;

$V_{R1} = 10.63$ V; $V_{R2} = 9.35$ V; $V_{R3} = 9.37$ V; $V_A = 9.37$ V

Answers may vary slightly due to rounding differences.

(Complete solutions to the Practice Exercises can be found in Appendix A.)

EXAMPLE 3–4   Analysis of a Combination DC Circuit

Given the combination circuit in Figure 3–8, determine the total equivalent resistance, the current through each resistor, and the voltage across each resistor. Verify KCL at node A. Verify KVL for each closed loop. Determine the voltage at node A and node B. Verify the solutions using computer simulation.

FIGURE **3–8**

Circuit for Example 3–4

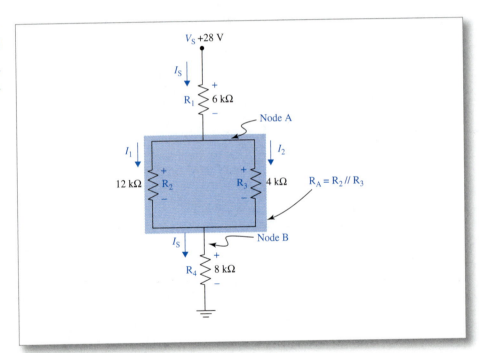

✓ SOLUTION

It is important to understand that the +28 V DC voltage indicated in Figure 3–8 is measured with respect to ground. The positive terminal of the 28 V DC voltage source is connected to the point labeled V_S on the circuit diagram; and the negative terminal of the voltage source is connected to circuit ground.

Observe that R_1 and R_2 are connected between node A and node B and are therefore connected in parallel. Let R_A represent the equivalent resistance of the parallel combination of R_2 and R_3. The source current I_S flows from the positive terminal of the voltage source V_S, through R_1, through the parallel combination R_A, and through R_4 to the negative terminal of the voltage source connected to circuit ground. Resistor R_1, and resistor R_4 have the same current (I_S) flowing through them and are therefore in series.

The total equivalent resistance of the circuit may be expressed as:

$$R_T = R_1 + R_A + R_4$$

where:
$$R_A = R_2 \,/\!/\, R_3 = 12 \text{ k}\Omega \,/\!/\, 4 \text{ k}\Omega = 3 \text{ k}\Omega$$

so that:
$$R_T = 6 \text{ k}\Omega + 3 \text{ k}\Omega + 8 \text{ k}\Omega = 17 \text{ k}\Omega$$

Ohm's law can be used to calculate the source current as:

$$I_S = \frac{V_S}{R_T} = \frac{28 \text{ V}}{17 \text{ k}\Omega} = 1.65 \text{ mA}$$

The current divider rule can be used to calculate the branch currents I_1 and I_2 as:

$$I_1 = I_S \frac{R_A}{R_1} = \frac{(1.65 \text{ mA})(3 \text{ k}\Omega)}{12 \text{ k}\Omega} = 0.41 \text{ mA}$$

$$I_2 = I_S \frac{R_A}{R_2} = \frac{(1.65 \text{ mA})(3 \text{ k}\Omega)}{4 \text{ k}\Omega} = 1.24 \text{ mA}$$

The current through each respective resistor may now be determined as:

$$I_S = I_{R1} = I_{R4} = 1.65 \text{ mA}$$

$$I_{R2} = 0.41 \text{ mA}$$

$$I_{R3} = 1.24 \text{ mA}$$

Ohm's law can be used to calculate the voltage across each respective resistor as:

$$V_{R1} = I_{R1}R_1 = (1.65 \text{ mA})(6 \text{ k}\Omega) = 9.9 \text{ V}$$

$$V_{R2} = I_{R2}R_2 = (0.41 \text{ mA})(12 \text{ k}\Omega) = 4.92 \text{ V}$$

$$V_{R3} = I_{R3}R_3 = (1.24 \text{ mA})(4 \text{ k}\Omega) = 4.96 \text{ V}$$

$$V_{R4} = I_{R4}R_4 = (1.65 \text{ mA})(8 \text{ k}\Omega) = 13.2 \text{ V}$$

KCL can be applied at node A to obtain:

$$I_S = I_1 + I_2 = 0.41 \text{ mA} + 1.24 \text{ mA} = 1.65 \text{ mA}$$

KVL can be expressed beginning at the positive terminal of V_S and traveling to circuit ground as:

$$V_S - V_{R1} - V_{R2} - V_{R4} = 0$$

so that:
$$28 \text{ V} - 9.9 \text{ V} - 4.92 \text{ V} - 13.2 \text{ V} = 0 \text{ (approximate)}$$

The voltage at node A may be determined as:

$$V_A = V_S - V_{R1} = 28 \text{ V} - 9.9 \text{ V} = 18.1 \text{ V}$$

or
$$V_A = V_{R4} + V_{R2} = 13.2 \text{ V} + 4.92 \text{ V} = 18.12 \text{ V}$$

The voltage at node B may be determined as:

$$V_B = V_{R4} = 13.2 \text{ V}$$

The MultiSIM simulation is shown in Figure 3–9.

MULTISIM

FIGURE **3–9**

Circuit Simulation
for Example 3–4

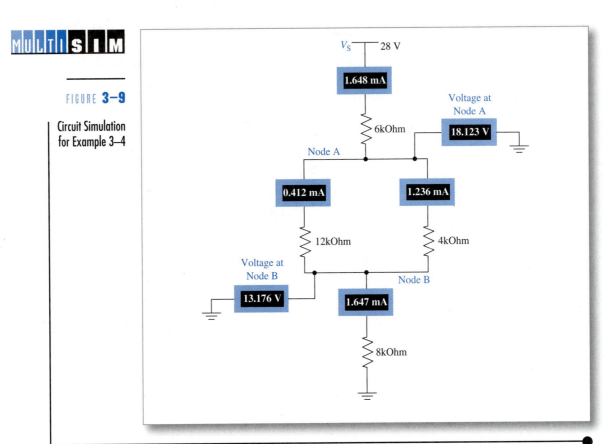

EXAMPLE **3–5**   Analysis of a Combination DC Circuit

Given the DC circuit in Figure 3–10, determine the current through each resistor and the voltage at node A.

FIGURE **3–10**

Circuit for
Example 3–5

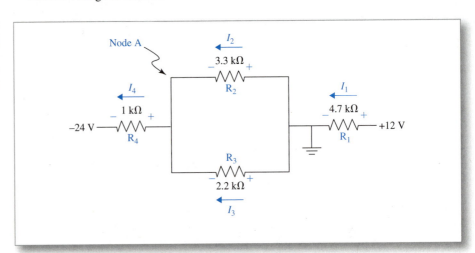

✓ SOLUTION

It is important to recall that all of the voltage values indicated on the circuit diagram in Figure 3–10 are referenced to circuit ground.

It is also important to recall that Ohm's law may be used to calculate current flow by dividing the voltage difference between any two points in a circuit by the total equivalent resistance between the same two points.

Observe that one end of resistor R_1 is connected to circuit ground and the other end is connected to a point whose voltage is 12 V with respect to ground. Therefore, the voltage across R_1 is 12 V.

Ohm's law can be used to calculate the current through R_1 as:

$$I_1 = \frac{12 \text{ V}}{4.7 \text{ k}\Omega} = 2.55 \text{ mA}$$

The direction of current flow through R_1 is from positive to negative as indicated in Figure 3–10.

Observe that the parallel combination of R_2 and R_3 is connected in series with resistor R_4. Observe also that the voltage across the combination of R_2, R_3, and R_4 is 24 V.

The equivalent resistance of the combination of R_2, R_3, and R_4 can be calculated as:

$$R_E = R_2 /\!/ R_3 + R_4 = 1.32 \text{ k}\Omega + 1 \text{ k}\Omega = 2.32 \text{ k}\Omega$$

Ohm's law can be used to calculate the current through R_4 as:

$$I_4 = \frac{24 \text{ V}}{2.32 \text{ k}\Omega} = 10.34 \text{ mA}$$

The direction of current flow through R_4 is from positive to negative as indicated in Figure 3–10.

The current divider rule may be used to calculate the current through R_2 and R_3 as:

$$I_2 = I_4 \frac{R_2 /\!/ R_3}{R_2} = 4.14 \text{ mA}$$

$$I_3 = I_4 \frac{R_2 /\!/ R_3}{R_3} = 6.20 \text{ mA}$$

The direction of current flow through R_2 and R_3 is from positive to negative as indicated in Figure 3–10.

The voltage at node A with respect to circuit ground may be determined as the voltage across R_2 as:

$$V_A = V_{R2} = I_3 R_2 = (4.14 \text{ mA})(3.3 \text{ k}\Omega) = -13.66 \text{ V}$$

The negative sign indicates that the voltage at node A is 13.66 V below circuit ground.

The voltage across R_4 may be calculated as:

$$V_{R4} = I_4 R_4 = (10.34 \text{ mA})(1 \text{ k}\Omega) = 10.34 \text{ V}$$

The voltage at node A may be determined as:

$$V_A = -24 \text{ V} + 10.34 \text{ V} = -13.66 \text{ V}$$

The MultiSIM simulation is shown in Figure 3–11.

FIGURE **3–11**

Circuit Simulation
for Example 3–5

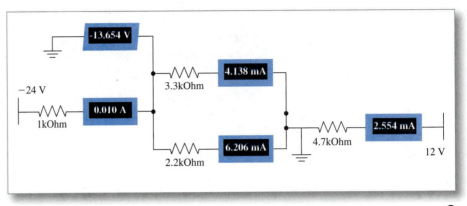

PRACTICE EXERCISE 3–2

Given the DC circuit in Figure 3–12, determine the current through each resistor and the voltage at node A.

FIGURE **3–12**

Circuit for Practice
Exercise 3–2

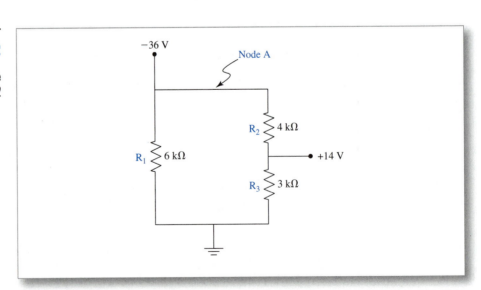

✔ **ANSWERS**

$I_{R1} = 6$ mA; $I_{R2} = 5.5$ mA; $I_{R3} = 4.67$ mA; $V_A = V_S = 36$ V

(Complete solutions to the Practice Exercises can be found in Appendix A.)

3.9 THE DC VOLTAGE DIVIDER CIRCUIT

Electronic systems typically require more than one DC voltage for proper operation. The purpose of a DC voltage divider circuit is to create multiple DC voltages from a single DC voltage source.

Most electronic circuits are designed to supply current and voltage to a designated load. A speaker in an audio system and a DC motor in a process control system are examples of circuit loads.

A circuit with no-load implies that the designated load terminals are open so that current to the load is zero.

EXAMPLE 3-6   Analysis of a DC Voltage Divider Circuit

Given the DC voltage divider circuit in Figure 3–13, determine the voltage at the external connection points labeled as A, B, and C.

FIGURE **3-13**

Circuit for
Example 3–6

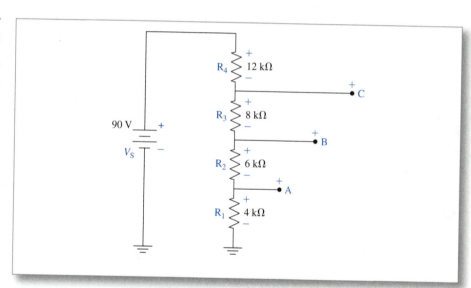

✔ A. SOLUTION USING OHM'S LAW AND KVL

Note that the output terminals of the voltage divider circuit in Figure 3–13 have no-loads connected.

Observing that the resistors and voltage source in Figure 3–13 are connected in series, the *total equivalent resistance* may be determined as:

$$R_T = R_1 + R_2 + R_3 + R_4 = 4 \text{ k}\Omega + 6 \text{ k}\Omega + 8 \text{ k}\Omega + 12 \text{ k}\Omega = 30 \text{ k}\Omega$$

Ohm's law can be used to calculate the source current as:

$$I_S = \frac{90 \text{ V}}{30 \text{ k}\Omega} = 3 \text{ mA}$$

Ohm's law can be used to calculate the voltage across each respective resistor as:

$$V_{R1} = I_S R_1 = (3 \text{ mA})(4 \text{ k}\Omega) = 12 \text{ V}$$

$$V_{R2} = I_S R_2 = (3 \text{ mA})(6 \text{ k}\Omega) = 18 \text{ V}$$

$$V_{R3} = I_S R_3 = (3 \text{ mA})(8 \text{ k}\Omega) = 24 \text{ V}$$

$$V_{R4} = I_S R_4 = (3 \text{ mA})(12 \text{ k}\Omega) = 36 \text{ V}$$

KVL can be used to determine the voltage at point A with respect to circuit ground as:

$$V_A = V_{R1} = 12 \text{ V}$$

KVL can be used to determine the voltage at point B with respect to circuit ground as:

$$V_B = V_{R1} + V_{R2} = 12 \text{ V} + 18 \text{ V} = 30 \text{ V}$$

KVL can be used to determine the voltage at point C with respect to circuit ground as:

$$V_C = V_{R1} + V_{R2} + V_{R3} = 12 \text{ V} + 18 \text{ V} + 24 \text{ V} = 54 \text{ V}$$

✔ B. SOLUTION USING THE VOLTAGE DIVIDER RULE

The voltage divider rule may be used to calculate the voltage at point A as:

$$V_A = V_S \frac{R_1}{R_T} = \frac{(90 \text{ V})(4 \text{ k}\Omega)}{30 \text{ k}\Omega} = 12 \text{ V}$$

The voltage divider rule may be used to calculate the voltage at point B as:

$$V_B = V_S \frac{(R_1 + R_2)}{R_T} = \frac{(90 \text{ V})(10 \text{ k}\Omega)}{30 \text{ k}\Omega} = 30 \text{ V}$$

The voltage divider rule may be used to calculate the voltage at point C as:

$$V_C = V_S \frac{(R_1 + R_2 + R_3)}{R_T} = \frac{(90 \text{ V})(18 \text{ k}\Omega)}{30 \text{ k}\Omega} = 54 \text{ V}$$

Note that the numerator in the voltage divider rule calculations is the sum of all of the resistors connected between circuit ground and the respective voltage point.

The MultiSIM simulation is shown in Figure 3–14.

FIGURE **3–14**

Circuit Simulation for Example 3–6

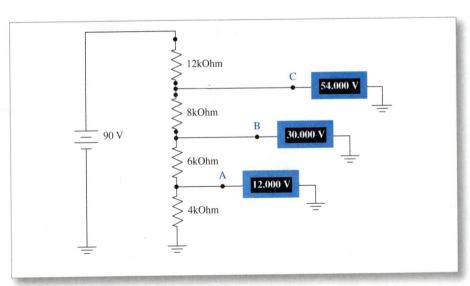

3.10   THE DC VOLTAGE DIVIDER CIRCUIT WITH LOAD

Connecting a load to an electronic circuit will change the voltages and currents that exist for the no-load condition. Circuit designers must be careful to ensure that the circuit will meet required specifications for both load and no-load conditions. Circuit designers also typically specify limitations and restrictions regarding the loads that may be connected to a circuit.

EXAMPLE 3–7   Analysis of a DC Voltage Divider Circuit with Load

The circuit shown in Figure 3–15 is the voltage divider circuit shown in Figure 3–13 with a load resistor (R_L) connected from the external connection V_B to circuit ground. Determine the voltage at the external connection points labeled as A, B, and C.

FIGURE **3–15**

A Parallel
Resistive Circuit

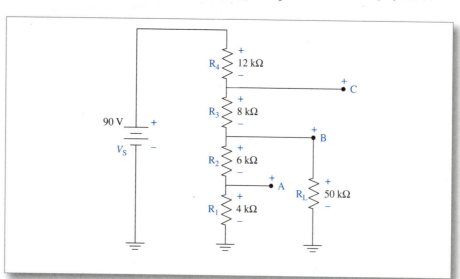

✓ SOLUTION

The voltage divider circuit with load shown in Figure 3–15 is a combination circuit. Note that the load resistor R_L is connected in parallel with the series combination of R_1 and R_2. Resistors R_3 and R_4 are connected in series with the combination of resistors R_1, R_2, and R_L. The total equivalent resistance may be calculated as:

$$R_T = [(R_1 + R_2)//R_L] + R_3 + R_4 = 8.33 \text{ k}\Omega + 8 \text{ k}\Omega + 12 \text{ k}\Omega = 28.33 \text{ k}\Omega$$

Ohm's law can be used to calculate the source current as:

$$I_S = \frac{90 \text{ V}}{28.33 \text{ k}\Omega} = 3.18 \text{ mA}$$

The current divider rule can be used to determine the current through the load resistor as:

$$I_L = I_S \frac{R_L // (R_1 + R_2)}{R_L} = 3.18 \text{ mA} \quad \frac{8.33 \text{ k}\Omega}{50 \text{ k}\Omega} = 0.53 \text{ mA}$$

The current divider rule can be used to determine the current through the series combination of R_1 and R_2 as:

$$I_{R1+R2} = I_S \frac{R_L // (R_1 + R_2)}{R_1 + R_2} = 3.18 \text{ mA} \quad \frac{8.33 \text{ k}\Omega}{10 \text{ k}\Omega} = 2.65 \text{ mA}$$

Ohm's law can be used to calculate the voltage across each respective resistor as:

$$V_{R1} = I_S R_1 = (2.65 \text{ mA})(4 \text{ k}\Omega) = 10.6 \text{ V}$$

$$V_{R2} = I_S R_2 = (2.65 \text{ mA})(6 \text{ k}\Omega) = 15.9 \text{ V}$$

$$V_{R3} = I_S R_3 = (3.18 \text{ mA})(8 \text{ k}\Omega) = 25.44 \text{ V}$$

$$V_{R4} = I_S R_4 = (3.18 \text{ mA})(12 \text{ k}\Omega) = 38.16 \text{ V}$$

KVL can be used to determine the voltage at point A with respect to circuit ground as:

$$V_A = V_{R1} = 10.6 \text{ V}$$

KVL can be used to determine the voltage at point B with respect to circuit ground as:

$$V_B = V_{R1} + V_{R2} = 10.6 \text{ V} + 15.9 \text{ V} = 26.5 \text{ V}$$

KVL can be used to determine the voltage at point C with respect to circuit ground as:

$$V_C = V_{R1} + V_{R2} + V_{R3} = 10.6 \text{ V} + 15.9 \text{ V} + 25.44 \text{ V} = 51.94 \text{ V}$$

Comparing the calculations in Example 3–6 and Example 3–7, we can observe that the addition of load resistor R_L has caused the output voltages (V_A, V_B, and V_C) to change in value. The voltage V_A for the unloaded voltage divider circuit is 12 V and is reduced to 10.58 V when the 50 kΩ load is connected. Similar reductions in V_B and V_C may also be observed.

The MultiSIM simulation is shown in Figure 3–16.

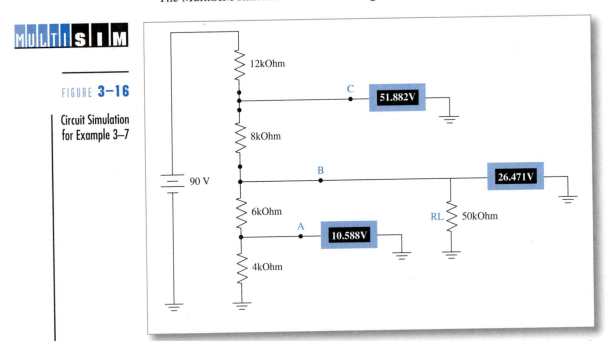

FIGURE **3–16**

Circuit Simulation for Example 3–7

3.11   ANALYSIS OF COMBINATION DC CIRCUITS WITH OPENS AND SHORTS

The most frequent failure in an electronic circuit results from either an unintentional short between two points in the circuit or from an unintentional open between two points.

A short typically results from insulation breakdown, or dirt buildup that establishes an undesired current-conducting path that has near zero resistance. An open typically results from component damage due to excessive heat buildup so that current can no longer flow through the damaged component.

The resistance between two points in a circuit that are shorted is considered to be zero. Current can flow through a short; however, the voltage across a short is zero. The resistance between two points in a circuit that are open is considered to be infinite. Current cannot flow through an open; however, voltage can exist across the open points.

EXAMPLE 3–8   Determining Open Circuit Voltage

Given the combination circuit in Figure 3–17, determine source current I_S and the open circuit voltage V_{open}. Verify the value of the open circuit voltage using computer simulation.

FIGURE **3–17**

Circuit for
Example 3–8

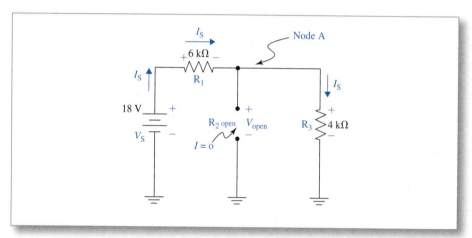

✔ **SOLUTION**

The source current in Figure 3–17 will flow through R_1 into node A. KCL requires that the current entering node A must equal the current leaving node A.

Since current cannot flow through the open, the current through R_3 (leaving node A) is also the source current (entering node A). Since the same current flows through R_1 and R_3, the resistors are in series.

The total equivalent resistance for the circuit in Figure 3–17 can be determined as:

$$R_T = R_1 + R_3 = 6 \text{ k}\Omega + 4 \text{ k}\Omega = 10 \text{ k}\Omega$$

Ohm's law can be used to calculate the source current as:

$$I_S = \frac{18 \text{ V}}{10 \text{ k}\Omega} = 1.8 \text{ mA}$$

Ohm's law can be used to calculate the voltage across each respective resistor as:

$$V_{R1} = I_S R_1 = (1.8 \text{ mA})(6 \text{ k}\Omega) = 10.8 \text{ V}$$

$$V_{R3} = I_S R_3 = (1.8 \text{ mA})(4 \text{ k}\Omega) = 7.2 \text{ V}$$

KVL can be used to determine the open circuit voltage beginning at the source ground as:

$$V_S - V_{R1} - V_{open} = 0$$

so that:

$$V_{open} = V_S - V_{R1} = 7.2 \text{ V}$$

KVL can also be used to determine the open circuit voltage beginning at the open ground as:

$$V_{open} = V_{R3} = 7.2 \text{ V}$$

The MultiSIM simulation is shown in Figure 3–18.

FIGURE **3–18**

Circuit Simulation for Example 3–8

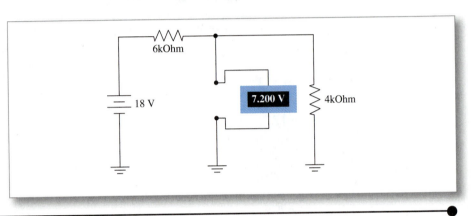

EXAMPLE 3–9   Determining Open Circuit Voltage

Given the combination circuit in Figure 3–19, determine source current I_S and the open circuit voltage V_{open}. Verify the value of the open circuit voltage using computer simulation.

FIGURE **3–19**

Circuit for Example 3–9

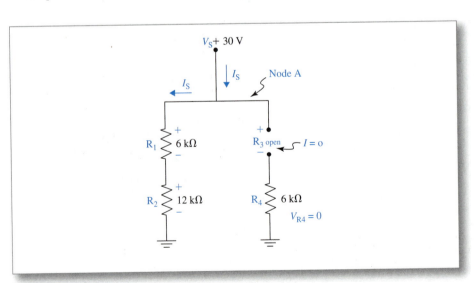

✓ SOLUTION

The source current in Figure 3–19 will flow into node A. KCL requires that the current entering node A must equal the current leaving node A. Since current cannot flow through the open, the source current (entering node A) will flow through the series combination of R_1 and R_2 (leaving node A).

The total equivalent resistance for the circuit can be determined as:

$$R_T = R_1 + R_3 = 6\ k\Omega + 12\ k\Omega = 18\ k\Omega$$

Note that the current through resistor R_4 is zero due to the open, so that R_4 has no effect on the total equivalent resistance.

Ohm's law can be used to calculate the source current as:

$$I_S = \frac{30\ V}{18\ k\Omega} = 1.67\ mA$$

Ohm's law can be used to calculate the voltage across each respective resistor as:

$$V_{R1} = I_S R_1 = (1.67\ mA)(6\ k\Omega) = 10.02\ V$$

$$V_{R2} = I_S R_2 = (1.67\ mA)(12\ k\Omega) = 20.04\ V$$

$$V_{R4} = (0\ mA)(6\ k\Omega) = 0\ V$$

KVL can be used to determine the open circuit voltage beginning at R_2 ground as:

$$V_{R2} + V_{R1} - V_{open} = 0$$

so that:

$$V_{open} = V_{R2} + V_{R1} = 30\ V$$

KVL can also be used to determine the open circuit voltage beginning at R_4 ground as:

$$V_{R3} + V_{open} - V_S = 0$$

so that:

$$V_{open} = V_S - V_{R4} = 30\ V - 0\ V = 30\ V$$

The MultiSIM simulation is shown in Figure 3–20.

FIGURE **3–20**

Circuit Simulation for Example 3–9

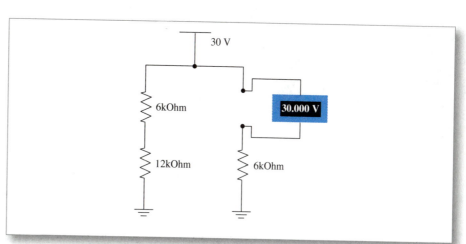

EXAMPLE 3–10   Determining Short Circuit Current

Given the circuit in Figure 3–21, assume that resistor R_4 has shorted. Determine the source current and the current through the short circuit. Verify the value of the short circuit current using computer simulation.

FIGURE **3-21**

Circuit for
Example 3–10

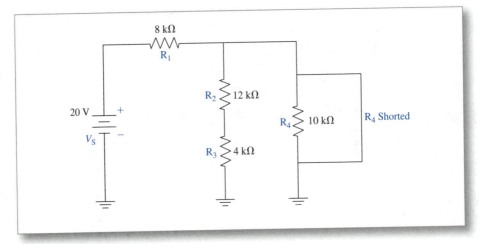

FIGURE **3-22**

Redrawn Circuit for
Example 3–10

✓ SOLUTION

The circuit in Figure 3–21 can be redrawn to indicate that R_4 is shorted as shown in Figure 3–22.

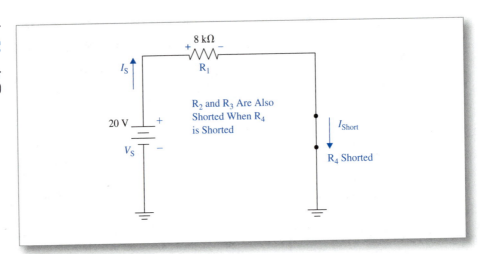

Note that the short across resistor R_4 has also shorted the series combination of R_2 and R_3 as indicated in Figure 3–22.

Considering R_2, R_3, and R_4 to be shorted, the total equivalent resistance may be determined as:

$$R_T = R_1 = 8 \text{ k}\Omega$$

Ohm's law can be used to calculate the source current as:

$$I_S = \frac{20 \text{ V}}{8 \text{ k}\Omega} = 2.5 \text{ mA}$$

The current through the short is equal to the source current so that:

$$I_{short} = I_S = 2.5 \text{ mA}$$

The MultiSIM simulation is shown in Figure 3–23.

FIGURE **3–23**

Circuit Simulation
for Example 3–10

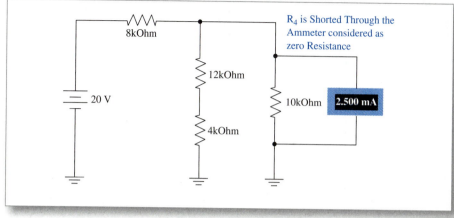

EXAMPLE 3–11   Determining Short Circuit Current

Given the circuit in Figure 3–24, assume that resistor R_3 has shorted. Determine the source current and the current through the short circuit. Verify the value of the short circuit current using computer simulation.

FIGURE **3–24**

Circuit for
Example 3–11

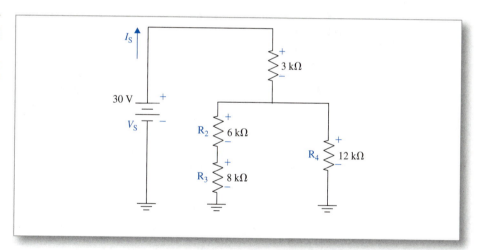

The circuit in Figure 3–24 can be redrawn to indicate that R_3 is shorted as shown in Figure 3–25.

FIGURE **3–25**

Redrawn Circuit
for Example 3–11

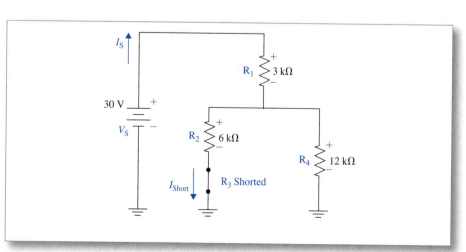

Note that shorting R_3 does not affect the remaining resistors in the circuit.

The total equivalent resistance may be determined as:

$$R_T = R_1 + (R_2 /\!/ R_4) = 3 \text{ k}\Omega + 4 \text{ k}\Omega = 7 \text{ k}\Omega$$

Ohm's law can be used to calculate the source current as:

$$I_S = \frac{30 \text{ V}}{7 \text{ k}\Omega} = 4.29 \text{ mA}$$

The current through R_2 and the short may be determined using the current divider rule as:

$$I_{\text{short}} = I_S \frac{(R_2 /\!/ R_4)}{R_2} = \frac{(4.29 \text{ mA})(4 \text{ k}\Omega)}{6 \text{ k}\Omega} = 2.86 \text{ mA}$$

The MultiSIM simulation is shown in Figure 3–26.

FIGURE 3-26

Circuit Simulation
for Example 3–11

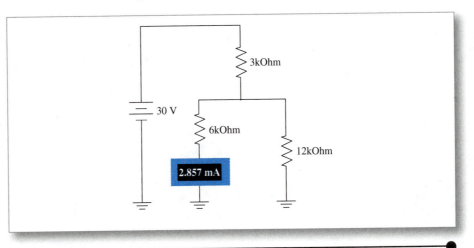

PRACTICE EXERCISE 3-3

a. Determine the open circuit voltage V_{open} for the circuit in Figure 3–27.

FIGURE 3-27

Circuit for
Practice Exercise 3–3a

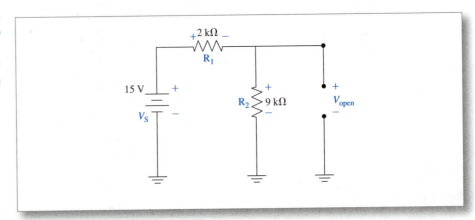

b. Determine the short circuit current for the circuit in Figure 3–28.

FIGURE **3–28**

Circuit for
Practice Exercise 3–3b

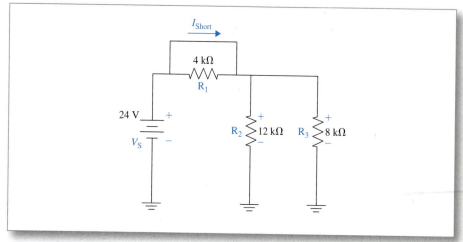

✔ANSWERS

a. $V_{open} = 12.27$ V

b. $I_{short} = 5$ mA

(Complete solutions to the Practice Exercises can be found in Appendix A.)

EXAMPLE 3–12   Analysis of a DC Combination Circuit

Given the circuit shown in Figure 3–29, determine the values of currents I_1, I_2, I_3, and I_4.

FIGURE **3–29**

Circuit for
Example 3–12

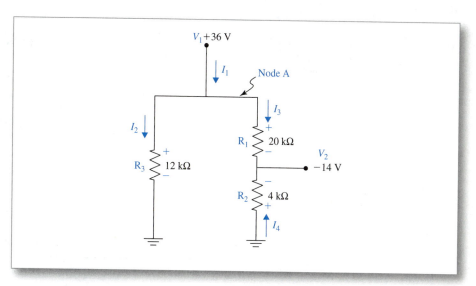

✔SOLUTION

The current I_1 is the current that flows through resistor R_1. Note that one end of R_1 is connected to $+36$ V with respect to circuit ground and the other end is connected to -14 V with respect to circuit ground. The voltage across R_1 can be determined as:

$$V_{R1} = +36 \text{ V} - (-)14 \text{ V} = 50 \text{ V}$$

so that:
$$I_3 = \frac{50 \text{ V}}{20 \text{ k}\Omega} = 2.5 \text{ mA}$$

In similar fashion, the current I_2 is the current through resistor R_3 and may be calculated as:
$$I_2 = \frac{36 \text{ V}}{12 \text{ k}\Omega} = 3 \text{ mA}$$

where the voltage across R_3 is determined as:
$$V_{R3} = +36 \text{ V} - 0 = 36 \text{ V}$$

The current I_4 is the current through resistor R_2 and may be calculated as:
$$I_4 = \frac{-14 \text{ V}}{4 \text{ k}\Omega} = -3.5 \text{ mA}$$

where the voltage across R_2 is determined as:
$$V_{R2} = -14 \text{ V} - 0 \text{ V} = -14 \text{ V}$$

The current I_1 can be determined by applying KCL at node A to obtain:
$$I_1 - I_2 - I_3 = 0$$

so that:
$$I_1 = I_2 + I_3 = 3 \text{ mA} + 2.5 \text{ mA} = 5.5 \text{ mA}$$

It is important to observe that the direction of current flow is always from $(+)$ to $(-)$.

The MultiSIM simulation is shown in Figure 3–30.

FIGURE **3–30**

Circuit
Simulation for
Example 3–12

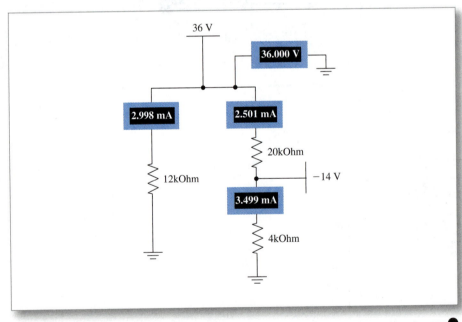

3.12   ANALYSIS OF THE BRIDGE CIRCUIT

The DC bridge circuit is shown in Figure 3–31. The DC bridge circuit is said to be balanced when the ratio of R_1 to R_2 is equal to the ratio of R_3 to R_4. The voltage across the balanced bridge circuit V_{open} is equal to zero. The voltage across an unbalanced bridge V_{open} is a nonzero voltage.

FIGURE **3-31**

The DC Bridge Circuit

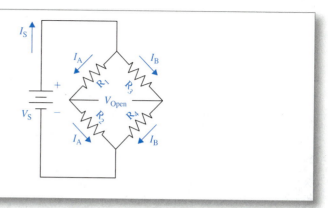

EXAMPLE 3-13   Analysis of an Unbalanced Bridge Circuit

Given the bridge circuit shown in Figure 3–32, determine the bridge voltage V_{open}. Verify the value of V_{open} using computer simulation.

FIGURE **3-32**

Circuit for Practice Exercise 3–13

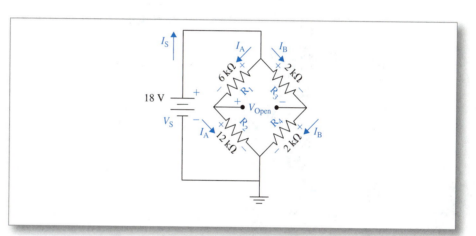

✓ **SOLUTION**

The bridge circuit consists of two parallel branches. One branch consists of R_1 in series with R_2, and the other consists of R_3 in series with R_4.

Ohm's law can be used to determine the current through each respective branch as:

$$I_A = \frac{18 \text{ V}}{18 \text{ k}\Omega} = 1 \text{ mA}$$

$$I_B = \frac{18 \text{ V}}{4 \text{ k}\Omega} = 4.5 \text{ mA}$$

Ohm's law can be used to calculate the voltage across each respective resistor as:

$$V_{R1} = (1 \text{ mA})(6 \text{ k}\Omega) = 6 \text{ V}$$

$$V_{R2} = (1 \text{ mA})(12 \text{ k}\Omega) = 12 \text{ V}$$

$$V_{R3} = (4.5 \text{ mA})(2 \text{ k}\Omega) = 9 \text{ V}$$

$$V_{R4} = (4.5 \text{ mA})(2 \text{ k}\Omega) = 9 \text{ V}$$

KVL may be used to determine the bridge voltage V_{open} as:

$$V_{R3} - V_{R1} - V_{open} = 0$$

so that: $V_{open} = 9 \text{ V} - 6 \text{ V} = 3 \text{ V}$ (polarity as indicated)

KVL can also be used to determine the bridge voltage V_{open} as:

$$V_{R2} - V_{open} - V_{R4} = 0$$

so that: $V_{open} = 12\ V - 9\ V = 3\ V$ (polarity as indicated)

Note that the bridge voltage for the circuit in this example is not equal to zero, indicating that the bridge circuit is unbalanced.

The ratio of R_1 to R_2 can be calculated as:

$$\frac{R_1}{R_2} = \frac{6\ k\Omega}{12\ k\Omega} = 0.5$$

and the ratio of R_3 to R_4 can be calculated as:

$$\frac{R_3}{R_4} = \frac{2\ k\Omega}{2\ k\Omega} = 1$$

The inequality of the ratios indicates that the bridge circuit is unbalanced.

The MultiSIM simulation is shown in Figure 3–33.

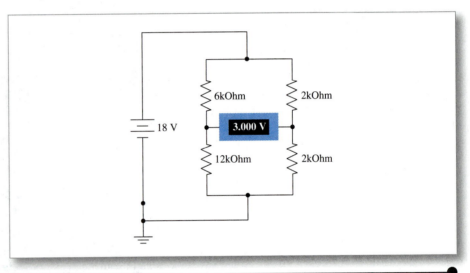

FIGURE **3–33**

Circuit Simulation for Example 3–13

EXAMPLE 3–14   Analysis of a Balanced Bridge Circuit

Given the bridge circuit shown in Figure 3–34, determine the bridge voltage V_{open}. Verify the value of V_{open} using computer simulation.

FIGURE **3–34**

Circuit for Example 3–14

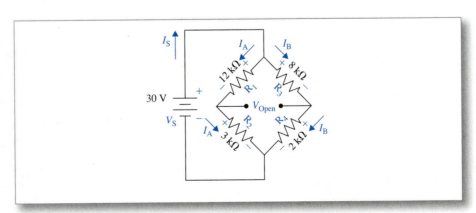

✓SOLUTION

Ohm's law can be used to determine the current through each respective branch in Figure 3–34 as:

$$I_A = \frac{30\text{ V}}{15\text{ k}\Omega} = 2\text{ mA}$$

$$I_B = \frac{30\text{ V}}{10\text{ k}\Omega} = 3\text{ mA}$$

Ohm's law can be used to calculate the voltage across each respective resistor as:

$$V_{R1} = (2\text{ mA})(12\text{ k}\Omega) = 24\text{ V} \qquad V_{R2} = (2\text{ mA})(3\text{ k}\Omega) = 6\text{ V}$$

$$V_{R3} = (3\text{ mA})(8\text{ k}\Omega) = 24\text{ V} \qquad V_{R4} = (3\text{ mA})(2\text{ k}\Omega) = 6\text{ V}$$

KVL can be used to determine the bridge voltage V_{open} as:

$$V_{R3} - V_{R1} - V_{open} = 0$$

so that:

$$V_{open} = 24\text{ V} - 24\text{ V} = 0$$

KVL can also be used to determine the bridge voltage V_{open} as:

$$V_{R2} - V_{open} - V_{R4} = 0$$

so that:

$$V_{open} = 6\text{ V} - 6\text{ V} = 0\text{ V}$$

Note that the bridge voltage for the circuit in this example is equal to zero, indicating that the bridge circuit is balanced.

The ratio of R_1 to R_2 can be calculated as:

$$\frac{R_1}{R_2} = \frac{12\text{ k}\Omega}{3\text{ k}\Omega} = 4$$

and the ratio of R_3 to R_4 can be calculated as:

$$\frac{R_3}{R_4} = \frac{8\text{ k}\Omega}{2\text{ k}\Omega} = 4$$

The equality of the ratios indicates that the bridge circuit is balanced.

The MultiSIM simulation is shown in Figure 3–35.

FIGURE **3–35**

Circuits Simulation for Example 3–14

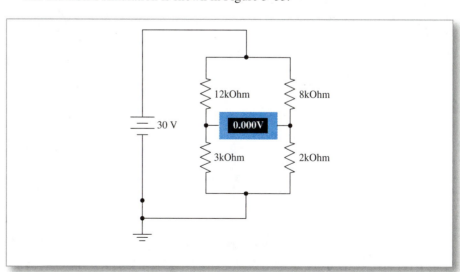

SUMMARY

Circuits that contain both series and parallel components or branches are referred to as *combination circuits*. Combination circuits may also be referred to as *series-parallel circuits*.

The resistance of a combination circuit may be reduced to a single value that is referred to as the *total equivalent resistance*. The current that flows from a voltage source in a combination circuit depends only upon the voltage of the source and the total equivalent resistance. When determining total equivalent resistance of a combination circuit, it is often beneficial to consider the circuit as a collection of branches rather than a collection of individual components.

Ohm's law for a combination circuit may be expressed as $I_S = \dfrac{V_S}{R_T}$, where I_S is the source current, V_S is the source voltage, and R_T is the total equivalent resistance. *Kirchhoff's voltage law* must be satisfied around every closed loop in a combination circuit. *Kirchhoff's current law* must be satisfied at every node in a combination circuit.

The *voltage divider rule* may be applied to any series components or series branches in a combination circuit. The *current divider rule* may be applied to any parallel components or parallel branches in a combination circuit.

A combination circuit is solved when the current through and voltage across each component has been determined.

ESSENTIAL LEARNING EXERCISES

(ANSWERS TO ODD-NUMBERED ESSENTIAL LEARNING EXERCISES CAN BE FOUND IN APPENDIX A.)

1. Given the circuit diagram in Figure 3–36, determine the total equivalent resistance and source current. Calculate the current through each resistor and verify KCL at node A. Verify KVL.

FIGURE **3–36**

Circuit for Essential Learning Exercise 1

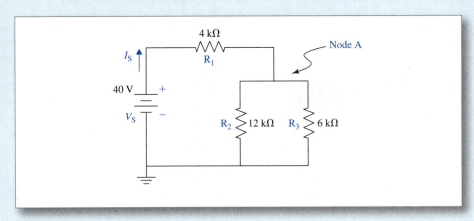

2. Given the circuit diagram in Figure 3–37, determine the total equivalent resistance and source current. Calculate the current through each resistor and verify KCL at node A. Verify KVL.

FIGURE **3–37**

Circuit for
Essential Learning
Exercise 2

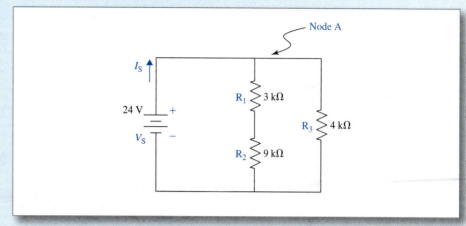

3. Given the circuit diagram in Figure 3–38, determine the total equivalent resistance and source current. Calculate the current through each resistor and verify KCL at node A. Verify KVL.

FIGURE **3–38**

Circuit for
Essential Learning
Exercise 3

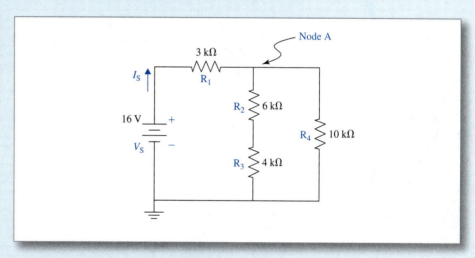

4. Given the circuit diagram in Figure 3–39, determine the total equivalent resistance and source current. Calculate the current through each resistor and verify KCL at node A. Verify KVL.

FIGURE **3–39**

Circuit for
Essential Learning
Exercise 4

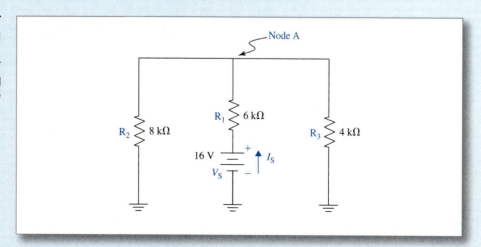

5. Given the circuit diagram in Figure 3–40, determine the total equivalent resistance and source current. Calculate the current through each resistor and verify KCL at node A. Verify KVL.

FIGURE **3–40**

Circuit for
Essential Learning
Exercise 5

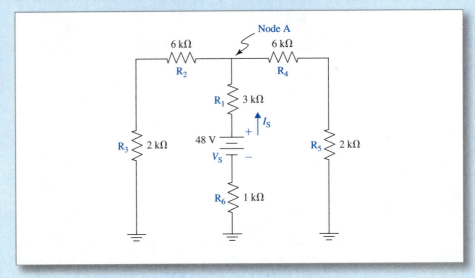

6. Given the circuit diagram in Figure 3–41, determine the total equivalent resistance and source current. Calculate the current through each resistor and verify KCL at node A. Verify KVL.

FIGURE **3–41**

Circuit for
Essential Learning
Exercise 6

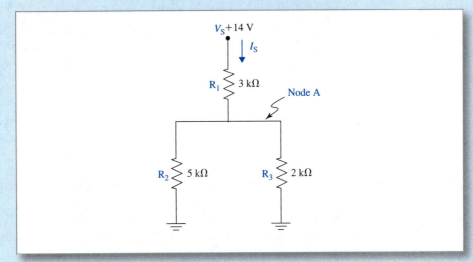

7. Given the circuit diagram in Figure 3–42, determine the total equivalent resistance and source current. Calculate the current through each resistor and verify KCL at node A. Verify KVL.

FIGURE **3–42**

Circuit for
Essential Learning
Exercise 7

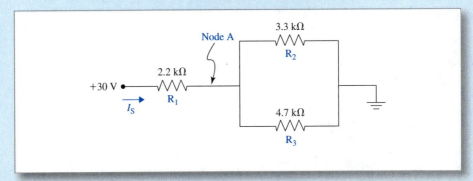

8. Given the circuit diagram in Figure 3–43, determine the total equivalent resistance and source current. Calculate the current through each resistor and verify KCL at node A. Verify KVL.

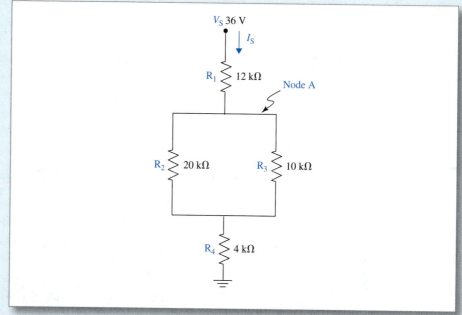

FIGURE **3–43**

Circuit for Essential Learning Exercise 8

9. Given the circuit diagram in Figure 3–44, calculate the current through each resistor and verify KCL at node A.

FIGURE **3–44**

Circuit for Essential Learning Exercise 9

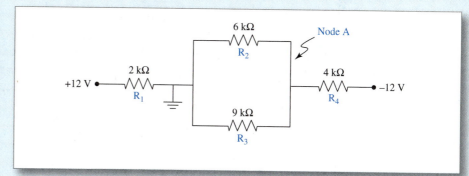

10. Given the circuit diagram in Figure 3–45, calculate the current through each resistor and verify KCL at node A.

FIGURE **3–45**

Circuit for Essential Learning Exercise 10

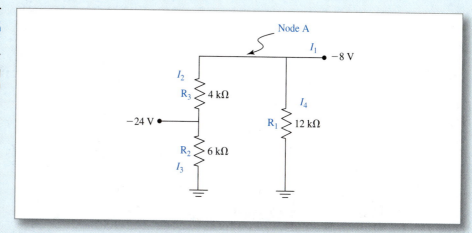

11. Given the voltage divider circuit diagram in Figure 3–46, determine the output voltage at points indicated as A, B, and C.

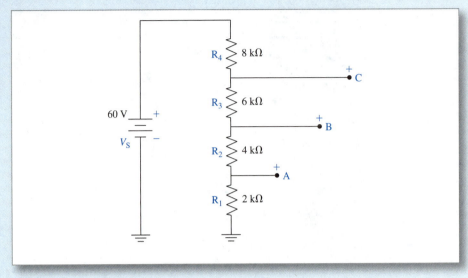

12. Given the voltage divider circuit diagram in Figure 3–47, determine the output voltage at points indicated as A and B with a 30 kΩ load as indicated. Also calculate the output voltage at points A and B with the 30 kΩ load removed. Calculate the percent difference between no-load and load voltage as:

$$\% \text{ Difference} = \frac{V_{\text{no-load}} - V_{\text{load}}}{V_{\text{no-load}}}$$

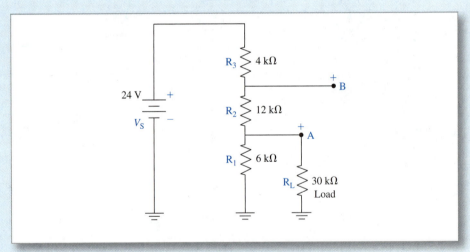

13. Given the circuit diagram in Figure 3–48, determine the voltage indicated as V_{open}.

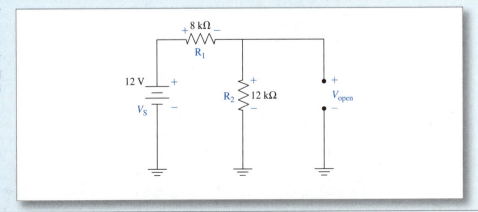

14. Given the circuit diagram in Figure 3–49, determine the voltage indicated as V_{open}.

FIGURE **3-49**

Circuit for
Essential Learning
Exercise 14

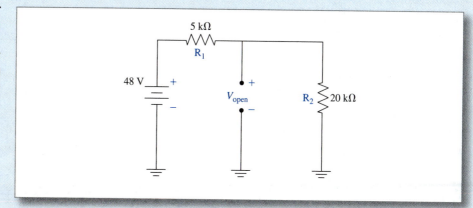

15. Given the circuit diagram in Figure 3–50, determine the voltage indicated as V_{open}.

FIGURE **3-50**

Circuit for
Essential Learning
Exercise 15

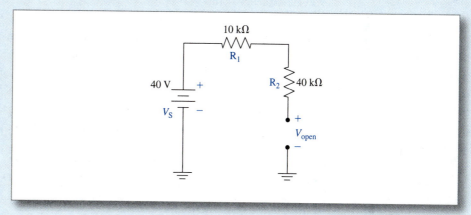

16. Given the circuit diagram in Figure 3–51, determine the current indicated as I_{short}.

FIGURE **3-51**

Circuit for
Essential Learning
Exercise 16

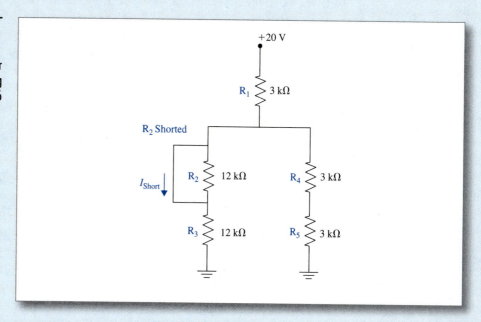

17. Given the circuit diagram in Figure 3–52, determine the current indicated as I_{short}.

FIGURE **3–52**

Circuit for
Essential Learning
Exercise 17

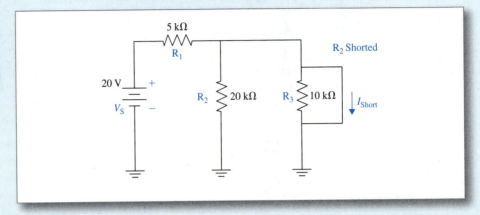

18. Given the bridge circuit diagram in Figure 3–53, determine the voltage indicated as V_{open}. Is the given circuit a balanced bridge?

FIGURE **3–53**

Circuit for
Essential Learning
Exercise 18

19. Given the bridge circuit diagram in Figure 3–54, determine the voltage indicated as V_{open}. Is the given circuit a balanced bridge?

FIGURE **3–54**

Circuit for
Essential Learning
Exercise 19

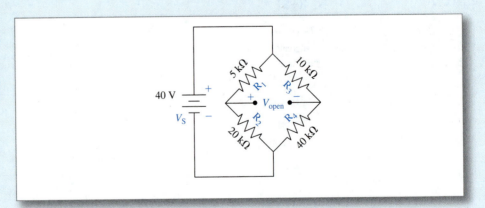

20. Given the circuit diagram in Figure 3–55 and given that $V_A = 12$ V, determine the value of R_4.

FIGURE **3–55**

Circuit for
Essential Learning
Exercise 20

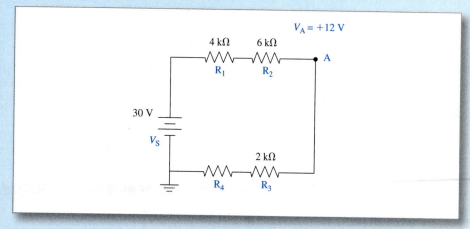

21. The current through R_1 in the circuit diagram in Figure 3–56 is 4 mA and the voltage across R_4 is 8 V as indicated. Determine the current through resistors R_2 and R_3.

FIGURE **3–56**

Circuit for
Essential Learning
Exercise 21

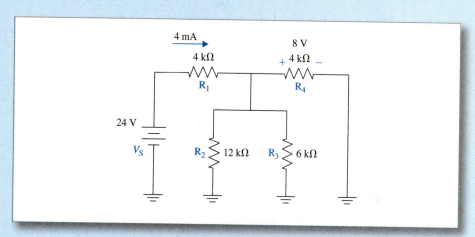

22. Given the circuit diagram in Figure 3–57, determine the value of R_X so that the current through each of the 6 kΩ resistors will be 3 mA.

FIGURE **3–57**

Circuit for
Essential Learning
Exercise 22

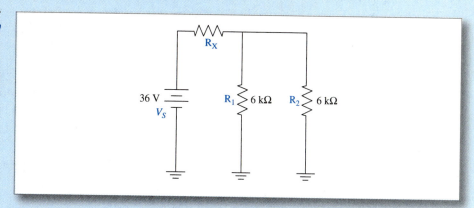

23. A 12 kΩ resistor and an 8 kΩ resistor are connected in parallel in an electronic circuit. The current through the 8 kΩ resistor is measured to be 3.5 mA. Determine the current through the 12 kΩ resistor.

24. Design a two-resistor DC voltage divider network so that the output voltage is 12.5 V. The voltage source is specified to be 36 V, and the resistor across which the output voltage is taken is specified to be 4.7 kΩ. Verify your design using computer simulation.

25. The connections for a certain electronic circuit are hidden from view. The components for the circuit are determined to be four 6 kΩ resistors. A measurement is taken and the total equivalent circuit resistance is determined to be 10 kΩ. Sketch a possible circuit diagram for the circuit.

CHALLENGING LEARNING EXERCISES

(ANSWERS TO SELECTED CHALLENGING LEARNING EXERCISES CAN BE FOUND IN APPENDIX A.)

1. Given the circuit diagram in Figure 3–58, determine the current through each resistor.

FIGURE **3–58**

Circuit for
Challenging Learning
Exercise 1

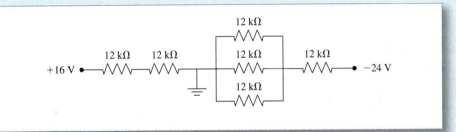

2. Given the circuit diagram in Figure 3–59, determine the value of R_3. Verify your solution using computer simulation.

FIGURE **3–59**

Circuit for
Challenging Learning
Exercise 2

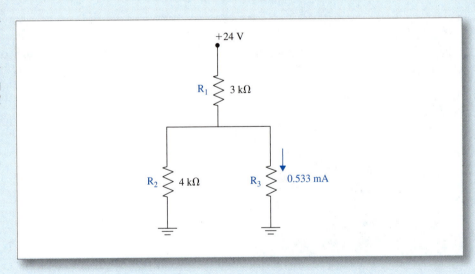

3. Given the circuit diagram in Figure 3–60, determine the voltage indicated as V_{open}.

FIGURE **3–60**

Circuit for
Challenging Learning
Exercise 3

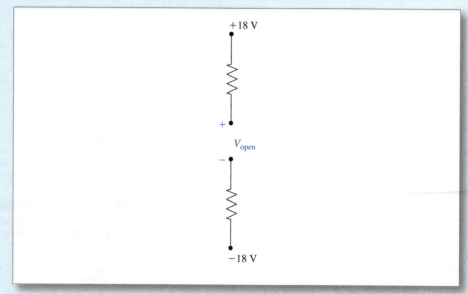

4. Given the circuit diagram in Figure 3–61, determine the current indicated as I_{short}.

FIGURE **3–61**

Circuit for
Challenging Learning
Exercise 4

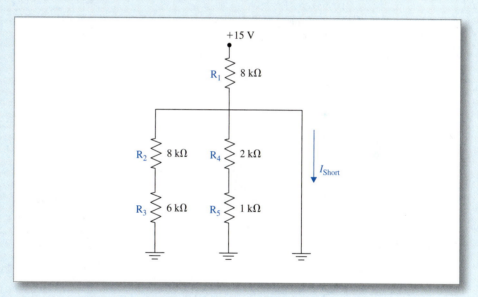

5. Given the circuit diagram in Figure 3–62, determine the voltage indicated as V_{out}.

FIGURE **3–62**

Circuit for
Challenging Learning
Exercise 5

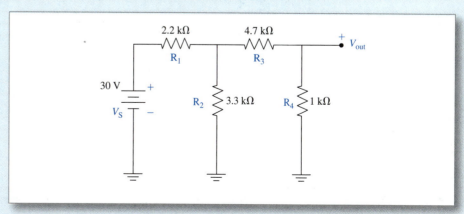

6. The circuit diagram in Figure 3–63 is referred to as a "ladder network." Determine the current through each resistor. Verify your solution using computer simulation.

FIGURE **3–63**

Circuit for
Challenging Learning
Exercise 6

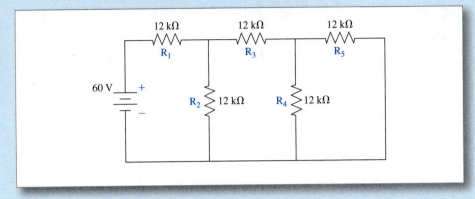

7. Given the bridge circuit in Figure 3–64, determine the voltage indicated as V_{open}. Is the circuit a balanced bridge?

FIGURE **3–64**

Circuit for
Challenging Learning
Exercise 7

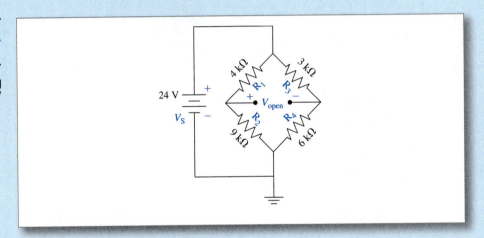

8. Given the circuit diagram in Figure 3–65, determine the current through each resistor. Determine the voltage at node A. Verify your solution using computer simulation.

FIGURE **3–65**

Circuit for
Challenging Learning
Exercise 8

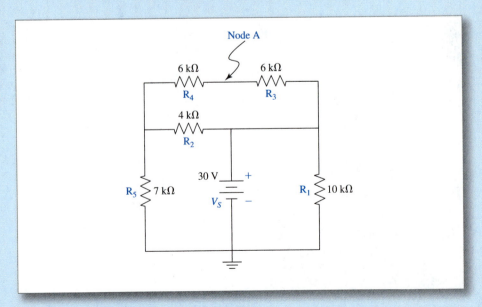

9. Given the voltage divider circuit diagram in Figure 3–66, determine the voltage at points A and B with no-load connected. Determine the voltage at points A and B with 30 kΩ and 100 kΩ loads connected as indicated.

FIGURE **3–66**

Circuit for
Challenging Learning
Exercise 9

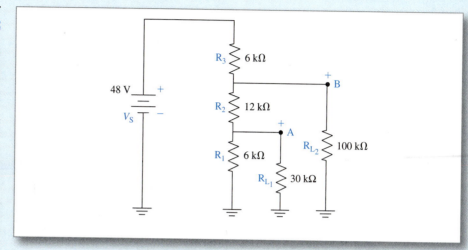

10. Design a DC voltage divider circuit to produce output voltages of 5 V, 12 V, and 24 V from a 50 V source. The source current is specified to be 3.5 mA. Verify your design using computer simulation.

TEAM ACTIVITY

The members of your study group are to design a two resistor DC voltage divider network to meet the following specifications:

1. The voltage source is specified to be 40 V.
2. The no-load output voltage is to be 25 V.

You are also required to determine the minimum load resistance that can be connected to the circuit so that the output voltage does not change more than 5% from the no-load voltage. Verify your design using computer simulation.

THE SINE WAVE AND LINEAR RESPONSE

> "ANYTHING ONE PERSON CAN IMAGINE,
> ANOTHER PERSON CAN MAKE REAL."
>
> —Jules Verne

CHAPTER OUTLINE

LEARNING OBJECTIVES

Upon successful completion of this chapter you will be able to:

- Express the sine wave as a function of time in the time domain.
- Determine the parameters of a sine wave.
- Calculate the instantaneous value of a sine wave.
- Express a sine wave as a phasor in the phasor domain.
- Convert a sine wave from the time domain to the phasor domain.
- Use an engineering calculator to perform mathematical operations in the phasor domain.

CHAPTER FOUR

- Express a phasor in polar or rectangular form.
- Calculate current and voltage for a resistor in the time domain.
- Calculate current and voltage for a resistor in the phasor domain.
- Calculate current and voltage for an inductor in the time domain.
- Calculate the reactance of an inductor at a give frequency.
- Calculate the reactance of a capacitor at a given frequency.
- Calculate current and voltage for an inductor in the phasor domain.
- Calculate current and voltage for a capacitor in the time domain.
- Calculate current and voltage for a capacitor in the phasor domain.
- Sketch a phasor diagram.

INTRODUCTION

The sinusoidal waveform is the most important electronic signal to be considered in this text, and the early sections of this chapter are devoted to the study of the sine wave. The sine wave is represented mathematically as a function of time in the time domain and as a phasor in the phasor domain. Sinusoidal voltage and current are typically expressed in the phasor domain to simplify mathematical calculations.

This chapter also introduces the linear components that are utilized by engineers and engineering technologists to design the electronic circuits that create, process, and transmit electronic signals. A linear component is one for which the current is always proportional to the voltage across the device. The resistor, inductor, and capacitor are linear components. This chapter will consider the physical nature of each component, derive its mathematical representation or mathematical model, and determine how each of the components responds to sinusoidal waveforms. Future chapters will demonstrate how these components are interconnected to create electronic circuits and how such circuits may be analyzed and designed.

4.1   THE SINE WAVE AS A FUNCTION OF TIME

The sine wave is the most important electronic signal waveform to be considered in the analysis and design of analog electronic circuits. The sine function is often presented in mathematics texts as a function of an independent variable often designated by the letter x.

The **amplitude** of a sine wave is the value of the function at the point of maximum displacement from the zero axis.

The sine function is typically expressed in mathematics texts as:

$$\mathrm{f}(x) = A \sin(x) \tag{4.1}$$

where A is the amplitude of the waveform and x is the independent variable.

A MATLAB plot of $\mathrm{f}(x) = A \sin(x)$ is shown in Figure 4–1.

MATLAB

FIGURE **4–1**

A Sine Wave as a Function of the Independent Variable "x"

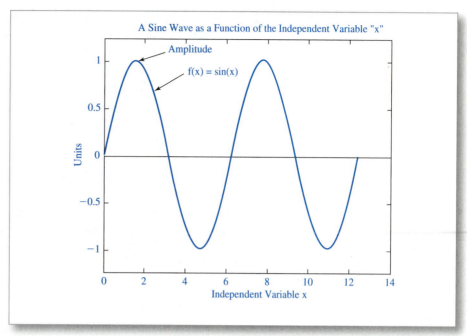

A waveform that is expressed as a function of time is said to be expressed in the **time domain.**

Figure 4–2 shows a sinusoidal voltage signal as a function of time represented in the time domain.

FIGURE **4–2**

A Sinusoidal Voltage as a Function of Time

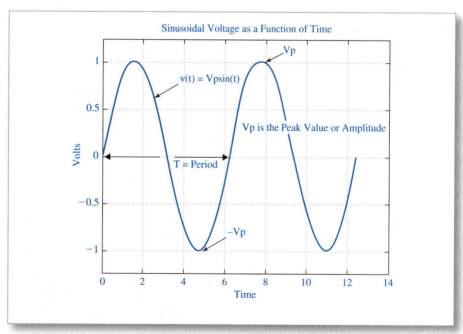

4.2 SINE WAVE PARAMETERS

Several important observations can be made regarding the sine wave in Figure 4–2 that are common to all sine waves. The sine wave repeats itself at regular intervals and is therefore said to be a **periodic waveform.** Each complete repetition is called a **cycle** of the waveform. The time required to complete one cycle of the waveform is called the **period** and is represented by the symbol T.

The **cyclic frequency** of a periodic waveform is represented by the symbol f and is defined as the number of cycles that occur in one second of time. Frequency (f) is measured in units of hertz (Hz) to honor the German physicist Heinrich Hertz (1857–1894).

The cyclic frequency of a sine wave is the reciprocal of the period such that:

$$f = \frac{1}{T} \tag{4.2}$$

Maximum displacement from the zero axis may be referred to as amplitude; however, electronic engineers typically refer to the maximum displacement as the **peak value** of the waveform. The terms *amplitude* and *peak value* describe the same quantity so that the terms may be used interchangeably.

Designating the peak value of a sinusoidal voltage waveform as V_P and expressing the voltage as a function of time, we obtain:

$$v(t) = V_P \sin(t) \tag{4.3}$$

Although the expression $v(t) = V_P \sin(t)$ in Equation (4.3) indicates that the wave shape is sinusoidal and indicates the peak value or amplitude, other important features or parameters of electronic signals are not included in this expression.

The frequency of an electronic signal is an important consideration in the analysis and design of analog electronic circuits. The expression in Equation (4.3) must be modified to indicate the frequency of the sine wave.

A **vector** is a mathematical term used to indicate quantities that are defined by both magnitude and direction or position.

The magnitude of a vector is proportional to its length, and its position is typically specified by an angular displacement from a given reference point. If a unit vector (magnitude = 1) is revolved counterclockwise about a fixed point, the projection of the vector (distance from the tip of the vector to the reference axis) describes the sine function as illustrated in Figure 4–3.

FIGURE **4–3**

A Revolving Unit Vector Generating a Sine Wave

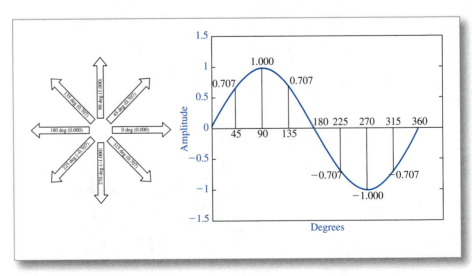

Notice that the projection of the unit vector through 360° of revolution generates one complete cycle of the sine wave, the point at which the waveform begins a new repetition. Although angular position may be measured in degrees, the radian provides an alternative unit of measure that is often preferred by engineers and scientists.

The *radian* is derived from the geometric relationship between the circumference and the radius of a circle. It is defined as the angle subtended by an arc equal to the length of the radius along the circumference, and that angle, regardless of the size of the circle, is 57.296° as indicated in Figure 4–4.

FIGURE **4–4**

One Radian is Equivalent
to 57.296 Degrees

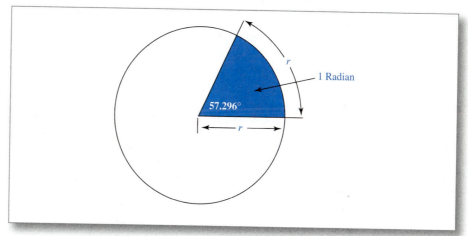

Since the circumference of a circle is 2π times the radius, it follows that 2π radians is equivalent to 360°.

$$2\pi \text{ radians} = 360°$$

(4.4)

Angular displacement, therefore, may be indicated in degrees or radians. The radian measure is traditionally chosen to represent the angular displacement of the unit vector as it revolves to generate the sine wave. Therefore, a unit vector that revolves through 2π radians will describe one complete cycle of the sine wave.

The sine wave shown in Figure 4–5 is shown as a function of angular displacement measured in both degrees and radians.

FIGURE **4–5**

A Sinusoidal Voltage as a
Function of Degrees
and Radians

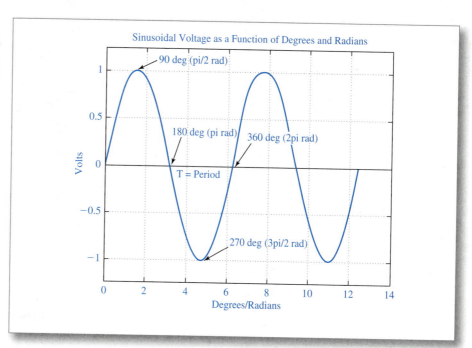

The equivalent values of degrees and radians at the points of maximum displacement and at the zero crossing points are shown below:

1. 2π radians $= 360°$ $(V = 0)$

2. π radians $= 180°$ $(V = 0)$

3. $\dfrac{\pi}{2}$ radians $= 90°$ $(V = V_P)$

4. $\dfrac{3\pi}{2}$ radians $= 270°$ $(V = -V_P)$

Our objective, however, is to represent the sinusoidal waveform as a function of time, not as a function of degrees or radians. We therefore need to devise a procedure that will indicate the angular position of the revolving vector illustrated in Figure 4–3 as a function of time.

Consider that if the angular velocity of the revolving vector in Figure 4–3 is known and the time that the vector has been revolving is also known, the angular position of the vector can be determined. This is analogous to the way that the distance traveled may be determined if the speed or velocity and the time of travel are known (distance = velocity × time).

The angular velocity of a revolving vector whose projection generates a sine wave is given the name omega and is represented by the symbol ω. Although omega (ω) is technically a measure of angular velocity having the units radians/second (rad/s), it is also referred to as angular frequency or radian frequency. The quantity omega will be referred to as **radian frequency** throughout this text.

The angular position of the revolving vector can be determined by multiplying the **radian frequency** (ω) of the revolving vector by **elapsed time** (t). The sinusoidal voltage waveform may therefore be expressed as a function of time (t) and omega (ω) to obtain:

$$v(t) = V_P \, \sin(\omega t) \tag{4.5}$$

Caution must be exercised to avoid confusing radian frequency (ω), which is measured in units of rad/s, with cyclic frequency (f), which is measured in units of hertz (Hz).

As noted, the revolving vector will travel 2π radians during each complete revolution, so that angular velocity or radian frequency may be defined as:

$$\omega = \frac{2\pi}{T} \ \text{(rad/s)} \tag{4.6}$$

An equivalent expression may be derived by substituting $f = \dfrac{1}{T}$ in Equation (4.6) to obtain:

$$\omega = 2\pi f \tag{4.7}$$

The sinusoidal voltage signal as a function of time and cyclic frequency may be expressed as:

$$v(t) = V_P \, \sin(2\pi f t) \tag{4.8}$$

where f is the cyclic frequency measured in Hz.

It is important to note that Equations (4.5) and (4.8) are expressions that indicate the peak value (amplitude) as well as the frequency of the waveform. Although the expressions in Equations (4.5) and (4.8) are equivalent, Equation (4.5) is the expression preferred by engineers and will be used most often throughout this text.

Figure 4–6 illustrates the relationship between omega, period, and cyclic frequency (Hz).

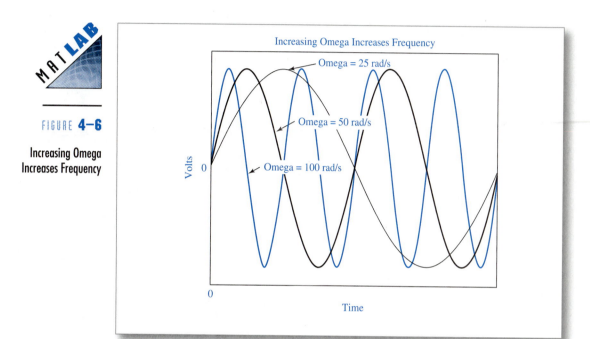

FIGURE **4–6**

Increasing Omega Increases Frequency

Observe in Figure 4–6 that increasing the radian frequency (ω) decreases the period (T) and increases the cyclic frequency (f).

Many circuits that are used to process electronic signals may cause the signals to be delayed or advanced in time. The analytical expression for a sinusoidal signal waveform should therefore include information regarding any time shifts that may occur. The shift may also be indicated in angular measurement of degrees, rather than a time shift measured in seconds, which is simply a matter of preference.

Phase angle is defined as the angular displacement of a waveform with respect to a specified reference. The phase angle of electronic signals is measured in degrees and designated by the symbol θ (theta). A positive value for θ has the effect of advancing the signal in time, and a negative value of θ delays the signal in time.

The phase angle of a waveform has meaning only if a reference is specified. The reference may be simply specified to be 0°. The phase angle may also be referenced to another sine wave that exists in the circuit.

Phase difference is a measure of the angular displacement of a wave from the specified reference wave. Phase difference is determined by subtracting the phase angle of a given waveform from the phase angle of the reference waveform. The phase angle of the reference is typically considered to be 0°.

The phase difference between the two sine waves shown in Figure 4–7 is 90° as indicated.

FIGURE **4–7**

Waveforms Whose Phase
Difference is 90 Degrees

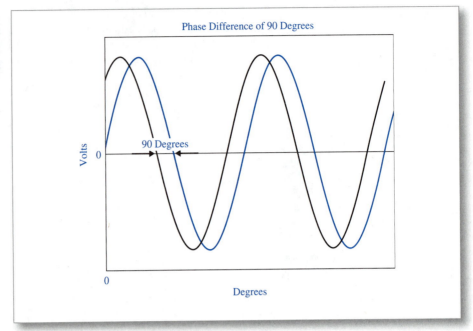

Note that the sine waves in Figure 4–7 do not cross the zero axis at the same point. The difference between the crossover points measured in degrees is 90° as indicated.

The general expression for a sinusoidal voltage waveform as a function of time that includes phase angle may be expressed as:

$$v(t) = V_P \, \sin(\omega t \pm \theta) \qquad (4.9)$$

The expression in Equation (4.9) represents the analytical expression of a sinusoidal voltage signal from which the peak value, period, frequency, and phase angle may be determined.

Quantities that vary over time are, by convention, represented by lowercase symbols, and quantities that are constant or do not vary over time are represented by uppercase symbols. This convention will be followed throughout the text. Therefore, v is understood to be equivalent to $v(t)$, and i is equivalent to $i(t)$. The expression in Equation (4.9) may therefore be written as:

$$v = V_P \, \sin(\omega t \pm \theta) \qquad (4.10)$$

and a similar expression for sinusoidal current as a function of time may be written as:

$$i = I_P \, \sin(\omega t \pm \theta) \qquad (4.11)$$

EXAMPLE 4–1   Expressing Voltage As a Function of Time

The peak value of a sinusoidal voltage waveform is given as 3 V. The phase angle of the waveform is 90°. Express the waveform as a function of time.

✔ SOLUTION

The peak value of the waveform is given as 3 V and the phase angle is given as 90°. Substituting the given values into the general expression in Equation (4.10), the waveform may be expressed as:

$$v = 3 \sin(\omega t + 90°) \text{ V}$$

The waveform is shown in the MATLAB plot in Figure 4–8.

FIGURE **4–8**

Waveform for Example 4–1

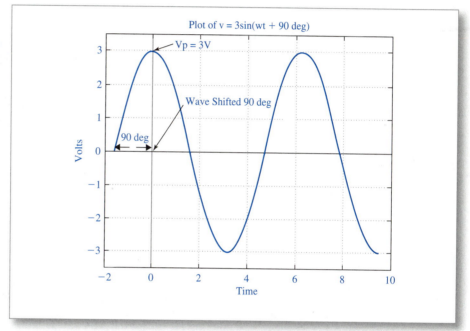

Note that the waveform in Figure 4–8 does not begin a new cycle at the origin (specified reference) at time ($t = 0$). The waveform has been shifted to the left by 90° (advanced in time). The 90° phase shift is also indicated by the given analytical expression $v = 3 \sin(\omega t + 90°)$. As noted, the phase shift of a waveform is typically expressed in degrees rather than units of time.

EXAMPLE 4–2   Determining Phase Difference between Two Waveforms

Determine the phase difference between the given voltage waveforms, designating v_1 as the reference waveform. $v_1 = 1.5 \sin(\omega t + 90°)$ V; $v_2 = 6 \sin(\omega t + 60°)$ V.

✔ SOLUTION

The phase difference is calculated as:

$$\theta = (\theta_1 - \theta_2) = (90° - 60°) = 30°$$

The two voltage waveforms are shown in Figure 4–9.

FIGURE **4–9**

Waveforms for
Example 4–2

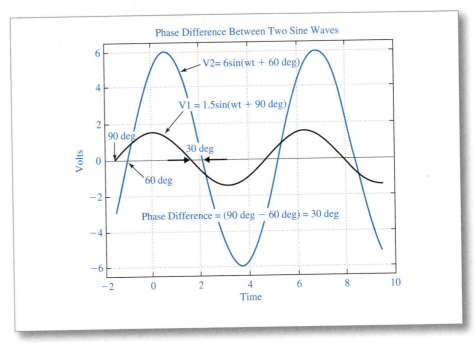

Phase Difference Between Two Sine Waves

V2= 6sin(wt + 60 deg)

V1 = 1.5sin(wt + 90 deg)

90 deg

30 deg

60 deg

Phase Difference = (90 deg − 60 deg) = 30 deg

Volts

Time

Note that v_1 has been shifted 90° to the left of the origin ($t = 0$) and v_2 has been shifted 60° to the left of the origin. The phase difference between v_1 and v_2 is 30° as indicated in Figure 4–9. Note that the positive peak of v_1 occurs before the peak of v_2.

We therefore may say that v_1 *leads* v_2 by 30°. *Note that the amplitude of the waveforms is not considered when determining phase difference.*

EXAMPLE 4–3   Determining the Parameters of a Sinusoidal Voltage Signal

A sinusoidal voltage signal is expressed as $v = 12 \sin(1000t)$ V. Determine the value of peak voltage (V_P), amplitude, radian frequency (ω), frequency (f), period (T), and phase angle (θ).

✓ SOLUTION

Comparing the given expression to the general expression of a sinusoidal voltage waveform expressed as:

$$v = V_P \sin(\omega t \pm \theta)$$

observe that the peak voltage or amplitude is 12 V. The radian frequency (ω) is 1000 rad/s, and the phase angle (θ) is 0°. The frequency (f) may be calculated as:

$$f = \frac{\omega}{2\pi} = \frac{1000}{2\pi} = 159.15 \text{ Hz}$$

In summary, the parameters of the voltage $v = 12 \sin(1000t)$ are:

$V_P = 12$ V; $\omega = 1000$ rad/s; $T = 6.283$ ms; $f = 159.15$ Hz;   and $\theta = 0°$.

A MATLAB plot of the given function is shown in Figure 4–10.

FIGURE **4–10**

Waveforms for
Example 4–3

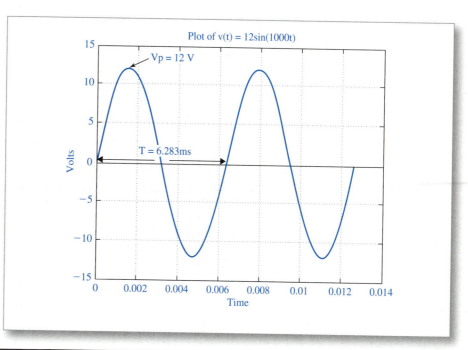

PRACTICE EXERCISES 4–1

a. Given $v = 25 \sin(1000t)$ V, determine the indicated parameters: V_P, A (amplitude), ω, f, T, and θ.

b. Given $v = 16 \sin(3000t + 81°)$ V, determine the indicated parameters: V_P, A (amplitude), ω, f, T, and θ.

✓ ANSWERS

a. $V_P = 25$ V, $A = 25$ V, $\omega = 1000$ rad/s, $f = 159.15$ Hz, $T = 6.28$ ms, $\theta = 0°$

b. $V_P = 16$ V, $A = 16$ V, $\omega = 3000$ rad/s, $f = 477.46$ Hz, $T = 2.09$ ms, $\theta = 81°$

(Complete solutions for the Practice Exercises can be found in Appendix A.)

4.3 INSTANTANEOUS VALUE

The sine wave is a continuous function of time and will have a unique value that can be determined for every instant of time.

The **instantaneous value** of a function of time is the unique value of the function at a specified time.

The instantaneous value of a sine wave can be either positive or negative, and the absolute value is less than or equal to the peak value of the waveform.

EXAMPLE 4–4   Calculating Instantaneous Values

Given $v = 12 \sin(1000t)$ V, calculate the instantaneous value of the voltage at $t = 1.57$ ms, $t = 2$ ms, and $t = 4.7$ ms. Obtain a MATLAB plot to verify the solution.

✓ SOLUTION

At $t = 1.57$ ms

Substituting the given value of time (1.57 ms) and evaluating the argument of the sine function, we obtain:

$$v = 12 \sin(1000 \times 1.57 \text{ ms}) = 12 \sin(1.57 \text{ rad})$$

With the calculator in *radian mode*, evaluate $\sin(1.57 \text{ rad}) = 0.999$ so that:

$$v = 12(0.99999) = 12 \text{ V (rounded)}$$

Note that at time $t = 1.57$ ms the function is at its *positive peak value* (12 V). This is true because 1.57 ms is one-fourth of the period (6.283 ms / 4 = 1.57 ms), which is the time at which the positive peak will occur. The argument of the sine function is 1.57 rad, which is equivalent to $\pi/2$ rad or 90°, which is the angle at which the positive peak value occurs.

At $t = 2$ ms

$$v = 12 \sin(1000 \times 2 \text{ ms}) = 12 \sin(2 \text{ rad}) = 12(0.9093) = 10.91 \text{ V}$$

At $t = 4.71$ ms

$$v = 12 \sin(1000 \times 4.71 \text{ ms}) = 12 \sin(4.71 \text{ rad}) = -12 \text{ V}$$

Note that at time $t = 4.71$ ms the function is at its *negative peak value* (-12 V). This is true because 4.71 ms is three-fourths of the period ($0.75\,T$), which coincides with the negative peak value for the given sine wave. Note that 4.71 rad = $3\pi/2$ rad = 270°.

Figure 4–11 illustrates the instantaneous values calculated in Example 4–4.

FIGURE **4–11**

Instantaneous Values for Example 4–4

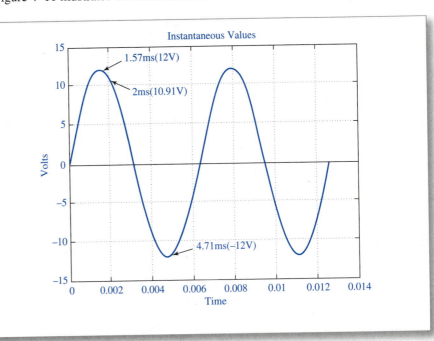

Determine the instantaneous value of the given sinusoidal signals at the specified times.

    a. $v = 6\sin(2500t)$ V at 1.3 ms

    b. $i = 9\sin(6000t)$ mA at 250 μs

    c. $v = 24\sin(4000t + 30°)$ V at 0.5 ms

✔ ANSWERS

    a. −0.65 V

    b. 8.98 mA

    c. 13.89 V

The calculator can be used to convert 30° to radians as required by Practice Exercise 4.2c by selecting RADIAN mode. Enter 30° into the display and press ENTER. The result indicates that 30° is equivalent to 523.6 × 10-3 rad. By selecting EXACT for the EXACT/APPROXIMATE mode, the calculator returns the exact answer π/6 rad.

(Complete solutions to the Practice Exercises can be found in Appendix A.)

Often engineers and scientists solve technical problems assisted by engineering calculators, as well as mathematically based computer software such as MATLAB and computer simulation programs such as MultiSIM. Calculators and computers are indispensable tools and are used extensively throughout this text. However, such tools are productive only when the user has a clear understanding and appreciation of the mathematics upon which the computations are based.

4.4 ADDING SINE WAVES IN THE TIME DOMAIN

During the analysis of electronic circuits it is necessary to perform mathematical operations involving sinusoidal voltages and currents. This section illustrates how the sum of two sine waves may be determined in the time domain using trigonometric identities.

EXAMPLE 4–5   Addition of Sine Waves Using Trigonometry

Determine the sum of the two sinusoidal voltages v_1 and v_2 using trigonometric identities.

$$v_1 = 1\sin(1000t + 60°); \quad v_2 = 1\sin(1000t + 30°)$$

✔ SOLUTION

The trigonometric identity that expresses the sum of two sine functions is:

$$\sin(\alpha) + \sin(\beta) = 2\sin\left(\frac{\alpha + \beta}{2}\right)\cos\left(\frac{\alpha - \beta}{2}\right)$$

This identity may be used to find the sum of the given sine waves as follows:

Let $\alpha = (1000t + 60°)$ and $\beta = (1000t + 30°)$, then:

$$\frac{(\alpha + \beta)}{2} = \frac{1}{2}[(1000t + 60°) + (1000t + 30°)] = 1000t + 45°$$

$$\frac{(\alpha - \beta)}{2} = \frac{1}{2}[(1000t + 60°) - (1000t + 30°)] = 15°$$

so that: $$2 \sin\frac{(\alpha + \beta)}{2} \cos\frac{(\alpha - \beta)}{2} = 2 \sin(1000t + 45°) \cos(15°)$$

Using the calculator, we find $\cos(15°) = 0.9659$

so that: $$2 \sin(1000t + 45°) \cos(15°) = 2 \sin(1000t + 45°)(0.9659)$$

and: $$v_1 + v_2 = 1.932 \sin(1000t + 45°) \text{ V}$$

Note that both of the given signals have the same peak value (1V). Adding sine waves that have different amplitudes using trigonometry greatly increases the complexity of the procedure.

4.5 THE SINE WAVE IN THE PHASOR DOMAIN

We have seen how sinusoidal expressions may be added in the time domain using trigonometric identities. Sine waves may also be added using a method that is based upon the use of *phasor mathematics* and an appropriate engineering calculator. Phasor mathematics, as we will see, is similar to the mathematics involving complex numbers.

Recall that the general expression of a sinusoidal voltage as a function of time is expressed as:

$$v = V_P \sin(\omega t \pm \theta)$$

Since we know that the waveform is sinusoidal, and if the frequency is also known, then the signal can be described using a shorthand mathematical expression:

$$\mathbf{V} = V_P \angle \theta_V \qquad (4.12)$$

The quantity in Equation (4.12) is referred to as a *phasor in polar form*. Observe that the phasor representation of the sine wave specifies only the peak value (amplitude) and phase angle. Throughout this text, phasors will be represented by bold-faced uppercase symbols.

Although phasors, vectors, and complex numbers are similar entities, the term phasor is preferred when referring to electronic voltage or current signals.

A signal expressed using phasor notation is said to be expressed in the **phasor domain.**

When converting waveforms to their phasor domain expressions, it is important to keep in mind that the phasor domain is meaningful only for waveforms that are sinusoidal and that have the same frequency.

EXAMPLE 4–6   Converting Signals from Time Domain to Phasor Domain

Convert the given voltage waveform from the time domain to the phasor domain and sketch a plot of the phasor in polar form in the phasor domain.

$$v = 12 \, \sin(8000t + 38°) \text{ V (time domain)}$$

✓ SOLUTION

The polar form in the phasor domain need indicate only the peak value and phase angle for the phasor. Therefore, the given sinusoidal voltage may be represented in the phasor domain as:

$$\mathbf{V} = 12 \angle 38° \text{ V (phasor domain)}$$

The plot of the phasor in polar form in the phasor domain is shown in Figure 4–12.

FIGURE **4–12**

Phasor
V = 12 ∠ 38° V
Plotted in Polar
Form in the
Phasor Domain

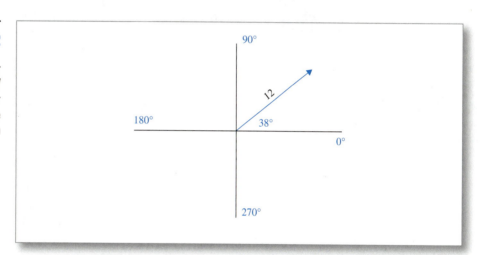

Note that the axes in the phasor domain diagram are labeled in degrees when expressing the phasors in polar form. Although the instantaneous values of a sinusoidal waveform cannot be determined from its phasor domain representation, this is a small price to pay for the advantages that are gained by performing mathematical computations in the phasor domain.

4.6 PHASOR MATHEMATICS

Mathematical operations in the phasor domain require that each voltage and current be represented in its phasor form.

An engineering calculator greatly simplifies calculations in the phasor domain, as illustrated in the following examples.

EXAMPLE 4–7   Adding Sine Waves in the Phasor Domain

Determine the sum of the sinusoidal voltages given in Example 4–5 using phasor mathematics. Express the final answer in the time domain. Sketch a phasor diagram of the two voltages and their sum.

$$v_1 = 1 \sin(1000t + 60°) \text{ V}; \quad v_2 = 1 \sin(1000t + 30°) \text{ V}$$

✓ SOLUTION

Each of the given sine waves must be converted from the time domain to the phasor domain as follows:

$$v_1 = 1 \sin(1000t + 60°) \text{ V (time domain)} \rightarrow \mathbf{V_1} = 1\angle 60° \text{ V (phasor domain)}$$

$$v_2 = 1 \sin(1000t + 30°) \text{ V (time domain)} \rightarrow \mathbf{V_2} = 1\angle 30° \text{ V (phasor domain)}$$

The following calculator keystrokes may be used for the TI-89 calculator.

Set the calculator COMPLEX FORMAT mode to POLAR and ANGLE mode to DEGREES. *Each phasor must be placed in parentheses in the calculator.* The calculator symbol for the angle of the phasor ($\angle$) is the second function of the ⌈EE⌉ key. The symbol for degree (°) is *not required*.

The calculation is keyed as: (⌈1⌉ $\angle$ ⌈6⌉⌈0⌉) ⌈+⌉ (⌈1⌉ $\angle$ ⌈3⌉⌈0⌉) ⌈ENTER⌉ to obtain: 1.932 $\angle$ 45.

The phasor 1.932 $\angle$ 45 V can be converted to the time domain if desired by reinserting the information that was eliminated when the signals were converted to the phasor domain to obtain:

$$v_1 + v_2 = 1.932 \sin(1000t + 45°) \text{ V}$$

Note that the sum of $v_1 + v_2$ obtained in Example 4–7 is the same result obtained in Example 4–5. The graphical representation of the given phasors and their sum is shown in the phasor diagram in Figure 4–13.

FIGURE **4–13**

Phasor Diagram for Example 4–7

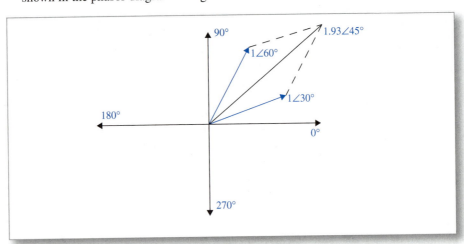

EXAMPLE 4–8   Adding Sine Waves in the Phasor Domain

Add the given sine waves using phasor mathematics. Express the final sum in polar form in the phasor domain and in the time domain. Sketch a phasor diagram of the two voltages and their sum.

$$v_1 = 18 \sin(10t + 36°) \text{ V}; \quad v_2 = 7 \sin(10t + 53°) \text{ V}$$

✓ SOLUTION

The given voltage signals may be converted to the phasor domain as:

$$v_1 = 18 \sin(10t + 36°) \text{ V (time domain)} \rightarrow \mathbf{V_1} = 18 \angle 36° \text{ V (phasor domain)}$$

$$v_2 = 7 \sin(10t + 53°) \text{ V (time domain)} \rightarrow \mathbf{V_2} = 7 \angle 53° \text{ V (phasor domain)}$$

The calculation is keyed as: $(\boxed{1}\boxed{8} \angle \boxed{3}\boxed{6}) \boxed{+} (\boxed{7} \angle \boxed{5}\boxed{3})$ **ENTER** to obtain: $24.78 \angle 40.74$

so that: $\mathbf{V_3} = \mathbf{V_1} + \mathbf{V_2} = 18 \angle 36° \text{ V} + 7 \angle 53° \text{ V} = 24.78 \angle 40.74° \text{ V}$

The sum of the two voltages can be converted to the time domain to obtain:

$$v_1 + v_2 = 24.78 \sin(10t + 40.74°) \text{ V}$$

The graphical representation of the given phasors and their sum is shown in the phasor diagram in Figure 4–14.

FIGURE **4–14**

Phasor Diagram for Example 4–8

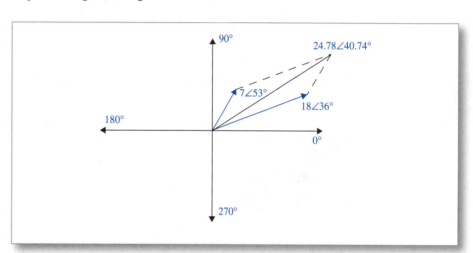

EXAMPLE 4–9   Phasor Mathematics Using the Engineering Calculator

a. Use the calculator and phasor math to perform the indicated phasor addition or subtraction. Represent the sum or difference in the phasor domain.

$$\mathbf{V} = 16 \angle -36° \text{ V} - 25 \angle 48° \text{ V}$$

✓ SOLUTION

Since the phasors are given in polar form, we place the calculator in polar form by selecting **MODE**...COMPLEX FORMAT... POLAR.

The calculation is keyed as: $(\boxed{1}\boxed{6} \angle \boxed{3}\boxed{6}) \boxed{-} (\boxed{2}\boxed{5} \angle \boxed{4}\boxed{8})$ **ENTER** to obtain: $28.24 \angle -97.7$

so that: $\mathbf{V} = 28.24 \angle -97.7°$

b. Use the calculator to perform the indicated phasor addition and subtraction. Represent the sum in the phasor domain.

$$\mathbf{V} = 18 \angle 24° + 8 \angle -21° - 14 \angle 40°$$

✓ SOLUTION

The calculation is keyed as: ($\boxed{1}\boxed{8}$ $\angle$ $\boxed{2}\boxed{4}$) $\boxed{+}$ ($\boxed{8}$ $\angle$ $\boxed{-}\boxed{2}\boxed{1}$) $\boxed{-}$ ($\boxed{1}\boxed{4}$ $\angle$ $\boxed{4}\boxed{0}$) $\boxed{\text{ENTER}}$ to obtain: $13.95 \angle -19.01$

so that: $\mathbf{V} = 13.95 \angle -19.01°$

c. Given $v_1 = 3.8 \sin(4000t + 25.6°)$ V and $v_2 = 7.3 \sin(4000t - 63.2°)$ V, use the calculator to determine the *product* of the two waveforms in the phasor domain.

✓ SOLUTION

$$v_1 = 3.8 \sin(4000t + 25.6°) \text{ V} \rightarrow \mathbf{V_1} = 3.8 \angle 25.6° \text{ V}$$

$$v_2 = 7.3 \sin(4000t - 63.2°) \text{ V} \rightarrow \mathbf{V_2} = 7.3 \angle -63.2° \text{ V}$$

The calculation is keyed as: ($\boxed{3}.\boxed{8}$ $\angle$ $\boxed{2}\boxed{5}.\boxed{6}$) $\boxed{\times}$ ($\boxed{7}.\boxed{3}$ $\angle$ $\boxed{-}\boxed{6}\boxed{3}.\boxed{2}$) $\boxed{\text{ENTER}}$ to obtain: $27.74 \angle -37.6$

so that: $(\mathbf{V_1})(\mathbf{V_2}) = 27.74 \angle -37.6° \text{ V}^2$

Converting the solution to the time domain, we obtain:

$$(v_1)(v_2) = 27.74 \sin(4000t - 37.6°) \text{ V}^2$$

PRACTICE EXERCISES 4-3

The following practice exercise calculations should be performed using phasor math and an appropriate engineering calculator. The calculations should be performed in polar form.

a. Add the given sinusoidal voltage waveforms in the phasor domain. Express the answer in polar form in the phasor domain.

$$v_1 = 6 \sin(3500t + 75°) \text{ V}; \quad v_2 = 11.5 \sin(3500t - 42°) \text{ V}$$

b. Multiply the given sinusoidal current waveforms. Express the answer in polar form in the phasor domain.

$$i_1 = 5.3 \sin(8000t + 53°) \text{ mA}; \quad i_2 = 2.6 \sin(8000t + 15°) \text{ mA}$$

c. Subtract v_2 from v_1 and express the answer in polar form in the phasor domain.

$$v_1 = 6.4 \sin(\omega t + 20°) \text{ V}; \quad v_2 = 7.8 \sin(\omega t - 39°) \text{ V}$$

✓ ANSWERS

a. $10.27 \angle -10.65° \text{ V}$

b. $13.78 \angle 68° \text{ mA}^2$

c. $7.09 \angle 90.38° \text{ V}$

Note that answers may vary slightly due to rounding differences.

(Complete solutions to the Practice Exercises can be found in Appendix A.)

4.7 THE PHASOR IN RECTANGULAR FORM

An alternate form of phasor representation in the phasor domain often used by engineers expresses a phasor as the sum of its rectangular components. The symbol j is used to indicate that a phasor has been revolved 90° counterclockwise.

The axes of the phasor diagram in rectangular form using the symbol j to indicate 90° of counterclockwise revolution can be labeled as indicated in Figure 4–15.

FIGURE **4–15**

The Phasor Domain with Axes Labeled in Rectangular Form

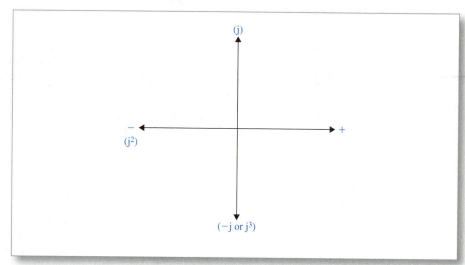

As indicated in Figure 4–15, the symbol j^2 could be used to indicate 180° of revolution; however, the minus sign is the more common notation because j^2 is defined to be equal to -1. Likewise, j^3 could be used to indicate 270°, however, $-j$ is the more common notation.

Consider the phasor **V** in Figure 4–16. Since the rectangular component a is shown in the positive direction along the horizontal axis, it is represented as **a** in rectangular notation. Since the rectangular component b *is* revolved 90° counterclockwise from the horizontal axis, it is represented in rectangular notations as jb. The phasor **V** is the sum of its rectangular components and may therefore be written as:

$$\mathbf{V} = a + j\text{b} \ (\text{rectangular form})$$

FIGURE **4–16**

A Phasor in Rectangular Form

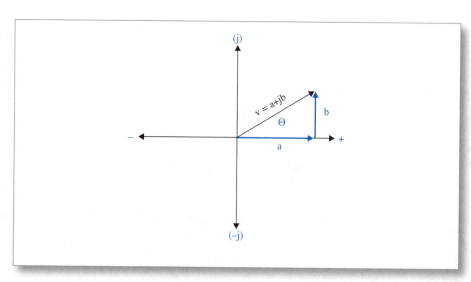

Right-angle trigonometry can be used to determine the peak value or amplitude of the voltage phasor **V** and its phase angle θ respectively as:

$$\mathbf{V} = \sqrt{a^2 + b^2} \tag{4.13}$$

and:

$$\theta = \tan^{-1}(b/a) \tag{4.14}$$

EXAMPLE 4-10   A Phasor in Rectangular Form and Polar Form

A voltage phasor is given in rectangular form as **V** = 6 + *j*8. Determine the peak value of the phasor and its phase angle. Express the phasor in polar form. Sketch the phasor in both rectangular form and polar form.

✔ **SOLUTION**

Applying Equations (4.13) and (4.14), the peak value (amplitude) and phase angle may be determined respectively as:

$$\mathbf{V_P} = \sqrt{6^2 + 8^2} = 10 \text{ V}$$

$$\theta = \tan^{-1}(b/a) = \tan^{-1}(8/6) = 53.13°$$

The phasor may be expressed in polar form as:

$$\mathbf{V} = 10\angle 53.13° \text{ V}$$

The phasor **V** = 6 + *j*8 is shown in rectangular form in Figure 4–17a and in polar form as 10∠53.13° in Figure 4–17b.

FIGURE **4-17**

V =
10 ∠ 53.13° V
in Polar Form is
Equivalent to
V = 6 + *j*8 in
Rectangular Form

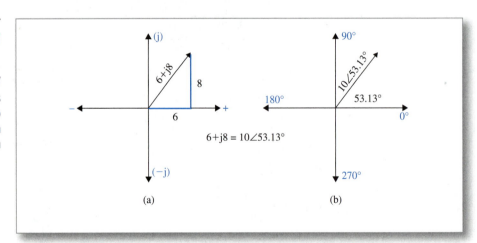

The phasors in Figure 4–17 a and b are equivalent in every way, but simply expressed in different form.

Converting a phasor from rectangular form to polar form can be performed using the TI-89 or similar calculator. The phasor given in Example 4–10 may be converted from the given rectangular form to polar form using the following keystrokes.

Select MODE...COMPLEX FORMAT...POLAR.

Key: (6+8*i*) ENTER to obtain: 10 ∠ 53.13°. Note the symbol *i* is keyed as: 2nd CATALOG.

Converting a phasor from polar form to rectangular form may also be performed as:

Select ⌈MODE⌉...COMPLEX FORMAT...RECTANGULAR.

Key: (⌈1⌉⌈0⌉ ∠ ⌈5⌉⌈3⌉.⌈1⌉⌈3⌉) ⌈ENTER⌉ to obtain: 6+8*i* (substituting *j* for *i*, we obtain 6 + *j*8).

Note: To avoid confusion, the symbol j is substituted for the calculator symbol i because i is the symbol chosen to represent current.

PRACTICE EXERCISES 4–4

Convert each phasor given in polar form to an equivalent phasor in rectangular form. Convert each phasor given in rectangular form to an equivalent phasor in polar form.

    a. $8\angle58°$ V

    b. $36\angle-35°$ mA

    c. $14 + j24$ V

    d. $16 - j12.5$ mA

✓ ANSWERS

    a. $4.236 + j6.78$ V

    b. $29.49 - j20.64$ mA

    c. $27.78\angle39.74°$ V

    d. $20.30\angle-37.99°$ mA

(Complete solutions to the Practice Exercises can be found in Appendix A.)

EXAMPLE 4–11   Addition of Phasors in Rectangular Form

Use the calculator and phasor math to add the given phasors in rectangular form. Also determine the sum in polar form.

$$\mathbf{V_1} = 6 + j8 \text{ V}; \quad \mathbf{V_2} = 3 + j5 \text{ V}$$

✓ SOLUTION

The TI-89 calculator can perform mathematical computations involving phasors expressed in either rectangular or polar form.

Select ⌈MODE⌉...COMPLEX FORMAT...RECTANGULAR.

Key: (⌈6⌉ ⌈+⌉ 8*i*) ⌈+⌉ (⌈3⌉ ⌈+⌉ 5*i*) ⌈ENTER⌉ to obtain: 9 + 13*i* (substituting *j* for *i*, we obtain 9 + *j*13).

Note that phasor addition in rectangular form is accomplished simply by adding the *respective parts* of the phasors (6 + 3 = 9 and 8*i* + 5*i* = 13*i*). Before the invention of the engineering calculator, phasors were required to be expressed in rectangular form in order to perform the operations of addition or subtraction.

Although the calculator has minimized the need to express phasors in rectangular form, many engineers continue to use the rectangular form to represent phasor quantities in electronic circuits.

4.8 PLOTTING SINE WAVES WITH MATLAB

It is often necessary to create a mathematical or analytical expression for a sinusoidal waveform that is observed on an oscilloscope, so that required mathematical calculations can be made. It is equally important to create a graphical representation of a sinusoidal waveform if its analytical expression is known.

The modern digital oscilloscope and computer plotting software such as MATLAB have eliminated much of the labor formerly required to create analytical expressions from graphical presentations. Digital oscilloscopes directly display waveform parameters such as frequency, period, peak voltage, and phase angles, while MATLAB can plot functions given appropriate parameters for the waveforms. Many engineering solutions typically combine both mathematical calculations and graphical analysis using software such as MATLAB.

EXAMPLE 4–12   Plotting a Sine Wave with MATLAB

Create a MATLAB plot for the given sinusoidal voltage waveform:

$$v = 3.2 \sin(50t + 45°) \text{ V}$$

✓ SOLUTION

The following commands will provide a basic plot of the given sine function. Note the phase angle (45°) is entered into MATLAB as pi/4 rad.

```
EDU ≫ t = linspace(−pi/50,4*pi/50,400);    (scales the horizontal axis)
EDU ≫ v=3.2*sin(50*t+pi/4);                (enters the desired function)
EDU ≫ plot(t,v)                            (plots v as a function of time)
```

The basic MATLAB plot is shown in Figure 4–18.

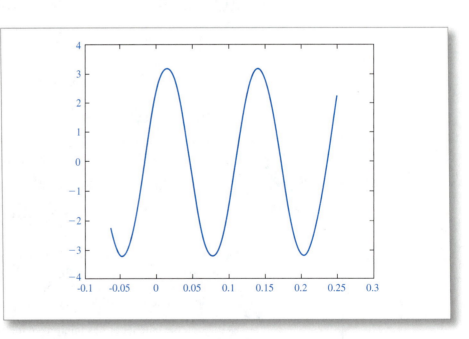

FIGURE 4–18

MATLAB Plot of $v = 3.2 \sin(50t + 45°)$ V

The command t = linspace(−pi/50,4*pi/50,400) scales the horizontal axis to cover the range in time that is equivalent to the angular range of $−\pi$ to 4π rad in 400 steps.

The first two terms of the linspace command are properly scaled by dividing by the radian frequency ($\omega = 50$ rad/s) as indicated in the linspace command.

Additional features such as a title, axes labels, text, arrows, and grid lines can be added to the basic MATLAB plot using the EDIT feature available on the plot figure menu. Several additional enhancements to the basic plot are shown in Figure 4–19.

FIGURE 4–19

A MATLAB Plot with Axes Labels, Title, and Text

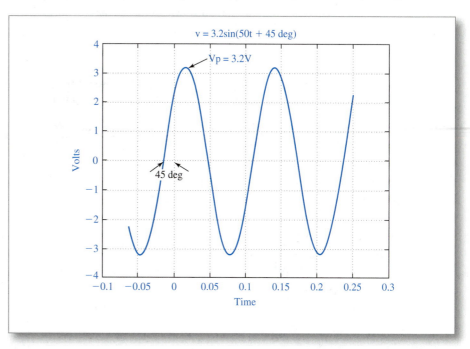

Note that the plot does not begin a cycle at $t = 0$. The given analytical expression indicates that the waveform is shifted to the left by 45° ($\pi/4$ rad) as indicated in Figure 4–19.

4.9 THE RESISTOR IN THE TIME DOMAIN

Assume that the voltage across the resistor in Figure 4–20 is sinusoidal and expressed in the time domain as:

$$v_R = V_P \sin(\omega t + \theta) \tag{4.15}$$

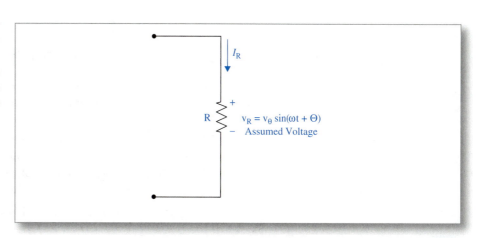

Ohm's law can be used to obtain an expression for the current flowing through the resistor as:

$$i_R = \frac{v_R}{R} = \frac{V_P \sin(\omega t + \theta)}{R} = \frac{V_P}{R} \sin(\omega t + \theta) \qquad (4.16)$$

The coefficient of the sine function in Equation (4.16) indicates that the peak value of the current flowing through the resistor is:

$$I_P = \frac{V_P}{R}$$

Equation (4.16) also indicates that the current through the resistor has the same radian frequency (ω) and the same phase angle (θ) as the voltage across the resistor.

Waveforms that have the same phase angle have a phase difference of zero and are said to be in phase.

The phase relationship between sinusoidal current and voltage for a resistor is illustrated in Figure 4–21.

FIGURE **4–21**

Resistor Current and Voltage are "In Phase"

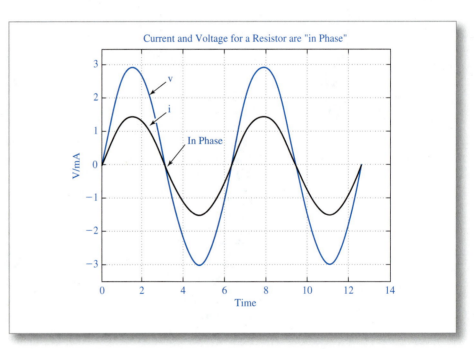

EXAMPLE 4–13 A Resistor in the Time Domain

The voltage across a resistor is given as $v_R = 16 \sin(\omega t)$ V, and the resistor value is given as 2 kΩ. Determine the current through the resistor in the time domain.

✓ **SOLUTION**

It is an excellent practice to sketch a schematic drawing that includes important information such as schematic symbols, component connections, voltage polarities, current flow directions, and the values of all circuit components.

The schematic drawing of the problem statement is shown in Figure 4–22.

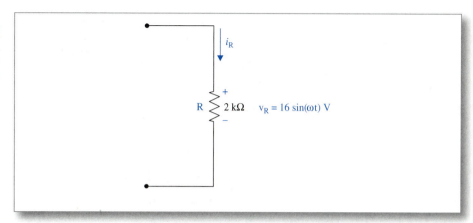

FIGURE **4-22**

Circuit Diagram
for Example 4–13

Ohm's law can be used to calculate the peak value of current as:

$$I_P = \frac{V_P}{R} = \frac{16 \text{ V}}{2 \text{ k}\Omega} = 8 \text{ mA}$$

The current as a function of time can be expressed as:

$$i_R = 8 \sin(\omega t) \text{ mA}$$

The phase difference between the current through and voltage across a resistor is $0°$. Noting that the given voltage in this example has a phase angle of $0°$, the current must also have a phase angle of $0°$.

A sinusoidal voltage will reverse polarity on every half cycle. As the voltage reverses polarity, the current through the resistor will reverse direction as well. The voltage polarities and current directions indicated on a schematic are references at an instant of time. It is understood that the voltage polarities and current directions will reverse with each successive half cycle.

The voltage and current waveforms are shown in the MATLAB plot in Figure 4–23.

FIGURE **4-23**

Voltage and
Current
Waveforms for
Example 4–13

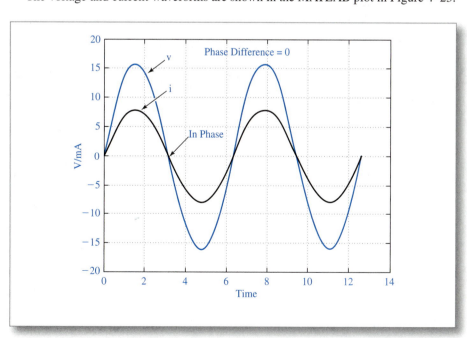

Note that the current i_R has the same radian frequency (ω) and cyclic frequency (f) as the applied voltage in Figure 4–23. Also note that the voltage and current waveforms cross the horizontal axis at the same time, indicating that the phase difference between the current and voltage is 0.

EXAMPLE 4–14   A Resistor in the Time Domain

The current through a resistor is given as $i = 20 \sin(50t + 45°)$ mA; and the resistance value is given as $R = 3$ kΩ. Determine the voltage across the resistor in the time domain.

✓ **SOLUTION**

The schematic is shown in Figure 4–24.

FIGURE **4–24**

Circuit Diagram for Example 4–14

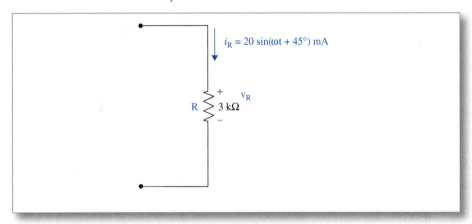

$i_R = 20 \sin(\omega t + 45°)$ mA

$+$ v_R

R 3 kΩ

$-$

Ohm's law can be used to calculate the peak voltage as:

$$V_P = I_P R = (20 \text{ mA})(3 \text{ kΩ}) = 60 \text{ V}$$

The voltage across the resistor can be expressed as a function of time as:

$$v_R = 60 \sin(50t + 45°) \text{ V}$$

Note that once again the resistor current and voltage have the same phase angle of 45°.

The waveforms are shown in the MATLAB plot in Figure 4–25.

FIGURE **4–25**

Voltage and Current Waveforms for Example 4–14

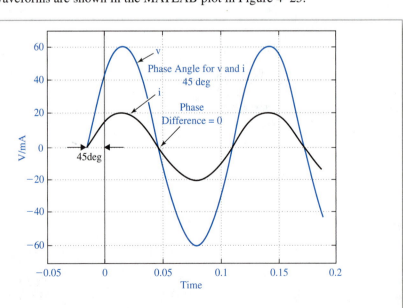

Note that the current and voltage waveforms *do not begin a cycle at t = 0*. Both waveforms have been shifted to the left from the origin. The analytical expressions for the current and voltage indicate that the shift is 45° ($\pi/4$ rad). The 45° phase shift is also indicated in Figure 4–24. *The phase angle of the current and voltage is 45° so that the phase difference between the two waveforms is 0°.*

PRACTICE EXERCISES 4–5

a. The current through a 4 kΩ resistor is given as $i_R = 5.3 \sin(300t + 18°)$ mA. Determine the expression for the voltage across the resistor as a function of time.

b. The voltage across a 10 kΩ resistor is given as $v_R = 24 \sin(2000t + 120°)$ V. Determine the expression for the current through the resistor as a function of time.

c. A voltage source is given as $v_S = 15 \sin(300t)$ V. Determine how much resistance is required to limit the source current to $i_S = 20 \sin(300t)$ mA.

✓ANSWERS

a. $21.2 \sin(300t + 18°)$ V

b. $2.4 \sin(2000t + 120°)$ mA

c. 750 Ω

(Complete solutions to the Practice Exercises can be found in Appendix A.)

4.10 THE RESISTOR IN THE PHASOR DOMAIN

Mathematical calculations involving sinusoidal waveforms are more easily performed in the phasor domain than in the time domain. In order to perform mathematical operations in the phasor domain, however, it is necessary to convert all sinusoidal voltages and currents from their time domain expressions to equivalent phasor domain expressions. In addition, a phasor representation for each circuit component must be determined as well. Voltages and currents may be expressed in the phasor domain in either polar or rectangular form. Circuit components also may be represented in either polar or rectangular form.

A resistor is represented in polar form in the phasor domain as:

$$\mathbf{R} = R \angle 0° \qquad (4.17)$$

A resistor is represented in rectangular form in the phasor domain as:

$$\mathbf{R} = R + j0 \qquad (4.18)$$

Note that the symbol R without bold-face print is the resistance value of the resistor.

Using the phasor notation $\mathbf{R} = R \angle 0°$ to represent a resistor, the proper phase relationship between current and voltage for the resistor is maintained when making calculations in the phasor domain.

EXAMPLE 4-15   The Resistor in the Phasor Domain

In Example 4–13, the voltage across a 2 kΩ resistor is given as $v = 16 \sin(\omega t)$ V. Calculate the current through the resistor using phasor mathematics, and compare the solution to the one obtained in Example 4–13.

✔ SOLUTION

Converting the given sinusoidal voltage to polar form in the phasor domain, we obtain:

$$v_R = 16 \sin(\omega t) \text{ V} \rightarrow \mathbf{V} = 16 \angle 0° \text{ V}$$

The resistor is represented in polar form in the phasor domain using Equation (4.17) as:

$$\mathbf{R} = 2 \angle 0° \text{ k}\Omega$$

Ohm's law can be used to determine the resistor current in polar form in the phasor domain as:

$$\mathbf{I} = \frac{\mathbf{V}}{\mathbf{R}} = \frac{16 \angle 0° \text{ V}}{2 \angle 0° \text{ k}\Omega} = 8 \angle 0° \text{ mA}$$

The calculation reveals that the current through the resistor has a magnitude (peak value) of 8 mA and that the phase angle of the current is 0°. Since the phase angle of the current and voltage are both 0°, the phase difference is 0°.

The current expressed in the phasor domain can be converted to the time domain as:

$$i_R = 8 \sin(\omega t) \text{ mA}$$

The expression for i_R is the same result obtained in Example 4–13 where the calculations were performed in the time domain.

EXAMPLE 4-16   The Resistor in the Phasor Domain

Given $i_R = 20 \sin(50t + 45°)$ mA and $R = 3$ kΩ, determine the voltage across the resistor as a function of time.

✔ SOLUTION

Converting i_R to polar form in the phasor domain, we obtain:

$$\mathbf{I} = 20 \angle 45° \text{ mA}$$

The resistor is represented in polar form in the phasor domain as:

$$\mathbf{R} = 3 \angle 0° \text{ k}\Omega$$

Ohm's law can be used to determine the voltage across the resistor as:

$$\mathbf{V} = \mathbf{IR} = (20 \angle 45° \text{ mA})(3 \angle 0° \text{ k}\Omega) = 60 \angle 45° \text{ V}$$

Converting the voltage phasor to the time domain, we obtain:

$$v_R = 60 \sin(50t + 45°) \text{ V}$$

The expression for v_R is the same result obtained in Example 4–14 where the calculations were performed in the time domain.

The phasor diagrams for the current and voltage phasors in Examples 4–15 and 4–16 are shown in Figure 4–26 a and b respectively.

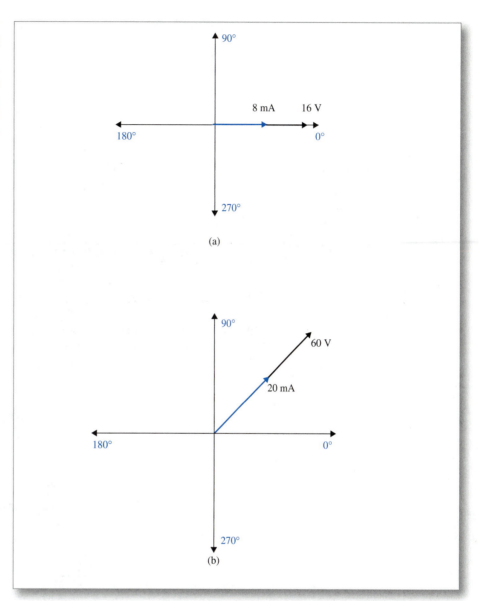

FIGURE 4–26

Phasor Diagrams for the Current and Voltage Phasors in Example 4–15 and Example 4–16

The phasor diagrams in Figure 4–26 clearly indicate that the voltage across a resistor has the same phase angle as the current through the resistor so that the phase difference is 0.

PRACTICE EXERCISES 4–6

a. The voltage across a 4.7 kΩ resistor is 13 sin(480*t* + 29°) V. Calculate the current through the resistor in polar form.

b. The current through a 500 Ω resistor is 41 sin(1000*t* + 68°) mA. Calculate the voltage across the resistor in polar form.

✓ ANSWERS

a. $\mathbf{I} = 2.77 \angle 29°$ mA

b. $\mathbf{V} = 20.5 \angle 68°$ V

(Complete solutions to the Practice Exercises can be found in Appendix A.)

The important characteristics of a resistor may be summarized as:

– The resistance of a resistor is independent of frequency.

– The magnitudes of current and voltage for a resistor are related by Ohm's law:

$$I_R = \frac{V_R}{R}; \ V_R = I_R R$$

– The phase difference between the voltage and current for a resistor is 0°.

– A resistor is represented in the phasor domain as $\mathbf{R} = R\angle 0°$ (polar form) or $\mathbf{R} = R + j0$ (rectangular form).

4.11 THE INDUCTOR IN THE TIME DOMAIN

An inductor is a circuit component that is constructed by winding turns of copper wire around an iron core, as illustrated in Figure 4–27. As current passes through an inductor, an expanding magnetic field will develop around the device. As the expanding magnetic field cuts across the windings of the inductor, a voltage is induced across the inductor, as illustrated in Figure 4–27.

FIGURE **4–27**

Increasing Current through an Inductor Produces an Expanding Magnetic Field around the Inductor

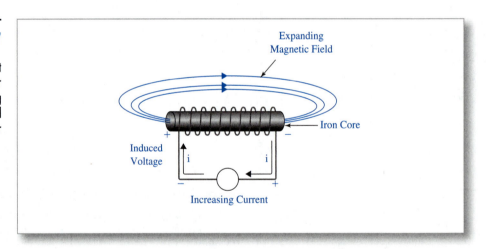

As the current diminishes, the magnetic field will begin to collapse, again cutting across the windings of the inductor. The collapsing magnetic field will also induce a voltage across the inductor, but with the opposite polarity. In both cases, *relative motion* exists between the inductor and the magnetic field surrounding the device, resulting in an induced voltage across the inductor.

A change in the current flowing through an inductor will produce either an expanding or collapsing magnetic field that will induce a voltage across the inductor. Since a sinusoidal current is continuously changing, it will induce a continuous voltage across the inductor. The induced voltage will also be sinusoidal and will have the same frequency as the current.

The polarity of the voltage across an inductor is determined by Lenz's law, which states that an induced effect will always oppose the cause that produced it. It follows that the polarity of the induced voltage across an inductor will always oppose a change of current through the device.

An important electrical characteristic of an inductor is called *inductance*, which is determined by the geometry of the device. Inductance is measured in units of henries (H) to honor Joseph Henry (1797–1878). The schematic symbol chosen for an inductor is the letter *L*. The inductance of an inductor is typically indicated on the device or can be obtained from a manufacturer's specification sheet.

If we are to analyze circuits that contain inductors in the time domain, then we must determine how an inductor responds to sinusoidal voltage and current. Experimentation has shown that an inductor responds or reacts to a change of current in two important ways.

The magnitude of the voltage induced across an inductor is directly proportional to its inductance (*L*) measured in henries and to the *rate of change of the current through the inductor*. In addition, in accordance with Lenz's law, the polarity of the induced voltage always opposes a change of current through the device.

The inductor voltage may be represented mathematically as:

$$v_L = L \frac{di_L}{dt} \tag{4.19}$$

where $\frac{di_L}{dt}$ is the mathematical notation for the rate of change of current (i_L) with respect to time (*t*), and *L* is inductance measured in henries (H), as indicated in Figure 4–28.

You may be familiar with the notation $\frac{d}{dt}$ and the mathematical process of differentiation that it represents. If you are not familiar with the mathematical process of differentiation, the following expression can be considered to be a definition.

The rate of change of a sine wave with respect to time is defined as:

$$\frac{d}{dt}(A \sin \omega t) = \omega A \cos(\omega t) \tag{4.20}$$

Assume that the current through an inductor is given as $i_L = I_L \sin(\omega t)$, as shown in Figure 4–28, where I_L is the peak value of current.

FIGURE **4–28**

An Inductor with Assumed
Current $i = I_L \sin(\omega t)$

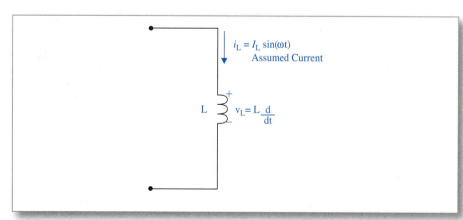

Using the definition given in Equation (4.20), we obtain:

$$v_L = L \frac{d}{dt}(I_L \sin \omega t) = \omega L I_L \cos(\omega t) \tag{4.21}$$

The cosine function is defined as the sine function shifted by 90° so that the expression in Equation (4.21) can expressed as:

$$v_L = \omega L I_L \sin(\omega t + 90°) \tag{4.22}$$

Equation (4.22) indicates that a sinusoidal current passing through an inductor will produce a sinusoidal voltage across the device that has the same frequency (ω) as the current. Equation (4.22) also indicates that the phase difference between the current through an inductor and the voltage across the inductor is 90°.

The coefficient of the sine function in Equation (4.22) is the peak value of the voltage across the device, and designating the peak voltage as V_L, we obtain:

$$V_L = \omega L I_L \tag{4.23}$$

Ohm's law can be generalized to state that the current that flows through any linear component is directly proportional to the voltage across the component and inversely proportional to the opposition to current flow offered by the component.

Letting X_L represent the opposition to current flow from the inductor, Ohm's law for the inductor can be expressed as:

$$X_L = \frac{V_L}{I_L} = \frac{\omega L I_L}{I_L} = \omega L \tag{4.24}$$

The opposition to current flow from an inductor is called **inductive reactance** and given the symbol X_L. Inductive reactance is calculated as ωL, or equivalently as $2\pi f L$, in units of ohms (Ω). It is important to note that inductive reactance varies directly with the frequency of the current through the device.

The relationship between the current and voltage for an inductor is shown in Figure 4–29.

FIGURE **4–29**

Inductor Voltage Leads
the Current by 90°

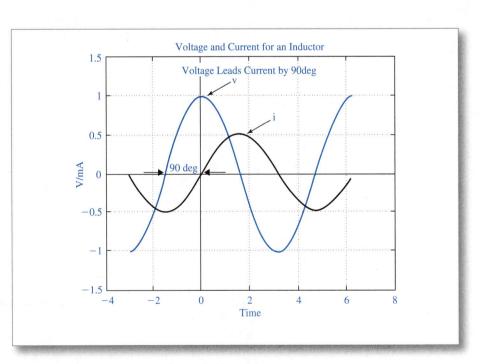

Note that the positive peak of the voltage waveform occurs 90° before the positive peak of the current waveform so that *the voltage leads the current by 90°*.

EXAMPLE 4–17   Calculating Inductive Reactance

a. The current through a 50 mH inductor is given as: $i_L = 26 \sin(1000t)$ mA. Determine the inductive reactance (X_L).

✓ **SOLUTION**

$$X_L = \omega L = (1000)(50 \text{ mH}) = 50 \ \Omega$$

b. The current through a 50 mH inductor is given as: $i_L = 26 \sin(2500t + 45°)$ mA. Determine the inductive reactance (X_L).

✓ **SOLUTION**

$$X_L = \omega L = (2500)(50 \text{ mH}) = 125 \ \Omega$$

Note that the inductive reactance increases as the frequency of the current waveform increases.

EXAMPLE 4–18   Calculating Inductive Current and Voltage

a. The current through a 50 mH inductor is given as: $i_L = 25 \sin(3000t)$ mA. Determine the voltage across the inductor in the time domain.

✓ **SOLUTION**

The inductive reactance (X_L) is calculated for $\omega = 3000$ rad/s as:

$$X_L = \omega L = (3000)(50 \text{ mH}) = 150 \ \Omega$$

Ohm's law can be used to calculate the peak voltage across the inductor as:

$$V_L = I_L X_L = (25 \text{ mA})(150 \ \Omega) = 3.75 \text{ V}$$

The expression for the voltage across the inductor as a function of time is obtained by adding 90° to the phase angle of the current to obtain:

$$v_L = 3.75 \sin(3000t + 90°) \text{ V}$$

Note that the voltage and current have the same radian frequency (3000 rad/s).

b. The current through a 150 mH inductor is given as: $i_L = 18 \sin(1500t + 26°)$ mA. Determine the voltage across the inductor in the time domain.

✓ **SOLUTION**

$$X_L = \omega L = (1500)(150 \text{ mH}) = 225 \ \Omega$$

$$V_L = I_L X_L = (18 \text{ mA})(225 \ \Omega) = 4.05 \text{ V}$$

so that: $$v_L = 4.05 \sin(1500t + 116°) \text{ V}$$

Again, the phase angle of the voltage across the inductor is obtained by adding 90° to the phase angle of the current.

The current and voltage waveforms for Example 4–18b are shown in Figure 4–30.

FIGURE **4–30**

Current and Voltage Waveform for Example 4–18b

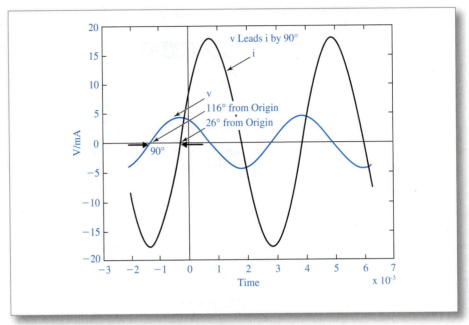

Note that the voltage across the inductor leads the current through the inductor by 90° in Figure 4–30.

4.12 THE INDUCTOR IN THE PHASOR DOMAIN

The analysis of the inductor in the time domain in the previous section indicated that the voltage across an inductor leads the current by 90°. The value of inductive reactance at a given frequency was determined to be $X_L = \omega L$ measured in units of ohms.

Inductive reactance is represented in the phasor domain at an angle of 90° so that:

$$\mathbf{X_L} = \omega L \angle 90° = 2\pi fL \angle 90° \text{ (polar form)} \qquad (4.25)$$

or equivalently as: $\mathbf{X_L} = 0 + j\omega L = 0 + j2\pi fL$ (rectangular form) (4.26)

Using the phasor notation $\mathbf{X_L} = \omega L \angle 90°$ to represent an inductor, the proper phase relationship between current and voltage for the inductor is maintained when making calculations in the phasor domain.

Inductive reactance is not technically a phasor quantity because it does not represent a sinusoidal function of time. However, it is a quantity that is represented in the phasor domain by indicating both a magnitude and an angle. Therefore, in this text, $\mathbf{X_L}$ will appear in bold-face print when making calculations in the phasor domain. The symbol X_L without bold-face print represents only the magnitude of inductive reactance.

The phasor diagram in Figure 4–31 shows the relationship between the current and voltage for an inductor in the phasor domain.

FIGURE **4-31**

Phasor Diagram Showing
Inductor Voltage Leading
Current by 90°

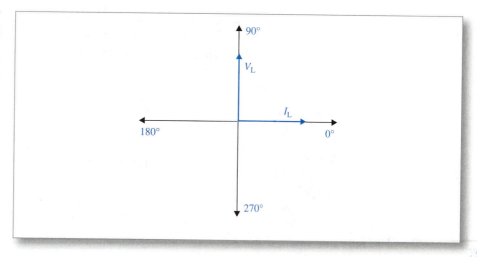

The phasor diagram in Figure 4–31 indicates that the voltage across an inductor *leads* the current by 90°.

EXAMPLE 4-19   The Inductor in the Phasor Domain

a. The current through a 50 mH inductor is given as 25 sin(500t) mA. Use phasor math to determine the voltage across the inductor in the phasor domain. Sketch a phasor diagram.

✔ SOLUTION

The inductive reactance is calculated as:

$$X_L = \omega L = (500)(50 \text{ mH}) = (500)(50 \text{ mH}) = 25 \ \Omega$$

The inductive reactance in polar form in the phasor domain is expressed as:

$$\mathbf{X_L} = 25 \angle 90° \ \Omega$$

The given current is expressed in polar form in the phasor domain as:

$$\mathbf{I_L} = 25 \angle 0° \text{ mA}$$

Ohm's law can be used to calculate the inductor voltage as:

$$\mathbf{V_L} = \mathbf{I_L} \ \mathbf{X_L} = (25 \angle 0° \text{ mA})(25 \angle 90° \ \Omega) = 0.625 \angle 90° \text{ V}$$

Note that representing the inductive reactance in the phasor domain at an angle of 90° has resulted in the correct phase angle for the voltage.

b. The voltage across a 125 mH inductor is given as 12 sin(300t + 35°) V. Determine the inductor current in the phasor domain. Sketch a phasor diagram.

✔ SOLUTION

The inductive reactance is calculated as:

$$X_L = \omega L = (300)(125 \text{ mH}) = 37.5 \ \Omega$$

so that: $$\mathbf{X_L} = 37.5 \angle 90° \ \Omega$$

Ohm's law can be used to calculate the inductor current to obtain:

$$\mathbf{I_L} = \frac{\mathbf{V_L}}{\mathbf{X_L}} = \frac{12 \angle 35° \text{ V}}{37.5 \angle 90° \ \Omega} = 320 \angle -55° \text{ mA}$$

Note again that the phase angle of the voltage is equal to the phase angle of the current plus 90° ($-55° + 90° = 35°$) so that *the voltage leads the current by 90°*.

The phasor diagrams for Example 4–19 a and b are shown in Figure 4–32a and b, respectively.

FIGURE **4–32**

Phasor Diagrams
for
Example 4–19a
and b

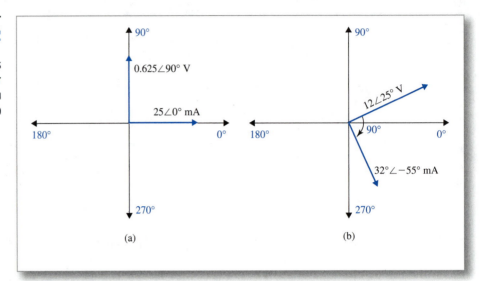

The phasor diagrams in Figure 4–32 indicates that the voltage across an inductor leads the current by 90°.

PRACTICE EXERCISES 4–7

Use phasor math and an appropriate engineering calculator to determine the indicated quantities:

a. The voltage across a 30 mH inductor is given as $v = 21 \sin(4000t + 34°)$ V. Determine the current through the inductor in the phasor domain.

b. The current through a 75 mH inductor is given as $i = 40 \sin(2500t - 41°)$ mA. Determine the voltage across the inductor in the phasor domain.

c. The peak voltage across a 25 mH inductor is given as $V_L = 36.9$ V, and the peak value of current through the inductor is given as $I_P = 12.3$ mA. Determine the value of $\mathbf{X_L}$.

✓ ANSWERS

a. $I = 175 \angle 124°$ mA

b. $V = 7.5 \angle 49°$ V

c. $\mathbf{X_L} = 3 \angle 90°$ kΩ

(Complete solutions to the Practice Exercises can be found in Appendix A.)

The important characteristics of the response of an inductor to sinusoidal excitation may be summarized as:

– The reactance of an inductor ($X_L = \omega L$) varies directly with frequency.

 – The magnitudes of voltage and current for an inductor are related by Ohm's law:

$$I_L = \frac{V_L}{X_L}; \; V_L = I_L X_L$$

 – The voltage across an inductor leads the current by 90°.
 – An inductor is represented in the phasor domain as $\mathbf{X_L} = X_L \angle 90°$ (polar form) or $\mathbf{X_L} = 0 + jX_L$ (rectangular form).

4.13 THE CAPACITOR IN THE TIME DOMAIN

The capacitor consists of two metallic plates that possess an abundance of loosely bound free electrons. The plates are physically and electrically separated by a nonconducting *dielectric* material, as indicated in Figure 4–33.

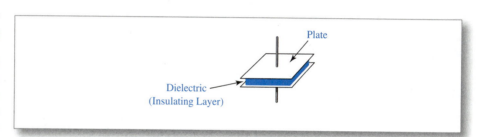

When a voltage is placed across the capacitor, negatively charged free electrons will flow from one of the capacitor plates toward the positive terminal of the voltage source. Electrons will simultaneously leave the negative terminal of the voltage source and be deposited on the other capacitor plate.

The nonconducting dielectric material that electrically separates the capacitor plates prevents current from flowing *through* the capacitor.

Capacitive current results from electrons being displaced from one capacitor plate to the other and is referred to as **displacement current.** Displacement current will continue to flow so long as the voltage across the capacitor is changing.

An important electrical characteristic of a capacitor is called *capacitance*, which is determined by the geometry of the device. Capacitance is measured in units of farads (F) to honor Michael Faraday (1791–1867). The schematic symbol chosen for a capacitor is the letter *C*. The capacitance of a capacitor is typically indicated on the device or can be obtained from a manufacturer's specification sheet.

In the years that have passed since the invention of circuit components such as the capacitor and the inductor, components have become much smaller both in their physical size and electrical measurement. Therefore, it would be unlikely that an *electronic signal processing circuit* would contain a capacitor of even one farad of capacitance. The capacitance of capacitors in *electronic signal processing circuits* is typically measured in units of microfarads (μF, 10^{-6}), nanofarads (nF, 10^{-9}), or picofarads (pF, 10^{-12}).

Since a sinusoidal voltage waveform is continuously changing, a continuous capacitive displacement current is sustained when a sinusoidal voltage is applied across the plates of a capacitor. The current produced by a sinusoidal voltage across a capacitor is also sinusoidal and has the same frequency as the applied voltage.

The capacitor is a reactive component and responds to oppose a change of voltage across its terminals. Experimentation has shown that a capacitor responds or reacts to a change of voltage in two important ways.

The magnitude of the displacement current is directly proportional to capacitance (C) measured in farads and to the rate of change of voltage across the capacitor. In addition, in accordance with Lenz's law, the direction of the displacement current opposes the voltage change that produced it.

Capacitor displacement current may be represented mathematically as:

$$i_C = C \frac{dv_C}{dt} \tag{4.27}$$

where C is the capacitance of the capacitor and $\dfrac{dv_C}{dt}$ represents the rate of change of voltage across the capacitor.

Assume that the voltage across a capacitor is given as $v_C = V_C \sin(\omega t)$, as shown in Figure 4–34, where V_C is the peak value of voltage.

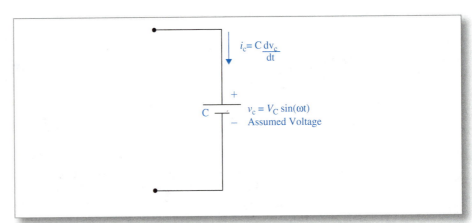

FIGURE 4–34

A Capacitor with Assumed Voltage $v = V_C \sin(\omega t)$

Using the definition introduced in Equation (4.20): $\dfrac{d}{dt}(A \sin \omega t) = \omega A \cos(\omega t)$, the capacitor displacement current i_C can be expressed as:

$$i_C = C \frac{d}{dt}(V_C \sin \omega t) \tag{4.28}$$

so that:

$$i_C = \omega C V_C \cos(\omega t)$$

Substituting the identity $\cos \theta = \sin(\theta + 90°)$, we obtain:

$$i_C = \omega C V_C \sin(\omega t + 90°) \tag{4.29}$$

Equation (4.29) indicates that a sinusoidal voltage across a capacitor produces the displacement current i_C that is also sinusoidal at the same frequency as the applied voltage.

Equation (4.29) also indicates that the phase difference between the capacitor voltage and current is 90° and that the capacitive current leads the voltage by 90°.

The coefficient of the sinusoidal current waveform expressed in Equation (4.29) is the peak value of current and can be expressed as:

$$I_C = \omega C V_C \tag{4.30}$$

Indicating the opposition to current flow of the capacitor as X_C and substituting the peak value of the current $I_C = \omega C V_C$ from Equation (4.30), Ohm's law may be applied to obtain:

$$X_C = \frac{V_C}{I_C} = \frac{V_C}{\omega C V_C}$$

so that:

$$X_C = \frac{1}{\omega C} \qquad (4.31)$$

The opposition to current flow of a capacitor is called **capacitive reactance** and given the symbol X_C. Capacitive reactance is calculated as $1/\omega C$, or equivalently $1/2\pi f C$, in units of ohms (Ω). It is important to note that capacitive reactance varies inversely with frequency.

The angle of capacitive current can be determined by adding 90° to the angle of the voltage across the capacitor.

The relationship between current and voltage for a capacitor is shown in Figure 4–35.

FIGURE **4–35**

Capacitor Current
Leads the
Voltage by 90°

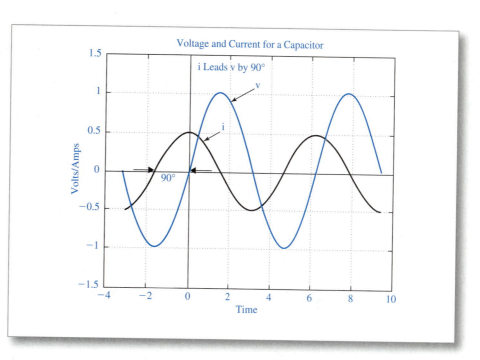

Voltage and Current for a Capacitor

i Leads v by 90°

EXAMPLE 4-20 Calculating Capacitive Reactance

a. The voltage across a 10 µF capacitor is given as $v_C = 8 \sin(2000t)$ V. Determine the capacitive reactance (X_C).

✔ **SOLUTION**

$$X_C = \frac{1}{\omega C} = \frac{1}{(2000)(10\ \mu\text{F})} = 50\ \Omega$$

b. The voltage across a 10 µF capacitor is given as $v_C = 8 \sin(4000t + 25°)$ V. Determine the capacitive reactance (X_C).

$$X_C = \frac{1}{\omega C} = \frac{1}{(4000)(10 \ \mu F)} = 25 \ \Omega$$

Note that the only difference in the calculations in this example is that the radian frequency (ω) of the applied voltage has increased from 2000 rad/s to 4000 rad/s. The calculations indicate that capacitive reactance varies inversely with frequency.

EXAMPLE 4–21 Calculating Capacitor Current in the Time Domain

The voltage across a 0.47 μF capacitor given as $v = 12 \sin(1000t)$ V. Determine the expression for the capacitor current in the time domain.

✔ **SOLUTION**

The capacitive reactance is calculated as:

$$X_C = \frac{1}{\omega C} = \frac{1}{(1000)(0.47 \ \mu F)} = 2.13 \ k\Omega$$

Ohm's law can be used to calculate the peak current as:

$$I_C = \frac{V_C}{X_C} = \frac{12 \ V}{2.13 \ k\Omega} = 5.64 \ mA$$

The expression for capacitor current can be expressed as:

$$i_C = 5.64 \sin(1000t + 90°) \ mA$$

Note that the phase angle for the capacitor current has been determined by adding 90° to the phase angle of the voltage across the capacitor.

The capacitor current and voltage are shown in the MATLAB plot in Figure 4–36.

FIGURE **4–36**

Capacitor Current and Voltage for Example 4–21

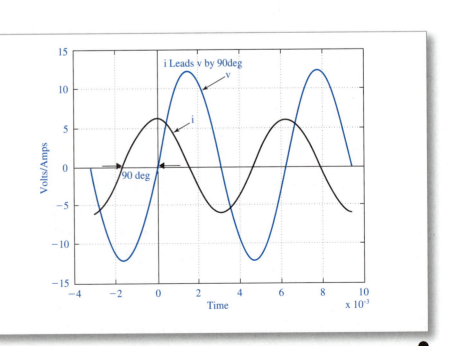

EXAMPLE 4–22   Calculating Capacitor Voltage in the Time Domain

The current for a 0.22 μF capacitor is given as $i_C = 21 \sin(2000t)$ mA. Determine the expression for the capacitor voltage in the time domain.

✓ SOLUTION

The capacitive reactance is calculated as:

$$X_C = \frac{1}{\omega C} = \frac{1}{(2000)(0.22 \ \mu F)} = 2.27 \text{ k}\Omega$$

Ohm's law can be used to calculate the peak voltage as:

$$V_C = I_C X_C = (21 \text{ mA})(2.27 \text{ k}\Omega) = 47.67 \text{ V}$$

The voltage across the capacitor can be expressed as:

$$v_C = 47.67 \sin(2000t - 90°) \text{ V}$$

Note that because the phase angle of the given current is 0°, the voltage must have a phase angle of −90°.

The current and voltage are shown in Figure 4–37.

FIGURE 4–37

Capacitor Current and Voltage for Example 4–22

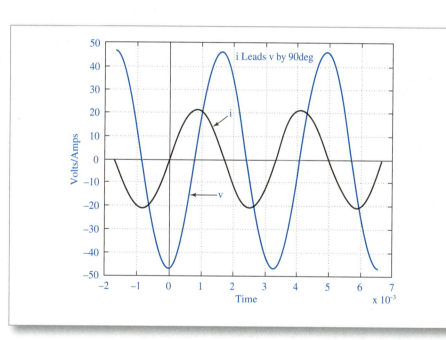

Note that the current in Figure 4–37 leads the voltage by 90°.

a. The voltage across a 0.47 μF capacitor is given as $v_C = 2.88 \sin(5000t - 18°)$ V. Determine the capacitive reactance ($\mathbf{X_C}$).

b. The voltage across a 0.22 μF capacitor is $v_C = 20 \sin(4000t + 25°)$ V. Determine the expression for the capacitor current in the time domain.

✓ **ANSWERS**

a. $425.53 \angle -90° \ \Omega$

b. $17.54 \sin(4000t + 115°)$ mA

(Complete solutions to the Practice Exercises can be found in Appendix A.)

4.14 THE CAPACITOR IN THE PHASOR DOMAIN

Capacitive reactance is represented in *polar form in the phasor domain* at an angle of $-90°$ so that:

$$\mathbf{X}_C = \frac{1}{\omega C} \angle -90° = \frac{1}{2\pi f C} \angle -90° \tag{4.32}$$

or equivalently in *rectangular form* as:

$$\mathbf{X}_C = \frac{1}{j\omega C} = \frac{1}{j2\pi f C} \tag{4.33}$$

Using the phasor notation $\mathbf{X}_C = \dfrac{1}{\omega C} \angle -90°$ to represent a capacitor, the proper phase relationship between current and voltage for the capacitor is maintained when making calculations in the phasor domain.

A phasor diagram showing the phase relationship between capacitor voltage and current is shown in Figure 4–38.

FIGURE **4–38**

Phasor Diagram Showing Capacitor Current Leading Voltage by 90°

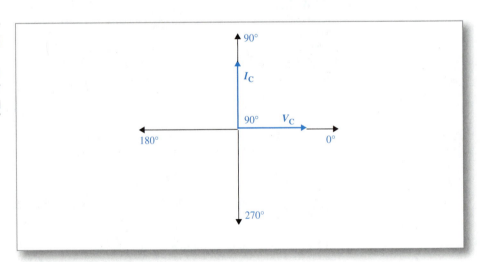

The phasor diagram in Figure 4–38 indicates that the capacitor current leads the voltage by 90°.

EXAMPLE 4–23   The Capacitor in the Phasor Domain

a. The voltage across a 0.2 µF capacitor is given as $12 \sin(500t)$ V. Use phasor math to determine the capacitor current in the phasor domain.

✔ SOLUTION

The capacitive reactance is calculated as:

$$\mathbf{X_C} = \frac{1}{\omega C}\angle -90° = \frac{1}{(500)(0.2\mu F)}\angle -90° = 10\angle -90° \text{ k}\Omega$$

Ohm's law can be used to calculate capacitor current as:

$$\mathbf{I_C} = \frac{\mathbf{V_C}}{\mathbf{X_C}} = \frac{12\angle 0°\text{V}}{10\angle -90°\text{k}\Omega} = 1.2\angle 90° \text{ mA}$$

Note that the phase angle of the current is 90° so that the current leads the voltage by 90°.

b. The voltage across a 5 μF capacitor is given as 3.6 sin(200t + 39°) V. Use phasor math to determine the capacitor current in the phasor domain.

$$\mathbf{X_C} = \frac{1}{\omega C}\angle -90° = \frac{1}{(200)(5\mu F)}\angle -90° = 1\angle -90° \text{ k}\Omega$$

$$\mathbf{I_C} = \frac{\mathbf{V_C}}{\mathbf{X_C}} = \frac{3.6\angle 39°\text{V}}{1\angle -90° \text{ k}\Omega} = 3.6\angle 129° \text{ mA}$$

The phasor diagrams for Example 4–23 a and b are shown in Figure 4–39.

FIGURE **4–39**

Phasor Diagrams
for
Example 4–23a
and b

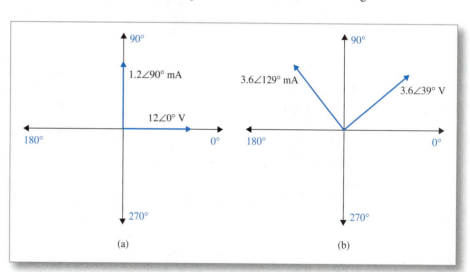

(a)   (b)

Observe that the capacitor current in Figure 4–39(b) leads the voltage by 90° (39° + 90° = 129°).

PRACTICE EXERCISES 4–9

a. The voltage across a 0.33 μF capacitor is given as v_C = 6 sin(2500t) V. Use phasor math to determine the capacitor current in the phasor domain.

b. The voltage across a 0.22 μF capacitor is given as v_C = 14 sin(1000t = 23°) V. Use phasor math to determine the capacitor current in the phasor domain.

✔ ANSWERS

a. $I_C = 4.96\angle 90°$ mA

b. $I_C = 3.08\angle 113°$ mA

(Complete solutions to the Practice Exercises can be found in Appendix A.)

The important characteristics of the response of a capacitor to sinusoidal excitation can be summarized as:

– The reactance of a capacitor $X_C = \dfrac{1}{\omega C}$ varies inversely with frequency.

– The magnitudes of voltage and current for a capacitor are related by Ohm's law:

$$I_C = \frac{V_C}{X_C}; \; V_C = I_C X_C$$

– The capacitor current leads the voltage across the capacitor by 90°.

– A capacitor is represented in the phasor domain as $\mathbf{X_C} = X_C \angle -90°$ (polar form) or $\mathbf{X_C} = 0 - jX_C$ (rectangular form).

SUMMARY

The **sine wave** is the most important electronic signal in the study of circuit analysis. A sinusoidal voltage is expressed in the **time domain** as: $v(t) = V_P \sin(\omega t \pm \theta)$. V_P is the **peak voltage**, ω is the **radian**, and θ is the **phase angle**. The time required to complete one cycle of a sine wave is referred to as the **period** of the waveform. The value of the voltage at a specified time is referred to as the instantaneous value of the sine wave.

A **phasor** is a shorthand representation of a sine wave that indicates only the peak value and phase angle of the waveform. A phasor may be represented in either polar form or rectangular form. A sine wave expressed as a phasor is said to be expressed in the **phasor domain**. Mathematical calculations involving sine waves are typically performed in the phasor domain.

The resistance of a **resistor** is independent of frequency. A resistor is represented in the phasor domain in polar form as $\mathbf{R} = R \angle 0°$. A quantity represented by bold-faced print must be represented by both a magnitude and an angle. The resistor current and voltage are in phase.

An **inductor** responds to sinusoidal current by inducing a sinusoidal voltage that has the same frequency as the current. The magnitude of the voltage is determined by the inductance of the inductor and the rate of change of the current. The opposition to current flow from an inductor is referred to as **inductive reactance** and is calculated as $X_L = \omega L$. The voltage across an inductor leads the current by 90°. An inductor is represented in the phasor domain in polar form as $\mathbf{X_L} = X_L \angle 90°$ and in rectangular form as $\mathbf{X_L} = 0 + j\omega L$.

A **capacitor** responds to sinusoidal voltage by producing a sinusoidal current that has the same frequency as the voltage. The magnitude of the current is determined by the capacitance of the capacitor and the rate of change of the voltage. The opposition to current flow from a capacitor is referred to as capacitive reactance and is calculated as $X_C = 1/\omega C$. The capacitor current leads the voltage by 90°. A capacitor is represented in the phasor domain in polar form as $\mathbf{X_C} = X_C \angle -90°$ and in rectangular form as $\mathbf{X_L} = 0 - j\omega L$.

ESSENTIAL LEARNING EXERCISES

(ANSWERS TO ODD-NUMBERED ESSENTIAL LEARNING EXERCISES CAN BE FOUND IN APPENDIX A.)

1. Write the symbol and indicate the proper units for the following quantities:

a. Peak value (voltage)

b. Peak value (current)

c. Frequency

d. Radian frequency

e. Phase angle

2. Write a brief definition of each of the terms given in Exercise 1.

3. An electronic signal is expressed as: $v = 12 \sin(2000t)$ V. Determine the value of each indicated parameter.

a. Peak value (amplitude)

b. Frequency

c. Radian frequency

d. Phase angle

4. Determine the instantaneous value of the sinusoidal voltage $v = 3.7 \sin(377t)$ V at the specified times: $t = 1$ ms, 2.4 ms, 9.4 ms, 12.62 ms, and 16.67 ms.

5. Given $v_1 = 3.19 \sin(1000t + 35°)$ V and $v_2 = 1.04 \sin(1000t + 16°)$ V, convert each voltage to an expression in the phasor domain.

6. Determine the sum of the two voltages given in Exercise 5 in polar form in the phasor domain.

7. Given the current phasor $\mathbf{I} = 3.6 \angle 25°$ mA, convert the phasor to an equivalent rectangular form in the phasor domain.

8. Given $\mathbf{V_1} = 6.5 + j\,8.2$ and $\mathbf{V_2} = 5.3 - j\,16.4$, determine the sum of the voltages in rectangular form in the phasor domain. Convert the sum from rectangular form to polar form in the phasor domain.

9. Sketch a plot of the voltage $v_1 = 6.5 \sin(500t + 36°)$ V in the time domain and in polar form in the phasor domain.

10. Sketch a plot of the current phasor $i = 14 \sin(1000t + 50°)$ mA in the time domain and in polar form in the phasor domain.

11. Use an engineering calculator to perform the indicated mathematical operations in polar form in the phasor domain.

a. $\dfrac{36 \angle 29°}{1.42 \angle -52°}$

b. $12.8 \angle 31° + 4.9 \angle 12° - 24.7 \angle -74°$

c. $(0.98 \angle 22°)(13.4 \angle -65°)$

12. The voltage across a 4.7 kΩ resistor is given as $v = 15 \sin(6000t)$ V. Determine the expression for the resistor current in the time domain.

13. The current through a 2.2 kΩ resistor is given as $i = 6.73 \sin(5500t + 38°)$ mA. Determine the expression for the voltage across the resistor in the time domain.

14. The current through a 2.8 mH inductor is given as $i = 12.7 \sin(12000t + 38°)$ mA. Calculate the reactance of the inductor.

15. The voltage across a 4.7 μF capacitor is given as $v = 8.8 \sin(500t - 12°)$ V. Calculate the reactance of the capacitor.

16. The current through a 3.5 mH inductor is given as $i = 1.5 \sin(8000t)$ mA. Determine the expression for the voltage across the inductor in the time domain.

17. The voltage a 150 mH inductor is given as $v = 24 \sin(5000t + 45°)$ V. Determine the expression for the inductor current in the time domain.

18. The displacement current for a 0.5 μF capacitor is given as $i = 0.38 \sin(800t)$ mA. Determine the expression for the voltage across the capacitor in the time domain.

19. The voltage a 150 nF capacitor is given as $v = 16 \sin(2000t - 30°)$ V. Determine the expression for the capacitor current in the time domain.

20. The voltage across a 10 kΩ resistor is given as $v = 23 \sin(550t)$ V. Determine the expression for the resistor current in polar form in the phasor domain. Sketch a phasor diagram showing the voltage and current in polar form.

21. The current through a 4 kΩ resistor is given as $i = 5.28 \sin(1200t - 68°)$ mA. Determine the expression for the voltage across the resistor in polar form in the phasor domain. Sketch a phasor diagram showing the voltage and current in polar form.

22. The current through a 200 mH inductor is given as $i = 0.98 \sin(6500t)$ mA. Determine the expression for the voltage across the inductor in polar form in the phasor domain. Sketch a phasor diagram showing the voltage and current in polar form.

23. The voltage across a 250 mH inductor is given as $v = 12.8 \sin(4500t - 20°)$ V. Determine the expression for the inductor current in polar form in the phasor domain. Sketch a phasor diagram showing the voltage and current in polar form.

24. The displacement current for a 0.01 μF capacitor is given as $i = 2.42 \sin(400t + 34°)$ mA. Determine the expression for the voltage across the capacitor in polar form in the phasor domain. Sketch a phasor diagram showing the voltage and current in polar form.

25. The voltage a 0.1 μF capacitor is given as $v = 20 \sin(350t + 80°)$ V. Determine the expression for the capacitor current in polar form in the phasor domain. Sketch a phasor diagram showing the voltage and current in polar form.

CHALLENGING LEARNING EXERCISES

(ANSWERS TO SELECTED CHALLENGING LEARNING EXERCISES CAN BE FOUND IN APPENDIX A.)

1. Which of the following circuit components will offer the most opposition to current flow at a frequency of 1.2 kHz?

   a. 5 kΩ resistor

   b. 2.3 μF capacitor

   c. 3.9 mH inductor

2. Determine the instantaneous value of the given sinusoidal voltage at 5 ms:

$$v = 25 \sin(500t + 45°) \text{ V}$$

3. Add the given sinusoidal voltages and express the answer in polar form in the phasor domain. Convert the sum from the phasor domain to the time domain.

$$v_1 = 6 \sin(5000t + 42°) \text{ V}; \quad v_2 + 9 \sin(5000t - 28°) \text{ V}; \quad v_3 = 5 \sin(5000t + 63°) \text{ V}$$

4. Given the MATLAB plot in Figure 4–40, determine the peak value of each wave, the frequency of each wave, and the approximate phase angle between the two waves.

FIGURE **4–40**

Waveforms for
Challenging
Exercise 4

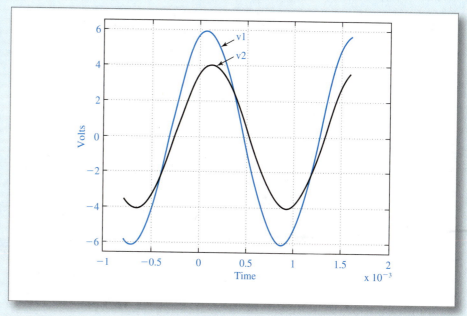

5. Create a spreadsheet document that will calculate the value of sin θ in 15° increments over 360°. Create a plot of the values using appropriate plotting software such as Microsoft Excel.

6. A current waveform is given as $i = 30 \sin(377t)$ mA. Sketch one complete cycle of the waveform as a function of time. Label the peak value points and the zero crossing point in degrees, radians, and seconds. Create a MATLAB plot of the given voltage waveform.

7. Create a MATLAB plot of the given voltage waveform and use the plot to determine the instantaneous value of the voltage at time $t = 1.5$ ms, $t = 3.9$ ms, and $t = 5.2$ ms. Calculate the instantaneous value of the voltage at the given times and compare to the values obtained from the plot.

$$v = 20 \sin(1000t) \text{ V}$$

TEAM ACTIVITY

Two voltage waveforms have a frequency of 20 kHz and are separated in time by 10 μs. Determine the phase difference between the two waveforms.

ESSENTIALS OF AC CIRCUIT ANALYSIS

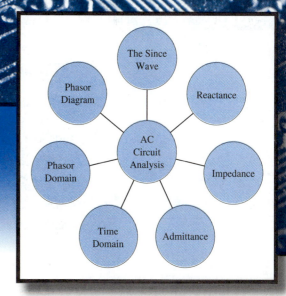

"LEARN TO REASON FORWARD AND BACKWARD
ON BOTH SIDES OF A QUESTION."

—Abraham Lincoln

CHAPTER OUTLINE

LEARNING OBJECTIVES

Upon successful completion of this chapter you will be able to:

- Calculate total equivalent impedance.
- Calculate total equivalent admittance.
- Sketch an impedance diagram.
- Convert sinusoidal waveforms from the time domain to the phasor domain.
- Apply Ohm's law in the phasor domain.
- Apply Kirchhoff's laws in the phasor domain.
- Apply the voltage divider and current divider rules in the phasor domain.
- Calculate current and voltage in a resistive, R-L, or R-C circuit.
- Sketch a phasor diagram.

INTRODUCTION

This chapter demonstrates how the circuit analysis laws and techniques learned in previous chapters are used to analyze circuits that contain resistors, capacitors, and inductors whose input signal is a sine wave. Since the sine wave voltage changes polarity every half cycle and the sine wave current changes direction every half cycle, the circuits are referred to as alternating current or AC circuits.

This chapter considers only circuits whose currents and voltages are sinusoidal. By applying Ohm's law, Kirchhoff's laws, and the voltage and current divider rules, we will be able to determine the magnitudes and phase relationships of the current and voltage in a variety of analog circuits.

5.1 THE SERIES RESISTIVE CIRCUIT IN THE PHASOR DOMAIN

All of the voltages and currents presented in this chapter are sinusoidal.

Since the sine wave voltage changes polarity every half cycle and the sine wave current changes direction every half cycle, the circuits are referred to as alternating current or AC circuits.

Recall that the resistance of a resistor is constant regardless of the frequency of the applied voltage, and the phase difference between the voltage across a resistor and the current through the resistor is zero.

Therefore, a resistor in the phasor domain is represented as:

$$\mathbf{R} = R \angle 0°$$

The bold-faced print indicates a quantity that is defined in the phasor domain by specifying both magnitude and angle. The angle of zero degrees indicates that the phase difference between the voltage across a resistor and the current through the resistor is zero.

Consider the circuit in Figure 5–1.

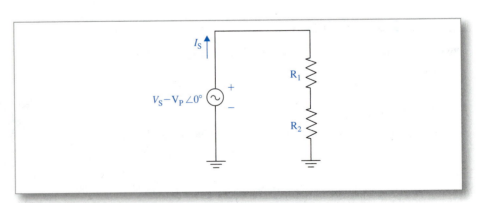

FIGURE **5–1**

A Series Resistive Circuit in the Phasor Domain

The signal source in Figure 5–1 is represented in the *phasor domain* as:

$$\mathbf{V_S} = V_P \angle 0°$$

Observe that the resistors in Figure 5–1 are connected in series with the signal source because the source current flows throughout the circuit. Recalling that total equivalent resistance for resistors connected in series is determined as the sum of the individual resistors, the total equivalent resistance in the phasor domain can be determined as:

$$\mathbf{R_T} = \mathbf{R_1} + \mathbf{R_2} = R_1 \angle 0° + R_2 \angle 0°$$

Ohm's law can be used to determine the source current in the phasor domain as:

$$\mathbf{I_S} = \frac{\mathbf{V_S}}{\mathbf{R_T}}$$

Ohm's law may also be used to calculate the voltage across each respective resistor in the phasor domain as:

$$\mathbf{V_{R1}} = \mathbf{I_{R1}}\ \mathbf{R_1}$$

$$\mathbf{V_{R2}} = \mathbf{I_{R2}}\ \mathbf{R_2}$$

where $\mathbf{I_S} = \mathbf{I_{R1}} = \mathbf{I_{R2}}$ in the series circuit.

KVL may be written in the phasor domain for the circuit in Figure 5–1 as:

$$\sum \mathbf{V}_{\text{closed loop}} = 0$$

so that:

$$\mathbf{V_S} = \mathbf{V_{R1}} + \mathbf{V_{R2}}$$

The voltage across each respective resistor can also be determined in the phasor domain using the voltage divide rule as:

$$\mathbf{V_{R1}} = \mathbf{V_S}\,\frac{\mathbf{R_1}}{\mathbf{R_1} + \mathbf{R_2}}$$

$$\mathbf{V_{R2}} = \mathbf{V_S}\,\frac{\mathbf{R_2}}{\mathbf{R_1} + \mathbf{R_2}}$$

EXAMPLE 5–1   The Series Resistive Circuit in the Phasor Domain

Given the circuit in Figure 5–2, determine the total equivalent resistance. Determine the source current. Calculate the voltage across each respective resistor using Ohm's law. Verify KVL for the circuit. Calculate the voltage across each respective resistor using the voltage divider rule. Sketch a phasor diagram for the circuit.

FIGURE **5–2**

Circuit for
Example 5–1

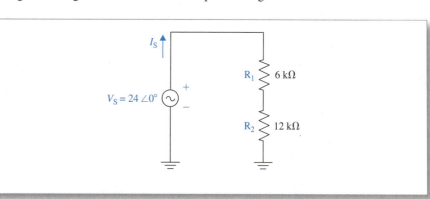

✓ SOLUTION

All calculations are performed in the phasor domain using an appropriate engineering calculator.

The signal source in Figure 5–2 is given in the phasor domain as:

$$\mathbf{V_S} = 24\angle 0°\ \text{V}$$

The total equivalent resistance is calculated as the sum of the individual resistors to obtain:

$$\mathbf{R_T} = 6\angle 0°\ \text{k}\Omega + 12\angle 0°\ \text{k}\Omega = 18\angle 0°\ \text{k}\Omega$$

The total equivalent resistance may be calculated as: 6 EE 3 ∠ 0 + 1 2 EE 3 ∠ 0 5 to obtain: 1 8 EE 3 ∠ 0. An angle is entered on the TI-89 calculator as 2nd EE. The degree symbol is not required.

Note that when resistors are summed in the phasor domain, the resistors must be represented by their phasor notation including the appropriate magnitude and angle.

Ohm's law can be used to calculate the source current in the phasor domain as:

$$\mathbf{I_S} = \frac{\mathbf{V_S}}{\mathbf{R_T}} = \frac{24\angle 0^\circ \text{ V}}{18\angle 0^\circ \text{ k}\Omega} = 1.33\angle 0^\circ \text{ mA}$$

Ohm' law can be used to calculate the voltage across each respective resistor as:

$$\mathbf{V_{R1}} = \mathbf{I_{R1}}\ \mathbf{R_1} = (1.33\angle 0^\circ \text{ mA})(6\angle 0^\circ \text{ k}\Omega) = 8\angle 0^\circ \text{ V}$$

$$\mathbf{V_{R2}} = \mathbf{I_{R2}}\ \mathbf{R_2} = (1.33\angle 0^\circ \text{ mA})(12\angle 0^\circ \text{ k}\Omega) = 16\angle 0^\circ \text{ V}$$

KVL can be expressed in the phasor domain as:

$$\mathbf{V_S} - \mathbf{V_{R1}} - \mathbf{V_{R2}} = 24\angle 0^\circ \text{ V} - 8\angle 0^\circ \text{ V} - 16\angle 0^\circ \text{ V} = 0$$

The voltage divider rule may also be used to calculate the voltage across each respective resistor as:

$$\mathbf{V_{R1}} = \mathbf{V_S}\frac{\mathbf{R_1}}{\mathbf{R_1}+\mathbf{R_2}} = \frac{(24\angle 0^\circ \text{ V})(6\angle 0^\circ \text{ k}\Omega)}{18\angle 0^\circ \text{ k}\Omega} = 8\angle 0^\circ \text{ V}$$

$$\mathbf{V_{R2}} = \mathbf{V_S}\frac{\mathbf{R_2}}{\mathbf{R_1}+\mathbf{R_2}} = \frac{(24\angle 0^\circ \text{ V})(12\angle 0^\circ \text{ k}\Omega)}{18\angle 0^\circ \text{ k}\Omega} = 16\angle 0^\circ \text{ V}$$

Note that the circuit laws and rules that were used to solve for current and voltage in the resistive DC circuits in Chapter 3 are equally valid in the phasor domain. It is important to remember, however, that all quantities represented in the phasor domain must be specified by a magnitude and angle.

A phasor diagram illustrates the phasors for the currents and voltages in an AC circuit indicating magnitude and phase angle.

The phasor diagram for the circuit in Figure 5–2 is shown in Figure 5–3.

FIGURE **5–3**

Phasor Diagram for Example 5–1

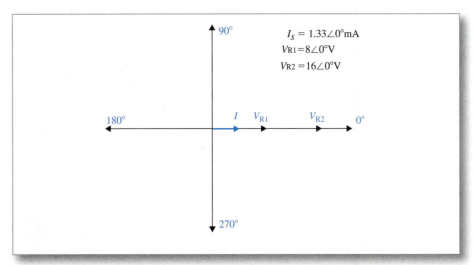

Note that the current is in phase with the voltage across the resistors and that the phase difference between the voltages in a purely resistive circuit is zero. The length of a phasor in a phasor diagram is proportional to the magnitude of the quantity represented. The phasor representing V_{R2} ($16\angle0°$ V) in Figure 5–3 is twice the length of the phasor representing V_{R1} ($8\angle0°$ V). The phasor representing current I_S is not proportional to the voltage phasors because the phasors are measured in different units (volts/amps).

The current and voltage for Example 5–1 are shown in the time domain in the MATLAB plot in Figure 5–4.

FIGURE **5–4**

Current and Voltage for Example 5–1 in the Time Domain

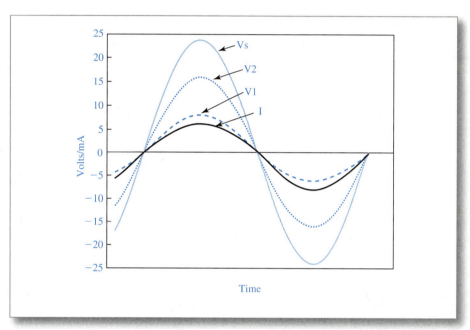

The time domain plot in Figure 5–4 and the phasor diagram in Figure 5–3 indicate that voltages and current are in phase for a purely resistive circuit.

5.2 THE SERIES R-L CIRCUIT IN THE PHASOR DOMAIN

Ohm's law can be generalized to state that current can be determined if the voltage that exists between two points in a circuit is known and if the total opposition to current flow that exists between the same two points can be determined. The total opposition to current flow in an AC circuit may include both resistive and reactive circuit components (capacitors and inductors).

The total opposition to current flow for circuits that contain both resistive and reactive components is referred to as **impedance** and is designated in the phasor domain by the symbol **Z**. The total impedance of a series circuit is the sum of the resistance and reactance of the individual components measured in units of ohms.

Impedance is not technically a phasor because it is not a sinusoidal time-varying quantity. However, since it is represented in the phasor domain by specifying both magnitude and angle, it will be treated in this text as a phasor quantity and represented by bold-faced uppercase print.

EXAMPLE 5-2   Analysis of a Series R-L Circuit

Given the series R-L circuit in Figure 5–5, determine the impedance of the circuit, the source current, and the voltage across each component. Verify that the voltages across the components are correct by writing KVL around the circuit. Use the voltage divider rule to calculate the voltage across each component. Sketch an impedance diagram and phasor diagram.

FIGURE **5-5**

Circuit for
Example 5–2

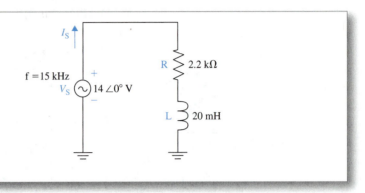

✓ **SOLUTION**

All calculations will be performed in the phasor domain using an appropriate engineering calculator.

The correct settings for the TI-89 calculator are:

- ANGLE = DEGREE
- COMPLEX FORMAT = POLAR
- EXACT/APPROXIMATE = APPROXIMATE

The magnitude of the inductive reactance at the given frequency can be calculated as:

$$X_L = 2\pi f L = 2\pi(15 \text{ kHz})(20 \text{ mH}) = 1.88 \text{ k}\Omega$$

Recalling that inductive reactance is represented in the phasor domain at an angle of $\angle 90°$, the inductive reactance can be expressed as:

$$\mathbf{X_L} = 1.88 \angle 90° \text{ k}\Omega$$

The impedance of the circuit is the sum of **R** and $\mathbf{X_L}$ so that:

$$\mathbf{Z} = \mathbf{R} + \mathbf{X_L} = 2.2 \angle 0° \text{ k}\Omega + 1.88 \angle 90° \text{ k}\Omega = 2.89 \angle 40.5° \text{ k}\Omega$$

Ohm's law can be used to calculate the source current as:

$$\mathbf{I_S} = \frac{\mathbf{V_S}}{\mathbf{Z}} = \frac{14 \angle 0° \text{ V}}{2.89 \angle 40.5° \text{ k}\Omega} = 4.84 \angle -40.5° \text{ mA}$$

Ohm's law may be used to calculate the voltage across each respective circuit component as:

$$\mathbf{V_R} = \mathbf{I_S}\mathbf{R} = (4.84 \angle -40.5° \text{ mA})(2.2 \angle 0° \text{ k}\Omega) = 10.65 \angle -40.5° \text{ V}$$

$$\mathbf{V_L} = \mathbf{I_S}\mathbf{X_L} = (4.84 \angle -40.5° \text{ mA})(1.88 \angle 90° \text{ k}\Omega) = 9.10 \angle 49.5° \text{ V}$$

KVL can be expressed as:

$$\mathbf{V_S} - \mathbf{V_R} - \mathbf{V_L} = 14 \angle 0° \text{ V} - 10.65 \angle -40.5° \text{ V} - 9.10 \angle 49.5° \text{ V} = 0 \text{ (approximate)}$$

The voltage divider rule can be used to calculate the voltage across each respective component as:

$$\mathbf{V_R} = \mathbf{V_S}\frac{\mathbf{R}}{\mathbf{Z}} = \frac{(14\angle 0°\text{ V})(2.2\angle 0°\text{ k}\Omega)}{2.89\angle 40.5°\text{ k}\Omega} = 10.65\angle -40.5°\text{ V}$$

$$\mathbf{V_L} = \mathbf{V_S}\frac{\mathbf{X_L}}{\mathbf{Z}} = \frac{(14\angle 0°\text{ V})(1.88\angle 90°\text{ k}\Omega)}{2.89\angle 40.5°\text{ k}\Omega} = 9.10\angle 49.5°\text{ V}$$

The impedance diagram for the circuit is shown in Figure 5–6. Note that **R** is at angle 0°, $\mathbf{X_L}$ is at angle 90°, and **Z** is at an angle of 40.5°.

FIGURE **5–6**

Impedance
Diagram
for
Example 5–2

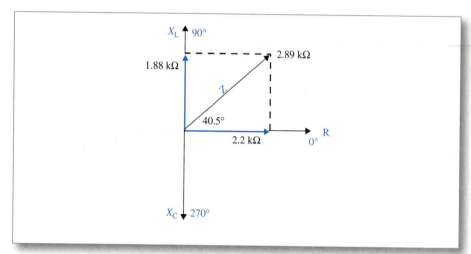

The phasor diagram for the circuit is shown in Figure 5–7.

FIGURE **5–7**

Phasor Diagram
for Example 5–2

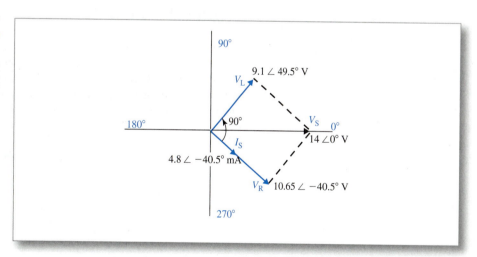

The phasor diagram in Figure 5–7 indicates the following important conclusions:

- The voltage across the resistor is in phase with the current through the resistor ($\mathbf{I_S}$).
- The voltage across the inductor leads the current by 90°.
- The sum of $\mathbf{V_R}$ and $\mathbf{V_L}$ is equal to the source voltage $\mathbf{V_S}$.

The axes of the phasor diagram in Figure 5–7 are labeled in degrees (polar form).

The current and voltages shown in the time domain in the MATLAB plot in Figure 5–8.

MATLAB

FIGURE **5-8**

Time Plot for
Example 5-2

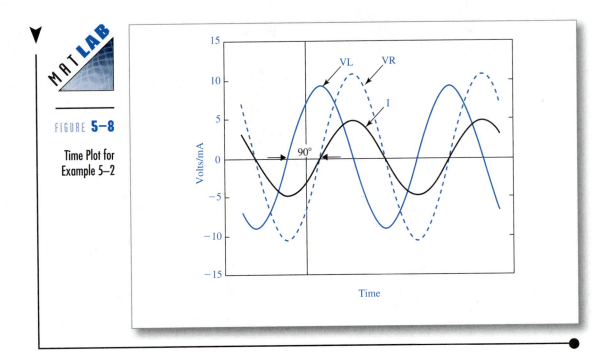

5.3 THE SERIES R-C CIRCUIT IN THE PHASOR DOMAIN

The analysis of the series R-C circuit is similar to the analysis of the series R-L circuit; however, capacitive reactance is represented at an angle of $-90°$ in the phasor domain.

EXAMPLE 5-3   Analysis of a Series R-C Circuit

Given the series R-C circuit in Figure 5–9, determine the impedance of the circuit, the source current, and the voltage across each component. Verify that the voltage across the components is correct by writing KVL around the circuit. Use the voltage divider rule to calculate the voltage across each component. Sketch an impedance diagram and phasor diagram.

FIGURE **5-9**

Circuit for
Example 5-3

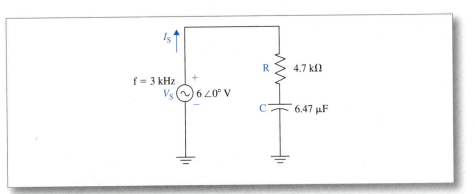

✓ SOLUTION

All calculations will be performed in the phasor domain using an appropriate engineering calculator.

The magnitude of the capacitive reactance may be calculated as:

$$X_C = \frac{1}{2\pi f C} = \frac{1}{2\pi(3\text{ kHz})(6.47\text{ nF})} = 8.2\text{ k}\Omega$$

so that:

$$\mathbf{X_C} = 8.2\angle -90°\text{ k}\Omega$$

The impedance of the circuit may be calculated as:

$$\mathbf{Z} = \mathbf{R} + \mathbf{X_C} = 4.7\angle 0°\text{ k}\Omega + 8.2\angle -90°\text{ k}\Omega = 9.45\angle -60.18°\text{ k}\Omega$$

Ohm's law can be used to calculate the source current as:

$$\mathbf{I_S} = \frac{\mathbf{V_S}}{\mathbf{Z}} = \frac{6\angle 0°\text{ V}}{9.45\angle -60.18°\text{ k}\Omega} = 634.92\angle 60.18°\text{ }\mu\text{A}$$

Ohm's law can be used to calculate the voltage across each respective component to obtain:

$$\mathbf{V_R} = \mathbf{I_S}\mathbf{R} = (634.92\angle 60.18°\text{ }\mu\text{A})(4.7\angle 0°\text{ k}\Omega) = 2.98\angle 60.18°\text{ V}$$

$$\mathbf{V_C} = \mathbf{I_S}\mathbf{X_C} = (634.92\angle 60.18°\text{ }\mu\text{A})(8.2\angle -90°\text{ k}\Omega) = 5.21\angle -29.82°\text{ V}$$

KVL may be expressed as:

$$\mathbf{V_S} - \mathbf{V_R} - \mathbf{V_C} = 6\angle 0°\text{ V} - 2.98\angle 60.18°\text{ V} - 5.21\angle -29.82°\text{ V} = 0\text{ (approximate)}$$

The voltage divider rule may be used to calculate the voltage across each respective component to obtain:

$$\mathbf{V_R} = \mathbf{V_S}\frac{\mathbf{R}}{\mathbf{Z}} = \frac{(6\angle 0°\text{ V})(4.7\angle 0°\text{ k}\Omega)}{9.45\angle -60.18°\text{ k}\Omega} = 2.98\angle 60.18°\text{ V}$$

$$\mathbf{V_C} = \mathbf{V_S}\frac{\mathbf{X_C}}{\mathbf{Z}} = \frac{(6\angle 0°\text{ V})(8.2\angle -90°\text{ k}\Omega)}{9.45\angle -60.18°\text{ k}\Omega} = 5.2\angle -29.82°\text{ V}$$

The impedance diagram for the circuit is shown in Figure 5–10.

FIGURE **5–10**

Impedance Diagram for Example 5–3

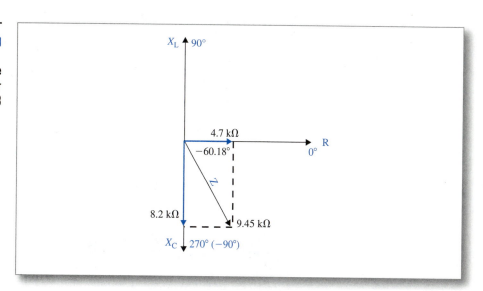

Note that **R** in the impedance diagram is at an angle of 0° and **X**_C is at an angle of −90°.

The phasor diagram is shown in Figure 5–11.

FIGURE **5–11**

Phasor
Diagram for
Example 5–3

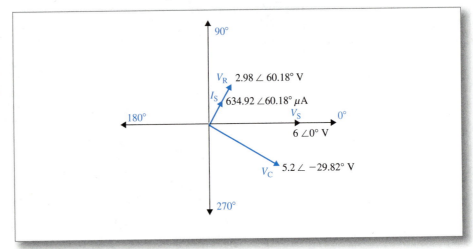

The phasor diagram in Figure 5–11 indicates the following important conclusions:

  – The voltage across the resistor is in phase with the current.

  – The current leads the voltage across the capacitor by 90°.

  – The sum of **V$_R$** and **V$_C$** is equal to the source voltage **V$_S$**.

The current and voltage for Example 5–3 are shown in the time domain in the MATLAB plot in Figure 5–12.

FIGURE **5–12**

Time Plot for
Example 5–3

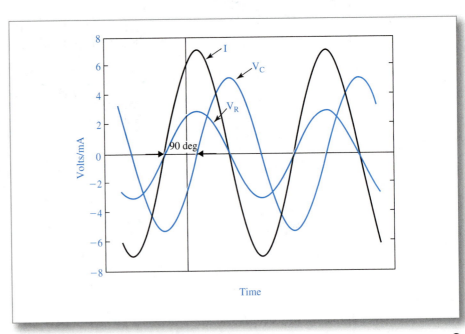

PRACTICE EXERCISES 5–1

a. Given the series resistive circuit of Figure 5–13, determine the total equivalent resistance, the source current, and voltage across each resistor using Ohm's law. Verify KVL around the circuit loop. Use the voltage divide rule to determine the voltage across each resistor. Sketch the phasor diagram for the circuit.

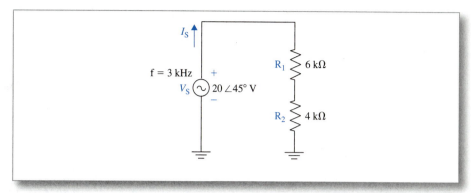

FIGURE **5-13**

Circuit for
Practice Exercise 5–1
Exercise a

b. Given the circuit in Figure 5–14, determine the circuit impedance, source current, and voltage across each component using Ohm's law. Determine the voltage across each component using the voltage divider rule. Verify KVL for the circuit. Sketch the impedance diagram and phasor diagram for the circuit.

FIGURE **5-14**

Circuit for
Practice Exercise 5–1
Exercise b

c. Given the circuit in Figure 5–15, determine the circuit impedance, current, and voltage across each component using Ohm's law. Determine the voltage across each component using the voltage divider rule. Verify KVL for the circuit. Sketch the impedance diagram and phasor diagram.

FIGURE **5-15**

Circuit for
Practice Exercise 5–1
Exercise c

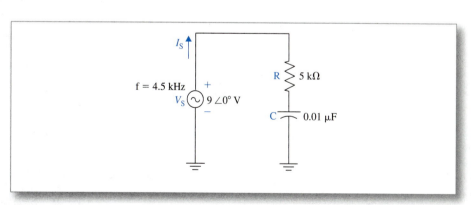

✓**ANSWERS**

a. $\mathbf{I_S} = 2\angle 45°$ mA; $\mathbf{V_{R1}} = 8\angle 45°$ V; $\mathbf{V_{R2}} = 12\angle 45°$ V;

KVL: $20\angle 45°$ V $- 8\angle 45°$ V $- 12\angle 45°$ V $= 0$

b. $\mathbf{Z} = 2.98 \angle 42.47° \text{ k}\Omega$; $\mathbf{I_S} = 4.04 \angle -42.27° \text{ mA}$;

   $\mathbf{V_R} = 8.89 \angle -42.27° \text{ V}$; $\mathbf{V_L} = 8.08 \angle 47.73° \text{ V}$

c. $\mathbf{Z} = 6.1 \angle -34.99° \text{ k}\Omega$; $\mathbf{I_S} = 1.48 \angle 34.99° \text{ mA}$;

   $\mathbf{V_R} = 7.4 \angle 34.99° \text{ V}$; $\mathbf{V_C} = 5.18 \angle -55.01° \text{ V}$

(Complete solutions to the Practice Exercises can be found in Appendix A.)

5.4 THE SERIES R-L-C CIRCUIT IN THE PHASOR DOMAIN

The series R-L-C circuit combines the effects of resistance as well as capacitive and inductive reactance.

EXAMPLE 5-4   Analysis of a Series R-L-C Circuit

Given the series R-L-C circuit in Figure 5–16, determine the impedance of the circuit, the source current, and the voltage across each component using Ohm's law. Determine the voltage across each component using the voltage divider rule. Verify KVL and sketch the impedance diagram and phasor diagram.

FIGURE **5–16**

Circuit for
Example 5–4

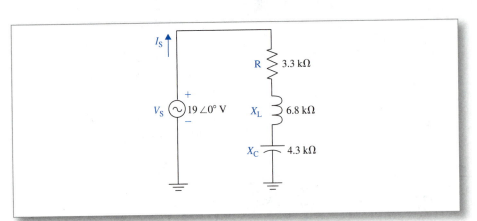

✔ SOLUTION

Note that the values of inductive and capacitive reactance are given directly in units of ohms so that the source frequency and reactive component values are not required in this example.

The circuit impedance is calculated as:

$$\mathbf{Z} = \mathbf{R} + \mathbf{X_C} + \mathbf{X_L}$$

$$\mathbf{Z} = 3.3 \angle 0° \text{ k}\Omega + 4.3 \angle -90° \text{ k}\Omega + 6.8 \angle 90° \text{ k}\Omega = 4.14 \angle 37.15° \text{ k}\Omega$$

Ohm's law can be used to calculate the source current as:

$$\mathbf{I_S} = \frac{\mathbf{V_S}}{\mathbf{Z}} = \frac{19 \angle 0° \text{ V}}{4.14 \angle 36.15° \text{ k}\Omega} = 4.59 \angle -37.15° \text{ mA}$$

Ohm's law can be used to calculate the voltage across each respective component as:

$$\mathbf{V_R} = \mathbf{I_S R} = (4.59\angle -37.15° \text{ mA})(3.3\angle 0° \text{ k}\Omega) = 15.15\angle -37.15° \text{ V}$$

$$\mathbf{V_C} = \mathbf{I_S X_C} = (4.59\angle -37.15° \text{ mA})(4.3\angle -90° \text{ k}\Omega) = 19.74\angle -127.15° \text{ V}$$

$$\mathbf{V_L} = \mathbf{I_S X_L} = (4.59\angle -37.15° \text{ mA})(6.8\angle 90° \text{ k}\Omega) = 31.21\angle 52.85° \text{ V}$$

KVL can be expressed as:

$$\mathbf{V_S} - \mathbf{V_R} - \mathbf{V_C} - \mathbf{V_L} = 0$$

so that:

$$19\angle 0° \text{ V} - 15.15\angle -37.15° \text{ V} - 19.74\angle -127.15° \text{ V} - 31.21\angle 52.85° \text{ V} = 0$$

The slight deviation from zero when evaluating KVL is due to rounding and is considered to be insignificant.

The impedance diagram is shown in Figure 5–17.

FIGURE **5–17**

Impedance
Diagram for
Example 5–4

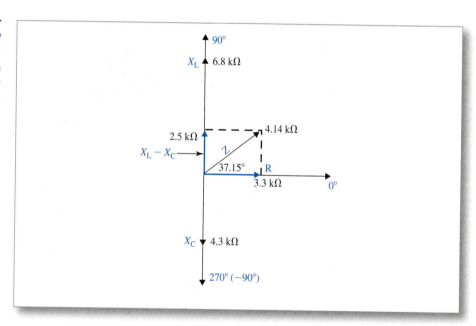

Note that the impedance in Figure 5–17 is the sum of $\mathbf{R}$, $\mathbf{X_C}$, and $\mathbf{X_L}$. Since $\mathbf{X_C}$ and $\mathbf{X_L}$ are separated by an angle of 180°, their sum is determined simply as the difference in magnitudes at the angle of the larger of the two quantities. Therefore, we can determine the sum of $\mathbf{X_L}$ and $\mathbf{X_C}$ as:

$$\mathbf{X_L} + \mathbf{X_C} = 6.8 \text{ k}\Omega - 4.3 \text{ k}\Omega = 2.5 \text{ k}\Omega$$

Since $\mathbf{X_L}$ is larger, the phasor representing the sum of $\mathbf{X_L}$ and $\mathbf{X_C}$ is at 90°.

The sum $\mathbf{X_L}$ and $\mathbf{X_C}$ could also be determined using the calculator as:

$$\mathbf{X_L} + \mathbf{X_C} = 6.8\angle 90° \text{ k}\Omega + 4.3\angle -90° \text{ k}\Omega = 2.5\angle 90° \text{ k}\Omega$$

The impedance as indicated in Figure 5–17 can be determined as:

$$\mathbf{Z} = 2.5\angle 90° \text{ k}\Omega + 3.3\angle 90° \text{ k}\Omega = 4.14\angle 37.15° \text{ k}\Omega$$

The phasor diagram is shown in Figure 5–18.

FIGURE **5–18**

Phasor Diagram
for Example 5–4

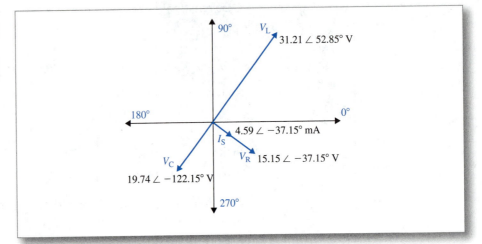

The phasor diagram in Figure 5–18 indicates the following important conclusions:

  – The voltage across the resistor is in phase with the current.

  – The inductor voltage leads the current by 90°.

  – The capacitor voltage lags the current by 90°.

  – The angle between the inductor voltage and capacitor voltage is 180°.

5.5 THE SERIES R–L–C CIRCUIT IN THE TIME DOMAIN

It is important to recall at this point that electronic signals are functions of time. The phasor domain permits convenient calculations of currents and voltages; however, we should remember that the phasors actually represent sinusoidal waveforms in the time domain.

The voltage and current phasors obtained in Example 5–4 may be converted to the time domain as:

$$v_S = 19 \sin(\omega t) \text{ V}$$

$$v_R = 15.15 \sin(\omega t - 37.15°) \text{ V}$$

$$v_C = 19.74 \sin(\omega t - 127.15°) \text{ V}$$

$$v_L = 31.26 \sin(\omega t + 52.85°) \text{ V}$$

$$i_S = 4.59 \sin(\omega t - 37.15°) \text{ mA}$$

Note again the use of lowercase symbols for quantities expressed in the time domain.

The voltage and current for Example 5–4 are shown in the time domain in the MATLAB plot in Figure 5–19.

MAT LAB

FIGURE **5-19**

Time Plot for
Example 5-4

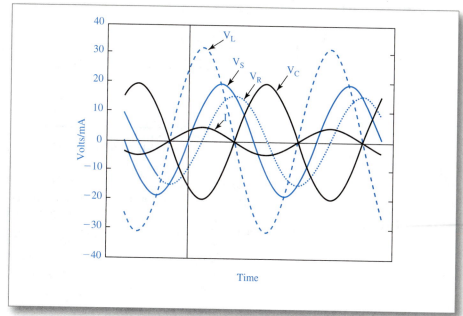

It is important to remember that the angles associated with currents and voltages obtained from the mathematical analysis of AC circuits result from time delays created by the capacitive and inductive components. It is always important to understand the physical implications of any mathematical calculation or notation.

PRACTICE EXERCISES 5-2

Given the circuit in Figure 5–20, determine the circuit impedance, source current, and voltage across each component using Ohm's law. Determine the voltage across each component using the voltage divider rule. Verify KVL. Sketch the impedance diagram and phasor diagram. Write an expression for V_R, V_C, and V_L in the time domain.

FIGURE **5-20**

Circuit for
Practice
Exercise 5–2

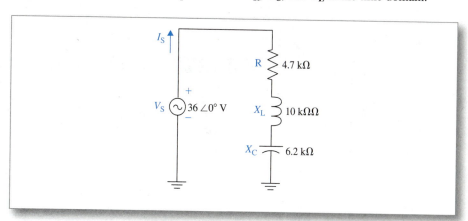

✓ **ANSWERS**

$$\mathbf{Z} = 6.04 \angle 38.96° \text{ k}\Omega; \quad \mathbf{I_s} = 5.96 \angle -38.96° \text{ mA};$$

$$\mathbf{V_R} = 28.01 \angle -38.96° \text{ V}; \quad \mathbf{V_L} = 59.6 \angle 51.04° \text{ V};$$

$$\mathbf{V_C} = 36.95 \text{ V} \angle -128.96°;$$

$$v_R = 28.01 \sin(\omega t - 38.96°) \text{ V};$$

$$v_C = 36.95 \sin(\omega t - 128.96°) \text{ V};$$

$$v_L = 59.6 \sin(\omega t + 51.04°) \text{ V}$$

(Complete solutions to the Practice Exercise can be found in Appendix A.)

5.6   THE PARALLEL R-L-C CIRCUIT IN THE PHASOR DOMAIN

Conductance is defined as the reciprocal of resistance ($\mathbf{G} = 1/\mathbf{R}$). The total equivalent conductance of resistors connected in parallel is calculated as the sum of conductances of each resistor. A similar approach may be used to analyze a parallel R-L-C circuit in the phasor domain.

The reciprocal of inductive reactance is called **inductive susceptance** and is given the symbol $\mathbf{B_L}$. The reciprocal of capacitive reactance is called **capacitive susceptance** and is given the symbol $\mathbf{B_C}$ so that:

$$\mathbf{G} = \frac{1}{\mathbf{R}}$$

$$\mathbf{B_L} = \frac{1}{\mathbf{X_L}}$$

$$\mathbf{B_C} = \frac{1}{\mathbf{X_C}}$$

Conductance and susceptance are measured in units of siemens (S).

The **admittance (Y)** of a parallel R-L-C circuit is defined as the reciprocal of impedance and can be expressed as:

$$\mathbf{Y} = \frac{1}{\mathbf{Z}} = \mathbf{G} + \mathbf{B_L} + \mathbf{B_C} \tag{5.1}$$

EXAMPLE 5-5   Analysis of a Parallel R-L-C Circuit

Given the circuit in Figure 5–21, determine the conductance or susceptance of each component. Calculate the admittance and impedance of the circuit and calculate the source current. Determine each branch current using the current divider rule. Verify KCL at node A.

FIGURE **5–21**

Circuit for
Example 5–5

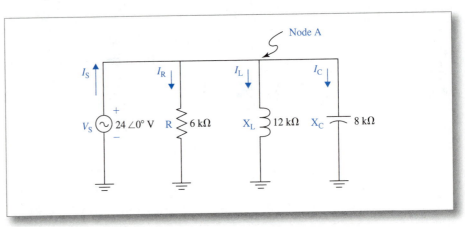

✓ **SOLUTION**

The conductance or susceptance of each respective parallel component (branch) is determined as:

$$\mathbf{G} = \frac{1}{\mathbf{R}} = \frac{1}{6 \angle 0° \text{ k}\Omega} = 166.67 \angle 0° \text{ } \mu S$$

$$\mathbf{B_L} = \frac{1}{\mathbf{X_L}} = \frac{1}{12 \angle 90° \text{ k}\Omega} = 83.33 \angle -90° \text{ } \mu S$$

$$\mathbf{B_C} = \frac{1}{\mathbf{X_C}} = \frac{1}{8 \angle -90° \text{ k}\Omega} = 125 \angle -90° \text{ } \mu S$$

The admittance of the circuit is calculated using Equation (5.1) as:

$$\mathbf{Y} = \mathbf{G} + \mathbf{B_L} + \mathbf{B_C}$$

so that:

$$\mathbf{Y} = 166.67 \angle 0° \text{ } \mu S + 83.33 \angle -90° \text{ } \mu S + 125 \angle 90° \text{ } \mu S = 171.79 \angle 14.04° \text{ } \mu S$$

The calculator can be used to determine the circuit admittance; however, care must be taken to observe proper parenthesis usage when making calculations. The calculation of admittance (**Y**) can be achieved by using the inverting keystrokes (^-1) or by using parentheses as shown:

Y may be keyed as: `( ( 1 / ( 6 EE 3 ∠ 0 ) ) + ( 1 / ( 12 EE ∠ 90 ) )` `+ ( 1 / ( 8 EE 3 ∠ 90 +/- ) ) ` ENTER to obtain: 171.79∠14.04 μS.

A second calculator option is:

Y may be keyed as: `( ( 6 EE 3 ∠ 0 ) ^ 1 +/- ) + ( ( 12 EE 3 ∠ 90 ) ^ 1` `+/- ) + ( ( 8 EE 3 ∠ 90 +/- ) ^ 1 ± )` ENTER to obtain: 171.79∠14.04 μS.

The impedance of the circuit can be determined as the reciprocal of admittance:

$$\mathbf{Z} = \frac{1}{\mathbf{Y}} = 1/171.79 \angle 14.04° \text{ } \mu S = 5.82 \angle -14.04° \text{ k}\Omega$$

Ohm's law can be used to calculate the source current as:

$$\mathbf{I_s} = \frac{\mathbf{V_s}}{\mathbf{Z}} = \frac{24 \angle 0° \text{ V}}{5.82 \angle -14.04° \text{ k}\Omega} = 4.12 \angle 14.04° \text{ mA}$$

Substituting $\mathbf{Y} = \frac{1}{\mathbf{Z}}$, Ohm's law can be expressed as:

$$\mathbf{I_s} = \mathbf{V_s} \mathbf{Y}$$

so that:

$$\mathbf{I_s} = \mathbf{V_s} \mathbf{Y} = (24 \angle 0° \text{ V})(171.79 \angle 14.04° \text{ } \mu S) = 4.12 \angle 14.04° \text{ mA}$$

The current divider rule may be used to calculate the branch currents as:

$$\mathbf{I_R} = \mathbf{I_s} \frac{\mathbf{Z}}{\mathbf{R}} = \frac{(4.12 \angle 14.04° \text{ mA})(5.82 \angle -14.04° \text{ k}\Omega)}{6 \angle 0° \text{ k}\Omega} = 4.00 \angle 0° \text{ mA}$$

$$\mathbf{I_L} = \mathbf{I_s} \frac{\mathbf{Z}}{\mathbf{X_L}} = \frac{(4.12 \angle 14.04° \text{ mA})(5.82 \angle -14.04° \text{ k}\Omega)}{12 \angle 90° \text{ k}\Omega} = 2.00 \angle -90° \text{ mA}$$

$$I_C = I_S \frac{Z}{X_C} = \frac{(4.12\angle 14.04° \text{ mA})(5.82\angle -14.04° \text{ k}\Omega)}{8\angle -90° \text{ k}\Omega} = 3.00\angle 90° \text{ mA}$$

KCL can be applied at node *A* to obtain:

$$I_S = I_R + I_L + I_C$$

$$I_S = 4.00\angle 0° \text{ mA} + 2.00\angle -90° \text{ mA} + 3.00\angle 90° \text{ mA} = 4.123\angle 14.04° \text{ mA}$$

EXAMPLE 5-6   An Alternate Method of Solution

Determine the branch currents, admittance, and impedance of the circuit in Figure 5–21 using an alternate approach. Sketch the admittance diagram and phasor diagram for the circuit.

✓ SOLUTION

Ohm's law can be used to obtain the branch currents directly as:

$$I_R = \frac{V_S}{R} = \frac{24\angle 0° \text{ V}}{6\angle 0° \text{ k}\Omega} = 4.00\angle 0° \text{ mA}$$

$$I_L = \frac{V_S}{X_L} = \frac{24\angle 0° \text{ V}}{12\angle 90° \text{ k}\Omega} = 2.00\angle -90° \text{ mA}$$

$$I_C = \frac{V_S}{X_C} = \frac{24\angle 0° \text{ V}}{8\angle 90° \text{ k}\Omega} = 3.00\angle 90° \text{ mA}$$

KCL can be used to determine the source current as the sum of the branch currents to obtain:

$$I_S = 4.00\angle 0° \text{ mA} + 2.00\angle -90° \text{ mA} + 3.00\angle 90° \text{ mA} = 4.123\angle 14.04° \text{ mA}$$

Ohm's law can be used to calculate the circuit impedance as:

$$Z = \frac{V_S}{I_S} = \frac{24\angle 0° \text{ V}}{4.123\angle 14.04° \text{ mA}} = 5.82\angle -14.04° \text{ k}\Omega$$

The circuit admittance may now be determined as the reciprocal of impedance to obtain:

$$Y = \frac{1}{Z} = \frac{1}{5.82\angle -14.04° \text{ k}\Omega} = 171.82\angle 14.04° \text{ μS}$$

Note that the solutions obtained in Example 5–6 are more direct and require less calculation than other methods that may have been chosen such as the ones demonstrated in Example 5–5. It is good engineering practice to check solutions by at least one independent method when possible to verify solutions.

The admittance diagram for the circuit in Figure 5–21 is shown in Figure 5–22.

FIGURE **5-22**

Admittance
Diagram for
Example 5-5

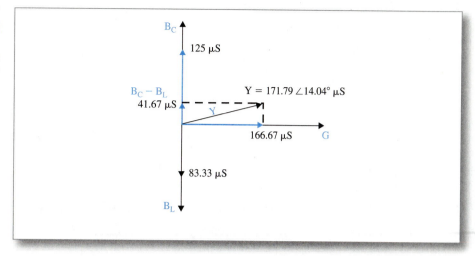

The phasor diagram for the circuit in Figure 5–21 is shown in Figure 5–23.

FIGURE **5-23**

Phasor
Diagram for
Example 5-5

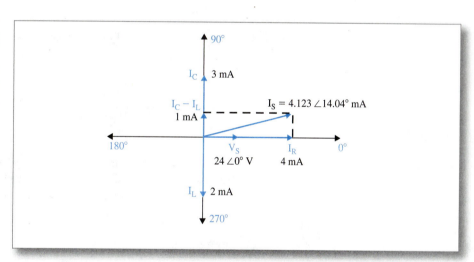

The phasor diagram in Figure 5–23 indicates the following important conclusions:

- The resistor current is in phase with the voltage across the resistor.

- The capacitor current leads the voltage by 90°.

- The inductor current lags the voltage by 90°.

- The sum of $\mathbf{I_R}$, $\mathbf{I_C}$, $\mathbf{I_L}$ is equal to the source current $\mathbf{I_S}$.

PRACTICE EXERCISES 5-3

Given the circuit in Figure 5–24, determine circuit admittance and impedance. Determine the values of the source current and each branch current. Apply KCL at node A.

FIGURE **5–24**

Circuit For
Practice
Exercise 5.3
Exercise 1

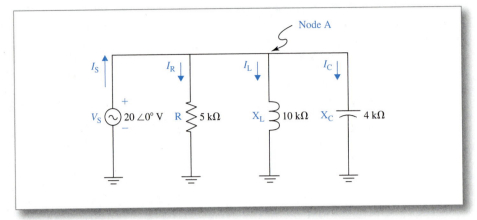

✓ **ANSWERS**

$$\mathbf{I_R} = 4\angle 0° \text{ mA}; \quad \mathbf{I_C} = 5\angle 90° \text{ mA}; \quad \mathbf{I_L} = 2\angle -90° \text{ mA};$$

$$\mathbf{I_S} = 5\angle 36.87° \text{ mA}; \quad \mathbf{Z} = 4\angle -36.87° \text{ k}\Omega;$$

$$\mathbf{Y} = 250\angle 36.87° \text{ μS}$$

(Complete solutions to the Practice Exercise can be found in Appendix A.)

5.7 THE COMBINATION R-L-C CIRCUIT IN THE PHASOR DOMAIN

The analysis of a combination circuit is accomplished by a careful application of Ohm's law, Kirchhoff's laws, and the current and voltage divider rules. All calculations involve phasor quantities that must be represented by a magnitude and angle.

EXAMPLE 5–7 Analysis of a Combination R-L-C Circuit in the Phasor Domain

Given the combination R-L-C circuit in Figure 5–25, determine the current and voltage for each component. Verify KCL at node A, and verify KVL around each closed loop. Determine the voltage at node A and node B with respect to ground.

FIGURE **5–25**

Circuit for
Example 5–7

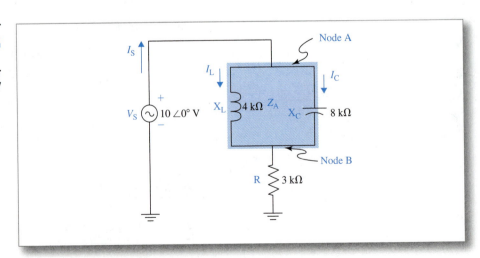

✓ SOLUTION

Observe that the inductor and capacitor in Figure 5–25 are both connected between node A and node B and therefore are connected in parallel. It is helpful to consider the parallel combination of X_L // X_C as a branch with equivalent impedance Z_A so that:

$$Z_A = X_L \text{ // } X_C$$

Also observe that the impedance Z_A is in series with the resistor R.

The total equivalent impedance of the circuit can therefore be expressed as:

$$Z_T = Z_A + R$$

The impedance Z_A can be calculated by using the product-over-sum method as:

$$Z_A = \frac{(X_L X_C)}{(X_L + X_C)} = \frac{(4\angle 90° \text{ k}\Omega)(8\angle -90° \text{ k}\Omega)}{(4\angle 90° \text{ k}\Omega + 8\angle -90° \text{ k}\Omega)} = 8\angle 90° \text{ k}\Omega$$

The total equivalent impedance can be determined as:

$$Z_T = R + Z_A = 3\angle 0° \text{ k}\Omega + 8\angle 90° \text{ k}\Omega = 8.54 \angle 69.44° \text{ k}\Omega$$

Ohm's law can be used to calculate the source current as:

$$I_S = \frac{V_S}{Z_T} = \frac{10\angle 0° \text{ V}}{8.5\angle 69.44° \text{ k}\Omega} = 1.17\angle -69.44° \text{ mA}$$

The current divider rule can be used to calculate the inductor current and capacitive current as:

$$I_L = I_T \frac{Z_A}{X_L} = \frac{(1.17\angle -69.44° \text{ mA})(8\angle 90° \text{ k}\Omega)}{4\angle 90° \text{ k}\Omega} = 2.34\angle -69.44° \text{ mA}$$

$$I_C = I_T \frac{Z_A}{X_C} = \frac{(1.17\angle -69.44° \text{ mA})(8\angle 90° \text{ k}\Omega)}{8\angle -90° \text{ k}\Omega} = 1.17\angle 110.56° \text{ mA}$$

It is important to note that Z_A in the numerator of the current divider rule represents the impedance of the parallel combination of X_L and X_C, not the total equivalent impedance of the circuit.

The current for each respective component has been determined as:

$$I_R = I_S = 1.17\angle -69.44° \text{ mA}$$

$$I_L = 2.34\angle -69.44° \text{ mA}$$

$$I_C = 1.17\angle 110.56° \text{ mA}$$

Verifying KCL at node A, we obtain:

$$I_S = I_L + I_C = 2.34\angle -69.44° \text{ mA} + 1.17\angle 110.56° \text{ mA} = 1.17\angle -69.44° \text{ mA}$$

Ohm's law can be used to calculate the voltage across each respective component as:

$$V_R = I_R R = (1.17\angle -69.44° \text{ mA})(3\angle 0° \text{ k}\Omega) = 3.51\angle -69.44° \text{ V}$$

$$V_L = I_L X_L = (2.34\angle -69.44° \text{ mA})(4\angle 90° \text{ k}\Omega) = 9.36\angle 20.56° \text{ V}$$

$$V_C = I_C X_C = (1.17\angle 110.56° \text{ mA})(8\angle -90° \text{ k}\Omega) = 9.36\angle 20.56° \text{ V}$$

Verifying KVL, we obtain:

$$\mathbf{V_S} - \mathbf{V_R} - \mathbf{V_C} = 10\angle 0° \text{ V} - 3.51\angle -69.44° \text{ V} - 9.36\angle 20.56° \text{ V} = 0 \text{ (approximate)}$$

The voltage at node A with respect to ground by observation is the source voltage:

$$\mathbf{V_A} = \mathbf{V_S} = 10\angle 0° \text{ V}$$

The voltage at node B with respect to ground is equal to the voltage across **R**:

$$\mathbf{V_B} = \mathbf{V_R} = 3.51\angle -69.44° \text{ V}$$

EXAMPLE 5-8   Analysis of a Combination R-L-C Circuit in the Phasor Domain

Given the combination R-L-C circuit in Figure 5–26, determine the current and voltage for each of the components. Verify KVL around each closed loop.

FIGURE **5-26**

Circuit for Example 5–8

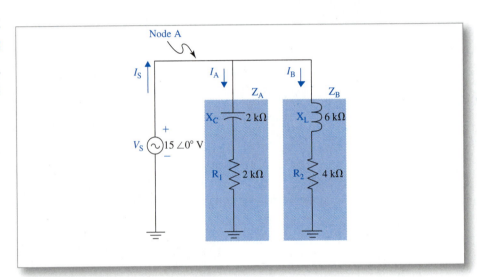

✓ SOLUTION

This circuit should be viewed as a parallel connection of two branches, rather than as four separate components. $\mathbf{Z_A}$ is the branch impedance of the series combination of $\mathbf{X_C}$ and $\mathbf{R_1}$; and $\mathbf{Z_B}$ is the branch impedance of the series combination of $\mathbf{X_L}$ and $\mathbf{R_2}$ so that:

$$\mathbf{Z_A} = \mathbf{X_C} + \mathbf{R_1} = 2\angle -90° \text{ k}\Omega + 2\angle 0° \text{ k}\Omega = 2.83\angle -45° \text{ k}\Omega$$

$$\mathbf{Z_B} = \mathbf{X_L} + \mathbf{R_2} = 6\angle 90° \text{ k}\Omega + 4\angle 0° \text{ k}\Omega = 7.21\angle 56.31° \text{ k}\Omega$$

A complex circuit can be solved using a variety of methods and approaches. We could, for example, calculate the admittance of the two parallel branches, use Ohm's law to determine the source current, and then apply the current divider rule to determine the current through each branch. We may also choose to determine the circuit impedance by applying the product-over-sum method to the parallel combination of $\mathbf{Z_A}$ and $\mathbf{Z_B}$.

We may also choose to calculate the branch currents directly using Ohm's law as:

$$\mathbf{I_A} = \frac{\mathbf{V_S}}{\mathbf{Z_A}} = \frac{15\angle 0° \text{ V}}{2.83\angle -45° \text{ k}\Omega} = 5.3\angle 45° \text{ mA}$$

$$\mathbf{I_B} = \frac{\mathbf{V_S}}{\mathbf{Z_B}} = \frac{15\angle 0°\ V}{7.21\angle 56.31°\ k\Omega} = 2.08\angle -56.31°\ mA$$

By selecting this approach, the number of calculations to obtain the branch currents changes from two steps (calculating admittance and then using the current divider rule) to a single Ohm's law calculation.

KCL can be applied at node A to determine the source current as:

$$\mathbf{I_S} = \mathbf{I_A} + \mathbf{I_B} = 5.3\angle 45°\ mA + 2.08\angle -56.31°\ mA = 5.3\angle 22.37°\ mA$$

The current for each respective component has been determined as:

$$\mathbf{I_C} = \mathbf{I_{R1}} = \mathbf{I_A} = 5.3\angle 45°\ mA$$

$$\mathbf{I_L} = \mathbf{I_{R2}} = \mathbf{I_B} = 2.08\angle -56.31°\ mA$$

Ohm's law can be used to determine the voltage across each respective component as:

$$\mathbf{V_C} = \mathbf{I_A}\mathbf{X_C} = (5.3\angle 45°\ mA)(2\angle -90°\ k\Omega) = 10.6\angle -45°\ V$$

$$\mathbf{V_{R1}} = \mathbf{I_A}\mathbf{R_1} = (5.3\angle 45°\ mA)(2\angle 0°\ k\Omega) = 10.6\angle 45°\ V$$

$$\mathbf{V_L} = \mathbf{I_B}\mathbf{X_L} = (2.08\angle -56.31°\ mA)(6\angle 90°\ k\Omega) = 12.48\angle 33.69°\ V$$

$$\mathbf{V_{R2}} = \mathbf{I_B}\mathbf{R_2} = (2.08\angle -56.31°\ mA)(4\angle 0°\ k\Omega) = 8.32\angle -56.31°\ V$$

Verifying KVL from source ground moving clockwise, we obtain:

$$\mathbf{V_S} - \mathbf{V_C} - \mathbf{V_{R1}} = 15\angle 0°\ V - 10.6\angle -45°\ V - 10.6\angle 45°\ V = 0$$

$$\mathbf{V_S} - \mathbf{V_L} - \mathbf{V_{R2}} = 15\angle 0°\ V - 12.46\angle 33.69°\ V - 8.32\angle -56.31°\ V = 0$$

PRACTICE EXERCISES 5–4

a. Given the combination R-L-C circuit in Figure 5–27, determine the source current and voltage for each component. Verify KCL a node A and verify KVL around each closed loop.

FIGURE **5–27**

Circuit for
Practice Exercise 5–4
Exercise a

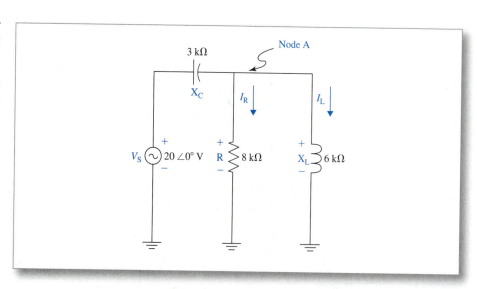

b. Given the combination R-L-C circuit in Figure 5–28, determine the source current and voltage for each component. Verify KCL at node A and verify KVL around each closed loop.

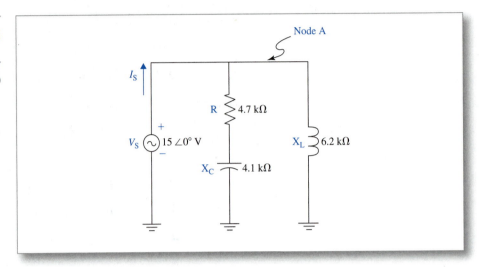

✓ ANSWERS

a. $I_C = 6.67 \angle -16.26°$ mA; $I_L = 5.34 \angle -53.13°$ mA;

   $I_R = 4.00 \angle 36.87°$ mA; $V_C = 20.01 \angle -106.26°$ V;

   $V_L = 32.22 \angle 36.87°$ V; $V_R = 32.22 \angle 36.87°$ V

b. $I_R = 2.4 \angle 41.1°$ mA; $I_C = 2.4 \angle 41.1°$ mA;

   $I_L = 2.42 \angle -90°$ mA; $V_R = 11.28 \angle 41.1°$ V;

   $V_C = 9.48 \angle -48.6°$ V; $V_L = 15 \angle 0°$ V

Note that answers may vary slightly due to rounding differences.

(Complete solutions to the Practice Exercise can be found in Appendix A.)

S U M M A R Y

A sinusoidal voltage or current is referred to as an alternating waveform. Circuits whose input signals are sinusoidal are referred to as alternating current or AC circuits.

The total equivalent **impedance** for components connected in series is the sum of the opposition to current flow (R, X_L, or X_C) for each respective component so that:

$$\mathbf{Z} = \mathbf{R} + \mathbf{X_L} + \mathbf{X_C}$$

where: $\mathbf{R} = R \angle 0°$; $\mathbf{X_L} = X_L \angle 90°$;   and $\mathbf{X_C} = X_C \angle -90°$

The total equivalent **admittance** of a parallel R-L-C circuit is determined as the sum of **conductance, inductive susceptance,** and **capacitive susceptance** and can be expressed as:

$$\mathbf{Y} = \mathbf{G} + \mathbf{B_L} + \mathbf{B_C}$$

where:

$$\mathbf{G} = \frac{1}{\mathbf{R}}; \quad \mathbf{B_L} = \frac{1}{\mathbf{X_L}}; \quad \text{and} \quad \mathbf{B_C} = \frac{1}{\mathbf{X_C}}$$

The impedance of a parallel R-L-C circuit may be calculated as the reciprocal of admittance:

$$\mathbf{Z} = \frac{1}{\mathbf{Y}}$$

The voltage across each component in an R-L-C circuit can be determined respectively using Ohm's law as:

$$\mathbf{V_R} = \mathbf{I_R R}; \quad \mathbf{V_L} = \mathbf{I_L X_L}; \quad \text{and} \quad \mathbf{V_C} = \mathbf{I_C X_C}$$

Ohm's law also states that the current that flows from a voltage source in an R-L-C circuit is directly proportional to the source voltage and inversely proportional to the total impedance and is expressed in the phasor domain as:

$$\mathbf{I}_S = \frac{\mathbf{V_S}}{\mathbf{Z}}$$

KVL states that the sum of voltages around a closed loop is equal to 0. KVL may be stated mathematically in the phasor domain as:

$$\sum \mathbf{V}_{\text{closed loop}} = 0$$

where the voltages are specified by a magnitude and angle.

KCL states that the sum of the currents entering a node must equal the sum of the currents leaving the node and may be stated mathematically in the phasor domain as:

$$\sum \mathbf{I}_{\text{entering node}} = \sum \mathbf{I}_{\text{leaving node}}$$

where the currents are specified by a magnitude and angle.

The voltage divider rule states that the voltage across a component in an R-L-C series circuit is directly proportional to the source voltage and to the ratio of the resistance or reactance value to the total impedance and can be stated mathematically in the phasor domain as:

$$\mathbf{V_R} = \mathbf{V_S} \frac{\mathbf{R}}{\mathbf{Z}}; \quad \mathbf{V_L} = \mathbf{V_S} \frac{\mathbf{X_L}}{\mathbf{Z}}; \quad \text{and} \quad \mathbf{V_C} = \mathbf{V_S} \frac{\mathbf{X_C}}{\mathbf{Z}}$$

The current divider rule states that the current for a component in an R-L-C parallel circuit is directly proportional to the total current flowing into the parallel circuit and inversely proportional to the ratio of the resistance or reactance value to the total impedance and may be stated mathematically in the phasor domain as:

$$\mathbf{I_R} = \mathbf{I_S} \frac{\mathbf{Z}}{\mathbf{R}}; \quad \mathbf{I_L} = \mathbf{I_S} \frac{\mathbf{Z}}{\mathbf{X_L}}; \quad \text{and} \quad \mathbf{I_C} = \mathbf{I_S} \frac{\mathbf{Z}}{\mathbf{X_C}}$$

A combination circuit consists of both series and parallel components or branches. The total equivalent impedance of a combination R-L-C circuit is determined by sequentially combining the series and parallel components or branches. Regardless of the complexity of an R-L-C circuit, Ohm's law and Kirchhoff's laws must be satisfied throughout the circuit.

The engineering calculator has greatly simplified the calculations involved in electronic circuit analysis, especially calculations in the phasor domain.

Creating impedance diagrams, phasor diagrams, and time domain plots using MATLAB offers a great deal of insight into the physical interpretation of complex mathematical calculations. In addition, computer simulation provides an excellent confirmation of analytical calculations.

ESSENTIAL LEARNING EXERCISES

(ANSWERS TO ODD-NUMBERED ESSENTIAL LEARNING EXERCISES CAN BE FOUND IN APPENDIX A.)

1. Determine the total equivalent impedance for the circuit in Figure 5–29 and sketch the impedance diagram.

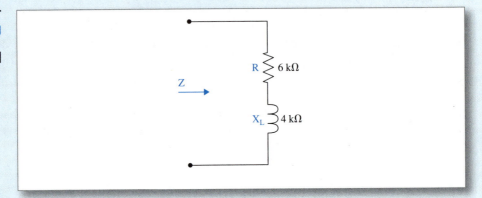

FIGURE **5–29**

Circuit for Essential Learning Exercise 1

2. Determine the total equivalnet impedance for the circuit in Figure 5–30 and sketch the impedance diagram.

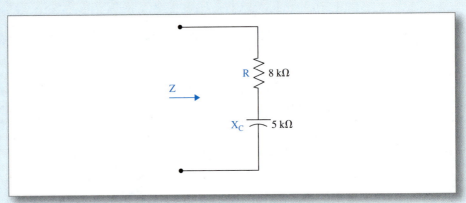

FIGURE **5–30**

Circuit for Essential Learning Exercise 2

3. Determine the total equivalent impedance for the circuit in Figure 5–31 and sketch the impedance diagram.

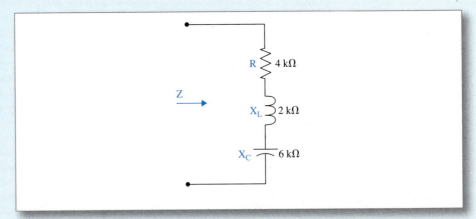

FIGURE **5–31**

Circuit for Essential Learning Exercise 3

4. Given the circuit in Figure 5–32, determine the total equivalent resistance. Determine the source current and the voltage across each respective resistor using Ohm's law. Determine the voltage across each respective resistor using the voltage divider rule. Verify KVL. All calculations should be performed in the phasor domain.

FIGURE **5–32**

Circuit for Essential Learning Exercise 4

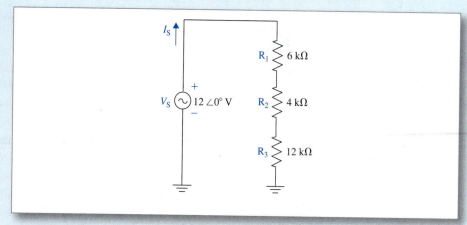

5. Given the circuit in Figure 5–33, determine the total equivalent resistance. Determine the source current and the voltage across each respective resistor using Ohm's law. Determine the voltage across each respective resistor using the voltage divider rule. Verify KVL. All calculations should be performed in the phasor domain.

FIGURE **5–33**

Circuit for Essential Learning Exercise 5

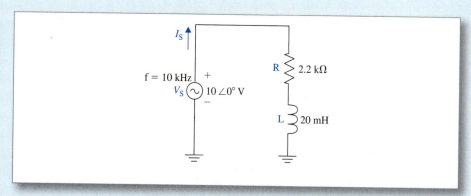

6. Given the circuit in Figure 5–34, determine the total equivalent impedance. Determine the source current and the voltage across each respective component using Ohm's law. Determine the voltage across each component using the voltage divider rule. Verify KVL. Sketch the impedance diagram and phasor diagram for the circuit. All calculations should be performed in the phasor domain.

FIGURE **5–34**

Circuit for Essential Learning Exercise 6

7. Given the circuit in Figure 5–35, determine the total equivalent impedance. Determine the current and voltage for each respective component using Ohm's law. Determine the voltage across each component using the voltage divider rule. Verify KVL. Sketch the impedance diagram and phasor diagram for the circuit. All calculations should be performed in the phasor domain.

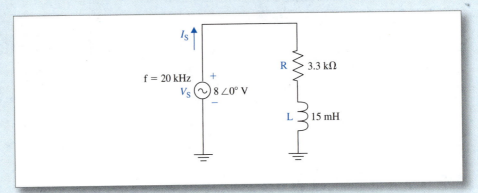

8. Given the circuit in Figure 5–36, determine the total equivalent impedance. Determine the current and voltage for each respective component using Ohm's law. Determine the voltage across each component using the voltage divider rule. Verify KVL. Sketch the impedance diagram and phasor diagram for the circuit. All calculations should be performed in the phasor domain.

(Figure 5–36: $f = 650$ Hz, V_S source $12 \angle 0°$ V, I_S, R = 2 kΩ, C = 0.1 μF)

9. Given the circuit in Figure 5–37, determine the total equivalent impedance. Determine the current and voltage for each respective component using Ohm's law. Determine the voltage across each component using the voltage divider rule. Verify KVL. Sketch the impedance diagram and phasor diagram for the circuit. All calculations should be performed in the phasor domain.

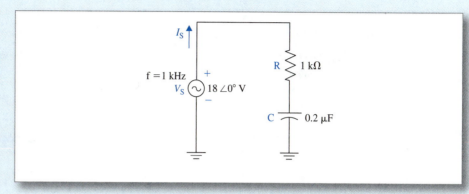

10. Given the circuit in Figure 5–38, determine the total equivalent impedance and total equivalent admittance. Determine the current for each component (branch) using Ohm's law. Determine the source current using KCL. Determine the current for each component (branch) using the current divider rule.

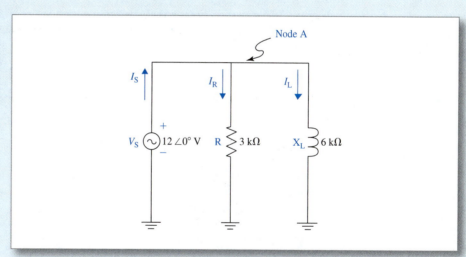

11. Given the circuit in Figure 5–39, determine the total equivalent impedance and total equivalent admittance. Determine the current for each component (branch) using Ohm's law. Determine the source current using KCL. Determine the current for each component (branch) using the current divider rule.

FIGURE **5–39**

Circuit for Essential Learning Exercise 11

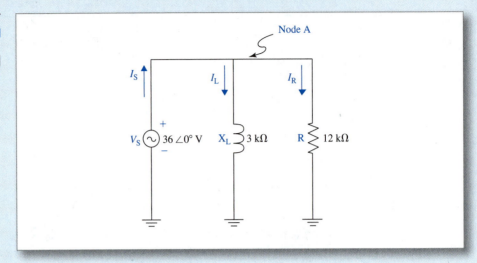

12. Given the circuit in Figure 5–40, determine the total equivalent impedance and total equivalent admittance. Determine the current for each component (branch) using Ohm's law. Determine the source current using KCL. Determine the current for each component (branch) using the current divider rule.

FIGURE **5–40**

Circuit for Essential Learning Exercise 12

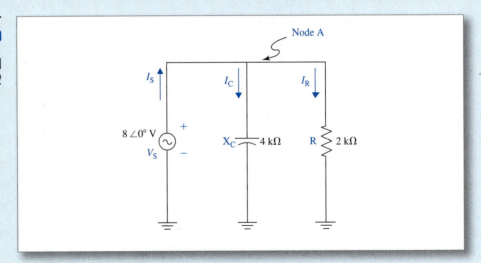

13. Given the circuit in Figure 5–41, determine the total equivalent impedance and total equivalent admittance. Determine the current for each component (branch) using Ohm's law. Determine the source current using KCL. Determine the current for each component (branch) using the current divider rule.

FIGURE **5–41**

Circuit for Essential Learning Exercise 13

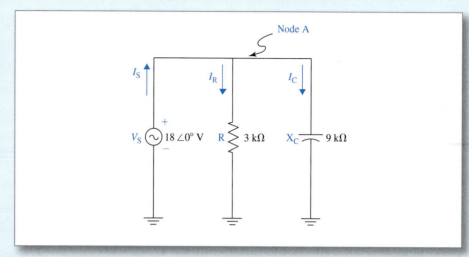

14. Given the circuit in Figure 5–42, determine the total equivalent impedance. Determine the current and voltage for each component. Verify KVL.

FIGURE **5–42**

Circuit for Essential Learning Exercise 14

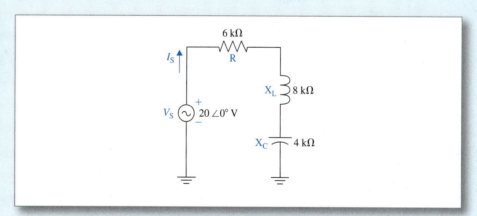

15. Given the circuit in Figure 5–43, determine the total equivalent impedance. Determine the current and voltage for each component. Verify KVL.

FIGURE **5–43**

Circuit for Essential
Learning Exercise 15

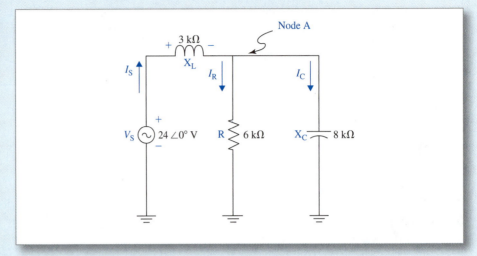

16. Given the circuit in Figure 5–44, determine the total equivalent impedance. Determine the current and voltage for each component. Verify KVL. Verify KCL at node A.

FIGURE **5–44**

Circuit for Essential
Learning Exercise 16

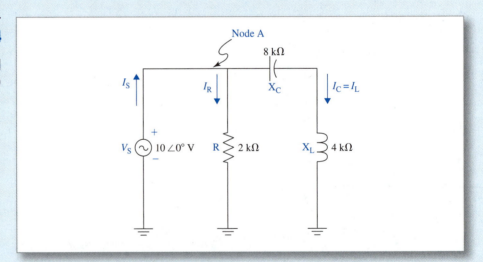

17. Given the circuit in Figure 5–45, determine the current and voltage for each component. Verify KVL around the closed loop containing V_R, V_C, and V_L.

FIGURE **5–45**

Circuit for Essential
Learning Exercise 17

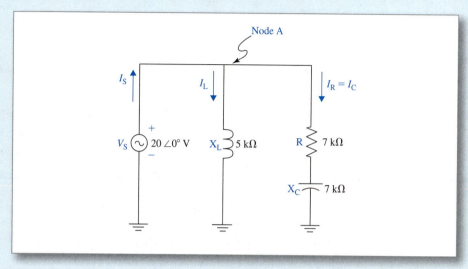

18. Given the circuit in Figure 5–46, determine the total equivalent impedance. Determine the current and voltage for each component. Verify KVL. Determine the output voltage V_{OUT}.

FIGURE **5–46**

Circuit for Essential
Learning Exercise 18

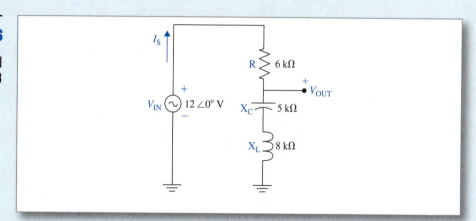

19. Given the circuit in Figure 5–47, use Ohm's law to determine the output voltage V_{OUT}. Determine the output voltage V_{OUT} using the voltage divider rule.

FIGURE **5–47**

Circuit for Essential
Learning Exercise 19

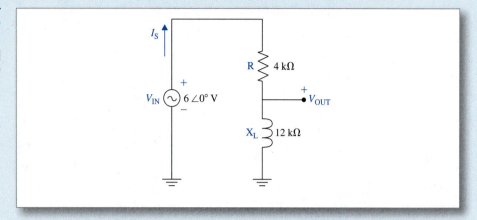

20. Given the circuit in Figure 5–48, determine the open circuit voltage V_X.

FIGURE **5–48**

Circuit for Essential
Learning Exercise 20

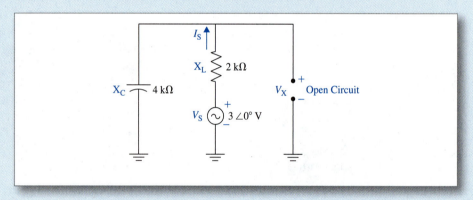

21. Given the circuit in Figure 5–49, use the voltage divider rule to determine the voltage across each component. Determine the voltage V_{AB}.

FIGURE **5–49**

Circuit for Essential
Learning Exercise 21

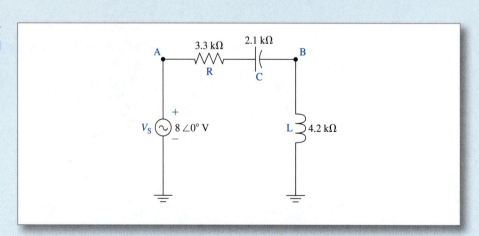

CHALLENGING LEARNING EXERCISES

(ANSWERS TO SELECTED CHALLENGING LEARNING EXERCISES CAN BE FOUND IN APPENDIX A.)

1. Given the circuit in Figure 5–50, determine the current I_X.

FIGURE 5–50

Circuit for
Challenging
Learning
Exercise 1

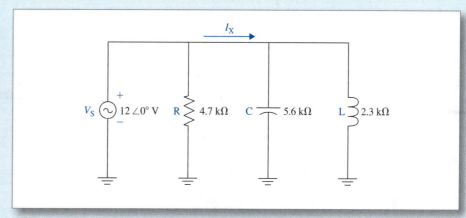

2. Given the circuit in Figure 5–51, determine the open circuit voltage V_{AB}.

FIGURE 5–51

Circuit for
Challenging
Learning
Exercise 2

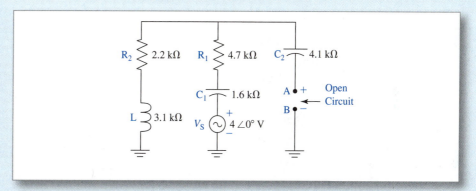

3. Given the circuit in Figure 5–52, determine the total equivalent impedance. Determine the voltage across each component expressed in the phasor domain.

FIGURE 5–52

Circuit for
Challenging
Learning
Exercise 3

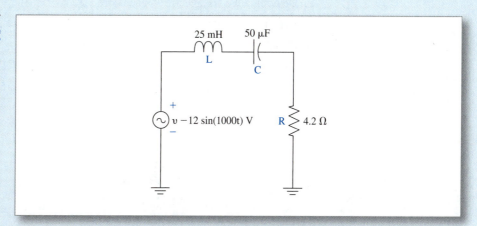

4. Given the circuit in Figure 5–53, determine the voltage across the inductor.

FIGURE **5–53**

Circuit for
Challenging
Learning
Exercise 4

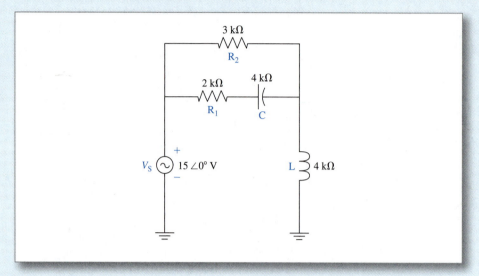

TEAM ACTIVITY

Given the circuit in Figure 5–54:

FIGURE **5–54**

Circuit for
TEAM ACTIVITY

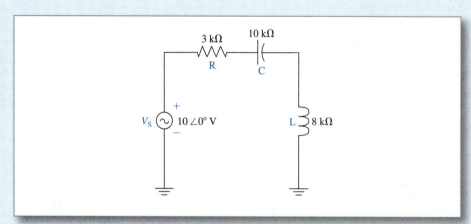

    a. Calculate the total impedance.

    b. Determine the voltage across each component.

    c. Sketch an impedance diagram.

    d. Sketch a phasor diagram.

    e. Verify the voltage across each component using computer simulation.

    f. If MATLAB is available, create a plot of the voltages in the circuit on the same time plot.

SIGNAL PROCESSING CIRCUIT ANALYSIS AND DESIGN

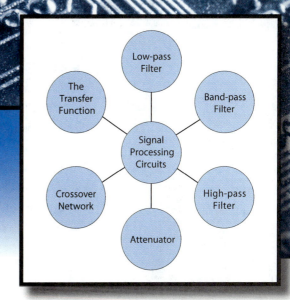

"DO WHAT YOU CAN WITH WHAT
YOU HAVE WHERE YOU ARE."

—Theodore Roosevelt

LEARNING OBJECTIVES

Upon successful completion of this chapter you will be able to:

- Describe the meaning of signal processing perspective.
- Determine the output signal of a resistive attenuator circuit.
- Determine the transfer function of a signal processing circuit.

CHAPTER SIX

- Define the frequency response curve plot and the Bode diagram.
- Calculate voltage gain and decibel voltage gain.
- Calculate the cutoff frequency for an R-C low-pass or high-pass filter.
- Sketch a Bode diagram for an R-C low-pass or high-pass filter.
- Calculate the cutoff frequency for an R-L low-pass or high-pass filter.
- Sketch a Bode diagram for an R-L low-pass or high-pass filter.
- Calculate the center frequency, high cutoff, and low cutoff frequency for a series resonant band-pass filter.
- Sketch the Bode diagram for a series resonant band-pass filter.
- Obtain a Bode diagram from MultiSIM.

INTRODUCTION

This chapter demonstrates how the circuit analysis laws and techniques learned in previous chapters are used to analyze several practical analog electronic signal processing circuits. By applying Ohm's law, Kirchhoff's laws, and the voltage and current divider rules, the characteristics of an output signal from a circuit, given a specified input signal, can be determined. Although much of contemporary signal processing circuit design utilizes integrated circuits, an appreciation of component-level circuit analysis is necessary to understand the fundamental operation of integrated circuits, as well as to design the interface networks required to connect integrated circuits together to create electronic signal processing systems.

6.1 AN INTRODUCTION TO SIGNAL PROCESSING CONCEPTS

Electronic signal processing circuits are designed to create or modify electronic signals. Signal processing circuits may be designed to increase or decrease signal amplitude or select only signals in a particular frequency range. Still other signal processing circuits are designed to convert analog signals to digital signals or digital signals to analog signals.

More sophisticated signal processing circuits are designed to create complex signals that contain information through the process of modulation. Demodulation circuits are designed to detect or recover information from complex signals. An example of the modulation-demodulation process is the common radio broadcast signal. Audio or sound information is added to a carrier wave (modulation) at the broadcast site, and the audio or sound is detected or recovered (demodulated) at the receiver site.

The purpose of an electronic circuit from a **signal processing perspective** is to process an input signal to produce a desired output signal. The objective of circuit analysis from a signal processing perspective is to determine the characteristics of an output signal for a given input signal.

The diagram in Figure 6–1 illustrates the signal processing perspective.

FIGURE **6–1**

The Signal Processing Perspective

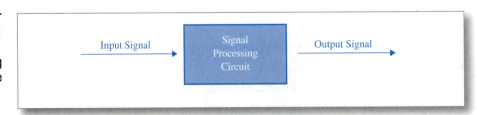

Input Signal → Signal Processing Circuit → Output Signal

A **transfer function** is a mathematical expression of the ratio of an output to an input signal such that:

$$\mathbf{T(f)} = \frac{\mathbf{V_{out}}}{\mathbf{V_{in}}} \qquad (6.1)$$

Equation (6.1) indicates that the transfer function is a function of frequency (f). The bold-faced print indicates that the transfer function is specified by a magnitude and angle.

If the mathematical expression for the input signal is known, the mathematical expression for the output signal may be determined by multiplying the input signal by the transfer function so that:

$$\mathbf{V_{out}} = \mathbf{V_{in}}\,\mathbf{T(f)} \qquad (6.2)$$

6.2 VOLTAGE GAIN AND THE DECIBEL

The ratio of the amplitude of the output signal to the amplitude of the input signal is an important consideration when viewing electronic circuits from a signal processing perspective.

An **amplifier** is an electronic circuit that produces an output signal whose amplitude is greater than the amplitude of the input signal. An **attenuator** is an electronic circuit that produces an output signal whose amplitude is less than the amplitude of the input signal.

Gain is defined as the ratio of the amplitude of the output signal of a circuit to the amplitude of the input signal. **Voltage gain** is given the symbol Av and is defined as:

$$Av = \frac{V_{out}}{V_{in}} \qquad (6.3)$$

Voltage gain is a ratio of similar units (volts) and is therefore a dimensionless quantity. Note that voltage is not bold-faced print, indicating that Av is simply the ratio of the amplitude of the output and input signals without regard to phase angles.

Comparing the expression of the transfer function in Equation (6.1) to the expression for voltage gain in Equation (6.3) indicates that voltage gain can be determined as the magnitude of the transfer function.

$$Av = T(f) \qquad (6.4)$$

Note that the symbol $T(f)$ without bold-faced print represents the magnitude of the transfer without regard to its angle.

EXAMPLE 6-1   Calculating Voltage Gain

The input signal to a signal processing circuit is given as $v_{in} = 3\sin(\omega t)$ V, and the output signal is given as $v_{out} = 12\sin(\omega t + 36°)$ V as illustrated in Figure 6-2. Determine the voltage gain for the circuit.

FIGURE **6–2**

Illustration for
Example 6–1

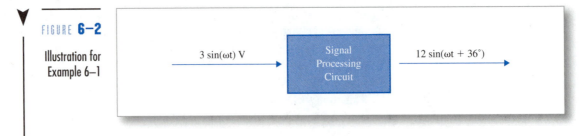

✓ SOLUTION

The voltage gain is calculated as the ratio of peak values of output and input voltage using Equation (6.3) to obtain:

$$Av = \frac{V_{out}}{V_{in}} = \frac{12\ V}{3\ V} = 4$$

Note that voltage gain is the ratio of amplitudes (peak values) with no consideration given to phase angles.

The **decibel** is a popular unit for expressing voltage gain because a logarithmic function best represents how the human ear responds to changes in sound intensity or volume, as well as to changes in frequency.

The decibel is a logarithmic quantity, and **decibel voltage gain** is defined as:

$$Av_{(dB)} = 20\ \log Av = 20\ \log \frac{V_{out}}{V_{in}} \qquad (6.5)$$

As an example of *logarithmic response*, the ear perceives the change of pitch from 400 Hz and 800 Hz as one octave. The change of pitch between 5 kHz and 10 kHz is also perceived as one octave even though the absolute frequency difference is 400 Hz in the first interval and 5 kHz in the second. The ear senses that the frequency has been doubled in each case. A response based upon interval change rather than absolute change can be described as a logarithmic response.

The decibel voltage gain for the circuit in Example 6–1 may be calculated as:

$$Av_{(dB)} = 20\ \log 4 = 12.04\ dB$$

The following table illustrates several common values of voltage gain and the approximate equivalent decibel voltage gain.

Voltage Gain	Decibel Voltage Gain
10	20 dB
2	6 dB
1	0 dB
0.7	−3 dB
0.5	−6 dB
0.1	−20 dB

Note that for voltage gain values less than 1, the decibel voltage gain is negative and for voltage gain values greater than 1, the decibel voltage gain is positive.

Electronic circuits that have a positive decibel voltage gain will *amplify* the input signal, and circuits with a negative decibel voltage gain will *attenuate* the input signal. A decibel voltage gain of 0 implies that the input signal and output signal have the *same amplitude*.

6.3 THE ATTENUATOR CIRCUIT

One of the most basic signal processing circuits is a voltage attenuator. The function of a voltage attenuator circuit is to reduce the amplitude of the input signal to a specified level. Consider the circuit in Figure 6–3.

FIGURE **6–3**

A Voltage
Attenuator Circuit

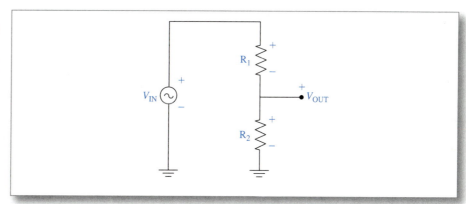

Observe that the output signal in Figure 6–3 is the voltage developed across R_2. The voltage divider rule may be used to determine the output voltage in the phasor domain as:

$$\mathbf{V}_{out} = \mathbf{V}_{in} \frac{\mathbf{R}_2}{\mathbf{R}_1 + \mathbf{R}_2}$$

The transfer function as defined in Equation (6.1) may be expressed as:

$$\mathbf{T(f)} = \frac{\mathbf{V}_{out}}{\mathbf{V}_{in}} = \frac{\mathbf{R}_2}{\mathbf{R}_1 + \mathbf{R}_2}$$

The voltage gain of the circuit may be determined as the magnitude of the transfer function:

$$Av = T(f)$$

The phase difference (θ) between the input and output signal is determined as the angle of the transfer function.

$$\theta = \text{The angle of } \mathbf{T(f)}$$

The decibel voltage gain may be calculated as:

$$Av_{(dB)} = 20 \log Av$$

Note that the transfer function for the circuit in Figure 6–3 is simply a ratio of resistor values. Since the value of a resistor does not depend upon frequency, the attenuator circuit in Figure 6–3 is referred to as *frequency independent*.

EXAMPLE 6–2   Analysis of an Attenuator Circuit

Given the attenuator circuit in Figure 6–4, determine the expression for the transfer function. Determine the voltage gain, decibel voltage gain, and phase difference (θ). Determine the amplitude (peak value) of the output voltage.

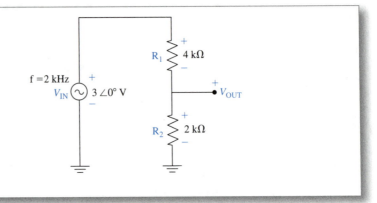

FIGURE **6-4**

Circuit for
Example 6-2

✔ **SOLUTION**

The voltage divider rule may be used to determine the output voltage in the phasor domain as:

$$\mathbf{V}_{out} = \mathbf{V}_{in}\frac{\mathbf{R}_2}{\mathbf{R}_1 + \mathbf{R}_2}$$

The transfer function is determined as:

$$\mathbf{T(f)} = \frac{\mathbf{V}_{out}}{\mathbf{V}_{in}} = \frac{\mathbf{R}_2}{\mathbf{R}_1 + \mathbf{R}_2}$$

The transfer function can be evaluated in the phasor domain as:

$$\mathbf{T(f)} = \frac{(4\angle 0°\ k\Omega)}{(4\angle 0°\ k\Omega + 2\angle 0°\ k\Omega)} = 0.67\angle 0°$$

The voltage gain of the circuit may be determined as the magnitude of the transfer function so that:

$$Av = T(f) = 0.67$$

The decibel voltage gain can be calculated as:

$$Av_{(dB)} = 20\ \log Av = 20\log 0.67 = -3.47\ dB$$

The phase difference between the input and output voltage is determined as the angle of the transfer function so that:

$$\theta = 0°$$

The phasor expression of the output signal may be determined as the phasor expression of the input signal multiplied by the transfer function so that:

$$\mathbf{V}_{out} = \mathbf{T(f)}\ \mathbf{V}_{in} = (0.67\angle 0°)(3\angle 0°\ V) = 2\angle 0°\ V \qquad (6.6)$$

Equation (6.6) indicates that the amplitude (peak value) of the output signal is 2 V and that the phase difference between the input and output signal is 0°.

Although the frequency is specified to be 2 kHz in Example 6–2, the frequency does not appear in any calculation, indicating the circuit is frequency independent.

PRACTICE EXERCISES 6-1

Given the attenuator circuit as in Figure 6–5, evaluate the transfer function. Determine the voltage gain, the decibel voltage gain, and the phase difference between the input and output signal. Determine the magnitude (peak value) and the frequency of the output signal.

FIGURE 6-5

Circuit for Practice
Exercise 6–1
Exercise 1

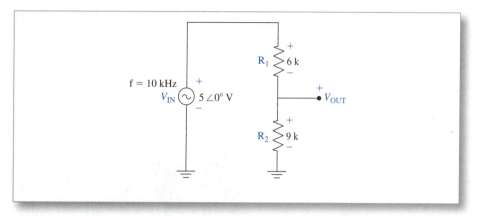

✓ ANSWERS

$T(f) = 0.60 \angle 0°$; $Av = 0.6$; $Av_{(dB)} = -4.44$ dB; $\theta = \angle 0°$;

$V_{out} = 3$ V; $f = 10$ kHz

(Complete solutions to the Practice Exercises can be found in Appendix A.)

6.4 CONCEPTS OF FREQUENCY RESPONSE

The introduction to signal processing concepts in this chapter indicated that one important signal processing function is to select only signals that fall within a particular frequency range.

The **frequency response diagram** of a circuit is a plot of the voltage gain of the circuit as a function of the frequency of the input signal. Frequency response diagrams may be obtained using software such as MATLAB or circuit simulator software such as MultiSIM.

Before software such as MultiSIM was readily available, approximations to actual frequency response diagrams were obtained using a procedure invented by Hendrik Bode (1905–1982).

A Bode diagram is a frequency response diagram with frequency plotted on a logarithmic scale and voltage gain plotted in decibels.

A typical Bode diagram indicating decibel gain as a function of frequency obtained from MultiSIM is shown in Figure 6–6.

MULTISIM

FIGURE 6-6

MultiSIM Bode Diagram
of Decibel Gain Versus
Frequency

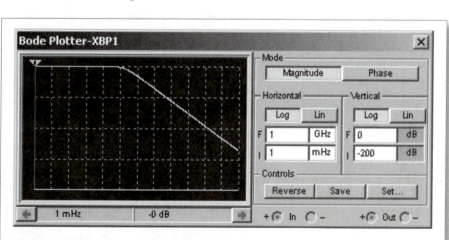

Note that the vertical axis of the Bode diagram in Figure 6–6 measures voltage gain in decibels (dB) and the horizontal axis measures signal frequency in Hz, plotted on a log scale.

The diagram in Figure 6–7 indicates the phase difference between the input and output signal as a function of frequency.

FIGURE **6–7**

MultiSIM Bode Diagram
of Phase Versus
Frequency

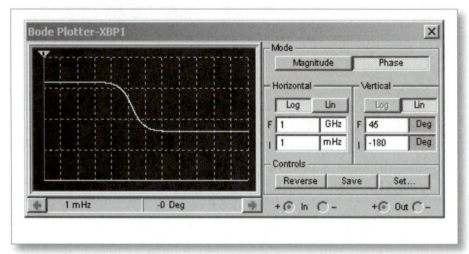

6.5 AN OVERVIEW OF FILTERS

A **filter** is an electronic circuit that is designed to pass signals whose frequency falls within a selected range and to reject or attenuate all other signals.

The most common electronic filters are the **low-pass, high-pass**, and **band-pass filters.** The Bode diagrams of a low-pass, a high-pass, and a band-pass filter are illustrated in Figure 6–8.

FIGURE **6–8**

An Overview
of Ideal and
Actual Filters

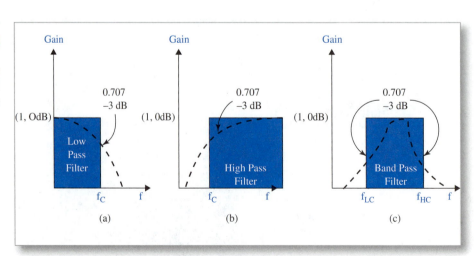

The highlighted areas of the frequency response illustrations in Figure 6–8 show the response of an ideal filter. As indicated, the response of an ideal filter changes instantaneously at the cutoff frequency (f_C) so that the value of f_C is clearly evident. The dashed lines in Figure 6–8 illustrate the frequency response of practical or actual filters.

Note that the gain of the practical filter does not change instantaneously at the cutoff frequency. It is therefore necessary to define where the cutoff frequency of a practical filter occurs.

The **cutoff frequency** of a practical filter is defined as the frequency at which the voltage gain (Av) is 0.707 ($\sqrt{2}/2$ exact), which corresponds to a decibel voltage gain ($Av_{(dB)}$) of $-3dB$.

Engineers may also refer to the cutoff frequency as the *critical frequency* or *break frequency*.

As illustrated in Figure 6–8, the low-pass filter will "pass" signals whose frequency is below the frequency designated as f_C and reject or attenuate frequencies above f_C. The high-pass filter will pass signals whose frequency is above f_C and reject or attenuate frequencies below f_C. The band-pass filter passes signals whose frequency lies between the low cutoff frequency f_{LC} and the high cutoff frequency f_{HC}.

6.6 THE R-C LOW-PASS FILTER

The circuit diagram of an R-C low-pass filter is shown in Figure 6–9.

FIGURE **6–9**

Prototype R-C
Low-Pass Filter

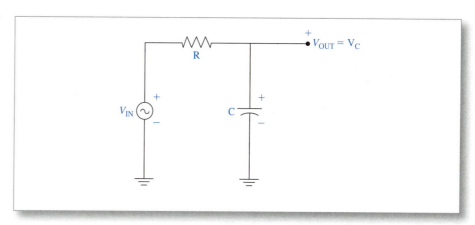

The resistance of the resistor in the low-pass filter circuit in Figure 6–9 remains constant regardless of the frequency of the input signal. The capacitive reactance, however, varies inversely with the input signal frequency as indicated by the expression:

$$X_C = \frac{1}{\omega C} = \frac{1}{2\pi f C}$$

The capacitor and the resistor in Figure 6–9 form a *voltage divider network*, and the output signal is the voltage that is developed across the capacitor as indicated.

As the capacitive reactance decreases, a smaller percentage of the output signal voltage will be developed across the capacitor. Since capacitive reactance varies inversely with frequency, the amplitude of the output signal will decrease as the frequency of the input signal increases.

The circuit will pass low-frequency signals to the output, while high-frequency signals will appear across the resistor and will not appear at the output. Such a circuit is, by definition, a low-pass filter.

The voltage divider rule can be used to obtain an expression for the output voltage in Figure 6–9 as:

$$\mathbf{V}_{out} = \mathbf{V}_{in}\frac{\mathbf{X}_C}{\mathbf{Z}} \tag{6.7}$$

The voltage gain at the cutoff frequency is defined to be $\frac{\sqrt{2}}{2}$ so that at cutoff:

$$Av = \frac{X_C}{Z} = \frac{\sqrt{2}}{2}$$

Recall that the magnitude of the impedance for an R-C series circuit can be expressed as:

$$Z = \sqrt{X_C^2 + R^2}$$

The voltage gain at the cutoff frequency can be expressed as:

$$Av = \frac{X_C}{\sqrt{X_C^2 + R^2}} = \frac{\sqrt{2}}{2} \tag{6.8}$$

Squaring both sides of Equation (6.8), we obtain:

$$\frac{X_C^2}{X_C^2 + R^2} = \frac{2}{4} = \frac{1}{2}$$

Simplifying, we obtain:

$$X_C^2 = \frac{1}{2}(X_C^2 + R^2)$$

$$2X_C^2 = (X_C^2 + R^2)$$

$$X_C^2 = R^2$$

so that: $$X_C = R \tag{6.9}$$

Equation (6.9) indicates that the cutoff frequency f_C for the R-C low-pass filter is the frequency at which X_C and R are equal.

Setting the magnitude of X_C equal to R, we obtain:

$$\frac{1}{2\pi f_C C} = R \tag{6.10}$$

Solving Equation (6.10) for f_C, we obtain an expression to calculate the cutoff frequency of an R-C low-pass filter as:

$$f_C = \frac{1}{2\pi RC} \tag{6.11}$$

The cutoff frequency f_C of the R-C low-pass filter may be determined by Equation (6.11) if the values of R and C are known. Recall that at the cutoff frequency f_C, the voltage gain of the low-pass filter is 0.707 ($\sqrt{2}/2$ exact) and the decibel voltage gain is -3dB.

The transfer function for the low-pass filter may be determined from Equation (6.7) and expressed as:

$$\mathbf{T(f)} = \frac{\mathbf{V}_{out}}{\mathbf{V}_{in}} = \frac{\mathbf{X}_C}{\mathbf{Z}} = \frac{\mathbf{X}_C}{\mathbf{X}_C + \mathbf{R}} \tag{6.12}$$

The transfer function can be evaluated at the cutoff frequency by setting the magnitudes of X_C and R equal to each other. The calculation in Equation (6.13) assumes the magnitude of X_C and R equal to 1 Ω to simplify the calculation as:

$$\mathbf{T(f)} = \frac{1\angle -90°\ \Omega}{1\angle -90°\ \Omega + 1\angle 0°\ \Omega} = 0.707\angle -45° \tag{6.13}$$

The voltage gain of the R-C low-pass filter can be determined as the magnitude of the transfer function in Equation (6.13) as:

$$Av = T(f) = 0.707$$

The phase difference between the input and output signal is determined as the angle of the transfer function in Equation (6.13) as:

$$\theta = -45°$$

The negative sign indicates that the output voltage lags the input voltage by 45°.

The results of the analysis of the R-C low-pass filter at the cutoff frequency f_C may be summarized as:

$$f_C = \frac{1}{2\pi RC}; \quad \mathbf{T(f)} = 0.707\angle -45°; \quad Av = 0.707; \quad Av_{(dB)} = -3dB;$$

$$\text{and} \quad \theta = -45°\,(V_{out} \text{ lags } V_{in})$$

Figure 6–10 shows the characteristics of the R-C low-pass filter at the cutoff frequency (f_C).

FIGURE **6-10**

Characteristics of
the R-C Low-Pass
Filter at Cutoff

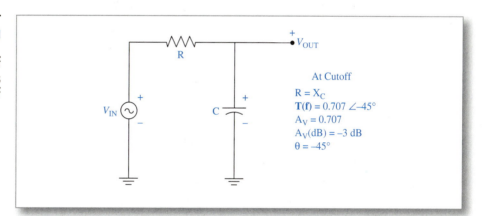

At Cutoff

$R = X_C$
$\mathbf{T(f)} = 0.707 \angle -45°$
$A_V = 0.707$
$A_V(dB) = -3\ dB$
$\theta = -45°$

EXAMPLE 6-3   Analysis of an R-C Low-Pass Filter

Given the R-C low-pass filter in Figure 6–11, calculate the cutoff frequency. Determine the voltage gain, decibel voltage gain, and phase difference θ at the cutoff frequency. Also determine the voltage gain, decibel voltage gain, and phase difference θ at the frequency of 2 kHz. Verify the decibel voltage gain and phase difference at the cutoff frequency f_C and at 2 kHz using computer simulation.

FIGURE **6-11**

R-C Low-Pass
Filter Circuit for
Example 6-3

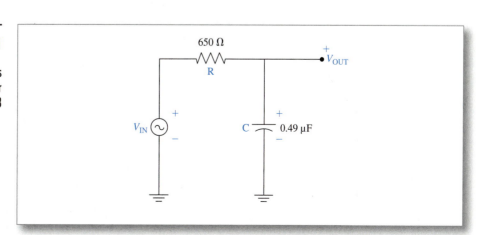

✔ **SOLUTION**

The cutoff frequency can be determined using Equation (6.11) to obtain:

$$f_C = \frac{1}{2\pi RC} = \frac{1}{2\pi(650\ \Omega)(0.49\ \mu F)} = 500\ \text{Hz}$$

The value of X_C at the cutoff frequency (500 Hz) may be calculated as:

$$X_C = \frac{1}{2\pi f_C C} = \frac{1}{2\pi(500\ \text{Hz})(0.49\ \mu F)} = 650\ \Omega$$

Note that the value of X_C at the cutoff frequency is equal to R (650 Ω).

Equation (6.13) indicates that the value of the transfer function at the cutoff frequency is:

$$\mathbf{T(f)} = 0.707\angle -45°$$

so that: $\quad Av = 0.707; \quad Av_{(dB)} = -3\ \text{dB}; \quad \text{and} \quad \theta = -45°$

It is important to understand that the value of the transfer function $T(f) = 0.707\angle -45°$ for the R-C low-pass filter is valid only at the cutoff frequency f_C.

To determine the value of voltage gain and phase difference at any frequency other than the cutoff frequency, the transfer function must be evaluated at the desire frequency.

The capacitive reactance at 2 kHz can be calculated as:

$$\mathbf{X_C} = \frac{1}{2\pi fC}\angle -90° = \frac{1}{2\pi(2\ \text{kHz})(0.49\ \mu F)}\angle -90° = 162.4\angle -90°\ \Omega$$

Note that R and X_C are not equal at 2 kHz.

The impedance of the circuit at 2 kHz can be calculated as:

$$\mathbf{Z} = \mathbf{R} + \mathbf{X_C} = 650\angle 0°\ \Omega + 162.4\angle -90°\ \Omega = 669.7\angle -14°\ \Omega$$

The transfer function can be determined using the voltage divider rule to obtain:

$$\mathbf{T(f)} = \frac{\mathbf{V_{out}}}{\mathbf{V_{in}}} = \frac{\mathbf{X_C}}{\mathbf{Z}}$$

The transfer function can be evaluated at 2 kHz to obtain:

$$\mathbf{T(f)} = \frac{\mathbf{X_C}}{\mathbf{Z}} = \frac{162.4\angle -90°\ \Omega}{669.7\angle -14°\ \Omega} = 0.242\angle -76° \tag{6.14}$$

The voltage gain can be determined as the magnitude of the transfer function so that:

$$Av = T(f) = 0.242$$

The decibel voltage gain is calculated as:

$$Av_{(dB)} = 20\ \log 0.242 = -12.3\ \text{dB}$$

The angle of the transfer function obtained in Equation (6.14) indicates that the phase difference between the input and output signal is:

$$\theta = -76°$$

The negative sign indicates that the output signal lags the input signal by 76°.

The results of the analysis of the R-C low-pass filter in Figure 6–11 at a frequency of 2 kHz may be summarized as:

$$\mathbf{T(f)} = 0.242 \angle -76°; \quad Av = T(f) = 0.242; \quad Av_{(dB)} = -12.3 \text{ dB};$$
$$\text{and} \quad \theta = -76° \, (V_{out} \text{ lags } V_{in})$$

The decibel gain and phase difference at the cutoff frequency (500 Hz) are shown in the MultiSIM Bode diagrams in Figure 6–12.

FIGURE **6–12**

MultiSIM Bode Diagrams for Example 6–3 at Cutoff (600 Hz)

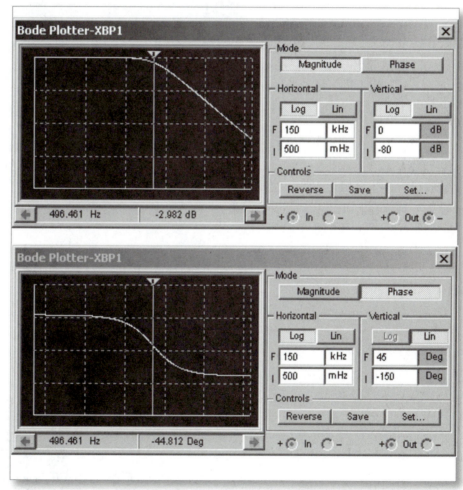

The decibel gain and phase angle at 2 kHz are shown in the MultiSIM Bode diagrams in Figure 6–13.

Note that *exact frequencies* cannot be obtained from MultiSIM Bode diagrams. However, the Bode diagram values approximate the values calculated in this example.

MULTISIM

FIGURE **6—13**

MultiSIM Bode
Diagrams for
Example 6—3
at 2 kHz

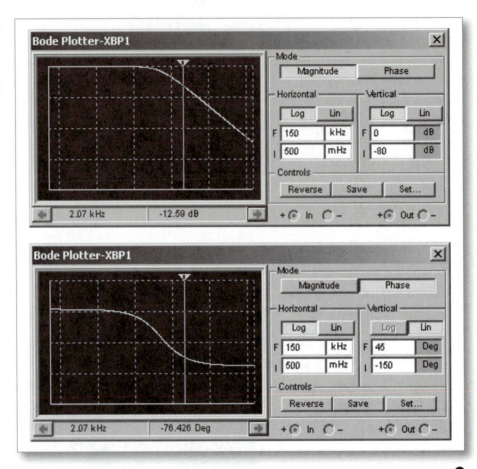

6.7 COMPUTER ANALYSIS OF THE R-C LOW-PASS FILTER

The frequency response of the R-C low-pass filter shown in Figure 6–11 can also be determined by evaluating the transfer function of the circuit over a range of frequencies. The transfer function can be evaluated by entering appropriate formulas into an electronic spreadsheet.

The transfer function of the circuit has been determined as:

$$T(f) = \frac{X_C}{Z} = \frac{X_C}{X_C + R}$$

The computer-generated table in Figure 6–14 contains the results of evaluating the transfer function at the indicated frequencies.

An important first observation of the data in the table in Figure 6–14 reveals that the decibel voltage gain of the circuit is near 0 at very low frequencies. A decibel voltage gain of 0 corresponds to a voltage gain of 1, which implies that the amplitude of the output signal is equal to the amplitude of the input signal.

The data in Figure 6–14 indicate that low-frequency signals will pass through the low-pass filter and appear at the output.

FIGURE **6-14**

Computer Generated
Values for the
R-C Low-Pass Filter

Frequency Hz	Av	Av$_{(dB)}$	Degrees
1	1.00	0.00	−0.11
2	1.00	0.00	−0.23
3	1.00	0.00	−0.34
10	1.00	0.00	−1.15
20	1.00	0.00	−2.29
30	1.00	0.00	−3.44
40	1.00	0.00	−4.58
50	1.00	0.00	−5.71
100	0.98	−0.17	−11.32
200	0.93	−0.65	−21.81
400	0.78	−2.15	−38.68
500	0.71	−3.01	−45.02
600	0.64	−3.88	−50.21
700	0.58	−4.72	−54.48
800	0.53	−5.52	−58.01
900	0.49	−6.28	−60.96
1000	0.45	−6.99	−63.45
2000	0.24	−12.31	−75.97
5000	0.10	−20.05	−84.29
50000	0.01	−40.01	−89.43

As the input signal frequency increases, the amplitude of the output signal begins to decrease. By definition, the cutoff frequency f_C occurs when the voltage gain is 0.707, corresponding to a decibel voltage gain of −3dB.

The data in Figure 6–14 indicate that the decibel voltage gain is −3dB at a frequency (f_C) of 500 Hz.

The data also indicate that the reactance of the capacitor X_C (650 Ω) is equal to the value of the resistance of the resistor R (650 Ω) at the cutoff frequency (f_C =500 Hz).

Lastly, the phase difference between the input and output signal ranges from 0° to −90° and is equal to −45° at the cutoff frequency f_C.

The results of the computer data in Figure 6–14 may be summarized as:

$$f_c = 500 \text{ Hz}$$

and at f_C: $Av = 0.707$; $Av_{(dB)} = -3 \text{ dB}$; $\theta = -45°$;   and $X_C = R$

which verify the results obtained in Example 6–3.

In addition, the table in Figure 6–14 indicates that the decibel voltage gain at a frequency of 5 kHz (10 f_C) is approximately −20 dB, and at 50 kHz (100 f_C) the dB gain is approximately −40dB. The frequency 5 kHz is ten times the cutoff frequency of 500 Hz and is referred to as one decade above the cutoff frequency. The frequency 50 kHz is one hundred times the cutoff frequency or two decades above the cutoff frequency.

The rate of attenuation of a filter is often referred to as **roll-off rate.** The roll-off rate of the R-C low-pass filter is −20 dB/decade above the cutoff frequency.

EXAMPLE 6-4   Analysis of an R-C Low-Pass Filter

Given the R-C low-pass filter in Figure 6–15, determine the cutoff frequency. Determine the voltage gain, decibel voltage gain, and phase difference (θ) at the cutoff frequency.

Evaluate the transfer function at a frequency of 4 kHz and determine the voltage gain, decibel voltage gain, and phase difference (θ) at 4 kHz.

Determine the amplitude of the output signal at cutoff and at 4 kHz if the input signal is given as 3 sin(5000*t*) V.

Verify the dB gain at cutoff and at 4 kHz using computer simulation.

FIGURE **6-15**

R-C Low-Pass
Filter Circuit for
Example 6–4

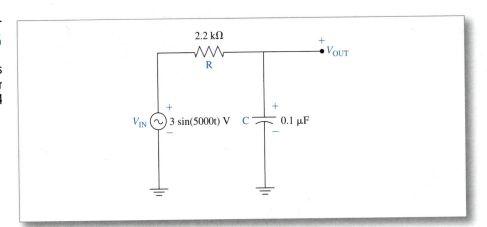

✓ **SOLUTION**

The cutoff frequency may be calculated as:

$$f_C = \frac{1}{2\pi RC} = \frac{1}{2\pi(2.2\ k\Omega)(0.1\ \mu F)} = 723.4\ Hz$$

At the cutoff frequency f_C:

$$\mathbf{T(f)} = 0.707\angle - 45°; \quad Av = 0.707; \quad Av_{(dB)} = -3dB; \quad and \quad \theta = -45°(V_{out}\ lags\ V_{in})$$

The gain and phase difference of the filter at any frequency other than the cutoff frequency is obtained by evaluating the transfer function of the circuit at the desired frequency.

The value of $\mathbf{X_C}$ at 4 kHz is calculated as:

$$\mathbf{X_C} = \frac{1}{2\pi fC}\angle - 90° = \frac{1}{2\pi(4\ kHz)(0.1\ \mu F)}\angle - 90° = 397.9\angle - 90°\ \Omega$$

The circuit impedance is calculated as:

$$\mathbf{Z} = 2.2\angle 0°\ k\Omega + 397.9\angle - 90°\ \Omega = 2.24\angle - 10°\ k\Omega$$

so that: $$\mathbf{T(f)} = \frac{\mathbf{X_C}}{\mathbf{Z}} = \frac{397.9\angle - 90°\ \Omega}{2.24\angle - 10°\ k\Omega} = 0.177\angle - 80°$$

The voltage gain can be determined as the magnitude of the transfer function as:

$$Av = T(f) = 0.177$$

The phase difference between the input and output signals can be determined from the transfer function as:

$$\theta = -80° (V_{out} \text{ lags } V_{in})$$

The decibel voltage gain can be calculated as:

$$Av_{(dB)} = 20 \log 0.177 = -15.04 \text{ dB}$$

The output signal can be determined in the phasor domain by multiplying the expression of the input signal by the transfer function so that:

$$\mathbf{V_{out}} = \mathbf{V_{in}} \ \mathbf{T(f)}$$

With the input signal given as $3 \sin(5000t)$ V, the output signal is calculated in the phasor domain at 4 kHz as:

$$\mathbf{V_{out}} = (3 \angle 0° \text{ V})(0.177 \angle -80°) = 0.531 \angle -80° \text{ V}$$

This voltage may be expressed in the time domain as:

$$v_{out} = 0.531 \sin(25132t - 80°) \text{ V}$$

where

$$\omega = 2\pi f = 2\pi(4000) = 25132 \text{ rad/s}$$

The MultiSIM Bode diagrams at the approximate cutoff frequency and at approximately 4 kHz are shown in Figures 6–16 and 6–17.

FIGURE **6–16**

MultiSIM Bode Diagrams for Example 6–4 at Cutoff

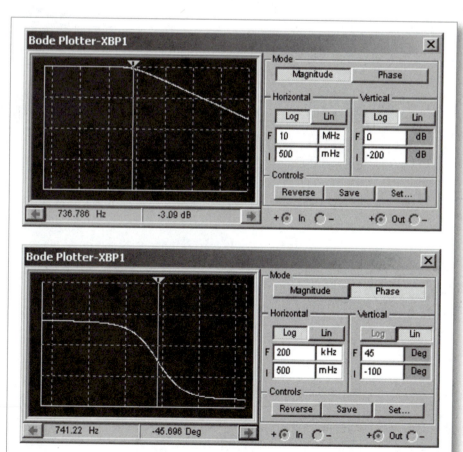

FIGURE **6–17**

MultiSIM Bode
Diagrams for
Example 6–4
at 4 kHz

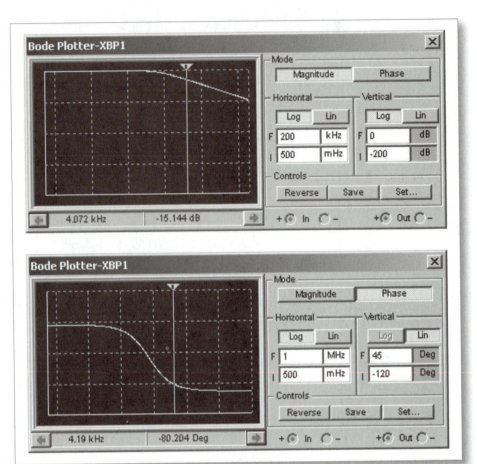

Note that exact frequencies cannot be obtained from MultiSIM Bode diagrams. However, the Bode diagram values approximate the values calculated in this example.

APPLICATION OF AN R-C LOW-PASS FILTER

The R-C low-pass filter may be used to remove undesired high-frequency "noise" from electronic circuits. High-frequency noise may be inadvertently created by circuit components or may enter the circuit from external sources. This application illustrates how an R-C low-pass filter can be designed to eliminate high-frequency noise.

SPECIFICATIONS: Design an R-C low-pass filter to eliminated high-frequency noise. The cutoff frequency of the filter is required to be 550 Hz. Verify the accuracy of the design using computer simulation.

✔ SOLUTION

A sketch of the prototype circuit for an R-C low-pass filter that will be used for the design is shown in Figure 6–18.

The cutoff frequency of the R-C low-pass filter can be expressed as:

$$f_C = \frac{1}{2\pi RC} \tag{6.15}$$

FIGURE **6-18**

Prototype
R-C Low-Pass Filter

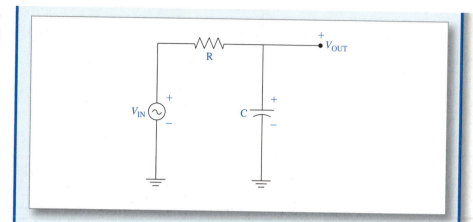

Equation (6.15) contains two unknown variables, R and C. Our only strategy at this point is to arbitrarily choose a value for either R or C and then calculate the appropriate value of the remaining variable.

We can arbitrarily choose R to be a 1 kΩ resistor and rearrange Equation (6.15) to calculate the value of C as:

$$C = \frac{1}{2\pi f_C R} = \frac{1}{2\pi(550\text{ Hz})(1\text{ k}\Omega)} = 289.37\text{ nF}$$

FIGURE **6-19**

Completed Circuit
Design and MultiSIM
Bode Diagrams

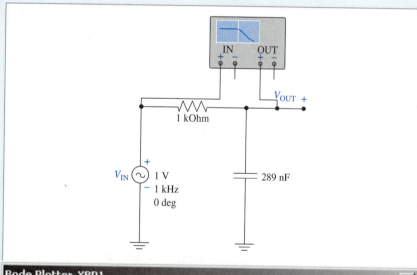

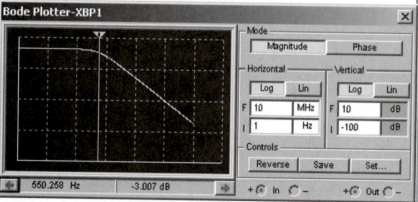

The completed circuit design is shown in Figure 6–19 along with the Bode diagram obtained from MultiSIM. Note that the Bode diagram verifies that the gain at the specified cutoff frequency of 550 Hz is -3 dB (approximate).

An arbitrary value of C could have been selected and the appropriate value of R calculated, rearranging Equation (6.16) as:

$$R = \frac{1}{2\pi f_C C}$$

The point to keep in mind is that as a circuit designer, arbitrary selection of one or more component values is often the key to successful design.

PRACTICE EXERCISES 6–2

Given an R-C low-pass filter with $R = 680\ \Omega$ and $C = 0.5\ \mu F$, determine the cutoff frequency of the filter. Determine the voltage gain, decibel voltage gain, and phase angle (θ) at the cutoff frequency.

Evaluate the transfer function at a frequency of 1 kHz and determine the voltage gain, the decibel voltage gain, and the phase difference between the input and output signal at 1 kHz.

If the input signal is given as $5\angle 0°$ V, determine the amplitude and phase of the output signal at 1 kHz.

✔ ANSWERS

$f_C = 468.1$ Hz;   at f_C: $Av = 0.707$; $Av_{(dB)} = -3$ dB; $\theta = -45°$;

at 1 kHz: $Av = 0.424$; $Av_{(dB)} = -7.45$ dB; $\theta = -64.9°$;

$V_{out} = 2.12\angle -64.9°$ V

(Complete solutions to the Practice Exercises can be found in Appendix A.)

6.8 THE R-C HIGH-PASS FILTER

The analysis of the R-C high-pass filter in Figure 6–20 is similar to the analysis of the R-C low-pass filter. Note that the output signal is developed across the resistor for the high-pass filter.

FIGURE 6–20

Prototype R-C High-Pass Filter

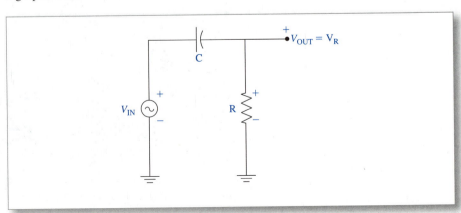

The capacitor and the resistor in Figure 6–20 form a voltage divider network with the output signal developed across the resistor. As the frequency of the input signal increases, the capacitive reactance will decrease and a larger percentage of the input signal will appear across the resistor as the output signal.

The circuit will attenuate low frequencies, while high-frequency signals will appear across the resistor at the output. Such a circuit is, by definition, a high-pass filter.

Considering the circuit in Figure 6–20, the voltage divider rule can be used to determine the output voltage as:

$$\mathbf{V}_{out} = \mathbf{V}_{in}\frac{\mathbf{R}}{\mathbf{Z}} \tag{6.16}$$

The transfer function can be expressed as:

$$\mathbf{T(f)} = \frac{\mathbf{V}_{out}}{\mathbf{V}_{in}} = \frac{\mathbf{R}}{\mathbf{Z}} \tag{6.17}$$

Analysis of the R-C high-pass filter is similar to the analysis of the R-C low-pass filter in the previous section. The cutoff frequency will occur at the frequency for X_C is equal to R such that:

$$X_C = R \tag{6.18}$$

Setting $X_C = R$, we obtain:

$$\frac{1}{2\pi f_C C} = R$$

Solving for f_C, we obtain the expression that can be used to calculate the cutoff frequency of the R-C high-pass filter as:

$$f_C = \frac{1}{2\pi RC}$$

Given that the magnitudes of $\mathbf{X}_C$ and $\mathbf{R}$ are equal at the cutoff frequency, the transfer function at the cutoff frequency can be determined as:

$$\mathbf{T(f)} = \frac{\mathbf{V}_{out}}{\mathbf{V}_{in}} = \frac{\mathbf{R}}{\mathbf{Z}} = \frac{\mathbf{R}}{\mathbf{R}+\mathbf{X}_C} = \frac{1\angle 0°\,\Omega}{1\angle 0°\,\Omega + 1\angle -90°\,\Omega} = 0.707\angle 45° \tag{6.19}$$

The magnitude of R and X_C in Equation (6.19) is chosen equal to $1\,\Omega$ to simplify the calculation.

The voltage gain of the filter at the cutoff frequency can be determined as the magnitude of the transfer function so that:

$$Av = T(f) = 0.707$$

The phase difference between the input and output signal may be determined as the angle of the transfer function so that:

$$\theta = 45°$$

The positive sign indicates that the output signal leads the input signal by 45°.

The phase difference between the input signal and output signal at cutoff for the high-pass filter is positive (+45°). Recall that the phase difference for the low-pass filter is negative (−45°) at cutoff.

The R-C high-pass filter can be summarized at the cutoff frequency as:

$$T(f) = 0.707\angle 45°; \quad Av = 0.707; \quad Av_{(dB)} = -3\text{dB}; \quad \text{and} \quad \theta = 45°$$

EXAMPLE 6-5   Analysis of an R-C High-Pass Filter

Given the R-C high-pass filter in Figure 6–21, determine the cutoff frequency. Determine the voltage gain, decibel voltage gain, and phase difference θ at the cutoff frequency.

Determine the voltage gain, decibel voltage gain, and phase difference θ at the frequency of 400 Hz.

Determine the expression for the output signal in the phasor domain at 400 Hz.

FIGURE **6–21**

R-C High-Pass
Filter Circuit for
Example 6–5

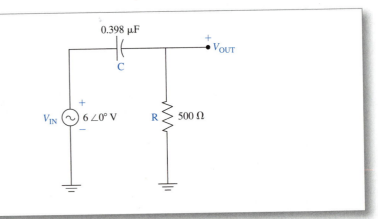

✓ **SOLUTION**

The cutoff frequency can be calculated as:

$$f_C = \frac{1}{2\pi RC} = \frac{1}{2\pi(500\ \Omega)(0.398\ \mu\text{F})} = 800\ \text{Hz}$$

At the cutoff frequency:

$$Av = 0.707; \quad Av_{(\text{dB})} = -3\ \text{dB}; \quad \text{and} \quad \theta = 45°$$

At the specified frequency of 400 Hz, the capacitive reactance may be calculated as:

$$\mathbf{X_C} = \frac{1}{2\pi fC} = \frac{1}{2\pi(400\ \text{Hz})(0.398\ \mu\text{F})} = 1\angle -90°\ \text{k}\Omega$$

$$\mathbf{Z} = \mathbf{R} + \mathbf{X_C} = 500\angle 0°\ \Omega + 1\angle -90°\ \text{k}\Omega = 1.12\angle -63.4°\ \text{k}\Omega$$

The transfer function of the R-C high-pass filter can be evaluated at 400 Hz as:

$$\mathbf{T(f)} = \frac{\mathbf{R}}{\mathbf{Z}} = \frac{500\angle 0°\ \Omega}{1.12\angle -63.4°\ \text{k}\Omega} = 0.466\angle 63.4°$$

The voltage gain of the filter is determined as the magnitude of the transfer function:

$$Av = 0.446$$

The decibel voltage gain can be calculated as:

$$Av_{(\text{dB})} = 20\ \log 0.446 = -7.01\ \text{dB}$$

The phase difference between the input and output signal is determined as the angle of the transfer function so that:

$$\theta = 63.4°$$

The positive sign indicates that the output signal leads the input signal by 63.4°.

The expression for the output voltage in the phasor domain is determined as:

$$\mathbf{V_{out}} = \mathbf{V_{in}} \, \mathbf{T(f)} = (6 \angle 0° \text{ V})(0.446 \angle 63.4°) = 2.68 \angle -63.4° \text{ V}$$

The MultiSIM Bode diagrams (magnitude plot only) at cutoff and at 400 Hz are shown in Figure 6–22 and Figure 6–23 respectively.

FIGURE **6–22**

MultiSIM Bode Diagram for Example 6–5 at Cutoff (800 Hz)

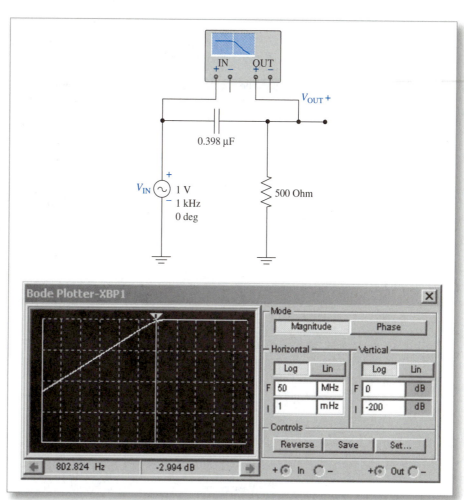

FIGURE **6–23**

MultiSIM Bode Diagram for Example 6–5 at 400 Hz

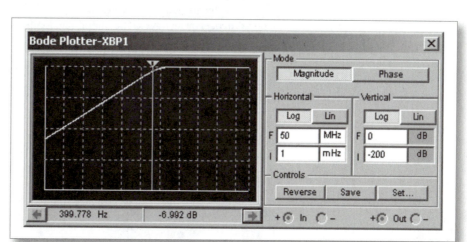

The Bode diagrams indicate a gain of -3 dB at 800 Hz (approximate) and a gain of -6.84 dB at 400 Hz (approximate). These values are close approximations of the values calculated in this example.

6.9 COMPUTER ANALYSIS OF THE R-C HIGH-PASS FILTER

The response of the R-C low-pass filter shown in Figure 6–21 can also be determined by evaluating the transfer function of the circuit over a range of frequencies.

The computer-generated spreadsheet shown in Figure 6–24 displays voltage gain and phase difference values for the high-pass filter shown in Figure 6–21 at the indicated frequencies.

FIGURE **6–24**

Computer Generated
Values for the
R-C High-Pass Filter

Frequency Hz	Av	$A_{V(dB)}$	Degrees
8	0.010	−40.00	89.43
10	0.013	−38.06	89.28
20	0.025	−32.04	88.57
50	0.062	−24.10	86.42
80	0.100	−20.04	84.29
100	0.124	−18.13	82.87
200	0.243	−12.30	75.96
300	0.351	−9.09	69.44
500	0.530	−5.51	57.99
600	0.600	−4.44	53.12
800	0.707	−3.01	44.99
1000	0.781	−2.15	38.65
2000	0.929	−0.64	21.80
3000	0.966	−0.30	14.93
4000	0.981	−0.17	11.31
6000	0.991	−0.08	7.59
10000	0.997	−0.03	4.57

We can observe from the data in Figure 6–24 that the cutoff frequency (-3 dB point) occurs at a frequency of 800 Hz. Also observe that frequencies below 800 Hz are attenuated, while frequencies above 800 Hz will pass through to the output (gain near 0 dB). The phase difference between the input and output signal is approximately 90° at low frequencies and approaches 0° as frequency increases. The roll-off rate for the R-C high-pass filter is -20 dB/decade.

APPLICATION OF AN R-C HIGH-PASS FILTER

Most electronic circuits require a DC voltage supply for proper operation. The DC voltage supply is often achieved by converting common utility voltage (a sinusoidal voltage at 60 Hz) into DC. This conversion process often produces low-frequency noise (60 Hz) that must be removed from the output signal.

A high-pass filter can be designed to remove low-frequency noise without affecting the amplitude of the higher signal frequencies.

The design of an R-C high-pass filter is similar to the design of the R-C low-pass filter. Recall that the formulta for calculating the cutoff frequency of an R-C filter is the same regardless of whether the filter is the low-pass or high-pass configuration.

SPECIFICATIONS: Design an R-C high-pass filter with a cutoff frequency of 60 Hz to remove low-frequency noise.

✔ SOLUTION

The prototype circuit for an R-C high-pass filter is shown in Figure 6–25.

FIGURE **6–25**

Prototype R-C
High-Pass Filter

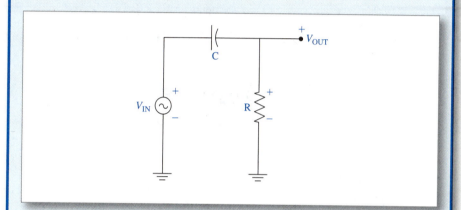

The cutoff frequency can be determined as:

$$f_c = \frac{1}{2\pi RC} \tag{6.20}$$

An arbitrary value for either R or C can be selected, and Equation (6.20) can be used to solve for the appropriate value of the remaining variable.

Arbitrarily selecting the value of C as 470 nF, the appropriate value of R may be determined by rearranging Equation (6.20) to obtain:

$$R = \frac{1}{2\pi f_c C} = \frac{1}{2\pi(60\ \text{kHz})(470\ \text{nF})} = 5643.8\ \Omega \tag{6.21}$$

Although choosing a value for C was an arbitrary selection, experience tells us that a 470 nF capacitor is a standard value that is readily available.

FIGURE **6–26**

Completed Circuit
Design and MultiSIM
Bode Diagrams

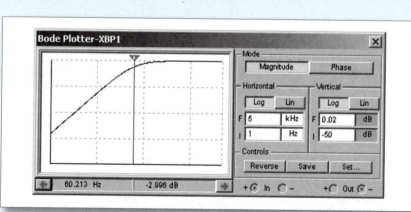

The completed circuit design is shown in Figure 6–26 along with the Bode diagram (magnitude plot only) from MultiSIM. Note that the Bode diagram verifies that the gain at the desired cutoff frequency of 60 Hz is −3 dB (approximate).

PRACTICE EXERCISES 6–3

Given an R-C high-pass filter with $R = 330\,\Omega$ and $C = 0.4\,\mu F$, determine the cutoff frequency. Determine the voltage gain, the decibel voltage gain, and θ at the cutoff frequency.

Evaluate the transfer function at a frequency of 3 kHz. Determine the voltage gain, the decibel voltage gain, and θ at 3 kHz.

If the input signal is given as $12\angle 0°$ V, determine the amplitude and phase of the output signal at 3 kHz.

✓ ANSWERS

$f_C = 1.21$ kHz;   at f_C : $Av = 0.707$; $Av_{(dB)} = -3$ dB; $\theta = 45°$;

at 3 kHz: $Av = 0.928$; $Av_{(dB)} = -0.65$ dB; $\theta = 21.9°$;

$V_{out} = 11.14\angle 21.9°$ V

(Complete solutions to the Practice Exercises can be found in Appendix A.)

6.10 THE R-L LOW-PASS FILTER

The prototype R-L low-pass filter is shown in Figure 6–27.

FIGURE **6–27**

Prototype R-L Low-Pass Filter

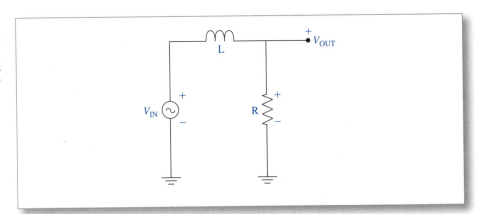

The inductor and resistor form a voltage divider network with the output signal taken across the resistor.

The inductive reactance increases with increasing frequency so that the high-frequency signals will appear across the inductor, while low-frequency signals will appear across the resistor at the output.

The voltage gain, decibel voltage gain, and phase difference between the input and output signal can be determined for any frequency for the R-L low-pass filter by evaluating the transfer function at the desired frequency.

The voltage divider rule can be used to express the output voltage as:

$$\mathbf{V}_{out} = \mathbf{V}_{in} \frac{\mathbf{R}}{\mathbf{Z}}$$

The transfer function of the R-L low-pass filter can be expressed as:

$$\mathbf{T(f)} = \frac{\mathbf{V}_{out}}{\mathbf{V}_{in}} = \frac{\mathbf{R}}{\mathbf{Z}}$$

where:

$$\mathbf{Z} = \mathbf{R} + \mathbf{X}_{L}$$

The cutoff frequency is defined as the frequency at which the real gain $Av = 0.707$ ($\sqrt{2}/2$ exact), which corresponds to a decibel gain of -3 dB.

Analysis of the filters in the previous sections indicates that the voltage gain will equal $\sqrt{2}/2$ when the reactance of the reactive component is equal to the resistance of the resistor. Setting inductive reactance equal to the resistance at the cutoff frequency, we obtain:

$$X_{L} = R$$

so that:

$$2\pi f_{c} L = R$$

The cutoff frequency (f_C) for the R-L low-pass filter can be determined as:

$$f_C = \frac{R}{2\pi L} \tag{6.22}$$

Setting the magnitudes of $\mathbf{R}$ and $\mathbf{X}_{L}$ equal to 1 Ω to simplify calculations, the transfer function at the cutoff frequency can be evaluated as:

$$\mathbf{T(f)} = \frac{\mathbf{R}}{\mathbf{Z}} = \frac{\mathbf{R}}{\mathbf{R} + \mathbf{X}_{L}} = \frac{1\angle 0^\circ\ \Omega}{(1\angle 0^\circ\ \Omega + 1\angle 90^\circ\ \Omega)} = 0.707\angle -45^\circ \tag{6.23}$$

The gain of the R-L low-pass filter at the cutoff frequency can be determined as the magnitude of the transfer function so that:

$$Av = 0.707$$

The phase difference between the output and input signal can be determined as the angle of the transfer function so that:

$$\theta = -45^\circ$$

The negative sign indicates that the output voltage lags the input voltage by 45°.

The results of the analysis of the R-L low-pass filter at the cutoff frequency can be summarized as:

$$f_C = \frac{R}{2\pi L}; \quad \mathbf{T(f)} = 0.707\angle -45^\circ; \quad A_v = 0.707; \quad A_{v(dB)} = -3\ \text{dB};$$

$$\theta = -45^\circ \ (V_{OUT} \text{ lags } V_{IN})$$

EXAMPLE 6-6   Analysis of the R-L Low-Pass Filter

Given the R-L low-pass filter in Figure 6–28, determine the cutoff frequency. Determine the voltage gain, the decibel voltage gain, and θ at the cutoff frequency.

Evaluate the transfer function at a frequency of 500 Hz. Determine the voltage gain, the decibel voltage gain, and θ at 500 Hz.

Verify the dB gain at cutoff and 500 Hz using computer simulation.

The input signal is $4\angle 0°$ V as indicated in Figure 6–28. Determine the amplitude and phase of the output signal at a frequency of 500 Hz.

FIGURE **6–28**

R-L Low-Pass
Filter Circuit for
Example 6–6

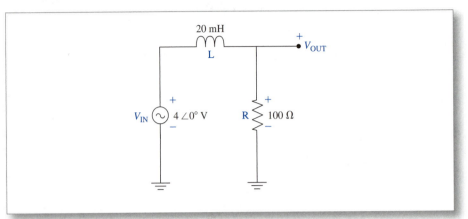

✓ **SOLUTION**

The cutoff frequency can be calculated as:

$$f_C = \frac{R}{2\pi L} = \frac{100\ \Omega}{2\pi(20\ \text{mH})} = 795.77\ \text{Hz}$$

At the cutoff frequency:

$$Av = 0.707; \quad Av_{(dB)} = -3\ \text{dB}; \quad \theta = -45°$$

The inductive reactance at 500 Hz can be calculated as:

$$\mathbf{X_L} = 2\pi fL\angle 90° = 2\pi(500\ \text{Hz})(20\ \text{mH})\angle 90° = 62.8\angle 90°\ \Omega$$

The impedance of the circuit at 500 Hz can be calculated as:

$$\mathbf{Z} = \mathbf{R} + \mathbf{X_L} = 100\angle 0°\ \Omega + 62.8\angle 90°\ \Omega = 118.1\angle 32.13°\ \Omega$$

The transfer function at 500 Hz is evaluated as:

$$\mathbf{T(f)} = \frac{\mathbf{V_{out}}}{\mathbf{V_{in}}} = \frac{\mathbf{R}}{\mathbf{Z}} = \frac{100\angle 0°\ \Omega}{118.1\angle 32.1°\ \Omega} = 0.847\angle -32.13°$$

The voltage gain is determined as the magnitude of the transfer function so that:

$$Av = 0.847$$

The decibel voltage gain can be calculated as:

$$Av_{(dB)} = 20\ \log 0.847 = -1.44\ \text{dB}$$

The phase difference between the input and output signal is determined as the angle of the transfer function so that:

$$\theta = -32.13°$$

The negative sign indicates that the output signal lags the input signal by 32.13°.

The output signal is determined as:

$$\mathbf{V_{out}} = \mathbf{V_{in}}\ \mathbf{T(f)} = (4\angle 0°\ \text{V})(0.847\angle -32.13°) = 3.39\angle -32.13°\ \text{V}$$

The MultiSIM Bode diagrams for this example are shown in Figure 6–29 and Figure 6–30.

Figure 6–29 indicates that the cutoff frequency (-3 dB) is approximately 800 Hz. Figure 6–30 indicates the gain near 500 Hz is approximately -1.44 dB as calculated.

FIGURE **6-29**

MultiSIM Bode
Diagram for
Example 6-6
at Cutoff (796 Hz)

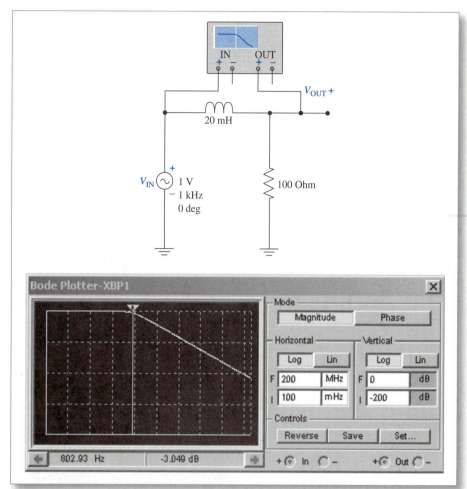

FIGURE **6-30**

MultiSIM Bode
Diagram for
Example 6-6
at 500 Hz

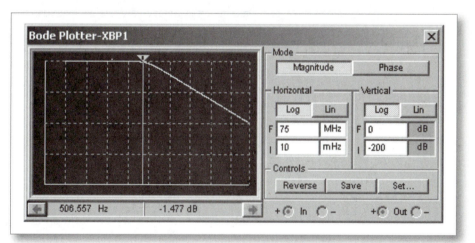

The computer-generated spreadsheet data displayed in Figure 6–31 are obtained by evaluating the transfer function $\mathbf{T(f)} = \mathbf{R/Z}$ for the R-L low-pass filter at the indicated frequencies.

The data in Figure 6–31 reveal that the frequency response of the R-L low-pass filter is similar to the response of the R-C low-pass filter. The filter passes low-frequency signals and attenuates frequencies higher than the cutoff frequency.

The gain at cutoff is 0.707 or −3 dB and θ is equal to −45°. The roll-off rate of the filter is −20 dB/decade.

Frequency Hz	Av	Av$_{(dB)}$	Degrees
25	1.000	0.00	− 1.8
50	0.998	− 0.02	− 3.6
100	0.992	− 0.07	− 7.2
150	0.983	− 0.15	−10.7
200	0.970	− 0.27	−14.1
250	0.954	− 0.41	−17.4
300	0.936	− 0.58	−20.7
350	0.915	− 0.77	−23.7
400	0.893	− 0.98	−26.7
500	0.847	− 1.45	−32.1
600	0.798	− 1.95	−37.0
700	0.751	− 2.49	−41.3
800	0.705	− 3.03	−45.2
1000	0.623	− 4.11	−51.5
2000	0.370	− 8.64	−68.3
3000	0.256	−11.82	−75.1
4000	0.195	−14.19	−78.7

PRACTICE EXERCISES 6–4

Given an R-L low-pass filter with R=150 Ω and L=18 mH, determine the cutoff frequency. Determine the voltage gain, the decibel voltage gain, and θ at the cutoff frequency.

Evaluate the transfer function at a frequency of 500 Hz. Determine the voltage gain, the dB gain, and θ at 500 Hz.

If the input signal is given as 8∠0° V, determine the amplitude and phase of the output signal at 500 Hz.

✓ANSWERS

f_C =1.33 kHz;   at f_C : Av = 0.707; $Av_{(dB)}$ =−3dB;   θ =−45°;

at 500 Hz: Av = 0.936; $Av_{(dB)}$ =−0.57 dB;   θ =−20.7°;

$\mathbf{V}_{out}$ = 7.49∠−20.7° V

(Complete solutions to the Practice Exercises can be found in Appendix A.)

6.11 THE R-L HIGH-PASS FILTER

The analysis of the prototype R-L high-pass filter in Figure 6–32 is similar to the analysis of the previous filters presented in this chapter.

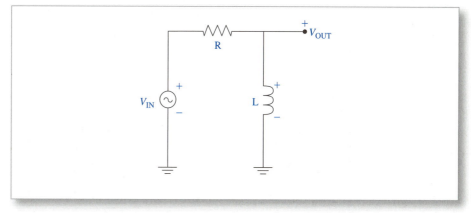

FIGURE **6–32**

FIGURE **6–32**

Prototype
R-L High-Pass
Filter

At the cutoff frequency:

$$f_C = \frac{R}{2\pi L}; \quad Av = 0.707; \quad Av_{(dB)} = -3 \text{ dB}; \quad \theta = 45°$$

The transfer function for an R-L high-pass filter can be expressed as:

$$\mathbf{T(f)} = \frac{\mathbf{X_L}}{\mathbf{Z}}$$

The voltage gain of the R-L high-pass filter can be obtained as the magnitude of the transfer function evaluated at any desired frequency. The phase difference between the input and output signal is determined as the angle of the transfer function.

EXAMPLE 6–7   Analysis of an R-L High-Pass Filter

Given the R-L high-pass filter in Figure 6–33, calculate the cutoff frequency. Determine the voltage gain, the decibel voltage gain, and θ at the cutoff frequency.

Determine the voltage gain, the decibel voltage gain, and θ at 1 kHz.

Verify the dB gain at cutoff and 1 kHz using computer simulation.

The input signal is $10\angle 0°$ V as indicated in Figure 6–33. Determine the amplitude and phase of the output signal at a frequency of 1 kHz.

FIGURE **6–33**

R-L High-Pass
Filter Circuit for
Example 6–7

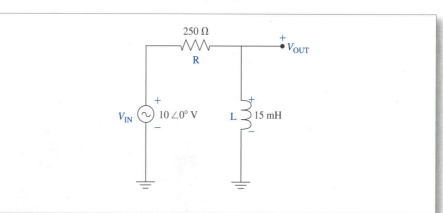

✔ SOLUTION

The cutoff frequency can be calculated as:

$$f_C = \frac{R}{2\pi L} = 2.65 \text{ kHz}$$

At the cutoff frequency:

$$Av = 0.707; \quad Av_{(dB)} = -3 \text{ dB}; \quad \text{and} \quad \theta = 45°$$

The output signal leads the input signal by 45°.

The inductive reactance at 1 kHz can be calculated as:

$$\mathbf{X_L} = 2\pi f L \angle 90° = 2\pi(1 \text{ kHz})(15 \text{ mH}) \angle 90° = 94.25 \angle 90° \text{ } \Omega$$

The impedance at 1 kHz can be calculated as:

$$\mathbf{Z} = \mathbf{R} + \mathbf{X_L} = 250 \angle 0° \text{ } \Omega + 94.25 \angle 90° \text{ } \Omega = 267.2 \angle 20.7° \text{ } \Omega$$

The transfer function can be evaluated at 1 kHz as:

$$\mathbf{T(f)} = \frac{\mathbf{X_L}}{\mathbf{Z}} = \frac{94.25 \angle 90° \text{ } \Omega}{267.2 \angle 20.7° \text{ } \Omega} = 0.353 \angle 69.3°$$

The voltage gain can be determined as the magnitude of the transfer function so that:

$$Av = T(f) = 0.353$$

The decibel voltage gain can be calculated as:

$$Av_{(dB)} = 20 \log 0.353 = -9.04 \text{ dB}$$

The phase difference between the input and output signal is determined as the angle of the transfer function so that:

$$\theta = 69.3°$$

The output signal leads the input signal by 69.3°.

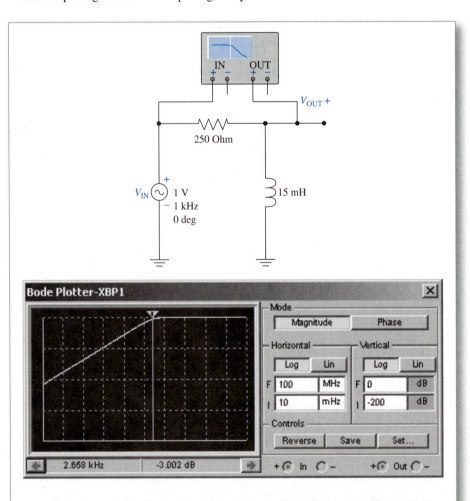

FIGURE **6-35**

MultiSIM Bode Diagram for Example 6–7 at 1 kHz

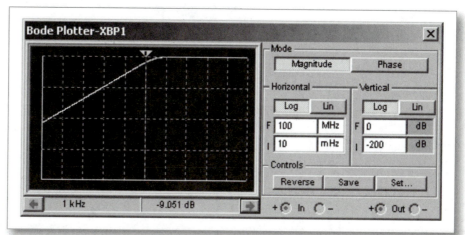

The MultiSIM Bode diagrams in Figure 6–34 and Figure 6–35 verify the calculated gain at the cutoff frequency of 2.65 kHz as -3 dB and at the specified frequency of 1 kHz as -9.04 dB.

PRACTICE EXERCISES 6–5

Given an R-L high-pass filter with $R = 300\ \Omega$ and $L = 10$ mH, determine the cutoff frequency of the filter. Determine the voltage gain, the decibel voltage gain, and θ at the cutoff frequency.

Evaluate the transfer function at a frequency of 500 Hz. Determine the voltage gain, the decibel gain, and θ at 500 Hz.

If the input signal is given as $24\angle 0°$ V, determine the amplitude and phase of the output signal at 500 Hz.

✓ ANSWERS

$f_C = 4.78$ kHz;   at f_C : $Av = 0.707$; $Av_{(dB)} = -3$ dB; $\theta = 45°$;

at 500 Hz: $Av = 0.104$; $Av_{(dB)} = -19.7$ dB; $\theta = 84°$;

$\mathbf{V}_{out} = 2.49\angle 84°$ V

(Complete solutions to the Practice Exercises can be found in Appendix A.)

6.12   THE SERIES RESONANT BAND-PASS FILTER

A series resonant band-pass filter is designed to pass a selected band of frequencies while rejecting or attenuating all others. Consider the band-pass filter circuit in Figure 6–36. The circuit contains *two reactive components* and is therefore referred to as a **second order filter.**

FIGURE **6-36**

Prototype
Series Resonant
Band-Pass Filter

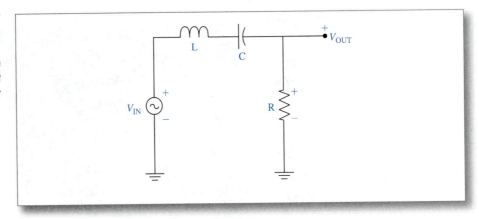

The series resonant band-pass filter can be analyzed using standard circuit analysis techniques to determine the frequency response of the circuit.

The voltage divider rule can be used to obtain an expression for the output voltage as:

$$\mathbf{V}_{out} = \mathbf{V}_{in} \frac{\mathbf{R}}{\mathbf{Z}} \tag{6.24}$$

where:

$$\mathbf{Z} = \mathbf{R} + \mathbf{X}_L + \mathbf{X}_C$$

The transfer function can be expressed as:

$$\mathbf{T(f)} = \frac{\mathbf{V}_{out}}{\mathbf{V}_{in}} = \frac{\mathbf{R}}{\mathbf{Z}} \tag{6.25}$$

The voltage gain of the filter can be determined as the magnitude of the transfer functions so that:

$$Av = Tf$$

The decibel voltage gain can be calculated as:

$$Av_{dB} = 20 \; \log Av$$

The phase difference (θ) between the input and out signal is determined as the angle of the transfer function.

The voltage gain of the series resonant band-pass filter may be expressed as:

$$A_v = \frac{\mathbf{R}}{\mathbf{Z}} = \frac{R}{\sqrt{R^2 + (X_L - X_C)^2}} \tag{6.26}$$

The denominator of the expression in Equation (6.26) represents the magnitude of the circuit impedance.

The resonant frequency of a series resonant band-pass filter is defined as the frequency at which the voltage gain is equal to 1 or 0 dB.

Setting the expression in Equation (6.26) equal to 1, we obtain:

$$\frac{R}{\sqrt{R^2 + (X_L - X_C)^2}} = 1$$

so that:

$$R = \sqrt{R^2 + (X_L - X_C)^2}$$

$$R^2 = R^2 + (X_L - X_C)^2$$

$$(X_L - X_C)^2 = 0$$

so that:

$$X_L = X_C \tag{6.27}$$

Equation (6.27) indicates that the resonant frequency of the series resonant band-pass filter is the frequency at which the inductive reactance and capacitive reactance are equal so that $X_L = X_C$.

Setting X_L equal to X_C, we obtain:

$$2\pi f_R L = \frac{1}{2\pi f_R C}$$

so that:

$$f_R^2 = \frac{1}{4\pi^2 LC} \tag{6.28}$$

and:

$$f_R = \frac{1}{2\pi\sqrt{LC}} \tag{6.29}$$

The expression in Equation (6.29) can be used to calculate the resonant frequency of the series resonant band-pass filter. Note that the resonant frequency depends only upon the values of the reactive components and is independent of the resistor value.

The low cutoff frequency (f_{LC}) and high cutoff frequency (f_{HC}) for the series resonant band-pass filter are defined as the frequencies at which the voltage gain is 0.707 ($\sqrt{2}/2$ exact) and the decibel voltage gain is -3 dB.

The gain of the series resonant band-pass filter will equal 0.707 (-3 dB) when the resistor value in Figure 6–36 is equal to the difference between the inductive and capacitive reactance such that:

$$R = (X_L - X_C) \tag{6.30}$$

Designating the upper cutoff radian frequency as ω_2 and substituting $X_L = \omega_2 L$ and $X_C = 1/\omega_2 C$, Equation (6.30) may be expressed as:

$$R = \omega_2 L - \frac{1}{\omega_2 C} \tag{6.31}$$

Multiplying Equation (6.31) by $\omega_2 C$, we obtain:

$$RC\omega_2 = LC\omega_2^2 - 1$$

so that:

$$\omega_2^2 - \frac{R}{L}\omega_2 - \frac{1}{LC} = 0 \tag{6.32}$$

The solution to Equation (6.32) may be obtained using the *standard quadratic equation as:*

$$\omega_2 = \frac{-b \pm \sqrt{b^2 - 4ac}}{2a}$$

where:

$$a = 1; \quad b = \frac{-R}{L}; \quad \text{and} \quad c = \frac{-1}{LC}$$

so that:

$$\omega_2 = \frac{R}{2L} \pm \frac{1}{2}\sqrt{\left(\frac{R}{L}\right)^2 + \frac{4}{LC}} \quad \text{rad/s} \tag{6.33}$$

Note that if X_C is greater than X_L, Equation (6.30) may be expressed as:

$$R = X_C - X_L$$

Designating the lower cutoff angular frequency as ω_1 and substituting $X_L = \omega_1 L$ and $X_C = 1/\omega_1 C$, we may express Equation (6.30) as:

$$R = \frac{1}{\omega_1 C} - \omega_1 L \tag{6.34}$$

Multiplying Equation (6.34) by $\omega_1 C$, we obtain:

$$RC\omega_1 = 1 - \omega_1^2 LC$$

so that:

$$\omega_1^2 + \frac{R}{L}\omega_1 - \frac{1}{LC} = 0 \qquad (6.35)$$

The solution to Equation (6.35) may be obtained using the standard quadratic equation as:

$$\omega_1 = \frac{-b \pm \sqrt{b^2 - 4ac}}{2a}$$

where:

$$a = 1; \quad b = \frac{R}{L}; \quad \text{and} \quad c = \frac{-1}{LC}$$

$$\omega_1 = \frac{R}{2L} + \frac{1}{2}\sqrt{\left(\frac{R}{L}\right)^2 + \frac{4}{LC}} \text{ rad/s} \qquad (6.36)$$

Subtracting Equation (6.36) from Equation (6.33), the similar parts of the equations whose signs are the same will cancel so that:

$$\omega_2 - \omega_1 = \frac{R}{2L} + \frac{R}{2L} = \frac{R}{L} \text{ rad/s} \qquad (6.37)$$

Substituting $f = \omega/2\pi$, Equation (6.37) may be expressed in terms of frequency as:

$$f_2 - f_1 = \frac{R}{2\pi L} \text{ Hz} \qquad (6.38)$$

The frequencies f_2 and f_1 in Equation (6.38) represent the upper and lower cutoff frequency respectively so that we may express Equation (6.38) as:

$$f_{HC} - f_{LC} = \frac{R}{2\pi L} \text{ Hz} \qquad (6.39)$$

The **bandwidth (BW)** of the series resonant band-pass filter is defined as the difference between the upper and lower cutoff frequencies ($f_{HC} - f_{LC}$) of the filter so that bandwidth may be expressed as:

$$BW = \frac{R}{2\pi L} \text{ Hz} \qquad (6.40)$$

The **quality factor (Q)** of a series resonant band-pass filter is an indication of the selectivity of the filter. A low Q filter will have a relatively broad bandwidth or low selectivity, and a high Q filter will have a relatively narrow bandwidth or high selectivity.

The quality factor Q of a series resonant band-pass filter is defined as the ratio of reactance to resistance at the resonant frequency such that:

$$Q = \frac{X_L}{R} \text{ (definition)} \qquad (6.41)$$

so that:

$$Q = \frac{2\pi L f_R}{R} \qquad (6.42)$$

where f_R is the resonant frequency as determined by Equation (6.29).

Substituting the expression for bandwidth (BW) from Equation (6.40) into Equation (6.42), we obtain:

$$Q = \frac{f_R}{BW} \qquad (6.43)$$

Equation (6.43) may be rearranged to obtain an expression for the filter bandwidth as:

$$BW = \frac{f_R}{Q} \qquad (6.44)$$

The low cutoff frequency of the series resonant band-pass filter is determined as:

$$f_{LC} = f_R - \frac{1}{2}BW \qquad (6.45)$$

The high cutoff frequency is determined as:

$$f_{HC} = f_R + \frac{1}{2}BW \qquad (6.46)$$

The important relationships for the series resonant band-pass filter may be summarized as:

$$f_R = \frac{1}{2\pi\sqrt{LC}}; \quad Q = \frac{X_L}{R} \text{ (calculated at } f_R\text{)}; \quad BW = \frac{f_R}{Q}; \quad f_{LC} = f_R - \frac{1}{2}BW;$$

$$\text{and } f_{HC} = f_R + \frac{1}{2}BW$$

EXAMPLE 6-8 Analysis of a Series Resonant Band-Pass Filter

Given the series resonant second order band-pass filter in Figure 6–37, calculate the resonant frequency, quality factor Q, bandwidth, low cutoff frequency, and high cutoff frequency. Verify the calculations using computer simulation.

FIGURE **6–37**

Series Resonant
Band-Pass Filter
Circuit for
Example 6–8

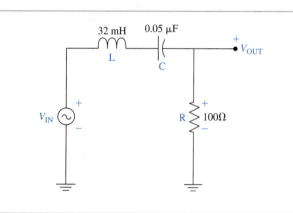

✓ **SOLUTION**

The resonant frequency may be calculated using Equation (6.29) to obtain:

$$f_R = \frac{1}{2\pi\sqrt{LC}} = \frac{1}{2\pi\sqrt{(32 \text{ mH})(0.05 \text{ μF})}} = 3978 \text{ Hz}$$

Note that X_L and X_C can be shown to be equal at the resonant frequency as:

$$X_L = 2\pi f_R L = 2\pi(3978 \text{ Hz})(32 \text{ mH}) = 800 \text{ Ω}$$

$$X_C = \frac{1}{2\pi f_R C} = \frac{1}{2\pi(3978 \text{ Hz})(0.05 \text{ μF})} = 800 \text{ Ω}$$

The quality factor Q can be calculated as the ratio of the inductive reactance at the resonant frequency to the resistance, as indicated in Equation (6.41), to obtain:

$$Q = \frac{X_L}{R} = \frac{800 \text{ Ω}}{100 \text{ Ω}} = 8$$

The bandwidth can be calculated using Equation (6.44) as:

$$BW = \frac{f_R}{Q} = \frac{3978 \text{ Hz}}{8} = 497 \text{ Hz}$$

The low cutoff frequency is determined as:

$$f_{LC} = f_R - \frac{1}{2}BW = 3978 \text{ Hz} - 249 \text{ Hz} = 3729 \text{ Hz}$$

The high cutoff frequency is determined as:

$$f_{HC} = f_R + \frac{1}{2}BW = 3978 \text{ Hz} + 249 \text{ Hz} = 4227 \text{ Hz}$$

The approximate resonant *frequency*, upper cutoff frequency, and lower cutoff frequency are shown in the MultiSIM Bode diagrams in Figure 6–38.

FIGURE **6–38**

MultiSIM Bode Diagram for Example 6–8 Showing f_R, f_{LC}, and f_{HC}

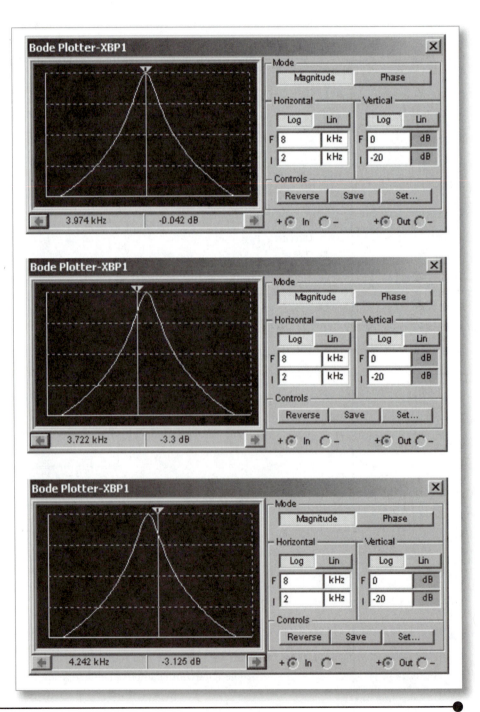

6.13 COMPUTER ANALYSIS OF THE SERIES RESONANT BAND-PASS FILTER

The computer-generated spreadsheet data displayed in Figure 6–39 are obtained by evaluating the transfer function as:

$$T(f) = \frac{R}{Z}$$

for the series resonant band-pass filter at the indicated frequencies.

Frequency Hz	Av	Av$_{(dB)}$	Degrees
100	0.003	−50.05	89.82
200	0.006	−44.01	89.64
500	0.016	−35.94	89.08
800	0.026	−31.64	88.50
1000	0.034	−29.50	88.08
1500	0.055	−25.22	88.85
2000	0.084	−21.54	85.19
2500	0.129	−17.81	82.60
3000	0.213	−13.42	77.68
3500	0.437	− 7.19	64.08
3600	0.529	− 5.53	58.05
3700	0.652	− 3.72	49.33
3800	0.805	− 1.88	36.36
3900	0.952	− 0.42	17.76
4000	0.996	− 0.03	− 4.84
4200	0.756	− 2.43	−40.89
4300	0.627	− 4.06	−51.19
4400	0.527	− 5.56	−58.20
4500	0.452	− 6.90	−63.13
4800	0.314	−10.05	−71.68
4900	0.286	−10.89	−73.41
5000	0.262	−11.64	−86.62
8000	0.082	−21.69	−87.95
10000	0.059	−24.58	−88.83

The analysis data in Figure 6–39 indicate that the resonant frequency for the band-pass filter occurs at a frequency near 4000 Hz where the $Av_{(dB)}$ is approximately equal to zero. The exact resonant frequency calculated in Example 6–8 was 3978 Hz.

The approximate value of the low cutoff frequency (−3dB) is determined from the data in Figure 6–39 as 3700 Hz (calculated in Example 6–8 as f_{LC} = 3729 Hz). The approximate value for the high cutoff frequency (–3dB) is determined from the data in Figure 6–39 as 4200 Hz (calculated in Example 6–8 as f_{HC} = 4227 Hz).

APPLICATION OF A SERIES RESONANT BAND-PASS FILTER

Band-pass filters are commonly used in audio equalizers and speaker crossover networks in home theater systems. The filters direct signals to various speakers depending upon the frequency of the signal.

This application illustrates how a series resonant band-pass filter can be designed to pass a specified range of frequencies.

SPECIFICATIONS: Design a series resonant band-pass filter to have a resonant frequency f_R of 4 kHz and a bandwidth of 500 Hz.

✓ SOLUTION

The prototype series resonant band-pass filter circuit is repeated in Figure 6–40.

FIGURE **6–40**

Prototype
Series Resonant
Band-Pass Filter

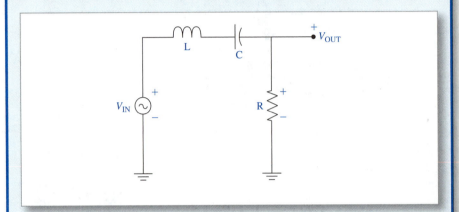

The resonant frequency of the series resonant band-pass filter is expressed as:

$$f_R = \frac{1}{2\pi\sqrt{LC}} \qquad (6.47)$$

Given the resonant frequency of 4 kHz, the value of the capacitor may be arbitrarily selected to be $C = 0.022$ µF. The value of C was selected to be 0.022 µF because this is a standard value for commercially available capacitors.

Equation (6.46) may be rearranged to obtain an expression to calculate L as:

$$f_R = \frac{1}{2\pi\sqrt{LC}}$$

$$f_R{}^2 = \frac{1}{(2\pi)^2\,LC}$$

Solving for the value of L:

$$L = \frac{1}{(2\pi)^2\,(C)(f_R{}^2)} = 72 \text{ mH}$$

The reactance of the capacitor at the resonant frequency may be calculated as:

$$X_C = \frac{1}{2\pi f_R C} = \frac{1}{2\pi(4\text{ kHz})(0.022\text{ µF})} = 1808\ \Omega \text{ (approximate)}$$

The reactance of the inductor at the resonant frequency may be calculated as:

$$X_L = 2\pi f_R L = 2\pi(4\text{ kHz})(72\text{ mH}) = 1808\ \Omega \text{ (approximate)}$$

The value of Q can be calculated as:

$$Q = \frac{f_R}{Bw} = \frac{4\ kHz}{500\ Hz} = 8$$

The value of the resistor R in the prototype may be calculated using the definition of Q as:

$$Q = \frac{X_L}{R}$$

so that:

$$R = \frac{X_L}{Q} = \frac{1808\ \Omega}{8} = 226\ \Omega$$

The design can be verified by substituting the calculated values into appropriate equations as:

$$f_R = \frac{1}{2\pi\sqrt{LC}} = \frac{1}{2\pi\sqrt{(72\ mH)(0.022\ \mu F)}} = 4\ kHz$$

$$Q = \frac{X_L}{R} = \frac{1808\ \Omega}{226\ \Omega} = 8$$

The bandwidth specification (500 Hz) may be verified as:

$$BW = \frac{f_R}{Q} = \frac{4\ kHz}{8} = 500\ Hz$$

The MultiSIM Bode diagrams for the design are shown in Figure 6–41.

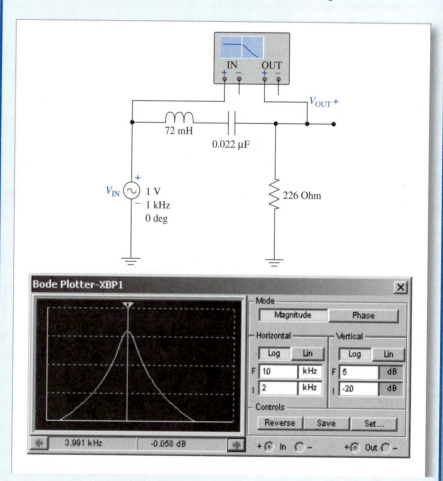

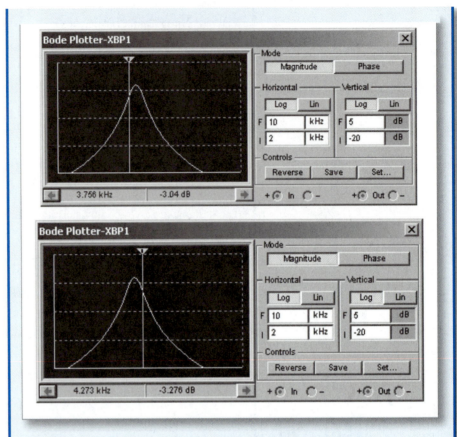

The Bode diagrams in Figure 6–41 indicate that the maximum gain for the circuit (0 dB) occurs at the frequency of 4 kHz (approximate), which is the specified resonant frequency of the filter. The Bode diagrams also indicate that the lower cutoff frequency (−3 dB) occurs at 3.75 kHz (approximate) and the upper cutoff frequency (−3 dB) occurs at 4.25 kHz (approximate). The total bandwidth is calculated as the difference between the lower and upper cutoff frequency or 4.25 kHz −3.75 kHz = 500 Hz, as required by the design specifications.

PRACTICE EXERCISES 6–6

Given a series resonant band-pass filter with $L = 25$ mH, $C = 0.03$ μF, and $R = 150$ Ω, determine the resonant frequency f_R, the quality factor Q, bandwidth BW, and the high and low cutoff frequency.

✓ ANSWERS

$f_R = 5.81$ kHz; $Q = 6.1$; $BW = 953$ Hz; $f_{LC} = 5.33$ kHz; $f_{HC} = 6.29$ kHz

(Complete solutions to the Practice Exercise can be found in Appendix A.)

6.14 A THREE-WAY CROSSOVER NETWORK

Most modern home theater systems typically include at least three separate speaker designs. Each speaker is designed for optimal response for a particular frequency range.

The woofer is designed for base response, the midrange speaker is designed for the middle range of frequencies, and the tweeter is designed to respond to high-frequency signals. The filter circuits that direct the proper frequencies to the appropriate speaker are referred to as a crossover network.

Signals whose frequency is less than 500 Hz are typically routed to the woofer, signals whose frequencies are between 500 Hz and 2 kHz are typically routed to the midrange speaker, and signals whose frequencies are greater than 2 kHz are typically routed to the tweeter.

Consider the three-way speaker system in Figure 6–42.

FIGURE **6–42**

A Three-Way Crossover
Speaker Network

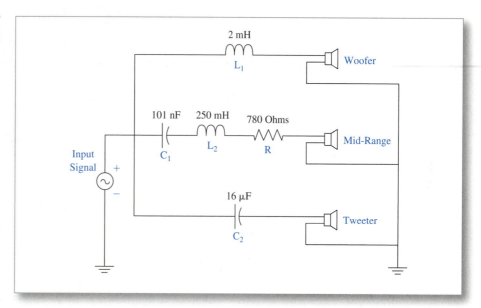

The resistance of each speaker is typically 8 Ω. The inductor L_1 and the 8 Ω resistance of the woofer form an R-L low-pass filter designed to pass low-frequency signals. The inductor L_2, capacitor C_1, resistor R_1, and the resistance of the midfrequency speaker form a series resonant band-pass filter designed to pass middle-range frequencies. The capacitor C_2 and the 8 Ω resistance of the high-frequency speaker form an R-C high-pass filter designed to pass high-frequency signals.

The frequency range for each of the filters may be determined as follows:

The inductor L_1 and the resistance of the woofer (8 Ω) form an R-L low-pass filter whose cutoff frequency may be calculated as:

$$f_C = \frac{R}{2\pi L_1} = \frac{8\ \Omega}{2\pi(2\ \text{mH})} = 637\ \text{Hz}$$

The capacitor C_1, inductor L_2, resistor R_1, and the resistance of the midfrequency speaker (8 Ω) form a series resonant band-pass filter whose resonant frequency may be calculated as:

$$f_1 = \frac{1}{2\pi\sqrt{C_1 L_2}} = \frac{1}{2\pi\sqrt{(101\ \text{nF})(250\ \text{mH})}} = 1\ \text{kHz}$$

The total equivalent resistance R_T for the band-pass filter is the sum of R_1 (780 Ω) and the resistance of the midrange speaker (8 Ω) so that the quality factor Q may be calculated as:

$$R_T = R_1 + 8\ \Omega = 780\ \Omega + 8\ \Omega = 788\ \Omega$$

$$Q = \frac{X_L}{R} = \frac{2\pi f_R L_2}{R_T} = \frac{1570\ \Omega}{788\ \Omega} = 1.99$$

The bandwidth may be calculated as:

$$BW = \frac{f_R}{Q} = \frac{1\ \text{kHz}}{1.99} = 503\ \text{Hz}$$

The low cutoff for the midrange speaker may be calculated as:

$$f_{LC} = f_R - \frac{1}{2}BW = 1\ \text{kHz} - 251.5\ \text{Hz} = 748.5\ \text{Hz}$$

The high cutoff frequency for the midrange speaker is:

$$f_{LC} = f_R + \frac{1}{2}BW = 1\ \text{kHz} + 251.5\ \text{Hz} = 1251.5\ \text{Hz}$$

The capacitor C_2 and the resistance of the high-frequency speaker form an R-C high-pass filter whose cutoff frequency may be calculated as:

$$f_C = \frac{1}{2\pi RC_2} = \frac{1}{2\pi(8\ \Omega)(16\ \mu F)} = 1198\ \text{Hz}$$

In summary, signal frequencies below 503 Hz will be routed to the low-frequency speaker, signal frequencies between 748.5 Hz and 1251.5 Hz will be routed to the midfrequency speaker, and signal frequencies above 1198 Hz will be routed to the high-frequency speaker.

6.15 THE PARALLEL RESONANT BAND-PASS FILTER

The equations that are used to solve a parallel AC circuit consisting of a resistor, capacitor, and inductor are similar to the equations that are used to solve a series AC circuit consisting of the same three components. However, certain interchanges must be made when comparing the analysis of the series and parallel circuits. For example, KVL is used in the analysis of the series circuit, while KCL is used in the analysis of the parallel circuit. In addition, a voltage source is typically considered as the input signal source for the series circuit, while a current source is often considered as the input signal source for the parallel circuit. Further interchanges include exchanging R, L, and C in equations for a series resonant circuit analysis with $1/R$, $1/C$, and $1/L$ respectively in the equations for a parallel circuit analysis.

Therefore, if the equations for the analysis of the series R-L-C circuit at resonance are known, the equivalent equations for the parallel R-L-C circuit at resonance can be determined by making the appropriate interchanges.

The series R-L-C circuit and the parallel R-L-C circuit are referred to as **dual** circuits, and the **principle of duality** can be used to convert equations for the series resonant circuit to equivalent equations for the parallel resonant circuit.

Consider the parallel circuit in Figure 6–43.

FIGURE **6–43**

Prototype
Parallel Resonant
Band-Pass Filter

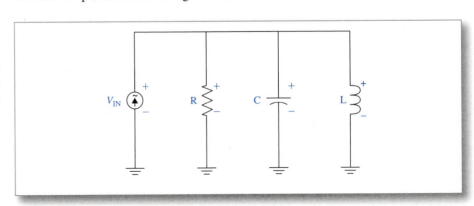

Note that the input signal source is a current source in keeping with the principal of duality.

The definition for the resonant condition is the same for the series or parallel resonant circuit. The resonant frequency is the frequency at which the inductive reactance and capacitive reactance are equal so that:

$$f_R = \frac{1}{2\pi\sqrt{LC}} \tag{6.48}$$

Selected equations that were derived for the series resonant band-pass filter are repeated in the following table for convenience.

The equivalent equations for the parallel resonant band-pass filter have been derived using appropriate substitutions as prescribed by the principle of duality. The symbols for R, L, and C in the equations for series resonance have been replaced by $1/R$, $1/C$, and $1/L$ respectively to obtain the equations for parallel resonance.

Equations for Series R-L-C Resonance		Equations for Parallel R-L-C Resonance	
$\omega_R = \dfrac{1}{\sqrt{LC}}; \quad f_R = \dfrac{1}{2\pi\sqrt{LC}}$	(6.29)	$\omega_R = \dfrac{1}{\sqrt{LC}}; \quad f_R = \dfrac{1}{2\pi\sqrt{LC}}$	(6.49)
$\omega_2 = \dfrac{R}{2L} \pm \dfrac{1}{2}\sqrt{\left(\dfrac{R}{L}\right)^2 + \dfrac{4}{LC}}$	(6.33)	$\omega_2 = \dfrac{1}{2RC} \pm \dfrac{1}{2}\sqrt{\left(\dfrac{1}{RC}\right)^2 + \dfrac{4}{LC}}$	(6.50)
$\omega_1 = \dfrac{R}{2L} \pm \dfrac{1}{2}\sqrt{\left(\dfrac{R}{L}\right)^2 + \dfrac{4}{LC}}$	(6.34)	$\omega_1 = \dfrac{1}{2RC} \pm \dfrac{1}{2}\sqrt{\left(\dfrac{1}{RC}\right)^2 + \dfrac{4}{LC}}$	(6.51)
$Q = \dfrac{X_L}{R} = \dfrac{X_C}{R}$	(6.41)	$Q = \dfrac{R}{X_L} = \dfrac{R}{X_C}$	(6.52)
$BW = \dfrac{f_R}{Q}$	(6.44)	$BW = \dfrac{f_R}{Q}$	(6.53)
$f_{LC} = f_R - \dfrac{1}{2}BW$	(6.45)	$f_{LC} = f_R - \dfrac{1}{2}BW$	(6.54)
$f_{HC} = f_R + \dfrac{1}{2}BW$	(6.46)	$f_{HC} = f_R + \dfrac{1}{2}BW$	(6.55)

EXAMPLE 6-9   Analysis of a Parallel Resonant Band-Pass Filter

Given the parallel circuit in Figure 6–44, determine the resonant frequency, quality factor, bandwidth, and upper and lower cutoff frequency.

FIGURE **6-44**

Parallel Resonant Band-Pass Filter for Example 6–9

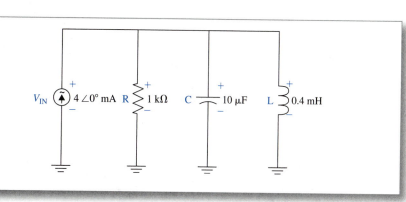

✓ SOLUTION

The resonant frequency can be calculated as:

$$f_R = \frac{1}{2\pi\sqrt{LC}} = \frac{1}{2\pi\sqrt{(0.4 \text{ mH})(10 \text{ }\mu\text{F})}} = 2.52 \text{ kHz}$$

The quality factor Q can be calculated as:

$$Q = \frac{R}{X_L} = \frac{1 \text{ k}\Omega}{(2\pi)(2.52 \text{ kHz})(0.4 \text{ mH})} = 158$$

The bandwidth can be calculated as:

$$BW = \frac{f_R}{Q} = \frac{2.52 \text{ kHz}}{158} = 16 \text{ Hz}$$

The approximate upper cutoff frequency may be determined as:

$$f_{HC} = f_R + \frac{1}{2}BW = 2.52 \text{ kHz} + 8 \text{ Hz} = 2.53 \text{ kHz}$$

The approximate lower cutoff frequency may be determined as:

$$f_{LC} = f_R - \frac{1}{2}BW = 2.52 \text{ kHz} - 8 \text{ Hz} = 2.51 \text{ kHz}$$

Note the narrow bandwidth of the filter in this example. A narrow bandwidth indicates that the filter is highly selective.

The parallel resonant band-pass filter can be designed to have extremely large values for Q, which results in a highly selective filter. The resonant frequency of the parallel circuit in Example 6–9 is shown in the MultiSIM Bode diagram shown in Figure 6–45.

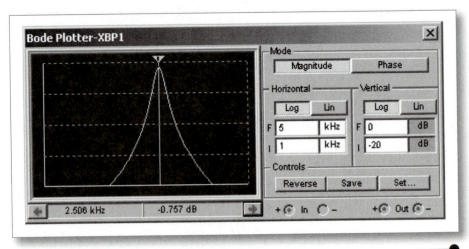

FIGURE **6–45**

MultiSIM Bode Diagram Showing Resonant Frequency for Example 6–9

An inductor is constructed by winding a length of copper wire into coils around an iron core. The copper wire has some small resistance that is typically negligible when analyzing the series resonant band-pass filter. However, small resistance in a parallel circuit often cannot be considered to be negligible. The circuit in Figure 6–46 is referred to as a tank circuit and is an important circuit in electronic communications systems.

Note that the inductor in Figure 6–46 is represented as an inductance in series with its resistance R_{LS}. The circuit in Figure 6–46 is actually a series-parallel circuit consisting of two branches, X_C and X_L in series with R_{LS}.

FIGURE **6-46**

Parallel Circuit Showing
the Resistance of
the Inductor R_{LS}

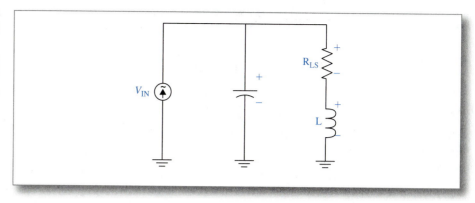

The admittance of the circuit in Figure 6–46 may be expressed in rectangular form:

$$Y = \frac{1}{R_{LS} + j\omega L} + j\omega C \tag{6.56}$$

Multiplying the denominator of the first term of Equation (6.54) by its complex conjugate, we obtain:

$$Y = \frac{1}{R_{LS} + j\omega L}\left(\frac{R_{LS} - j\omega L}{R_{LS} - j\omega L}\right) + j\omega C = \frac{R_{LS} - j\omega L}{R_{LS}^2 + \omega^2 L^2} + j\omega C \tag{6.57}$$

The first term of Equation (6.57) can be separated into its real part and imaginary part so that Equation (6.57) may be expressed as:

$$Y = \frac{R_{LS}}{R_{LS}^2 + \omega^2 L^2} - j\frac{\omega L}{R_{LS}^2 + \omega^2 L^2} + j\omega C \tag{6.58}$$

At resonance, the imaginary part of Equation (6.58) must be equal to zero so that:

$$\omega_R C = \frac{\omega_R L}{R_{LS}^2 + \omega_R^2 L^2} \tag{6.59}$$

Equation (6.59) can be divided by ω_R to obtain:

$$C = \frac{L}{R_{LS}^2 + \omega_R^2 L^2} \tag{6.60}$$

Using appropriate algebraic manipulation, Equation (6.60) can be solved for ω_R to obtain:

$$\omega_R = \sqrt{\frac{1}{LC} - \left(\frac{R_{LS}}{L}\right)^2} \tag{6.61}$$

Factoring $\dfrac{1}{\sqrt{LC}}$ from Equation (6.61), we obtain:

$$\omega_R = \frac{1}{\sqrt{LC}}\sqrt{1 - \frac{R_{LS}^2 C}{L}} \quad \text{rad/s} \tag{6.62}$$

$$f_R = \frac{1}{2\pi\sqrt{LC}}\sqrt{1 - \frac{R_{LS}^2 C}{L}} \quad \text{Hz} \tag{6.63}$$

Equation (6.63) indicates that the resonant frequency of the parallel resonant band-pass filter when the resistance of the inductor is taken into account is less the resonant frequency when the inductor resistance is neglected.

EXAMPLE 6-10   The Effect of Inductor Resistance on the Parallel Resonant Frequency

Given the circuit in Figure 6–47, determine the resonant frequency neglecting the effects of the inductor resistance R_{LS}. Determine the resonant frequency taking into account the inductor resistance R_{LS}.

FIGURE **6-47**

Circuit for
Example 6–10

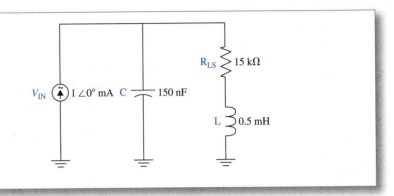

✔ **SOLUTION**

The resonant frequency, neglecting the effects of the inductor resistance R_{LS}, may be calculated as:

$$f_R = \frac{1}{2\pi\sqrt{LC}} = \frac{1}{2\pi\sqrt{(0.5 \text{ mH})(150 \text{ nF})}} = 18.4 \text{ kHz}$$

The resonant frequency, taking into account the effects of the inductor resistance R_{LS}, may be calculated as:

$$f_R = \frac{1}{2\pi\sqrt{LC}}\sqrt{1 - \frac{R_{LS}^2 C}{L}} = 18.4 \text{ kHz}\sqrt{1 - \frac{(15 \text{ }\Omega^2)(150 \text{ nF})}{0.5 \text{ mH}}} = 17.8 \text{ kHz}$$

Commercial radio and television stations are assigned a particular band of frequencies by the Federal Communications Commission. Parallel band-pass filters are often used to ensure that the signals broadcast from each station contain only frequencies within the specified bandwidth. Television and radio receivers use band-pass filters called tuned circuits to select only the band of frequencies transmitted from a desired station.

It is noteworthy that the impedance of a series resonant band-pass filter is at its minimum value at the resonant frequency. It follows that the circuit current is maximum at the resonant frequency. By contrast, the impedance of a parallel resonant band-pass filter is at its maximum value at the resonant frequency. It follows that the circuit current is minimum at the resonant frequency. The high-impedance characteristic of the parallel resonant band-pass filter is typically found in signal processing applications where minimum load current is desired.

S U M M A R Y

Electronic circuits are often viewed from a **signal processing perspective.** The primary consideration of the signal processing perspective is to determine the output signal of a circuit of a given input.

The **transfer function** of a circuit is defined as the ratio of the output signal to the input signal: $T(f) = V_{out}/V_{in}$. The transfer function is a function of frequency and is specified by a magnitude and angle.

Voltage gain is defined as the ratio of the amplitude of the output signal voltage to the amplitude of the input signal voltage: $Av = V_{out}/V_{in}$. The voltage gain is determined as the magnitude of the transfer function.

The phase difference between the input signal and output signal is determined as the angle of the transfer function.

The **decibel voltage gain** is defined as $Av_{(dB)} = 20 \log Av$.

An **attenuator** circuit is designed to reduce the amplitude of an input signal to a desired level without changing the shape of the input signal.

Filters are electronic signal processing circuits that are designed to pass signals in a particular frequency range and to attenuate all other signals. The most common filters are low-pass, high-pass, and band-pass filters.

The **cutoff frequency** (f_C) of an R-C or R-L low-pass or high-pass filter is defined as the frequency at which the voltage gain is 0.707 and the decibel voltage gain is -3dB.

A **low-pass filter** is designed to attenuate frequencies above the cutoff frequency f_C and to pass signal frequencies below the cutoff frequency to the output.

A **high-pass filter** is designed to attenuate frequencies below the cutoff frequency f_C and to pass signal frequencies above the cutoff frequency to the output.

A **band-pass filter** is designed to pass signal frequencies that lie between the low and high cutoff frequencies f_{LC} and f_{HC}. The band-pass filter may be a series resonant or a parallel resonant circuit.

The gain and phase difference between the input and output signal of a filter is determined by evaluating the circuit transfer function at any desired frequency. The magnitude of the transfer function is the voltage gain Av, and the phase difference between the input and output signal is the angle of the transfer function.

ESSENTIAL LEARNING EXERCISES

(ANSWERS TO ODD-NUMBERED ESSENTIAL LEARNING EXERCISES CAN BE FOUND IN APPENDIX A.)

1. Given the voltage attenuator circuit in Figure 6–48, write an expression for the transfer function.

FIGURE **6–48**

Circuit for
Learning Exercise 1

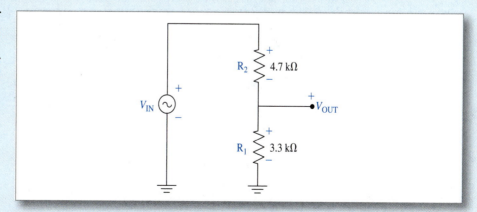

2. Evaluate the transfer function for the attenuator circuit in Figure 6–48 at the frequency of 1 kHz. Determine the voltage gain Av, the decibel voltage gain $Av_{(dB)}$, and the phase difference (θ) between the input and output signal at 1 kHz.

3. Determine the amplitude (peak value) and frequency of the output signal in Figure 6–48 if the input signal is given as $v = 24 \sin(500t)$ V.

4. A voltage attenuator circuit similar to the circuit in Figure 6–48 is designed with $R_1 = 6.1$ kΩ and $R_2 = 2.2$ kΩ with the output taken across the 6.1 kΩ resistor. Write an expression for the transfer function of the circuit. Determine the amplitude and frequency of the output signal if the input signal is given as $v = 24 \sin(6000t)$ V.

5. a. Determine the cutoff frequency (f_C) for the R-C filter in Figure 6–49. Determine the value of Av, $Av_{(dB)}$, and θ at the cutoff frequency. Verify the value of f_C, Av, $Av_{(dB)}$, and θ using computer simulation.

FIGURE 6–49

Circuit for Essential Learning Exercise 5

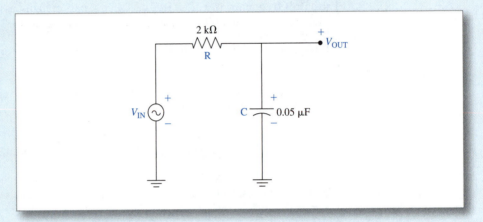

b. Determine the amplitude of the output signal at the cutoff frequency if the input signal is given as $v = 6 \sin(\omega t)$ V.

c. Write an expression for the transfer function of the circuit.

d. Evaluate the transfer function at the frequency of 3 kHz. Determine the value of Av, $Av_{(dB)}$, and θ at 3 kHz.

6. a. An R-C low-pass filter consists of a 4 kΩ resistor and a 0.033 μF capacitor. Determine the cutoff frequency for the filter. Determine the value of Av, $Av_{(dB)}$, and θ at the cutoff frequency. Verify the value of f_C, Av, $Av_{(dB)}$, and θ using computer simulation.

b. Determine the amplitude of the output signal at the cutoff frequency if the input signal is given as $v = 10 \sin(\omega t)$ V.

c. Write an expression for the transfer function of the circuit.

d. Evaluate the transfer function at the frequency of 6 kHz. Determine the value of Av, $Av_{(dB)}$, and θ at 6 kHz.

7. a. Determine the cutoff frequency (f_C) for the R-C filter in Figure 6–50. Determine the value of Av, $Av_{(dB)}$, and θ at the cutoff frequency. Verify the value of f_C, Av, $Av_{(dB)}$, and θ using computer simulation.

FIGURE **6–50**

Circuit for
Essential Learning
Exercise 7

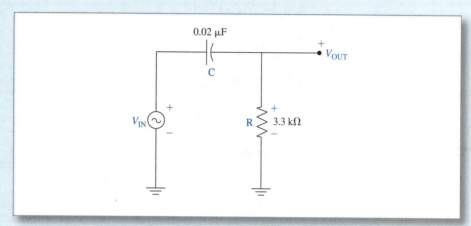

0.02 µF

C

V_{OUT}

V_{IN}

R   3.3 kΩ

b. Determine the amplitude of the output signal at the cutoff frequency if the input signal is given as $v = 4 \sin(\omega t)$ V.

c. Write an expression for the transfer function of the circuit.

d. Evaluate the transfer function at the frequency of 10 kHz. Determine the value of Av, $Av_{(dB)}$, and θ at 10 kHz.

8. a. An R-C high-pass filter consists of a 10 kΩ resistor and a 0.01 µF capacitor. Determine the cutoff frequency for the filter. Determine the value of Av, $Av_{(dB)}$, and θ at the cutoff frequency. Verify the value of f_C, Av, $Av_{(dB)}$, and θ using computer simulation.

b. Determine the amplitude of the output signal at the cutoff frequency if the input signal is given as $v = 20 \sin(\omega t)$ V.

c. Write an expression for the transfer function of the circuit.

d. Evaluate the transfer function at the frequency of 6 kHz. Determine the value of Av, $Av_{(dB)}$, and θ at 6 kHz.

9. a. Determine the cutoff frequency for the R-L filter in Figure 6–51. What are the theoretical values of Av, $Av_{(dB)}$, and θ at the cutoff frequency? Verify the value of f_C, Av, $Av_{(dB)}$, and θ using computer simulation.

FIGURE **6–51**

Circuit for
Essential Learning
Exercise 9

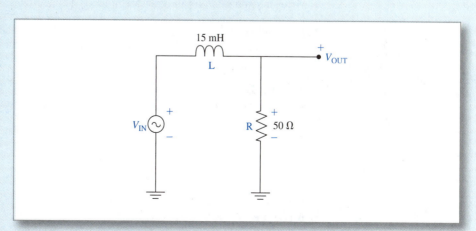

15 mH

L

V_{OUT}

V_{IN}

R   50 Ω

b. Determine the amplitude of the output signal at the cutoff frequency if the input signal is given as $v = 6 \sin(\omega t)$ V.

c. Write an expression for the transfer function of the circuit.

d. Evaluate the transfer function at the frequency of 2 kHz. Determine the value of Av, $Av_{(dB)}$, and θ at 2 kHz.

10. a. An R-L low-pass filter consists of a 100 Ω resistor and a 10 mH inductor. Determine the cutoff frequency for the filter. What are the theoretical values of Av, $Av_{(dB)}$, and θ at the cutoff frequency? Verify the value of f_C, Av, $Av_{(dB)}$, and θ using computer simulation.

b. Determine the amplitude of the output signal at the cutoff frequency if the input signal is given as $v = 10 \sin(\omega t)$ V.

c. Write an expression for the transfer function of the circuit.

d. Evaluate the transfer function at the frequency of 2 kHz. Determine the value of Av, $Av_{(dB)}$, and θ at 2 kHz.

11. a. Determine the cutoff frequency for the R-L filter in Figure 6–52. What are the theoretical values of Av, $Av_{(dB)}$, and θ at the cutoff frequency? Verify the value of f_C, Av, $Av_{(dB)}$, and θ using computer simulation.

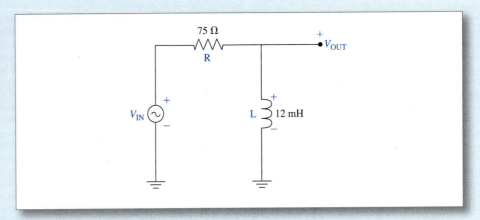

FIGURE **6–52**

Circuit for
Essential Learning
Exercise 11

b. Determine the amplitude of the output signal at the cutoff frequency if the input signal is given as $v = 15 \sin(\omega t)$ V.

c. Write an expression for the transfer function of the circuit.

d. Evaluate the transfer function at the frequency of 3 kHz. Determine the value of Av, $Av_{(dB)}$, and θ at 3 kHz.

12. a. An R-L high-pass filter consists of a 90 Ω resistor and a 10 mH inductor. Determine the cutoff frequency for the filter. What are the theoretical values of Av, $Av_{(dB)}$, and θ at the cutoff frequency? Verify the value of f_C, Av, $Av_{(dB)}$, and θ using computer simulation.

b. Determine the amplitude of the output signal at the cutoff frequency if the input signal is given as $v = 3.5 \sin(\omega t)$ V.

c. Write an expression for the transfer function of the circuit.

d. Evaluate the transfer function at the frequency of 200 Hz. Determine the value of Av, $Av_{(dB)}$, and θ at 200 Hz.

13. a. Determine the resonant frequency of the series resonant band-pass filter in Figure 6–53.

FIGURE **6–53**

Circuit for
Essential Learning
Exercise 13

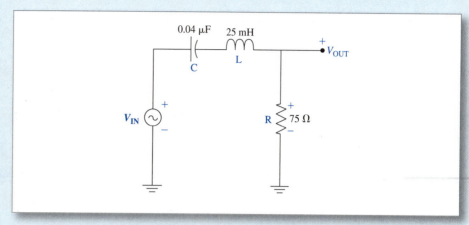

b. Determine the quality factor (Q) for the circuit.

c. Determine the bandwidth of the filter.

d. Determine the upper and lower cutoff frequency for the filter.

e. Sketch a Bode diagram (magnitude only) for the filter.

f. Verify the resonant frequency and the upper and lower cutoff frequency for the filter using computer simulation.

14. a. A series resonant band-pass filter consists of a 0.33 µF capacitor, a 30 mH inductor, and a 70 Ω resistor. Determine the resonant frequency of the filter.

b. Determine the quality factor (Q) for the circuit.

c. Determine the bandwidth of the filter.

d. Determine the upper and lower cutoff frequency for the filter.

e. Sketch a Bode diagram (magnitude only) for the filter.

f. Verify the resonant frequency and the upper and lower cutoff frequency for the filter using computer simulation.

15. a. Determine the resonant frequency of the parallel resonant band-pass filter in Figure 6–54.

FIGURE **6–54**

Circuit for
Essential Learning
Exercise 15

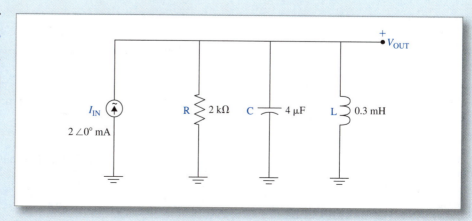

b. Determine the quality factor (Q) for the circuit.

c. Determine the bandwidth of the filter.

d. Determine the upper and lower cutoff frequency for the filter.

16. a. A parallel resonant band-pass filter consists of a 2.2 μF capacitor, a 0.5 mH inductor, and a 6 kΩ resistor. Determine the resonant frequency of the filter.

b. Determine the quality factor (Q) for the circuit.

c. Determine the bandwidth of the filter.

d. Determine the upper and lower cutoff frequency for the filter.

17. Determine the resonant frequency of the parallel circuit in Figure 6–55.

FIGURE **6–55**

Circuit for
Essential Learning
Exercise 17

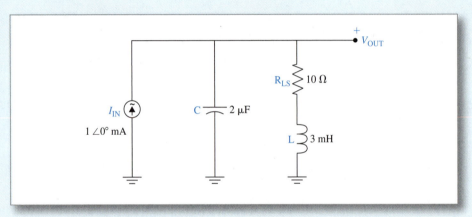

18. Determine the resonant frequency of the parallel resonant circuit in Figure 6–56.

FIGURE **6–56**

Circuit for
Essential Learning
Exercise 18

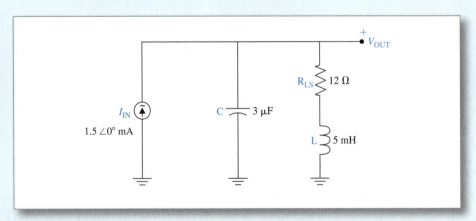

19. The cutoff frequency for an R-C low-pass filter is defined as the frequency at which the capacitive reactance X_C is equal to R. Set X_C equal to R to derive an expression that may be used to calculate the cutoff frequency for the filter.

20. Design an R-C low-pass filter to have a cutoff frequency of 1 kHz. The resistor for the design is specified to be 470 Ω. Verify the design using computer simulation.

21. Design an R-C low-pass filter to have a cutoff frequency of 600 Hz. The capacitor for the design is specified to be 0.22 μF. Verify the design using computer simulation.

22. Design an R-C high-pass filter to have a cutoff frequency of 450 Hz. The resistor for the design is specified to be 2.2 kΩ. Verify the design using computer simulation.

23. Design an R-C high-pass filter to have a cutoff frequency of 800 Hz. The capacitor for the design is specified to be 0.1 μF. Verify the design using computer simulation.

24. Design an R-L low-pass filter to have a cutoff frequency of 400 Hz. The inductor for the design is specified to be 8 mH. Verify the design using computer simulation.

25. Design an R-L low-pass filter to have a cutoff frequency of 300 Hz. The resistor for the design is specified to be 65 Ω. Verify the design using computer simulation.

26. Design an R-L high-pass filter to have a cutoff frequency of 1 kHz. The inductor for the design is specified to be 10 mH. Verify the design using computer simulation.

27. Design an R-L high-pass filter to have a cutoff frequency of 150 Hz. The resistor for the design is specified to be 25 Ω. Verify the design using computer simulation.

CHALLENGING LEARNING EXERCISES

(ANSWERS TO SELECTED CHALLENGING LEARNING EXERCISES CAN BE FOUND IN APPENDIX A.)

1. a. Given the voltage attenuator circuit in Figure 6–57, write an expression for the transfer function if the output is taken from point A to ground.

FIGURE **6–57**

Circuit for
Challenging Learning
Exercise 1

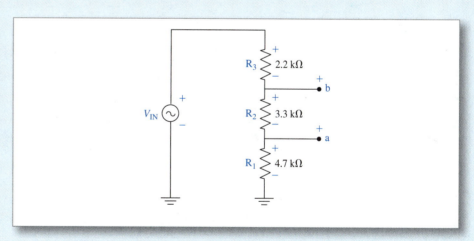

b. Determine the voltage gain and the decibel voltage gain of the circuit if the output is taken from point A to ground.

c. Write an expression for the transfer function if the output is taken from point A to ground.

d. Determine the voltage gain and the decibel voltage gain of the circuit if the output is taken from point B to ground.

2. a. Write an expression for the transfer function for the R-C filter in Figure 6–58.

FIGURE **6–58**

Circuit for
Challenging Learning
Exercise 2

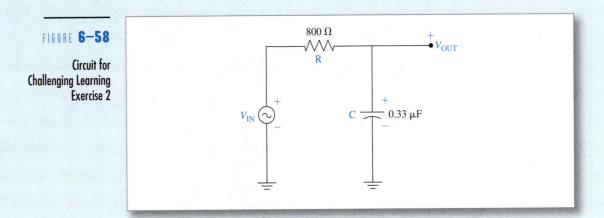

b. Determine the voltage gain and the decibel voltage gain at the frequency of 3 kHz. Note: 3 kHz is not the cutoff frequency.

c. Determine the phase difference between the input and output signal at the frequency of 3 kHz.

3. a. Write an expression for the transfer function for the R-C filter in Figure 6–59.

FIGURE **6–59**

Circuit for
Challenging Learning
Exercise 3

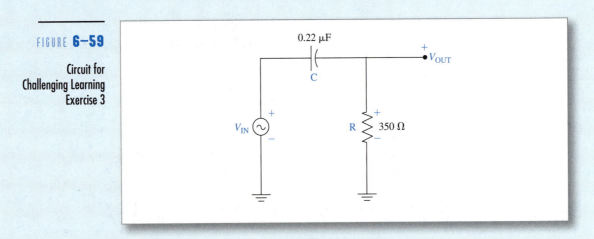

b. Determine the voltage gain and decibel voltage gain at the frequency of 4 kHz. Note: 4 kHz is not the cutoff frequency.

c. Determine the phase difference between the input and output signal at the frequency of 4 kHz.

4. Design an R-C low-pass filter to have a cutoff frequency of 2.5 kHz. Verify the design using computer simulation.

5. Design an R-C high-pass filter to have a cutoff frequency of 2.5 kHz. Verify the design using computer simulation.

6. Design an R-L low-pass filter to have a cutoff frequency of 850 Hz. Verify the design using computer simulation.

7. Design an R-L high-pass filter to have a cutoff frequency of 2.2 kHz. Verify the design using computer simulation.

8. Design a series resonant band-pass filter to have a resonant frequency of 5 kHz and a bandwidth of 500 Hz. Verify the design using computer simulation

9. The cutoff frequency for a series resonant R-L-C band-pass filter is defined as the frequency at which the capacitive reactance X_C is equal to the inductive reactance X_L. Derive an expression for the transfer function of the filter if the output is taken across the resistor R.

TEAM ACTIVITY

1. Use the computer to simulate an R-C low-pass filter consisting of a 500 Ω resistor and a 2 μF capacitor. Select the function generator from the instrument toolbar as the input signal to the filter. Connect the input terminal of the filter to channel one of the oscilloscope and the output of the filter to channel two. Expand the oscilloscope to maximum size.

Set the signal generator for a *sinusoidal waveform* at a frequency of 1 kHz as the input to the filter. Observe the input and output signals on the oscilloscope.

Set the signal generator for a *triangular waveform* at a frequency of 1 kHz as the input to the filter. Observe the input and output signals on the oscilloscope.

Set the signal generator for a *square waveform* at a frequency of 1 kHz as the input to the filter. Observe the input and output signals on the oscilloscope.

Discuss among your study group the results of your activity.

2. Design an R-C high-pass filter with a cutoff frequency of 5 kHz. Check your design using computer circuit simulation software to ensure a cutoff frequency (-3dB) of 5 kHz.

Now design an R-C low-pass filter with a cutoff frequency of 7 kHz. Check your design using computer circuit simulation software to ensure a cutoff frequency (-3dB) of 7 kHz.

What kind of filter do you think would be created by connecting both of the filters together? Connect both of the filters together using computer simulation. What do you think the Bode diagram (magnitude) should look like? Make a Bode diagram sketch and compare it to the Bode diagram obtained using computer simulation.

TRANSIENT CIRCUIT ANALYSIS

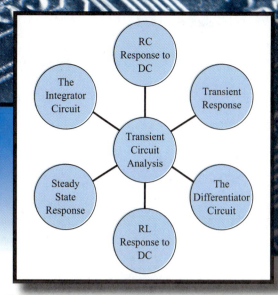

"THAT'S WHAT EDUCATION MEANS-TO DO WHAT
YOU'VE NEVER DONE BEFORE."

—George Herbert Palmer

LEARNING OBJECTIVES

Upon successful completion of this chapter you will be able to:

- Explain the meaning of transient response.
- Explain the meaning of steady-state response.
- Calculate transient voltage and current for an R-C circuit.
- Calculate transient voltage and current for an R-L circuit.
- Calculate the time constant for an R-C or R-L circuit.
- Determine the parameters of a rectangular pulse waveform.
- Determine the output of an integrator circuit.
- Determine the output of a differentiator circuit.
- Determine the response of an R-C or R-L circuit through computer-assisted analysis.

CHAPTER SEVEN

INTRODUCTION

Chapters 5 and 6 in this text are dedicated to the analysis of circuits whose current and voltage signals are sinusoidal waveforms. Many other waveforms are common in electronic circuits although the sine wave is the most important signal to be considered. It is important to remember that the concepts of reactance, impedance, and calculations in the phasor domain are valid only for R-L-C circuits when considering sinusoidal excitation.

This chapter considers the transient and steady-state response of the R-C and R-L circuit to DC excitation. Many practical electronic circuits such as amplifiers contain both DC and AC waveforms so that an appreciation of DC response is necessary to understand the behavior of such circuits.

7.1 STEADY-STATE AND TRANSIENT RESPONSE

It is sometimes necessary to consider signal conditions that exist only for brief moments of time. Such conditions are called *transients,* and the response of the circuit to these signals is called **transient response.**

Transients are often caused by undesired disturbances to the circuit such as noise spikes that may be generated by circuit components or introduced from external sources such as atmospheric disturbances or static electrical discharge. Transients are present only for a short duration. As transient conditions pass, the currents and voltages assume their "normal" values and the response of the circuit is called **steady-state response.**

7.2 THE SWITCHED DC WAVEFORM

As indicated in Figure 7–1, the voltage of the switched DC waveform is either 0 or V_S. To simplify our discussion, we will assume that the transition from 0 to V_S occurs at time $t = 0$ as indicated in Figure 7–1. Following the transition from 0 to V_S at $t = 0$, the waveform remains at V_S for an indefinite time. It should also be noted that the switch in Figure 7–1 is closed at time $t = 0$, as indicated.

FIGURE **7–1**

(a) A Switched DC Voltage
Source (b) A Switched DC
Voltage as a Function of Time

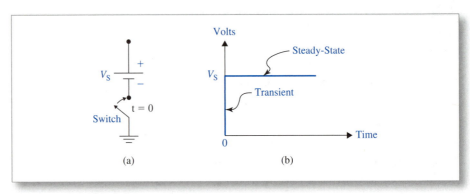

The switched DC waveform in Figure 7–1 consists of two distinct conditions. The transition from 0 to V_S may be considered a transient condition because the transition exists only for a brief moment of time. Following the transient condition, the voltage remains at V_S for an indefinite period of time, which is the steady-state value.

7.3 THE CAPACITOR STORING ENERGY

A capacitor has the ability to store energy in an electric field that resides between the capacitor plates. As the capacitor absorbs energy from the source, the voltage across the capacitor will increase and the capacitor is said to be **charging.** The amount of stored energy is calculated as:

$$E = \tfrac{1}{2} C V_C^{\,2} \tag{7.1}$$

where E is energy in joules, C is capacitance in farads, and V_C is the voltage across the capacitor in volts.

Consider the circuit in Figure 7–2.

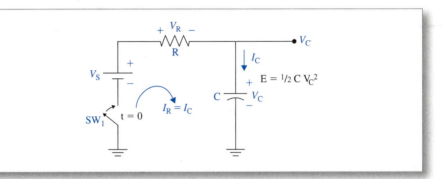

FIGURE 7–2

A Capacitor Storing Energy

Circuit Considerations for Time $t = 0^+$

Assume that the switch in Figure 7–2 is closed at time $t = 0$. The symbol $t = 0^+$ represents the briefest moment of time following time $t = 0$. During this brief period, the switched DC signal transitions from 0 to V_S. The capacitor, however, will not permit sudden changes of voltage across its terminals.

Assuming that the voltage across the capacitor (V_C) is initially 0, the voltage across the capacitor will remain at 0 at time $t = 0^+$.

KVL may be used to determine the voltage across the resistor at time $t = 0^+$ as:

$$V_R = V_S - V_C = V_S - 0 = V_S$$

Ohm's law can be used to determine the circuit current as:

$$I_R = I_C = \frac{V_R}{R} = \frac{V_S}{R}$$

The voltage and current at time $t = 0^+$ can be summarized as:

$$V_C = V_{out} = 0; \quad I_C = I_R = I_S = \frac{V_S}{R}; \quad \text{and} \quad V_R = V_S$$

as indicated in Figure 7–3.

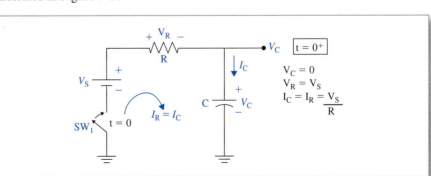

FIGURE 7–3

Voltage and Current at Time $t=0^+$

Circuit Considerations for Time $0^+ < t < t_{ss}$

The symbol t_{ss} represents the time required for the circuit to reach steady-state conditions. As time advances beyond $t = 0^+$, negatively charged free electrons will flow from the upper plate of the capacitor toward the positive terminal of the DC voltage source V_S.

The voltage source must maintain the constant voltage of V_S across its terminals. This voltage is created by an imbalance of electrons between the positive and negative terminals of the source. As electrons from the upper capacitor plate arrive at the positive terminal of the voltage source, an equal number of electrons must leave the negative terminal of the DC source. The number of electrons arriving and leaving must be equal so that the voltage of the DC source will remain constant. The electrons that leave the negative terminal of the DC source are deposited on the lower plate of the capacitor.

As electrons leave the upper plate of the capacitor, the upper plate becomes more positive. As electrons are deposited on the lower plate of the capacitor, the lower plate becomes more negative. Due to the imbalance of electrons (charge) between the capacitor plates, an increasing voltage is developed across the capacitor. The capacitor is said to be *charging* as its voltage increases.

KVL indicates that the sum of V_C and V_R must equal the constant source voltage V_S at all times. It follows that the voltage across the resistor must decrease as the voltage across the capacitor increases. Ohm's law ($I = V_R/R$) indicates that the current must also decrease as the voltage across the resistor decreases.

The voltage and current during the time interval $0^+ < t < t_{ss}$ can be summarized as:

$$V_C \text{ is increasing;} \quad I_C \text{ is decreasing;} \quad V_R \text{ is decreasing}$$

as indicated in Figure 7–4.

FIGURE 7–4

Voltage and Current for the Time Period $0^+ < t < t_{ss}$

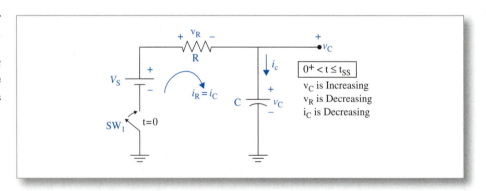

Circuit Considerations for Time $t \geq t_{ss}$ (steady state)

The voltage across the capacitor continues to increase until it reaches the maximum limit of V_S. *KVL* indicates that if the voltage across the capacitor is V_S, then the voltage across the resistor must be 0. Ohm's law indicates that the circuit current becomes 0 as well because $I_C = I_R = V_R/R$, where $V_R = 0$.

The circuit voltages and current for the steady-state condition can be summarized as:

$$V_C = V_S; \quad V_R = 0; \quad I = 0$$

as indicated in Figure 7–5.

FIGURE **7–5**

Steady-State
Voltage and Current

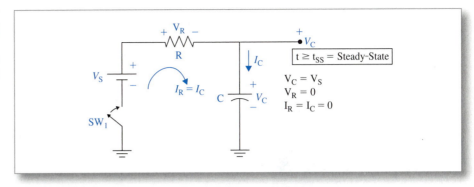

The qualitative analysis of the R-C response to switched DC reveals much about the behavior of the circuit and the nature of the voltage developed across the capacitor. Specifically, the voltage across the capacitor will begin at 0 and increase over time to a limiting value of V_S.

The qualitative analysis does not reveal, however, the manner in which the capacitor voltage transitions from 0 to V_S or how much time is required to reach the steady-state condition at which $V_C = V_S$. The answer to these questions requires a quantitative analysis of the circuit.

Consider the circuit in Figure 7–6.

FIGURE **7–6**

The Mathematical
Model for a Capacitor

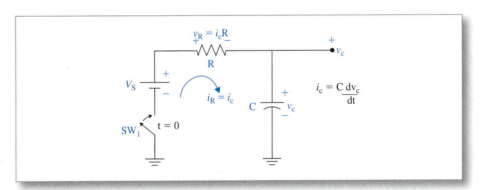

Recall that capacitive current is directly proportional to the capacitance value of the capacitor and to the rate of change of voltage across the capacitor terminals. This statement can be written mathematically as:

$$i_C = C\frac{dv_C}{dt} \tag{7.2}$$

as shown in Figure 7–6.

The voltage across the resistor in Figure 7–6 can be written using the expression in Equation (7.2) as:

$$v_R = iR = RC\frac{dv_C}{dt} \tag{7.3}$$

KVL for the circuit in Figure 7–6 can be written as:

$$V_S = v_R + v_C$$

Substituting the expression in Equation (7.3), we obtain:

$$V_S = RC\frac{dv_C}{dt} + v_C \tag{7.4}$$

Solving Equation (7.4) for $\dfrac{dv_C}{dt}$, we obtain:

$$\frac{dv_C}{dt} = \frac{(V_S - v_C)}{RC} \tag{7.5}$$

MATLAB can be used to solve Equation (7.5) for the capacitor voltage as a function of time to obtain:

EDU ≫ dsolve('DVc=(Vs −Vc)/RC','Vc(0)=0')

ans = Vs−exp(−1/RC*t)*Vs   (7.6)

Equation (7.6) expresses the voltage across the capacitor as a function of time and may be written in the more familiar form as:

$$v_C(t) = V_S(1 - e^{-t/RC}) \tag{7.7}$$

The voltage across the resistor may be determined using KVL as $V_S - v_C$, so that:

$$v_R(t) = V_S - V_S(1 - e^{-t/RC}) = V_S - V_S + V_S e^{-t/RC}$$

so that:
$$v_R(t) = V_S e^{-t/RC} \tag{7.8}$$

Ohm's law can be used to determine the circuit current as a function of time as:

$$i_C(t) = \frac{v_R(t)}{R} e^{-t/RC} = \frac{V_S}{R} e^{-t/RC} \tag{7.9}$$

The voltages and current for the R-C circuit as the capacitor stores energy (charges) can be summarized as:

$$v_C = V_S(1 - e^{-t/RC})$$

$$v_R(t) = V_S e^{-t/RC}$$

$$i_C(t) = \frac{V_S}{R} e^{-t/RC}$$

Plots of the voltage across the capacitor, voltage across the resistor, and capacitor current as functions of time are shown in the MATLAB plot in Figure 7–7.

FIGURE 7–7

Voltage and Current Plots as the Capacitor Stores Energy

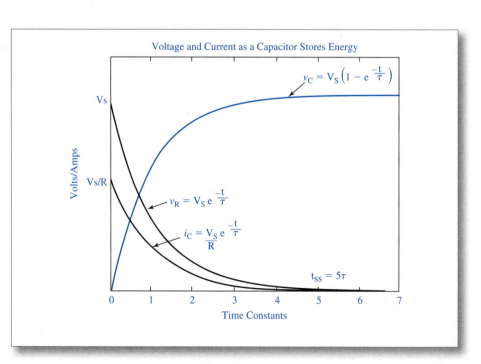

Note that Equation (7.7) indicates that an infinite amount of time would be required for the voltage across the capacitor to reach V_S.

A more practical solution is achieved by defining the quantity RC as one time constant represented by the Greek letter τ.

$$\tau = RC$$

The product RC is expressed in units of time (seconds, milliseconds, etc.).

It is commonly accepted that the steady-state voltage across the capacitor is reached after five time constants of time have passed, so that:

$$t_{SS} = 5\tau$$

This convention is indicated on the time axis in Figure 7–7.

EXAMPLE 7–1   The Charging Capacitor Storing Energy

Given the R-C circuit in Figure 7–8, determine the capacitor voltage and current and the voltage across the resistor for each of the indicated times. Also determine the time required for the capacitor voltage to reach steady-state voltage. Assume that the capacitor is initially uncharged (0 V) and that the switch is closed at time $t = 0$.

FIGURE **7–8**

R-C Circuit
for Example 7–1

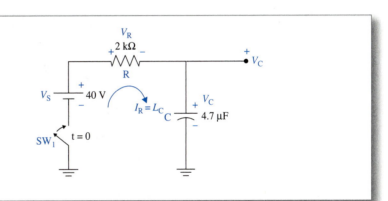

✔ **SOLUTION**

The time constant τ is calculated as:

$$\tau = RC = (2 \text{ k}\Omega)(4.7 \text{ μF}) = 9.4 \text{ ms}$$

At $t = 10$ ms:

The voltage across the capacitor can be calculated as:

$$V_C = 40(1 - e^{-10 \text{ ms}/9.4 \text{ ms}}) = 40(1 - 0.3451) = 26.19 \text{ V}$$

The voltage across the resistor can be calculated as:

$$V_R = V_S \, e^{-10 \text{ ms}/9.4 \text{ ms}} = 40(0.3451) = 13.81 \text{ V}$$

Note that $V_C + V_R = 26.19 \text{ V} + 13.81 \text{ V} = 40 \text{ V}$, as required by KVL.

The capacitor current can be calculated as:

$$I_C = I_R = \frac{V_S}{R} e^{-10 \text{ ms}/9.4 \text{ ms}} = 20(0.727) = 6.90 \text{ mA}$$

Ohm's law can also be used to calculate the capacitor current as:

$$I_C = I_R = \frac{V_R}{R} = \frac{13.81 \text{ V}}{2 \text{ k}\Omega} = 6.90 \text{ mA}$$

At $t = 20$ ms:

The voltage across the capacitor can be calculated as:

$$V_C = 40(1 - e^{-20 \text{ ms}/9.4 \text{ ms}}) = 40(1 - 0.119) = 35.24 \text{ V}$$

The voltage across the resistor can be calculated as:

$$V_R = 40 e^{-20 \text{ ms}/9.4 \text{ ms}} = 40(0.119) = 4.76 \text{ V}$$

Note that $V_C + V_R = 35.24 \text{ V} + 4.76 \text{ V} = 40 \text{ V}$, as required by KVL.

The capacitor current can be calculated as:

$$I_C = \frac{V_S}{R} e^{-20 \text{ ms}/9.4 \text{ ms}} = 20(0.119) = 2.38 \text{ mA}$$

Ohm's law can also be used to calculate the capacitor current as:

$$I_C = I_R = \frac{V_R}{R} = \frac{4.76 \text{ V}}{2 \text{ k}\Omega} = 2.38 \text{ mA}$$

At $t = 30$ ms:

The voltage across the capacitor can be calculated as:

$$V_C = 40(1 - e^{-30 \text{ ms}/9.4 \text{ ms}}) = 40(1 - 0.528) = 38.36 \text{ V}$$

The voltage across the resistor can be calculated as:

$$V_R = 40 e^{-30 \text{ ms}/9.4 \text{ ms}} = 40(0.0411) = 1.64 \text{ V}$$

Note that $V_C + V_R = 38.36 \text{ V} + 1.64 \text{ V} = 40 \text{ V}$, as required by KVL.

The capacitor current can be calculated as:

$$I_C = \frac{V_S}{R} e^{-30 \text{ ms}/9.4 \text{ ms}} = 20(0.0411) \text{ mA} = 0.82 \text{ mA}$$

Ohm's law can also be used to calculate the capacitor current as:

$$I_C = I_R = \frac{V_R}{R} = \frac{1.64 \text{ V}}{2 \text{ k}\Omega} = 0.82 \text{ mA}$$

The time required to reach steady state is calculated as:

$$t_{SS} = 5\tau$$

where: $\qquad\qquad \tau = RC = (2 \text{ k}\Omega)(4.7 \text{ } \mu\text{F}) \text{ ms} = 9.4 \text{ ms}$

so that: $\qquad\qquad t_{SS} = 5(9.4 \text{ ms}) = 47 \text{ ms}$

For engineering considerations, the voltage across the capacitor is assumed to be 40 V at time $t = t_{SS} = 47$ ms.

The actual value of the capacitor voltage at t_{SS} may be calculated as:

$$V_C = V_S (1 - e^{-t/\tau})$$

$$V_C = 40(1 - e^{-47 \text{ ms}/9.4 \text{ms}}) \text{ V}$$

so that: $\qquad\qquad V_C = 40(0.993) = 39.72 \text{ V}$

The small difference between the assumed value (40 V) and actual voltage (39.72 V) is considered negligible for most practical applications.

The MATLAB plot of the capacitor voltage and current and resistor voltage as functions of time for Example 7–1 is shown in Figure 7–9.

FIGURE 7–9

Voltage and
Current
Plots for
Example 7–1

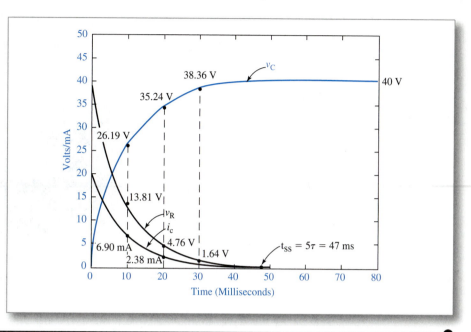

EXAMPLE 7–2   The Charging Capacitor Storing Energy

Given the R-C circuit in Figure 7–10, determine the capacitor voltage at the indicated times. Also determine the time required for the capacitor voltage to reach its steady-state voltage. Assume that the switch is closed at time $t = 0$. Verify the calculated values of capacitor voltage using computer simulation.

FIGURE 7–10

R-C Circuit for
Example 7–2

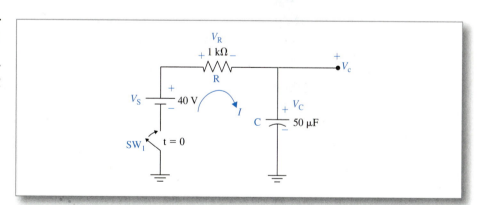

✓ SOLUTION

The time constant for the R-C circuit in Figure 7–10 can be calculated as:

$$\tau = RC = (1\ \text{k}\Omega)(50\ \mu\text{F}) = 50\ \text{ms}$$

At $t = 50$ ms:

The value of the voltage at $t = 50$ ms can be calculated as:

$$V_C = V_S(1 - e^{-t/\tau}) = 40(1 - e^{-50\ \text{ms}/50\ \text{ms}}) = 25.28\ \text{V}$$

At $t = 100$ ms:

The value of the voltage at $t = 100$ ms can be calculated as:

$$V_C = V_S(1 - e^{-t/\tau}) = 40(1 - e^{-100\ \text{ms}/50\ \text{ms}}) = 34.59\ \text{V}$$

The MultiSIM simulation of the voltage across the capacitor is shown in Figure 7–11.

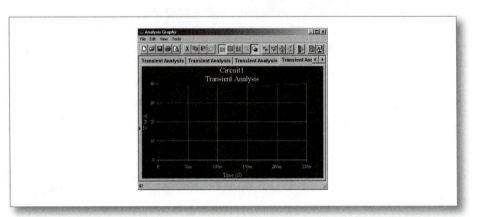

FIGURE **7–11**

MultiSIM 7
Simulation of
V_C for
Example 7–2

The plot in Figure 7–11 indicates that V_C at $t = 50$ ms is approximately 25 V and at $t = 100$ ms, V_C is approximately 34.5 V.

The MATLAB plot of the voltage across the capacitor in Example 7–2 is shown in Figure 7–12.

FIGURE **7–12**

MATLAB Plot of
V_C for
Example 7–2

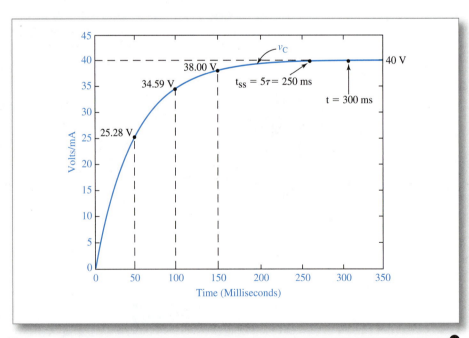

PRACTICE EXERCISES 7–1

    a. Determine the voltage across the capacitor in Figure 7–13 at times $t = 3.5$ ms, 6 ms, and 10 ms. How much time is required to reach steady-state conditions?

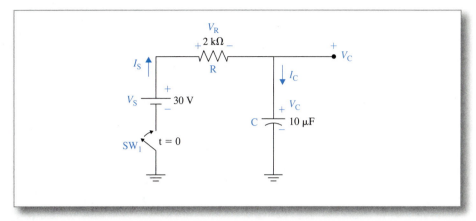

FIGURE 7-13

R-C Circuit for
Practice Exercises 7-1

b. Calculate the voltage across the resistor in Figure 7–13 at time $t = 3.5$ ms, 6 ms, and 10 ms. Add the voltage across the capacitor from exercises A and the voltage across the resistor at each of the specified times. Is KVL satisfied?

✓ANSWERS

a. At 3.5 ms: $V_C = 4.8$ V;   At 6 ms: $V_C = 7.8$ V;
At 10 ms: $V_C = 11.8$ V; $t_{ss} = 5\tau = 100$ ms
b. At 3.5 ms: $V_R = 25.2$ V;   At 6 ms: $V_R = 22.2$ V;
At 10 ms: $V_R = 18.2$ V;

KVL at 3.5 ms: 4.8 V + 25.2 V = 30 V;

KVL at 6.0 ms: 7.8 V + 22.2 V = 30 V;

KVL at 10 ms: 11.8 V + 18.2 V = 30 V

(Complete solutions to the Practice Exercises can be found in Appendix A.)

7.4 THE CAPACITOR RELEASING ENERGY

As noted in the previous section, a charging capacitor absorbs energy from a source that is temporarily stored in an electric field between the capacitor plates. As the capacitor releases or returns energy back to the circuit, the voltage across the capacitor will decrease and the capacitor is said to be **discharging.**

Consider the circuit in Figure 7–14.

FIGURE 7-14

Voltage and Current
at Time $t=0^+$

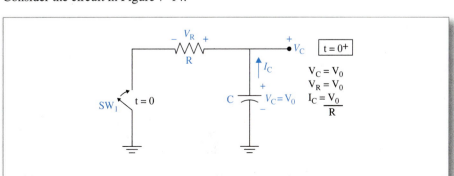

The voltage across the charged capacitor at time $t = 0$ in Figure 7–14 is V_0, as indicated. The voltage is created by energy stored in an electric field that exists between the capacitor plates. When the switch in the circuit is closed, the energy stored by the capacitor is returned to the circuit as the capacitor discharges. Note that there is *no voltage source* in Figure 7–14. Current flow in the circuit is produced only by the discharging capacitor returning stored energy to the circuit.

Circuit Considerations for Time $t = 0^+$

KVL indicates that the capacitor and resistor voltage in Figure 7–14 must have equal magnitudes and opposite polarity, so that:

$$V_C = -V_R = V_0$$

Ohm's law may be used to calculate the current as:

$$I_C = I_R = \frac{V_R}{R} = \frac{V_0}{R}$$

The voltage and current at time $t = 0^+$ can be summarized as:

$$V_C = -V_R = V_0; \quad I_C = \frac{V_0}{R}$$

as indicated in Figure 7–14.

Circuit Considerations for Time $0^+ < t < t_{ss}$

As time continues to pass, the capacitor voltage will decrease as the stored energy is returned to the circuit. The capacitor current may be expressed as a function of time using the mathematical model of the capacitor as:

$$i_C = C \frac{dv_C}{dt}$$

The voltage across the resistor may be expressed as:

$$v_R = i_C R = RC \frac{dv_C}{dt} = v_C$$

Solving for $\dfrac{dv_C}{dt}$, we obtain:

$$\frac{dv_C}{dt} = \frac{v_C}{RC} \tag{7.10}$$

MATLAB may be used to obtain the solution to Equation (7.10) as:

EDU ≫ dsolve('DVc=−Vc/RC','Vc(0)=Vc(0)')

ans = Vc(t)=Vc(0)*exp(−1/RC*t)

The MATLAB solution may be written in the more familiar form as:

$$v_C = V_0 e^{-t/\tau} \tag{7.11}$$

where V_0 is the initial voltage across the capacitor, t is the time during which the capacitor has been discharging, and $\tau = RC$ is the circuit time constant. KVL indicates that the voltage across the resistor in Figure 7–13 is equal to the voltage across the capacitor with the indicated polarity, so that:

$$v_C = -v_R = V_0 e^{-t/\tau} \tag{7.12}$$

The current can be determined using Ohm's law to obtain:

$$i_C = \frac{v_R}{R} = \frac{V_0}{R} e^{-t/\tau}$$

The capacitor and resistor voltage and current during the time interval $0^+ < t < t_{ss}$ can be summarized as:

$$v_R = -v_C = V_0\, e^{-t/\tau}; \quad i_C = \frac{V_0}{R} e^{-t/\tau}$$

as indicated in Figure 7–15.

FIGURE **7–15**

Voltage and Current
for the Time
Period $0^+ < t < t_{ss}$

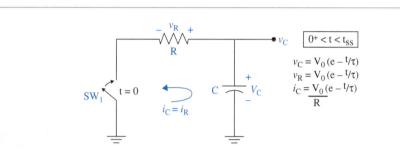

Circuit Considerations for Time t > t_ss (steady state)

The current and voltages will reach steady-state conditions after five time constants have passed. The capacitor is assumed to be completely discharged after five time constants, so that the voltage across the capacitor and resistor and capacitor current all equal zero, as indicated in Figure 7–16.

$$V_C = V_R = I_C = 0$$

FIGURE **7–16**

Steady-State
Voltage and Current

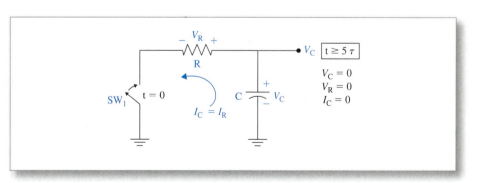

The voltage and current for the R-C circuit as the capacitor releases energy (discharges) can be summarized as:

$$v_C(t) = V_0\, e^{-t/RC}$$

$$v_R(t) = -V_0\, e^{-t/RC}$$

$$i_C(t) = \frac{V_0}{R} e^{-t/RC}$$

The plots of the voltage and current as functions of time are shown in the MATLAB plot in Figure 7–17.

FIGURE **7–17**

Voltage and Current Plots as the Capacitor Releases Energy

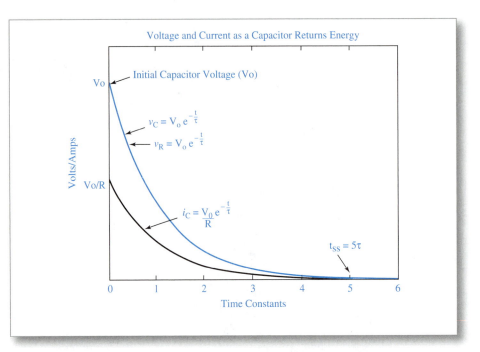

Voltage and Current as a Capacitor Returns Energy

Initial Capacitor Voltage (Vo)

Vo

$v_C = V_0 e^{-\frac{t}{\tau}}$

$v_R = V_0 e^{-\frac{t}{\tau}}$

Vo/R

$i_C = \frac{V_0}{R} e^{-\frac{t}{\tau}}$

$t_{SS} = 5\tau$

Volts/Amps

Time Constants

EXAMPLE 7–3   The Discharging Capacitor Releasing Energy

The initial voltage V_0 across the capacitor in Figure 7–18 is given as 60 V. Calculate the capacitor voltage and current and the resistor voltage for each of the specified times. Also determine the time required for the capacitor voltage to reach its steady-state value.

FIGURE **7–18**

R-C Circuit for Example 7–3

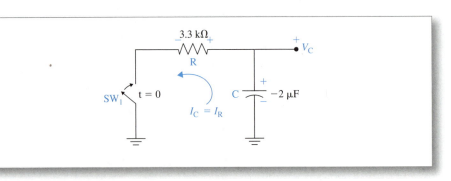

✓ SOLUTION

The time constant may be calculated as:

$$\tau = RC = (2\ \mu F)(3.3\ k\Omega) = 6.6\ ms$$

At $t = 5$ ms:

The voltage across the capacitor can be calculated as:

$$V_C = V_0\,e^{-t/\tau} = 60e^{-5\ ms/6.6\ ms} = 60(0.468) = 28.08\ V$$

KVL indicates that the voltage across the resistor is equal to the voltage across the capacitor, so that:

$$V_C = V_R = 28.08 \text{ V} \ (V_C \text{ and } V_R \text{ have opposite online polarity}).$$

The capacitor current can be calculated as:

$$I_C = \frac{60 \text{ V}}{3.3 \text{ k}\Omega} e^{-5 \text{ ms}/6.6 \text{ ms}} = 18.18(0.468) = 8.51 \text{ mA}$$

Ohm's law can also be used to calculate the capacitor current as:

$$I_C = I_R = \frac{V_R}{R} = \frac{28.08 \text{ V}}{3.3 \text{ k}\Omega} = 8.51 \text{ mA}$$

At $t = 10$ ms:

$$V_C = 60e^{-10 \text{ ms}/6.6 \text{ ms}} = 60(0.219) = 13.14 \text{ V}$$

$$V_R = V_C = 13.14 \text{ V}$$

$$I_C = \frac{60 \text{ V}}{3.3 \text{ k}\Omega} e^{-10 \text{ ms}/6.6 \text{ ms}} = 18.18(0.468) = 3.99 \text{ mA}$$

Ohm's law can also be used to calculate the capacitor current as:

$$I_C = I_R = \frac{V_R}{R} = \frac{13.14 \text{ V}}{3.3 \text{ k}\Omega} = 3.99 \text{ mA}$$

At $t = 15$ ms:

$$V_C = 60e^{-15 \text{ ms}/6.6 \text{ ms}} = 60(0.103) = 6.18 \text{ V}$$

$$V_R = V_C = 6.18 \text{ V}$$

$$I_C = \frac{60 \text{ V}}{3.3 \text{ k}\Omega} e^{-15 \text{ ms}/6.6 \text{ ms}} = 18.18(0.103) = 1.87 \text{ mA}$$

Ohm's law can also be used to calculate the capacitor current as:

$$I_C = I_R = \frac{V_R}{R} = \frac{6.18 \text{ V}}{3.3 \text{ k}\Omega} = 1.87 \text{ mA}$$

The steady-state value of the voltage across the capacitor will be 0 when the capacitor is completely discharged. The time required to reach steady state may be determined as:

$$t_{SS} = 5\tau = 5(6.6 \text{ ms}) = 33 \text{ ms}$$

For engineering considerations, the voltage across the capacitor is assumed to be 0 at time $t = t_{ss} = 33$ ms. The exact value of the capacitor voltage can be calculated as:

$$V_C = V_0 e^{-t/\tau} = 60e^{-33 \text{ ms}/6.6 \text{ ms}} = 0.40 \text{ V}$$

The small difference between the approximate and exact steady-state voltage is considered negligible for most practical applications.

The MATLAB plot of the voltage across the discharging capacitor in Example 7–3 is shown in Figure 7–19.

FIGURE **7–19**

Voltage and Current Plots for Example 7–3

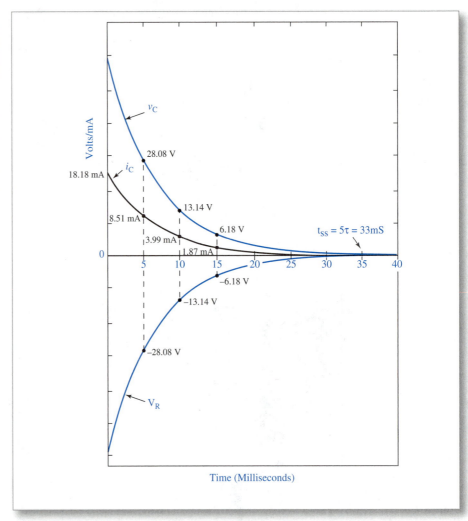

Time (Milliseconds)

PRACTICE EXERCISES 7–2

A 5 μF capacitor is charged to a voltage of 45 V. The capacitor is discharged through a 2 kΩ resistor. Calculate the voltage across the capacitor and resistor at time $t = 4$ ms, 8 ms, and 35 ms. How much time is required to reach steady-state conditions?

✓ ANSWERS

At 4 ms: $V_C = 30.2$ V;   At 8 ms: $V_C = 20.2$ V;

At 35 ms: $V_C = 1.4$ V; $t_{ss} = 5\tau = 5(10$ ms$) = 50$ ms

(Complete solutions to the Practice Exercises can be found in Appendix A.)

APPLICATION OF AN R-C TIMING CIRCUIT

The R-C circuit can be used in many applications as a timing circuit. The relay is a device whose internal contacts will be open or closed depending upon the magnitude of the voltage across the device. The relay is used in many industrial control applications where on and off control is required.

Consider the illustration in Figure 7–20.

FIGURE **7–20**

An R-C Timing
Circuit Application

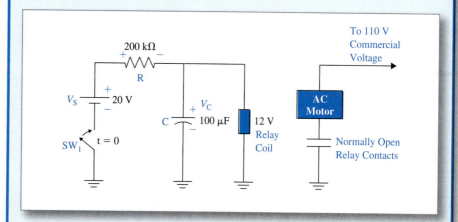

SPECIFICATIONS: The relay contacts in Figure 7–20 will close when the voltage across the relay coil reaches 12 V. The closed contacts will apply the 110 V commercial voltage to the motor. Determine how much time will pass before the commercial voltage is applied to the motor after the switch is closed.

✓ SOLUTION

The relay coil is connected directly across the capacitor, so the voltage across the relay coil is equal to the capacitor voltage. We can express the voltage across the capacitor as:

$$V_C = V_S(1 - e^{-t/\tau}); \quad \tau = RC$$

Substituting the given information, we obtain:

$$12 \text{ V} = 20 \text{ V}(1 - e^{-t/\tau})$$

$\tau = RC$ may be evaluated as:

$$\tau = RC = (200 \text{ k}\Omega)(100 \text{ }\mu\text{F}) = 20 \text{ s}$$

so that:
$$12 \text{ V} = 20 \text{ V}(1 - e^{-t/20 \text{ s}}) \tag{7.14}$$

The value of t in Equation (7.14) can be determined by the following procedure:

Divide both sides of Equation (7.14) by 20 V to obtain:

$$0.6 = (1 - e^{-t/20 \text{ s}}) \tag{7.15}$$

Subtract 1 from each side of Equation (7.15) to obtain:

$$-0.4 = -e^{-t/20 \text{ s}}$$

Multiply both sides of Equation (7.15) by (−1) to obtain:

$$0.4 = e^{-t/20 \text{ s}} \tag{7.16}$$

Take the natural logarithm of both sides of Equation (7.16) to obtain:

$$\ln(0.4) = \ln(e^{-t/20\,s}) \qquad (7.17)$$

One of the rules governing logarithms is expressed as:

$$\log_{(base\ A)}(A^x) = X$$

Since e is the base of the natural logarithm (ln), the rule may be applied to the right-hand side of the expression in Equation (7.17) to obtain:

$$\ln(0.4) = -\frac{t}{20\ s}$$

Use the calculator to determine the value of ln(0.4) to obtain:

$$\ln(0.4) = -0.916$$

so that:
$$-0.916 = -\frac{t}{20\ s}$$

$$t = 18.32\ s$$

The solution can be verified as:

$$V_C = V_S(1 - e^{-t/RC}) = 20(1 - e^{-18.32\,s/20\,s}) = 12\ V\ \text{(approximate)}$$

The solution to Equation (7.14) may also be obtained using the TI-89 calculator as:

Solve(12=20(1−e^(−x/(100EE−6*200E3))),x) ENTER to obtain 18.32 s

The MATLAB plot in Figure 7–21 shows the capacitor voltage reaching the value of 12 V in 18.32 s.

FIGURE **7–21**

The Capacitor Voltage Reaches 12 V in 18.32 ms

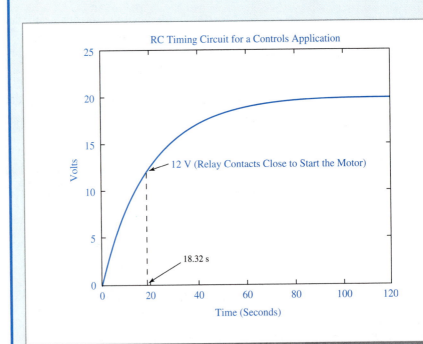

Determine the time required for the voltage across the capacitor in Figure 7–22 to reach 8 V. Solve using the natural logarithm as demonstrated in the previous application and also by using an engineering calculator.

FIGURE **7-22**

R-C Circuit for
Practice Exercise 7-3

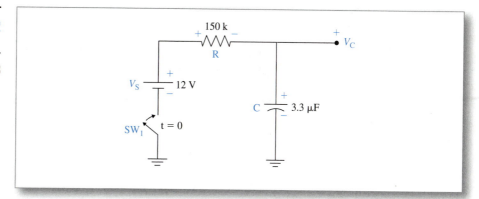

✔ **ANSWER**

$t = 0.544$ s

(Complete solutions to the Practice Exercises can be found in Appendix A.)

7.5 THE CAPACITOR STORING AND RELEASING ENERGY

As noted, a capacitor can absorb and store electrical energy from a source and release or return that energy to the circuit.

Consider the circuit in Figure 7–23.

FIGURE **7-23**

A Capacitor Storing and
Releasing Energy

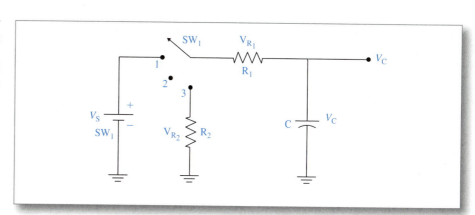

The switch SW$_1$ shown in Figure 7–23 is a three-position switch.

Switch in Position 1

When the switch is in position 1, the capacitor is in series with R_1 and the voltage supply V_S. With the switch in position 1, the capacitor will charge to V_S in five time constants.

The voltage across the capacitor with the switch in position 1 can be expressed as:

$$v_C = V_S(1 - e^{-t/\tau})$$

where:

$$\tau = R_1 C$$

Switch in Position 2

When the switch is in position 2, the circuit is open so that the capacitor can neither charge nor discharge. The capacitor voltage with the switch in position 2 remains constant at the final voltage that existed across the capacitor in position 1.

The voltage across the capacitor with the switch in position 2 may be expressed as:

$$V_C = V_{F1}$$

where V_{F1} is a constant equal to the final voltage that the capacitor reached in position 1.

Switch in Position 3

When the switch is in position 3, the capacitor will discharge through resistors R_1 *and* R_2 and reach a final value of 0 in five time constants. The initial voltage for the discharge phase is the voltage that existed across the capacitor in position 2 or V_{F1}.

The voltage across the capacitor with the switch in position 3 may be expressed as:

$$v_C = V_{F1} e^{-t/\tau}$$

where:

$$\tau = (R_1 + R_2)C$$

EXAMPLE 7–4   **The Capacitor Storing and Releasing Energy**

The switch in the circuit in Figure 7–24 is placed in position 1 at time $t = 0$ and remains there for 15 ms. The switch is then moved position 2 and remains there for the next 5 ms. The switch is then moved to position 3 and remains there indefinitely. Determine the voltage across the capacitor at $t = 8$ ms, 15 ms, 18 ms, 45 ms, and 60 ms. Sketch a plot of the voltage across the capacitor as a function of time for time $t = 0$ until time $t = 145$ ms.

FIGURE **7–24**

Circuit for
Example 7–4

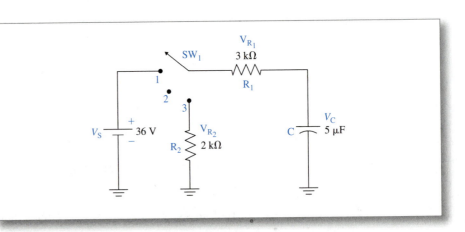

✓ SOLUTION

At $t = 8$ ms:

At time $t = 8$ ms, the switch is in position 1 and the capacitor has been charging for 8 ms, so the voltage across the capacitor can be calculated as:

$$v_C = V_S (1 - e^{-t/\tau})$$

where:

$$\tau = R_1 C = (3 \text{ k}\Omega)(5 \text{ }\mu\text{F}) = 15 \text{ ms}$$

so that:

$$v_C = 36(1 - e^{-8 \text{ ms}/15 \text{ ms}}) = 14.88 \text{ V}$$

At $t = 15$ ms:

At time $t = 15$ ms, the switch is in position 1 and the capacitor has been charging for 15 ms, so the voltage across the capacitor is calculated as:

$$v_C = 36(1 - e^{-15 \text{ ms}/15 \text{ ms}}) = 22.76 \text{ V}$$

Note that 22.76 V is the final voltage reached in position 1.

At $t = 18$ ms:

At time $t = 18$ ms, the switch is in position 2 and the capacitor voltage is constant at the final voltage reached in position 1 or 22.76 V. As long as the switch remains in position 2, the capacitor can neither charge nor discharge and the voltage will remain constant at 22.76 V.

$$V_C = 22.76 \text{ V}$$

At $t = 45$ ms:

At time $t = 45$ ms, the switch is in position 3 so that the capacitor will discharge through resistors R_1 and R_2. It is important to note that at time $t = 45$ ms, the switch has been in position 3 only for 25 ms (45 ms − 20 ms). The voltage across the capacitor can be calculated as:

$$v_C = V_S e^{-t/\tau}$$

where:

$$\tau = (R_1 + R_2) C = (3 \text{ k}\Omega + 2 \text{ k}\Omega) (5 \text{ }\mu\text{F}) = 25$$

so that:

$$V_C = 22.76 e^{-25 \text{ ms}/25 \text{ ms}} = 8.37 \text{ V}$$

At $t = 60$ ms:

At $t = 60$ ms, the switch has been in position 3 for 40 ms (60 ms − 20 ms), so the voltage across the capacitor is calculated as:

$$V_C = 22.76 e^{-40 \text{ ms}/25 \text{ ms}} = 4.59 \text{ V}$$

The voltage across the capacitor will reach 0 when the switch has been in position 3 for five time constants or 125 ms. The total time from time $t = 0$ is calculated as:

$$t_{\text{total}} = t_{\text{pos 1}} + t_{\text{pos 2}} + t_{\text{pos 3}} = 15 \text{ ms} + 5 \text{ ms} + 125 \text{ ms} = 145 \text{ ms}$$

The sketch of the voltage across the capacitor from $t = 0$ to $t = 145$ ms is shown in Figure 7–25.

FIGURE **7–25**

Plot of v_C for Example 7–4

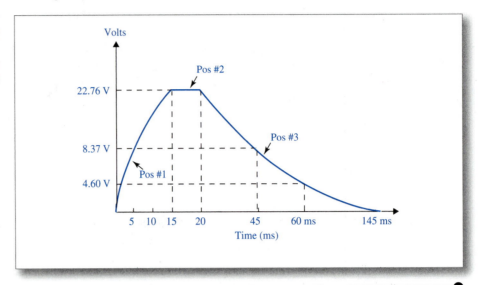

7.6 TOTAL EQUIVALENT CAPACITANCE

Capacitors that are connected in *parallel* effectively increase the total capacitance. The total equivalent capacitance of capacitors connected in parallel is calculated as the sum of the individual capacitances, so that:

$$C_T = C_1 + C_2 + C_3 + \cdots + C_N \tag{7.18}$$

as indicated in Figure 7–26.

Capacitors that are connected in *series* effectively decrease the total capacitance. The total equivalent capacitance of capacitors connected in series is calculated as:

$$\frac{1}{C_T} = \frac{1}{C_1} + \frac{1}{C_2} + \frac{1}{C_3} + \cdots + \frac{1}{C_N} \tag{7.19}$$

as indicated in Figure 7–26.

FIGURE **7–26**

Equivalent Capacitance for Series and Parallel Capacitors

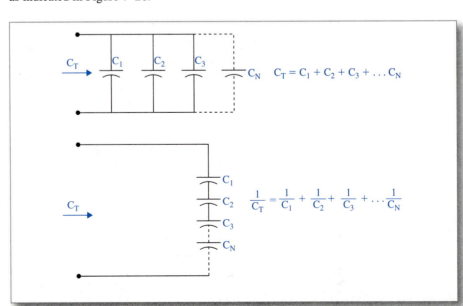

7.7 THE INDUCTOR STORING ENERGY

Like the capacitor, the inductor can absorb and store energy and release energy back to the circuit. The energy is stored temporarily in a magnetic field that surrounds the inductor. The magnetic field expands as the inductor absorbs energy and collapses as energy is returned to the circuit. As the magnetic field collapses or expands, the relative motion between the coils of the inductor and the changing magnetic field will induce a voltage across the device. The inductor will oppose any sudden change of current through the device. The amount of energy stored by an inductor is calculated as:

$$E = \frac{1}{2} L I_L^2$$

where E is energy in joules, L is inductance in henries, and I_L is the current through the inductor in amperes.

Consider the circuit in Figure 7–27.

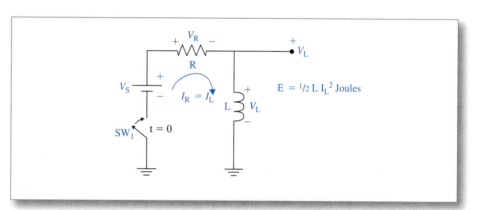

FIGURE **7–27**

An Inductor
Storing Energy

Circuit Considerations for Time $t = 0^+$

As previously noted, the symbol $t = 0^+$ represents the briefest moment of time following time $t = 0$. During this brief moment, the switched DC signal transitions from 0 to V_S. An inductor, however, will not permit sudden changes of current through the device. Assuming that the current through the inductor is initially 0, the current through the inductor will remain at 0 at time $t = 0^+$. Since I_L is 0, the current through the resistor and the voltage across the resistor are both 0. The voltage across the inductor may be determined by KVL as:

$$V_L + V_R - V_S = 0$$

Since $V_R = 0$, then:

$$V_L = V_S$$

The polarity of the voltage across the inductor in accordance with Lenz's law will oppose the polarity of the source voltage.

The voltage and current at time $t = 0^+$ can be summarized as:

$$I_L = I_R = 0; \quad V_R = 0; \quad V_L = V_S$$

as indicated in Figure 7–28.

FIGURE **7–28**

Voltage and
Current at
Time $t=0^+$

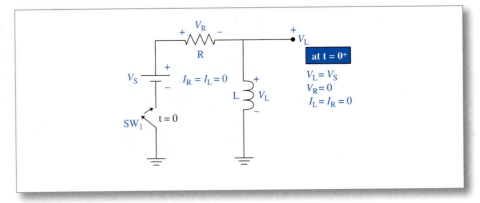

Circuit Considerations for Time $0^+ < t < t_{ss}$

As time advances from $t = 0^+$, the current will begin to increase, causing more voltage to be dropped across the resistor. The increase in voltage across the resistor will result in a decrease in voltage across the inductor to satisfy KVL.

The voltage and current during the time interval $0^+ < t < t_{ss}$ can be summarized as:

I_L is increasing from 0; V_R is increasing from 0; V_L is decreasing from V_S

as indicated in Figure 7–29.

FIGURE **7–29**

Voltage and Current
for the Time
Period $0^+ < t < t_{ss}$

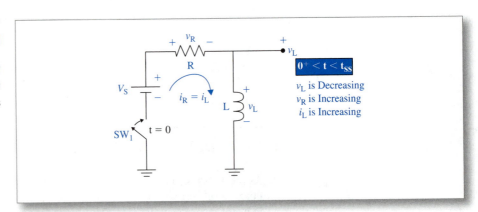

Circuit Considerations for Time $t \geq t_{ss}$ (steady state)

As the current continues to increase, the voltage across the resistor will increase until it reaches the *steady-state value* of V_s at t_{ss}. The voltage across the inductor reaches its steady-state value of 0, and the current will reach its steady-state value of $\dfrac{V_s}{R}$.

The voltages and current for the steady-state condition can be summarized as:

$$V_L = 0; \quad V_R = V_S; \quad I_L = I_R = \frac{V_S}{R}$$

as indicated in Figure 7–30.

FIGURE **7-30**

Steady-State
Voltage and Current

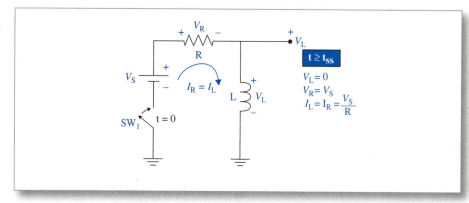

FIGURE **7-30**

Steady-State
Voltage and Current

The voltage across an inductor may be expressed as a function of time as:

$$v_L = L\frac{di_L}{dt}$$

where L is inductance measured in henries and $\dfrac{di_L}{dt}$ represents the rate of change of current through the device, as indicated in Figure 7–31.

FIGURE **7-31**

The Mathematical
Model for an Inductor

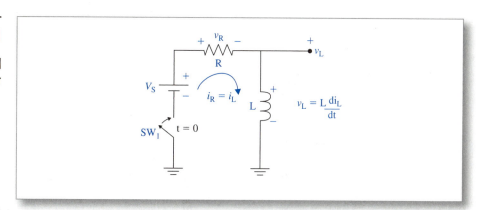

KVL can be expressed for the circuit in Figure 7–31 as:

$$V_S - i_L R - L\frac{di_L}{dt} = 0 \tag{7.20}$$

MATLAB can be used to obtain the solution to Equation (7.20) as:

EDU ≫ dsolve('DI=(V−I*R)/L','I(0)=0')

ans = V/R−exp(−1/L*R*t)*V/R

The MATLAB solution for the inductor current may be written in the more familiar form as:

$$i_L = \frac{V_S}{R}(1 - e^{-t/\tau}) \tag{7.21}$$

The time constant for the R-L circuit is calculated as:

$$\tau = \frac{L}{R} \tag{7.22}$$

The voltage across the resistor can be determined as:

$$v_R = i_L R = V_S (1 - e^{-t/\tau}); \quad \tau = \frac{L}{R} \qquad (7.23)$$

The voltage across the inductor can be determined as $V_S - v_R$ to obtain:

$$v_L = V_S - V_S (1 - e^{-t/\tau}) = V_S e^{-t/\tau} \qquad (7.24)$$

The steady-state condition is reached at five time constants, so that:

$$t_{SS} = 5\tau \qquad (7.25)$$

The voltage and current for the R-L circuit as the inductor stores energy (increasing current) can be summarized as:

$$v_L = V_S e^{-t/\tau}$$

$$v_R = V_S (1 - e^{-t/\tau})$$

$$i_L = \frac{V_S}{R} (1 - e^{-t/\tau})$$

where:

$$\tau = \frac{L}{R}$$

The plots of voltage and current are shown in Figure 7–32.

FIGURE **7–32**

Voltage and Current Plots as the Inductor Stores Energy

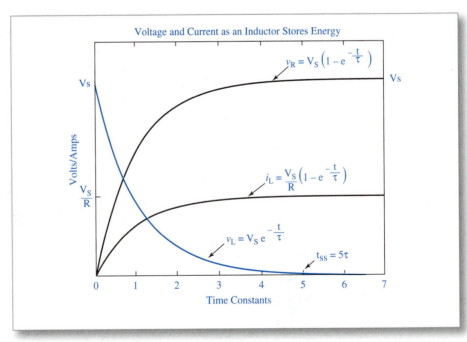

EXAMPLE 7–5   The Inductor Storing Energy

Determine the inductor voltage and current and resistor voltage in Figure 7–33 at $t = 1$ ms, 3 ms and 5 ms. How long will it take the circuit to reach steady-state conditions?

FIGURE **7-33**

RL Circuit for
Example 7-5

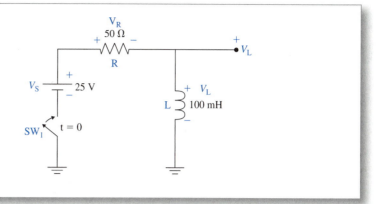

✓ **SOLUTION**

The time constant may be evaluated as:

$$\tau = \frac{L}{R} = \frac{100 \text{ mH}}{50 \text{ } \Omega} = 2 \text{ ms}$$

At t = 1 ms:

$$V_L = 25e^{-1 \text{ ms}/2 \text{ ms}} = 15.15 \text{ V}$$

$$V_R = 25(1 - e^{-1 \text{ ms}/2 \text{ ms}}) = 9.85 \text{ V}$$

$$I_L = \frac{25 \text{ V}}{50 \text{ } \Omega}(1 - e^{-1 \text{ ms}/2\text{ms}}) = 500(0.606) = 197 \text{ mA}$$

Note that $V_L + V_R = 15.15 \text{ V} + 9.85 \text{ V} = 25 \text{ V}$, as required by KVL.

At t = 3 ms:

$$V_L = 25e^{-3 \text{ ms}/2 \text{ ms}} = 5.58 \text{ V}$$

$$V_R = 25(1 - e^{-3 \text{ ms}/2 \text{ ms}}) = 19.42 \text{ V}$$

$$I_L = \frac{25 \text{ V}}{50 \text{ } \Omega}(1 - e^{-3 \text{ ms}/2 \text{ ms}}) = 388.5 \text{ mA}$$

Note that $V_L + V_R = 5.58 \text{ V} + 19.42 \text{ V} = 25 \text{ V}$, as required by KVL.

At t = 5 ms:

$$V_L = 25e^{-5 \text{ ms}/2 \text{ ms}} = 2.05 \text{ V}$$

$$V_R = 25(1 - e^{-5 \text{ ms}/2 \text{ ms}}) = 22.95 \text{ V}$$

$$I_L = \frac{25 \text{ V}}{50 \text{ } \Omega}(1 - e^{-5 \text{ ms}/2 \text{ ms}}) = 459 \text{ mA}$$

Note that $V_L + V_R = 2.05 \text{ V} + 22.95 \text{ V} = 25 \text{ V}$, as required by KVL.

The time required to reach steady-state conditions is five time constants, so that:

$$t_{ss} = 5\tau = 5(2 \text{ ms}) = 10 \text{ ms}$$

The MATLAB plot in Figure 7-34 shows the inductor and resistor voltage and the current as functions of time.

FIGURE **7-34**

Voltage and
Current
Plots for
Example 7-5

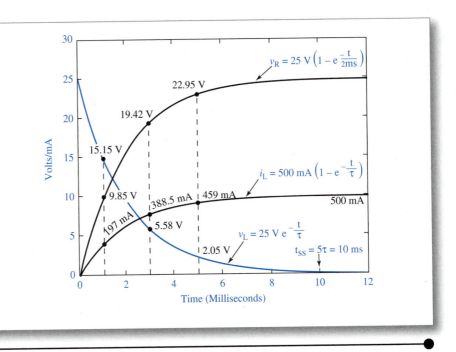

PRACTICE EXERCISES 7-4

Determine the voltage and current for the inductor in Figure 7–35 at times $t = 0.5$ ms and 3 ms. How much time is required to reach steady-state conditions?

FIGURE **7-35**

RL Circuit for
Practice Exercise 7-4

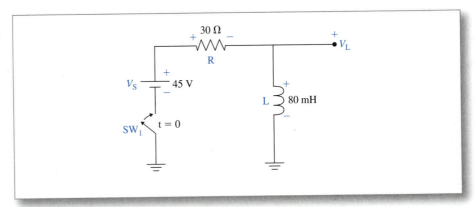

✓**ANSWERS**

At $t = 0.5$ ms: $V_L = 37.3$ V; $I_L = 0.26$ A;

At $t = 3$ ms: $V_L = 14.6$ V; $I_L = 1$ A;

$t_{ss} = 5\tau = 5(2.67 \text{ ms}) = 13.4$ ms

(Complete solutions to the Practice Exercises can be found in Appendix A.)

7.8 THE INDUCTOR RELEASING ENERGY

The inductor releases the energy stored in the magnetic field that surrounds the device as the magnetic field begins to collapse. As the collapsing magnetic field cuts across the windings of the inductor, a voltage is induced across the inductor, so that the inductor becomes a source of energy in the circuit.

An inductor opposes any change of current, so that the voltage induced across an inductor as it releases energy will become whatever voltage is required to maintain the current through the inductor.

Consider the circuit in Figure 7–36.

FIGURE **7-36**

An Inductor
Releasing Energy

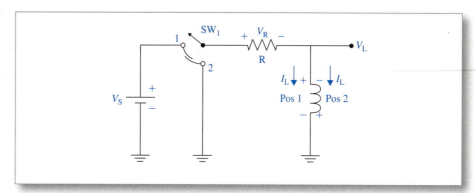

With the switch in position 1 as indicated, the inductor will absorb energy from the source. The voltage and current while the switch is in position 1 may be calculated as:

$$v_L = V_S e^{-t/\tau}; \quad v_R = V_S(1 - e^{-t/\tau}); \quad \text{and} \quad i_L = \frac{V_S}{R}(1 - e^{-t/\tau})$$

where:
$$\tau = \frac{L}{R}$$

The switch in Figure 7–36 is referred to as a *make-before-break switch,* implying that contact is made in position 2 before the contact in position 1 is broken. With the switch in position 2 and contact broken from position 1, the magnetic field that surrounds the inductor will begin to collapse, thereby inducing a voltage across the inductor.

The inductor becomes the source of energy in the circuit, causing current to flow. As indicated in Figure 7–36, the voltage across the inductor will reverse polarity as the energy from the inductor is released. Note that the current through the resistor flows in the same direction whether the switch is in position 1 or position 2. Recall that an inductor opposes any change of current, including change in both magnitude and direction.

The voltage and current for the R-L circuit as the inductor releases energy can be summarized as:

$$v_L = V_0 \, e^{-t/\tau}$$

$$v_R = -V_0 \, e^{-t/\tau}$$

$$i_L = \frac{V_0}{R} e^{-t/\tau}$$

where:
$$\tau = \frac{L}{R}$$

Note that V_0 is the voltage required to maintain the current that is flowing through the inductor. The magnitude of V_0 can be much larger than the supply voltage V_S.

The plots of the inductor and resistor voltage and the current as functions of time as the inductor releases energy are shown in Figure 7–37.

MATLAB

FIGURE **7–37**

Voltage and Current Plots as an Inductor Releases Energy

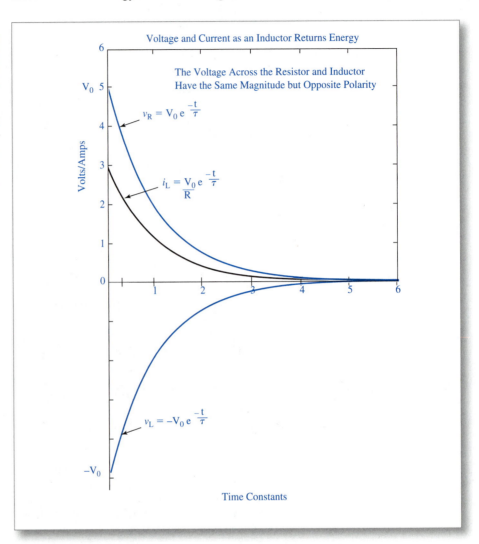

Voltage and Current as an Inductor Returns Energy

The Voltage Across the Resistor and Inductor Have the Same Magnitude but Opposite Polarity

$v_R = V_0\, e^{\frac{-t}{\tau}}$

$i_L = \dfrac{V_0}{R}\, e^{\frac{-t}{\tau}}$

$v_L = -V_0\, e^{\frac{-t}{\tau}}$

Time Constants

EXAMPLE 7–6   The Inductor Releasing Energy

The make-before-break switch in Figure 7–38 has been in position 1 for five time constants. Assume that the switch is moved to position 2 at time $t = 0$. Determine the voltage across the inductor, the voltage across each resistor, and the inductor current $t = 0^+, 3$ ms, and 8 ms.

FIGURE **7–38**

Circuit for Example 7–6

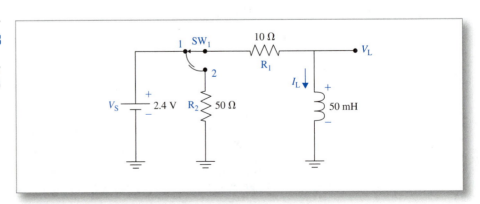

At $t = 0^+$:

A $t = 0^+$, the current that was flowing in position 1 will be momentarily sustained by the inductor, so that:

$$I_L = \frac{2.4 \text{ V}}{10 \text{ }\Omega} = 240 \text{ mA}$$

The inductor voltage required to maintain 240 mA of current as the switch is moved to position 2 is calculated as:

$$V_0 = I_L(R_1 + R_2) = (240 \text{ mA})(10 \text{ }\Omega + 50 \text{ }\Omega)$$

so that:

$$V_0 = 14.4 \text{ V}$$

The voltage divider rule can be used to determine the voltage across each respective resistor as:

$$V_{R1} = V_0 \frac{R_1}{R_1 + R_2} = 14.4 \text{ V} \frac{10 \text{ }\Omega}{60 \text{ }\Omega} = 2.4 \text{ V}$$

$$V_{R2} = V_0 \frac{R_2}{R_1 + R_2} = 14.4 \text{ V} \frac{50 \text{ }\Omega}{60 \text{ }\Omega} = 12 \text{ V}$$

Note that the initial voltage across the inductor in position 2 ($V_0 = 14.4$ V) is much larger than the source voltage ($V_S = 2.4$ V) in position 1.

At $t = 3$ ms:

$$\tau = \frac{L}{R} = \frac{50 \text{ mH}}{10 \text{ }\Omega} = 5 \text{ ms}$$

$$V_L = V_0 e^{-t/\tau} = 14.4(e^{-3 \text{ ms}/5 \text{ ms}}) = 7.9 \text{ V}$$

Using the voltage divider rule, the voltage across each resistor is calculated as:

$$V_{R1} = 7.9 \text{ V} \frac{10 \text{ }\Omega}{60 \text{ }\Omega} = 1.32 \text{ V}$$

$$V_{R2} = 7.9 \text{ V} \frac{50 \text{ }\Omega}{60 \text{ }\Omega} = 6.58 \text{ V}$$

$$I_L = 240 e^{-3 \text{ ms}/5 \text{ ms}} = 131.71 \text{ mA}$$

At $t = 8$ ms:

$$V_L = V_0 e^{-t/\tau} = 14.4 e^{-8 \text{ ms}/5 \text{ ms}} = 2.91 \text{ V}$$

$$V_{R1} = 2.91 \text{ V} \frac{10 \text{ }\Omega}{60 \text{ }\Omega} = 0.485 \text{ V}$$

$$V_{R2} = 2.91 \text{ V} \frac{50 \text{ }\Omega}{60 \text{ }\Omega} = 2.425 \text{ V}$$

$$I_L = 240 e^{-8 \text{ ms}/5 \text{ ms}} = 48.45 \text{ mA}$$

At $t = t_{ss}$:

At $t = t_{ss}$, the voltage and current will equal 0 as the total energy stored by the inductor has been depleted, so that:

$$V_L = V_{R1} = V_{R2} = I_L = 0$$

Figure 7–39 shows the voltage and current plots for Example 7–6.

FIGURE **7-39**

Voltage and
Current Plots
for Example 7–6

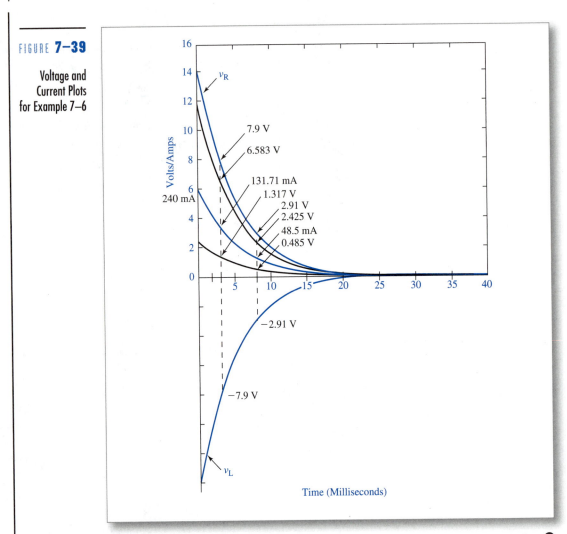

7.9 TOTAL EQUIVALENT INDUCTANCE

Inductors that are connected in *series* effectively increase the total inductance. The total equivalent inductance of inductors connected in series is calculated as the sum of each individual inductance, so that:

$$L_T = L_1 + L_2 + L_3 + \cdots + L_N \tag{7.26}$$

as indicated in Figure 7–40.

Inductors that are connected in *parallel* effectively decrease the total inductance. The total equivalent inductance of inductors connected in parallel is calculated as:

$$\frac{1}{L_T} = \frac{1}{L_1} + \frac{1}{L_2} + \frac{1}{L_3} + \cdots + \frac{1}{L_N} \tag{7.27}$$

as indicated in Figure 7–40.

FIGURE **7–40**

Equivalent Inductance
for Series and
Parallel Inductors

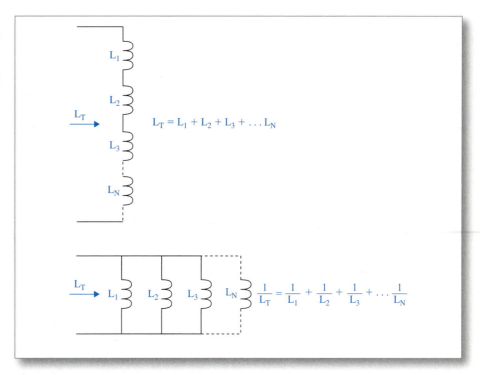

$$L_T = L_1 + L_2 + L_3 + \ldots L_N$$

$$\frac{1}{L_T} = \frac{1}{L_1} + \frac{1}{L_2} + \frac{1}{L_3} + \ldots \frac{1}{L_N}$$

7.10 AN INTRODUCTION TO THE RECTANGULAR PULSE WAVEFORM

The rectangular pulse waveform can be considered to be a DC voltage that is switched off and on at regular intervals. In practice, pulse waveforms are typically created by circuits that electronically switch a DC supply voltage at a predetermined frequency. The pulse waveform is used in many electronic systems applications including clock timing pulses, stepper motor control circuits, and digital logic circuits.

The rectangular pulse in Figure 7–41 is idealized so that the "edges" of the pulse are considered to be perfectly vertical and the "tops" of the pulse are considered to be perfectly flat.

FIGURE **7–41**

An Ideal Rectangular
Pulse Waveform

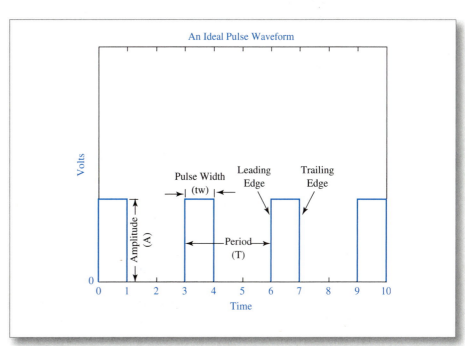

The important parameters of the pulse waveform indicated in Figure 7–41 are defined as follows:

- *Leading Edge:* The edge created as the pulse transitions from its low voltage level (typically 0) to its high voltage level.

- *Trailing Edge:* The edge created as the pulse transitions from its high voltage level to its low voltage level.

- *Amplitude (A):* The voltage difference between the low voltage level and the high voltage level.

- *Pulse Width:* The time duration of a single pulse from its leading edge to its trailing edge, indicated in Figure 7–41 as t_w.

- *Period:* The elapsed time between adjacent leading edges, or between adjacent trailing edges, indicated as T in Figure 7–41.

- *Pulse Frequency:* The number of pulses that occur in one second.

$$f_P = \frac{1}{T}$$

- *Duty Cycle:* Ratio of the pulse width to the period of the pulse expressed as a percent.

$$D = \frac{t_w}{T} \times 100\%$$

EXAMPLE 7–7   Determining the Parameters of a Rectangular Pulse Waveform

Given the rectangular pulse in Figure 7–42, determine the amplitude, pulse width, period, pulse frequency, and duty cycle of the pulse.

FIGURE **7–42**

Pulse Waveform for Example 7–7

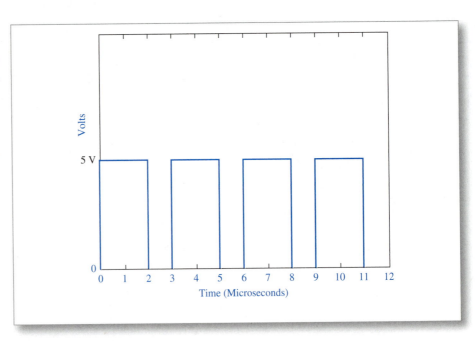

✓ SOLUTION

The amplitude (A) of the pulse is determined as:

$$A = V_{high} - V_{low} = 5\ V - 0\ V = 5\ V$$

The pulse width is determined as the time between any single pulse leading and trailing edge, so that:

$$t_w = 2\ \mu s - 0\ \mu s = 2\ \mu s \text{ (for the first pulse)}$$

$$t_w = 5\ \mu s - 3\ \mu s = 2\ \mu s \text{ (for the second pulse)}$$

The period is determined as the time between adjacent leading or trailing edges, so that:

$$T = 3\ \mu s - 0\ \mu s = 3\ \mu s \text{ (between leading edges of first and second pulses)}$$

$$T = 6\ \mu s - 3\ \mu s = 3\ \mu s \text{ (between leading edges of second and thhird pulses)}$$

The pulse frequency is determined as the reciprocal of the period:

$$f_p = \frac{1}{T} \tag{7.28}$$

so that:

$$f_p = \frac{1}{3\ \mu s} = 333.33 \text{ kHz}$$

The duty cycle is determined as the ratio of pulse width to the period expressed as a percent:

$$D = \frac{t_w}{T} \times 100\% \tag{7.29}$$

so that:

$$D = \frac{2\ \mu s}{3\ \mu s} \times 100\% = 66.7\%$$

PRACTICE EXERCISES 7-5

Given the rectangular pulse in Figure 7–43, determine the amplitude, pulse width, period, pulse frequency, and duty cycle of the pulse.

FIGURE **7–43**

Pulse Waveform for Practice Exercise 7–5

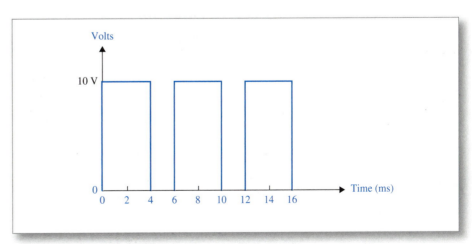

✓ ANSWERS

$A = 10$ V; $t_w = 4$ ms; $T = 6$ ms; $f_p = 166.7$ kHz; $D = 66.7\%$

(Complete solutions to the Practice Exercises can be found in Appendix A.)

7.11 THE INTEGRATOR AND DIFFERENTIATOR

The most common pulse signal is the **square wave,** which may be defined as the special case of the rectangular pulse in which the duty cycle is 50%. It follows from this definition that the pulse width (t_w) is equal to one-half of the period (T), so that:

$$t_W = \frac{T}{2} \tag{7.30}$$

The square wave is shown in Figure 7–44.

FIGURE **7–44**

A Square Wave (Duty Cycle = 50%)

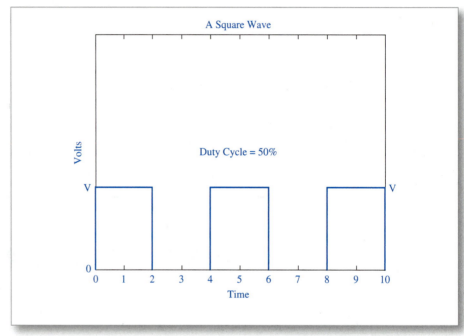

Virtually all electronic circuit designs contain some resistors and some capacitors. In addition, connecting wires and printed circuit board conducting tracks introduce small amounts of resistance and capacitance as well.

It is therefore important to consider how an R-C circuit affects switched DC pulse signals, in particular, the square wave pulse signal.

The output signal of an R-C circuit with a square wave input signal is determined by the relationship between the pulse width (t_w) and the time constant ($\tau = RC$) of the R-C circuit.

Since the square wave may be considered to be a DC voltage switched off and on at regular intervals, the equations developed in the previous sections can be used to analyze R-C circuit response to a square wave input signal as:

$$v_C = V_S (1 - e^{-t/RC}); \quad v_R = V_S (e^{-t/RC})$$

as the capacitor charges;

and as:

$$v_C = V_S (e^{-t/RC}); \quad v_R = -V_S (e^{-t/RC})$$

as the capacitor discharges.

It is important to recall that for practical analysis, a capacitor is considered to reach full charge (steady state) or full discharge after five time constants have passed.

Case 1: Output Taken across the Resistor and τ Much Less Than Pulse Width [t_w]

In this case, the voltage across the resistor will begin at V at the leading edge of the pulse and decay to 0 in five time constants, as defined by the equation $v_R = V_S(e^{-t/RC})$. The voltage across the resistor will remain at 0 until the trailing edge of the pulse appears.

At the trailing edge transition, the voltage across the resistor will become $-V$ and decay to 0 in five time constants, as defined by the equation $v_R = -V_S(e^{-t/RC})$. The voltage across the resistor becomes negative at the trailing edge because the current flowing through the resistor reverses direction as the capacitor discharges, thus reversing the polarity of the voltage across the resistor. The input and output waveforms for the output taken across the resistor are shown in Figure 7–45.

FIGURE **7–45**

The Differentiator Circuit

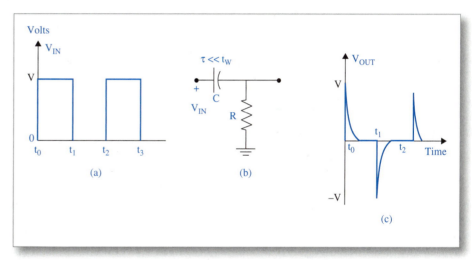

Note that when the output signal is taken across the resistor, the output signal indicates the presence of a leading edge of the pulse with a positive going spike and indicates the presence of a trailing edge with a negative going spike. This circuit has many useful applications and is referred to as a **differentiator circuit.**

Case 2: Output Taken across the Capacitor and τ Much Greater Than Pulse Width [t_w]

If the RC time constant (τ) for the circuit is much greater than the pulse width (t_w), the capacitor will not discharge significantly. The steady-state capacitor voltage will be approximately $V/2$, moving only slightly above and below this value as the input pulse waveform varies from 0 to V.

The input and output waveforms with the output taken across the capacitor are shown in Figure 7–46.

FIGURE 7–46

The Integrator Circuit

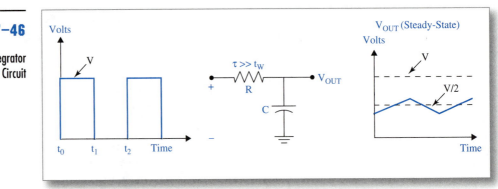

The average value of the output voltage across the capacitor is $V/2$, as shown in Figure 7–45. The area contained under the input pulse waveform over one *complete cycle* is also $V/2$. Integration is the mathematical process that calculates the area contained under a curve over a specified interval. The output signal of the circuit in Figure 7–46 is the integral of the input signal, and the circuit is referred to as an **integrator circuit.**

7.12 COMPUTER ANALYSIS OF THE R-C RESPONSE TO A PULSE WAVEFORM

The R-C circuit shown in Figure 7–47 has an RC time constant of:

$$\tau = RC = (2\ \text{k}\Omega)(0.05\ \mu\text{F}) = 100\ \mu\text{s}$$

FIGURE 7–47

The Differentiator Circuit ($\tau \ll t_W$)

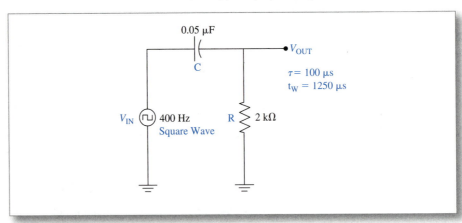

The pulse width (t_W) of the square wave input signal may be calculated using EQ (7.30) as:

$$t_W = \frac{T}{2}$$

Substituting $T = \dfrac{1}{f_p}$, we obtain:

$$t_W = \frac{1}{2f_p} = \frac{1}{800\ \text{Hz}} = 1250\ \mu\text{s}$$

where the frequency of the input square wave signal in Figure 7–47 is given as 400 Hz.

The time constant of the circuit (100 μs) can be considered to be much less than the pulse width (1250 μs). Also, the output signal is taken across the resistor, so the circuit can be considered to be a differentiator circuit. The input and output waveforms are shown in the MultiSIM simulation in Figure 7–48.

FIGURE **7-48**

Computer Simulation
of the Differentiator

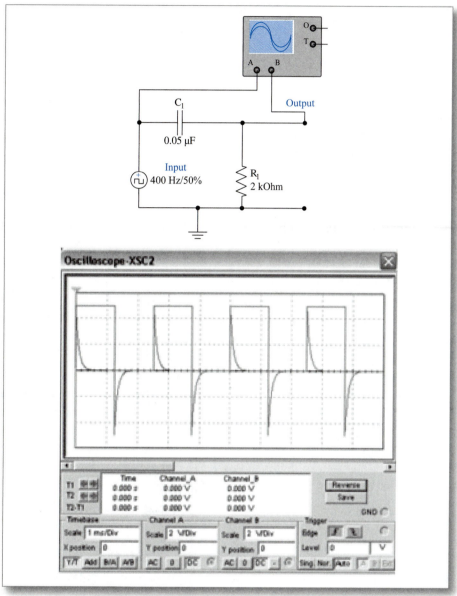

Note that the positive spikes of the output waveform occur at the leading edges of the square wave input signal and the negative spikes occur at the trailing edges of the input signal.

The time constant of the R-C circuit in Figure 7–49 may be calculated as:

$$\tau = RC = (3 \text{ k}\Omega)(3 \text{ } \mu\text{F}) = 9 \text{ ms}$$

FIGURE **7-49**

The Integrator
Circuit ($t_W \ll \tau$)

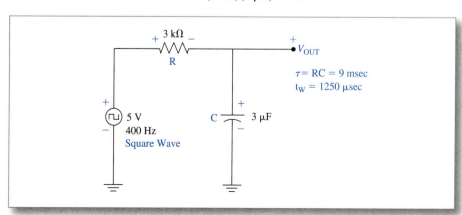

The pulse width (t_w) of the 400 Hz square wave input signal as previously calculated is 1250 μs. The time constant for the circuit can be considered to be much larger than the pulse width. Also, the output is taken across the capacitor, so that the circuit can be considered to be an integrator circuit. The input and output waveforms are shown in the MultiSIM simulation in Figure 7–50.

FIGURE **7–50**

Computer Simulation of the Integrator

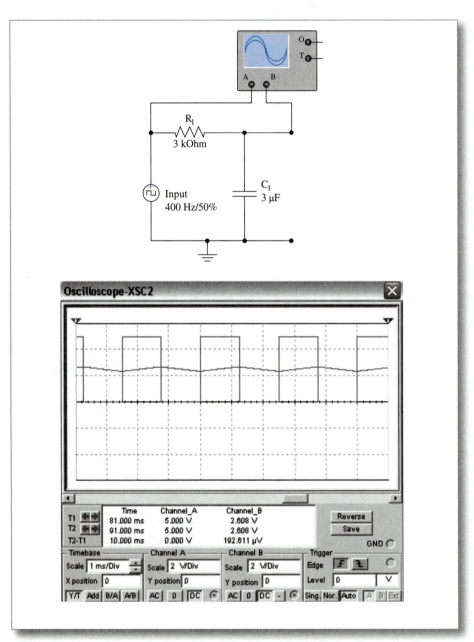

Note that the output signal is nearly constant at an average value of $\dfrac{V}{2}$.

SUMMARY

Transient response in electronic circuits exists only for brief moments after which the signals reach **steady-state** or normal values. Switched DC and pulse signals are examples of waveforms whose transitions create transient voltages and currents.

Capacitive current as a function of time is directly proportional to the capacitance (C) and to the rate of change of voltage across the capacitor expressed mathematically as:

$$i_C = C\frac{dv}{dt}$$

The transient voltage across a capacitor in an R-C circuit when a switched DC voltage (V_S) is applied increases exponentially as $v_C = V_S(1 - e^{-t/\tau})$, where $\tau = RC$ in units of time. The **charging** capacitor absorbs energy from the source that is temporarily stored in an electric field that exists between the capacitor plates. The amount of stored energy is calculated as $E = \frac{1}{2}CV^2$, where E is the stored energy in joules, C is the capacitance in farads, and V is the voltage across the capacitor in volts.

The transient voltage across a **discharging** capacitor decreases exponentially as $v_C = V_0 e^{-t/\tau}$, where V_0 is the initial voltage across the capacitor and $\tau = RC$ in units of time. The discharging capacitor returns stored energy to the circuit.

The **time constant** for an R-C circuit is calculated as $\tau = RC$.

For practical considerations, the transient voltage and current in the R-C circuit will reach **steady-state** values after five time constants have passed.

The steady-state model for a capacitor in a DC circuit is an open circuit.

The voltage across an inductor as a function of time is directly proportional to the inductance (L) and to the rate of change of current through the inductor expressed mathematically as:

$$v_L = L\frac{di}{dt}$$

The transient current through an inductor in an R-L circuit when a switched DC voltage (V_S) is applied increases exponentially as $i_L = (V_S/R)(1 - e^{-t/\tau})$, where $\tau = L/R$ in units of time. The inductor absorbs energy from the source that is temporarily stored in a magnetic field that surrounds the device. The amount of stored energy is calculated as $E = \frac{1}{2}LI^2$, where E is the stored energy in joules, L is the inductance in henries, and I is the current through the inductor in amperes.

The transient current through the inductor as the inductor returns energy to the circuit decreases exponentially as $i_L = I_0 e^{-t/\tau}$, where I_0 is the initial current through the inductor and $\tau = L/R$.

The time constant for the R-L circuit is calculated as $\tau = L/R$.

For practical considerations, the inductor transient voltage and current will reach its steady-state value after five time constants have passed.

The steady-state model for an inductor in a DC circuit is a short circuit.

The **pulse waveform** alternates between two voltage levels at regular intervals. One of the voltage levels is typically 0. The pulse waveform finds use in many electronic applications such as a clock pulse that coordinates signal processing operations. The important parameters of the pulse wave are amplitude, pulse width, period, frequency, and duty cycle. The **square wave** is a pulse waveform having a duty cycle of 50%.

The **differentiator circuit** is an R-C circuit for which the time constant (τ) is much (at least ten times) less than the pulse width of the input pulse signal. The output of the differentiator circuit with a pulse input signal is a positive spike at the leading edge of the pulse and a negative spike at the trailing edge. The differentiator circuit is often referred to as an "Edge Detector" circuit.

The **integrator circuit** is an R-C circuit for which the time constant (τ) is much (at least ten times) greater than the pulse width of the input pulse signal. The output of the integrator circuit with a pulse input signal is a nearly constant voltage equal to the average value of the input pulse signal.

The voltage across a charging capacitor in an R-C circuit for often used in control systems. The circuit is designed so that the capacitor voltage reaches a predetermined value at a specified time. The capacitor voltage may be used to change the state of a relay that in turn may start or stop a motor, conveyor belt, or other mechanical device.

ESSENTIAL LEARNING EXERCISES

(ANSWERS TO ODD-NUMBERED ESSENTIAL LEARNING EXERCISES CAN BE FOUND IN APPENDIX A.)

1. Write the equations for capacitor voltage and current as a function of time as the capacitor absorbs energy (charges) in an R-C series DC circuit.

2. Write the equations for capacitor voltage and current as a function of time as the capacitor releases energy (discharges) in an R-C series DC circuit.

3. Given the circuit in Figure 7–51, determine the voltage across the capacitor and the capacitor current at time $t = 3$ ms, $t = 10$ ms, and $t = 15$ ms. Assume that the switch is closed at time $t = 0$. Sketch the waveform for the voltage across the capacitor and the capacitor current as a function of time.

FIGURE **7–51**

Essential Learning
Exercise 3

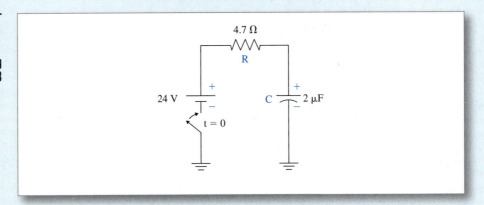

4. Given the circuit in Figure 7–52, determine the voltage across the capacitor and the capacitor current at time $t = 2$ ms and $t = 8$ ms. Assume that the switch is closed at time $t = 0$. Sketch the waveform for the voltage across the capacitor and the capacitor current as a function of time.

FIGURE **7–52**

Essential Learning
Exercise 4

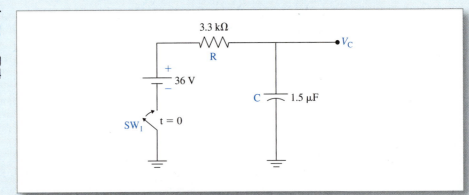

5. How much time is required for the capacitor in Figure 7–51 to reach its steady-state voltage? Determine the steady-state current for the circuit in Figure 7–51.

6. How much time is required for the capacitor in Figure 7–52 to reach its steady-state voltage? Determine the steady-state current for the circuit in Figure 7–52.

7. Given the circuit in Figure 7–53, determine the voltage across the capacitor and the resistor at time $t = 3.5$ ms and $t = 6$ ms. Assume that the switch is closed at time $t = 0$. Sketch the waveform for the voltage across the capacitor and the resistor as a function of time. Verify KVL at each of the specified times.

FIGURE **7–53**

Essential Learning
Exercise 7

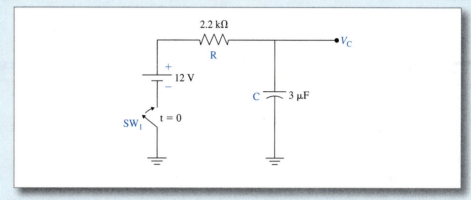

8. Given the circuit in Figure 7–54, determine the voltage across the capacitor and the resistor at time $t = 5.2$ ms and $t = 18$ ms. Assume that the switch is closed at time $t = 0$. Sketch the waveform for the voltage across the capacitor and the resistor as a function of time. Verify KVL at each of the specified times.

FIGURE **7–54**

Essential Learning
Exercise 8

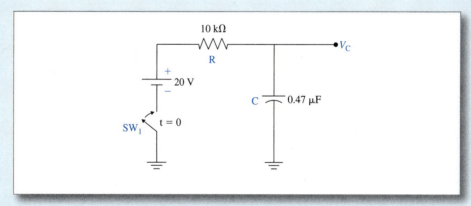

9. The capacitor in Figure 7–55 has been charged to 50 V as indicated. Determine the voltage across the capacitor and capacitor current at time $t = 4$ ms and $t = 12$ ms. Assume that the switch is closed at time $t = 0$. Sketch the waveform of the voltage across the capacitor and the capacitor current as a function of time.

FIGURE **7–55**

Essential Learning
Exercise 9

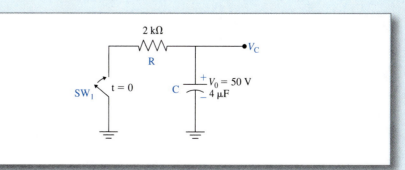

10. The capacitor in Figure 7–56 has been charged to 30 V as indicated. Determine the voltage across the capacitor and capacitor current at time $t = 6$ ms and $t = 25$ ms. Assume that the switch is closed at time $t = 0$. Sketch the waveform of the voltage across the capacitor and the capacitor current as a function of time.

FIGURE **7–56**

Essential Learning
Exercise 10

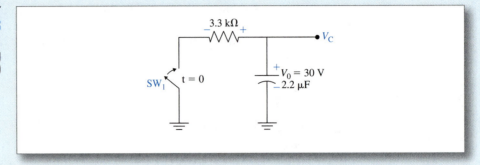

11. Determine the voltage across the resistor at the indicated times for the circuit in Essential Learning Exercise 9. Verify KVL at each of the indicated times.

12. Determine the voltage across the resistor at the indicated times for the circuit in Essential Learning Exercise 10. Verify KVL at each of the indicated times.

13. Write the equations for the inductor voltage and the inductor current as a function of time as the inductor absorbs energy in an R-L series DC circuit.

14. Write the equations for the inductor voltage and the inductor current as a function of time as the inductor releases energy in an R-L series DC circuit.

FIGURE **7–57**

Essential Learning
Exercise 15

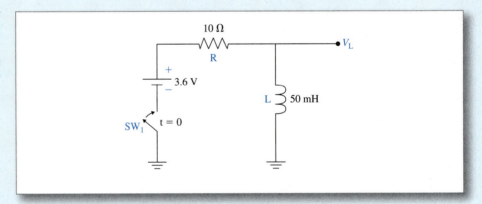

15. Given the circuit in Figure 7–57, determine the voltage across the inductor and inductor current at time $t = 4$ ms and $t = 13$ ms. Assume that the switch is closed at time $t = 0$.

FIGURE **7–58**

Essential Learning
Exercise 16

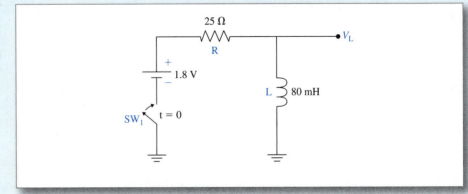

16. Given the circuit in Figure 7–58, determine the voltage across the inductor and inductor current at time $t = 2.8$ ms and $t = 10$ ms. Assume that the switch is closed at time $t = 0$.

17. How much time is required for the inductor in Essential Learning Exercise 15 to reach its steady-state voltage? Determine the steady-state current for the circuit in Essential Learning Exercise 15.

18. How much time is required for the inductor in Essential Learning Exercise 16 to reach its steady-state voltage? Determine the steady-state current for the circuit in Essential Learning Exercise 16.

19. Given the circuit in Figure 7–59, determine the voltages across the inductor and resistor at time $t = 100$ μs and $t = 500$ μs . Assume that the switch is closed at time $t = 0$. Verify KVL at each of the given times.

FIGURE **7–59**

Essential Learning
Exercise 19

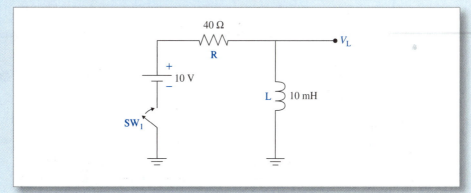

20. Given the circuit in Figure 7–60, determine the voltage across the inductor and resistor at time $t = 500$ μs and $t = 1500$ μs. Assume that the switch is closed at time $t = 0$. Verify KVL at each of the given times.

FIGURE **7–60**

Essential Learning
Exercise 20

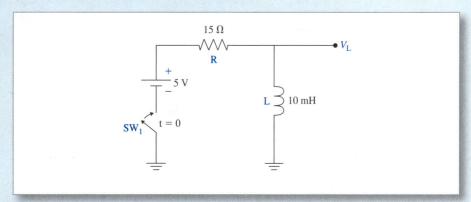

21. Given that the make-before-break switch in Figure 7–61 has been in position 1 for ten time constants and then moved to position 2 at time $t = 0$, determine the voltage across the inductor and inductor current at time $t = 350$ μs and $t = 900$ μs.

FIGURE **7–61**

Essential Learning
Exercise 21

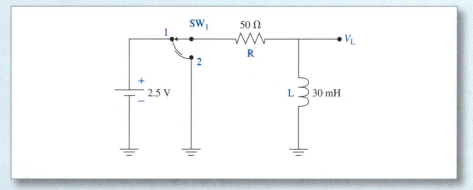

22. Given that the make-before-break switch in Figure 7–62 has been in position 1 for ten time constants and then moved to position 2 at time $t = 0$, determine the voltage across the inductor and inductor current at time $t = 100$ μs and $t = 600$ μs.

FIGURE **7–62**

Essential Learning
Exercise 22

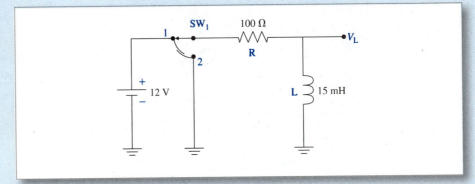

23. Calculate the voltage across the resistor at the indicated times in Essential Learning Exercise 21. Verify KVL at each of the indicated times.

24. Calculate the voltage across the resistor at the indicated times in Essential Learning Exercise 22. Verify KVL at each of the indicated times.

25. Determine the total equivalent capacitance of three 10 μF capacitors connected in series. Determine the total equivalent capacitance of three 10 μF capacitors connected in parallel.

26. Determine the total equivalent inductance of three 1 mH inductors connected in series. Determine the total equivalent inductance of three 1 mH inductors connected in parallel.

27. Determine the period, frequency, pulse width, and duty cycle for the pulse waveform shown in Figure 7–63.

FIGURE **7–63**

Essential Learning
Exercise 27

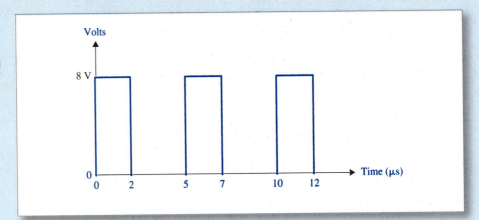

28. Determine the period, frequency, pulse width, and duty cycle for the pulse waveform shown in Figure 7–64.

FIGURE **7–64**

Essential Learning
Exercise 28

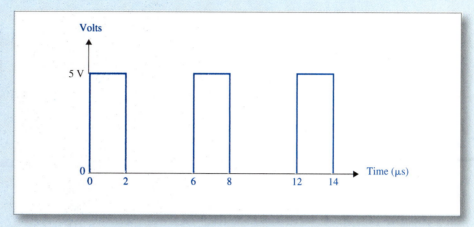

29. If the pulse waveform given in Essential Learning Exercise 28 is the input signal to a differentiator circuit, sketch the output waveform of the differentiator.

30. If the pulse waveform given in Essential Learning Exercise 28 is the input signal to an integrator circuit, sketch the output waveform of the integrator.

31. If the pulse waveform given in Essential Learning Exercise 29 is the input signal to a differentiator circuit, sketch the output waveform of the differentiator.

32. If the pulse waveform given in Essential Learning Exercise 29 is the input signal to an integrator circuit, sketch the output waveform of the integrator.

CHALLENGING LEARNING EXERCISES

(ANSWERS TO SELECTED CHALLENGING LEARNING EXERCISES CAN BE FOUND IN APPENDIX A.)

1. Given the circuit in Figure 7–65, $R = 100$ kΩ, and $C = 15$ μF, determine how much time is required for the capacitor to reach 12 V. Assume the switch is closed at time $t = 0$.

FIGURE **7–65**

Challenging Learning
Exercise 1

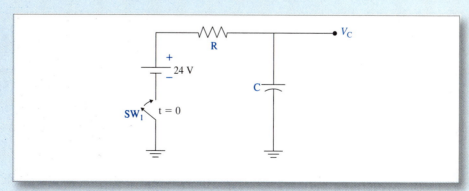

2. Given the circuit in Figure 7–66, determine the value of *R* so that the capacitor voltage will reach 10 V in 2.5 ms after the switch is closed.

FIGURE **7–66**

Challenging Learning
Exercise 2

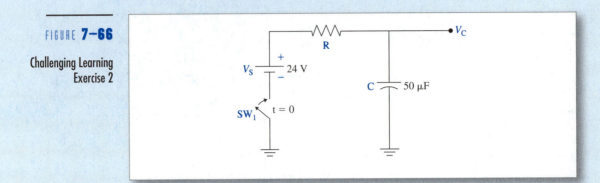

3. The switch in the circuit in Figure 7–67 is placed in position 1 at time *t* = 0 and remains there for 10 ms. The switch is then placed in position 2 and remains there indefinitely. Determine the voltage across the capacitor at the following times: *t* = 5 ms, *t* = 8 ms, *t* = 15 ms, and *t* = 25 ms. Sketch a plot of the capacitor voltage versus time from *t* = 0 to *t* = 75 ms.

FIGURE **7–67**

Challenging Learning
Exercise 3

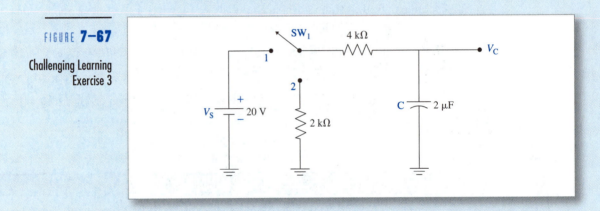

4. The make-before-break switch in Figure 7–68 is placed in position 1 at time *t* = 0 and remains there for 25 ms. The switch is then placed in position 2 and remains there for an indefinite period of time. Determine the voltage across the inductor and the inductor current at the following times: *t* = 1 ms, *t* = 2 ms, and *t* = 30 ms. Sketch a plot of the inductor voltage and the inductor current versus time from *t* = 0 to *t* = 35 ms.

FIGURE **7–68**

Challenging Learning
Exercise 4

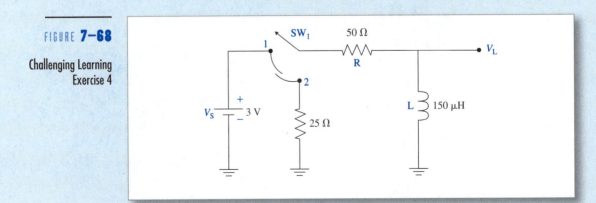

5. Show that the circuit in Figure 7–69 meets the requirements of a differentiator circuit. Sketch the output signal if the input signal is given as a 5 V square wave with a pulse width of 3 ms. Verify your sketch using computer simulation.

FIGURE **7–69**

Challenging Learning
Exercise 5

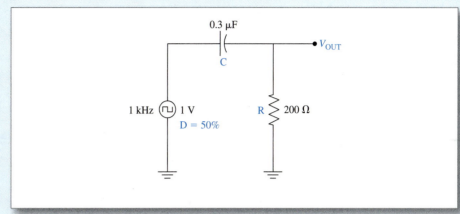

6. The mathematical model for a capacitor may be expressed as $i_C = C\dfrac{dv_C}{dt}$. Use the mathematical model to explain why a capacitor is an open circuit for steady-state DC considerations.

7. The mathematical model for an inductor may be expressed as $v_L = L\dfrac{di_L}{dt}$. Use the mathematical model to explain why an inductor is a short circuit for steady-state DC considerations.

TEAM ACTIVITY

A 24 V relay is to be used to start a conveyor belt in a process that delivers electronic components to a printed circuit board assembly site. Design an R-C timing circuit that will start the motor 10 s after a master control switch is closed and stop the motor 25 s after the master control switch is closed.

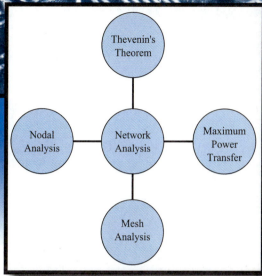

NETWORK ANALYSIS AND SELECTED THEOREMS

"THE MORE WE DO, THE MORE WE CAN DO."

—William Hazlitt

CHAPTER OUTLINE

LEARNING OBJECTIVES

Upon successful completion of this chapter you will be able to:

- Solve a complex network using mesh analysis.
- Solve a complex network using nodal analysis.
- Derive a Thevenin equivalent circuit.
- Derive a Norton equivalent circuit.
- Determine network current and voltage using the superposition theorem.
- Solve a complex network using alternate methods of analysis.

INTRODUCTION

This chapter introduces methods that are typically used to analyze networks that contain large numbers of components or branches. Although the networks presented in the examples in this chapter are not overly complex, the methods demonstrated are most suitable for analysis of larger networks. In addition, the theorems that are introduced can often simplify analysis, especially under changing load conditions.

The analysis methods and techniques presented thus far in this text have been limited to circuits that contain only one signal source. Although many electronic circuits are designed to process only one input signal at any given time, there are occasions in which more than one signal must be processed simultaneously. The theorems presented in this chapter may be applied to networks that contain any number of independent sources.

8.1 FORMAT MESH ANALYSIS

Mesh analysis is a systematic application of KVL around each loop in a given network. The resulting system of equations can then be solved to determine the values of the loop currents from which the current through each network branch or component can be determined.

Consider the circuit in Figure 8–1.

FIGURE 8–1

Assuming Loop Current I_1 and I_2 for Mesh Analysis

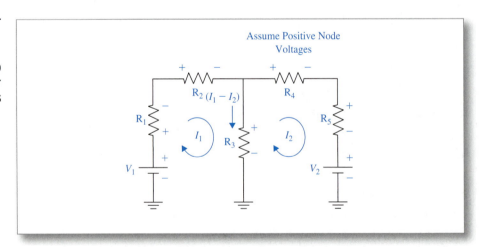

The currents I_1 and I_2 are referred to as loop currents, and both loop currents are assumed to be flowing clockwise as indicated in Figure 8–1. Note that each loop current is assumed to flow clockwise regardless of the polarity of the voltage sources V_1 and V_2.

Loop current I_1 flows through resistors R_1, R_2, and R_3, and loop current I_2 flows through resistors R_4 and R_5 and also through R_3. Both loop currents I_1 and I_2 flow through R_3. Noting the indicated current directions in Figure 8–1, I_1 flows down through R_3, and I_2 flows up through R_3. The net current through R_3 is therefore determined as the *difference* between I_1 and I_2 as indicated in Figure 8–1.

The polarity of the voltage across each resistor is determined by the direction of current flow through each resistor. The polarity of voltage across R_3 is uncertain because the polarity will depend upon which of the loop currents is larger.

The polarity of the voltage across each resistor is assumed to be the polarity indicated on the schematic in Figure 8–1. The mathematical solutions to the analysis will indicate whether or not the assumptions are correct.

Recalling that the voltage across a resistor may be written as IR, KVL may be written around loop 1 moving clockwise from V_1 source ground as:

$$V_1 - I_1 R_1 - I_1 R_2 - (I_1 - I_2) R_3 = 0 \qquad (8.1)$$

KVL around loop 2 moving clockwise from R_3 ground may be written as:

$$(I_1 - I_2) R_3 - I_2 R_4 - I_2 R_5 - V_2 = 0 \qquad (8.2)$$

The voltage transitions from $(+)$ to $(-)$ moving clockwise around a loop are entered into KVL as negative values, and voltages that transition from $(-)$ to $(+)$ are entered as positive values.

Removing the parenthesis in Equation (8.1), we obtain:

$$V_1 - I_1 R_1 - I_1 R_2 - I_1 R_3 + I_2 R_3 = 0 \qquad (8.3)$$

Collecting like terms in Equation (8.3), we obtain:

$$V_1 - I_1 (R_1 + R_2 + R_3) + I_2 R_3 = 0 \qquad (8.4)$$

Equation (8.4) may be rearranged as:

$$I_1 (R_1 + R_2 + R_3) - I_2 R_3 = V_1 \qquad (8.5)$$

Similar algebraic operations can be performed on Equation (8.2) to obtain:

$$-I_1 R_3 + I_2 (R_3 + R_4 + R_5) = -V_2 \qquad (8.6)$$

The KVL equations in Equations (8.5) and (8.6) are referred to as format mesh equations.

Format mesh equations may be determined using the following procedure:

1. Assume clockwise loop currents for each loop, designating the loop currents as $I_1, I_2, I_3 \ldots I_n$.

2. The left-hand side of each equation contains a term for each loop current. The loop current for the loop being considered is the only positive term. All other terms on the left-hand side are negative.

3. The coefficient of the term for the loop being considered is the sum of all of the resistances (self-resistance) in the loop.

4. The coefficient for the remaining terms is the mutual resistance between the two loops being considered.

5. The right-hand side of the equation is the algebraic sum of all sources in the respective loop. A source that transitions from $(+)$ to $(-)$ moving clockwise around the loop is entered as a negative quantity. A source that transitions from $(-)$ to $(+)$ moving clockwise around the loop is entered as a positive quantity.

The format mesh equations for the circuit in Figure 8–1 can be obtained by applying the procedure outlined about as:

1: Assume that all loop currents flow clockwise as indicated in Figure 8–1.

2: The format mesh equation for loop 1 will contain a term for each loop current variable with I_1 as the only positive term to obtain:

$$+I_1(\quad) - I_2(\quad) =$$

3: The coefficient for the I_1 term is the sum of all resistors (self-resistance) in loop 1 to obtain:

$$I_1(R_1 + R_2 + R_3) - I_2(\quad) =$$

4: The coefficient of the I_2 term is the mutual resistance between loop 1 and 2 to obtain:

$$I_1(R_1 + R_2 + R_3) - I_2(R_3) =$$

5: The right-hand side of the equation is the algebraic sum of the sources in loop 1, so that the format mesh equation for loop 1 may be expressed as:

$$I_1(R_1 + R_2 + R_3) - I_2 R_3 = +V_1 \qquad (8.7)$$

Note that voltage source V_1 transitions from $(-)$ to $(+)$ moving clockwise around loop 1 and is therefore entered as a positive quantity in the format mesh equation.

The format mesh equation for loop 2 can be obtained using a similar procedure.

1: Assume that all loop currents flow clockwise as indicated in Figure 8–1.

2: The format mesh equation for loop 2 will contain a term for each loop current variable with I_2 as the only positive term to obtain:

$$-I_1(\ \ ) + I_2(\ \ ) =$$

3: The coefficient for the I_2 term is the sum of all resistors (self-resistance) in loop 2 to obtain:

$$-I_1(\ \ ) + I_2(R_3 + R_4 + R_5) =$$

4: The coefficient of the I_1 term is the mutual resistance between loop 2 and 1 to obtain:

$$-I_1(R_3) + I_2(R_3 + R_4 + R_5) =$$

5: The right-hand side of the equation is the algebraic sum of the sources in loop 2, so that the format mesh equation for loop 2 may be expressed as:

$$-I_1(R_3) + I_2(R_3 + R_4 + R_5) = -V_2 \qquad (8.8)$$

Note that voltage source V_2 transitions from $(+)$ to $(-)$ moving clockwise around loop 2 and is therefore entered as a negative quantity in the format mesh equation.

Equations (8.5) and (8.6) are identical to Equations (8.7) and (8.8) obtained by using the procedure of format mesh analysis.

EXAMPLE 8–1   Format Mesh Analysis of a DC Resistive Network

Given the resistive network in Figure 8–2, write the format mesh equations and solve for the loop currents. Determine the current that flows through each resistor in the network.

FIGURE **8–2**

Circuit for Example 8–1 Indicating Loop Currents

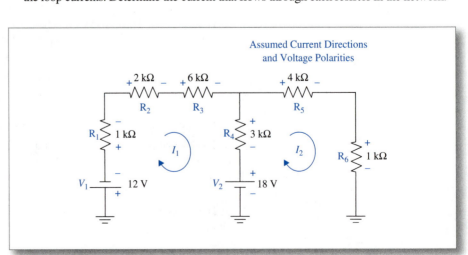

✓ SOLUTION

The format mesh equations can be obtained using the procedure outlined as:

Loop 1: $\qquad\qquad\qquad I_1(R_1 + R_2 + R_3 + R_4) - I_2 R_4 = -V_1 - V_2$

Loop 2: $\qquad\qquad\qquad -I_1 R_4 + I_2(R_4 + R_5 + R_6) = V_2$

Substituting numerical values from Figure 8–2, we obtain:

Loop 1: $\quad I_1(1\text{ k}\Omega + 2\text{ k}\Omega + 6\text{ k}\Omega + 3\text{ k}\Omega) - I_2(3\text{ k}\Omega) = -12\text{ V} - 18\text{ V} = -30\text{ V}$

Loop 2: $\quad -I_1(3\text{ k}\Omega) + I_2(3\text{ k}\Omega + 4\text{ k}\Omega + 1\text{ k}\Omega) = 18\text{ V}$

Collecting like terms, we obtain:

Loop 1: $\qquad\qquad\qquad I_1(12\text{ k}\Omega) - I_2(3\text{ k}\Omega) = -30\text{ V}$ $\qquad$ (8.9)

Loop 2: $\qquad\qquad\qquad -I_1(3\text{ k}\Omega) + I_2(8\text{ k}\Omega) = 18\text{ V}$ $\qquad$ (8.10)

Equations (8.9) and (8.10) may be solved using a variety of methods. Manual methods of solution include "adding and subtracting equations" and "the substitution method," both of which are demonstrated in most college algebra texts. The solutions for I_1 and I_2 can also be obtained by using the TI-89 calculator or MATLAB.

The following calculator solution is obtained using the TI-89 calculator.

Select 2nd Custom F3, Menu item 4 to obtain on the display: solve (and , (x, y)
Key as: solve (12x − 3y = −30 and −3x + 8y = 18, {x, y}) ENTER
Answers: $\qquad$ x = I_1 = −2.14 mA; y = I_2 = 1.45 mA

The MATLAB solution is obtained as:

EDU ≫ syms x y
EDU ≫ [a1 a2]=solve(12*x −3*y +30, −3*x +8*y −18)

to obtain:

$\qquad$ a1 = I_1 = 62/29 = −2.14 mA; a2 = I_2 = 42/29 = 1.45 mA

When applying format mesh analysis, all loop currents are always assumed to be clockwise. The negative sign associated with I_1 in the solution of Example 8–1 indicates that the assumed clockwise current direction must be reversed, so that loop current I_1 is actually flowing counterclockwise.

The circuit diagram in Figure 8–3 shows the actual loop current directions and the voltage polarities indicated by the mathematical solutions.

FIGURE 8–3

Actual Loop Current
Directions for the
Circuit in Figure 8–2

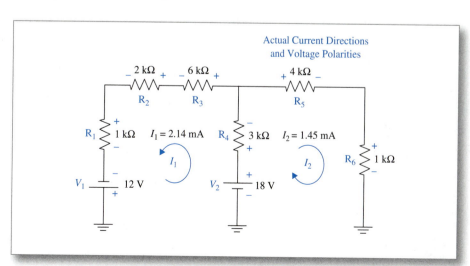

Observe that loop current I_1 flows through resistors R_1, R_2, and R_3 in the circuit diagram in Figure 8–3. Note that the magnitude of loop current I_1 is 2.14 mA and that the current is shown flowing counterclockwise because the mathematical solution for I_1 is negative.

Observe that loop current I_2 flows through resistors R_5 and R_6 in the circuit diagram in Figure 8–3. Note that the magnitude of loop current I_2 is 1.45 mA and that the current is shown flowing clockwise because the mathematical solution for I_2 is positive.

Recall that the loop current I_1 is actually flowing counterclockwise, so that both I_1 and I_2 are flowing up through R_4. The current through R_4 is therefore the sum of the loop currents to obtain the current through R_4 as:

$$I_{R4} = I_1 + I_2 = 2.14 \text{ mA} + 1.45 \text{ mA} = 3.59 \text{ mA}$$

It is important to realize that even though the current through R_4 is determined as the sum of two currents, this is only a mathematical concept. In reality, only one current flows through R_4, which has been determined to be 3.59 mA.

The resistor currents may be summarized as:

$$I_{R1} = 2.14 \text{ mA} \downarrow; \quad I_{R2} = I_{R3} = 2.14 \text{ mA} \leftarrow; \quad I_{R4} = 3.59 \text{ mA} \uparrow; \quad I_{R5} = 1.45 \text{ mA} \rightarrow;$$
$$I_{R6} = 1.45 \text{ mA} \downarrow$$

Ohm's law can be used to calculate the voltage across each resistor in Figure 8–3 respectively as:

$$V_{R1} = (1 \text{ k}\Omega)(2.14 \text{ mA}) = 2.14 \text{ V}$$

$$V_{R2} = (2 \text{ k}\Omega)(2.14 \text{ mA}) = 4.28 \text{ V}$$

$$V_{R3} = (6 \text{ k}\Omega)(2.14 \text{ mA}) = 12.84 \text{ V}$$

$$V_{R4} = (3 \text{ k}\Omega)(3.59 \text{ mA}) = 10.77 \text{ V}$$

$$V_{R5} = (4 \text{ k}\Omega)(1.45 \text{ mA}) = 5.80 \text{ V}$$

$$V_{R6} = (1 \text{ k}\Omega)(1.45 \text{ mA}) = 1.45 \text{ V}$$

The circuit in Figure 8–4 indicates the voltage across each resistor. The polarity of each resistor voltage is determined by the direction of current flow through the resistor.

FIGURE 8–4

Verifying KVL for the Circuit in Figure 8–2

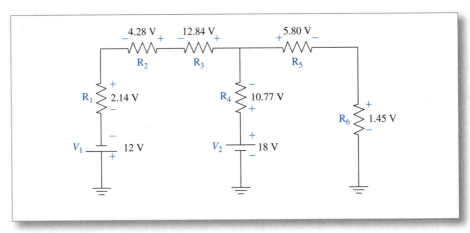

The results of the analysis can be verified by writing KVL around each loop in the circuit as:

Loop 1: $-12 \text{ V} + 2.14 \text{ V} + 4.28 \text{ V} + 12.84 \text{ V} + 10.77 \text{ V} - 18 \text{ V} = 0$

Loop 2: $18 \text{ V} - 10.77 \text{ V} - 5.80 \text{ V} - 1.45 \text{ V} = 0$

Slight inaccuracies are expected due to rounding.

The MultiSIM simulation is shown in Figure 8–5.

FIGURE **8–5**

MultiSIM Simulation
of the Circuit in
Figure 8–2

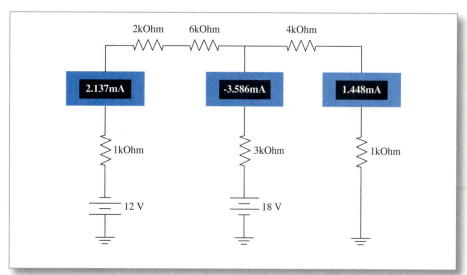

EXAMPLE 8–2   Format Mesh Analysis of a DC Resistive Network

Given the resistive network in Figure 8–6, write the format mesh equations and solve for the loop currents. Determine the current that flows through each resistor in the network. Verify the solution using computer simulation.

FIGURE **8–6**

Circuit for
Example 8–2

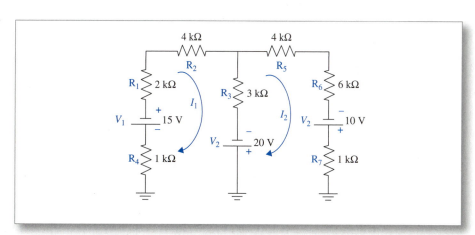

✓ **SOLUTION**

Assume loop currents I_1 and I_2 are flowing clockwise.

The format mesh equations are written as:

Loop 1: $\qquad I_1(R_1 + R_2 + R_3 + R_4) - I_2 R_3 = -V_1 - V_2$

Loop 2: $\qquad -I_1 R_3 + I_2(R_3 + R_5 + R_6) = V_2 + V_3$

Substituting numerical values from Figure 8–6, we obtain:

Loop 1: $I_1(2\text{ k}\Omega + 4\text{ k}\Omega + 3\text{ k}\Omega + 1\text{ k}\Omega) - I_2(3\text{ k}\Omega) = -15\text{ V} - 20\text{ V} = -35\text{ V}$

Loop 2: $-I_1(3\text{ k}\Omega) + I_2(3\text{ k}\Omega + 4\text{ k}\Omega + 6\text{ k}\Omega + 1\text{ k}\Omega) = 20\text{ V} + 10\text{ V} = 30\text{ V}$

Collecting like terms, we obtain:

Loop 1: $I_1(10\text{ k}\Omega) - I_2(3\text{ k}\Omega) = -35\text{ V}$

Loop 2: $-I_1(3\text{ k}\Omega) + I_2(14\text{ k}\Omega) = 30\text{ V}$

Key as: solve(10E3 x −3E3 y = −35 and − 3E3 x + 14E3 y = 30, {x,y}) ENTER
Answers: x = I_1 = −3.05 mA;   y = I_2 = 1.49 mA
Since the solution for I_1 is negative, the assumed direction of I_1 (clockwise) must be reversed (counterclockwise).

The loop current I_1 = 3.05 mA flows through resistors R_1, R_2, and R_4.

The loop current I_2 = 1.49 mA flows through resistors R_5, R_6, and R_7.

The current through resistor R_3 is $I_1 + I_2$ = 4.54 mA.

The MultiSIM simulation is shown in Figure 8–7.

FIGURE 8–7

MultiSIM
Simulation for
Example 8–2

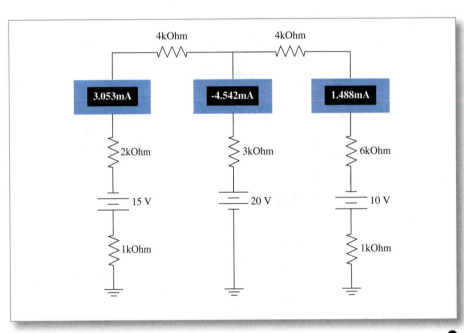

EXAMPLE 8-3   Format Mesh Analysis of a Resistive Network

Write format mesh equations for the resistive network in Figure 8–8. Determine the values of the loop currents. Verify the values of the loop currents using computer simulation.

✓ SOLUTION

Loop 1: $I_1(R_1 + R_2 + R_3 + R_4) - I_2 R_3 - I_3 R_2 = -V_1$

Loop 2: $-I_1 R_3 + I_2(R_3 + R_5 + R_6 + R_7) - I_3 R_5 = -V_2$

Loop 3: $-I_1 R_2 - I_2 R_5 + I_3(R_2 + R_5 + R_8) = -V_3$

Substituting numerical values from Figure 8–8, we obtain:

Loop 1: $I_1(1\text{ k}\Omega + 2.2\text{ k}\Omega + 4.7\text{ k}\Omega + 1\text{ k}\Omega) - I_2(4.7\text{ k}\Omega) - I_3(2.2\text{ k}\Omega) = -8\text{ V}$

FIGURE **8–8**

Circuit for
Example 8–3
Indicating Loop
Currents

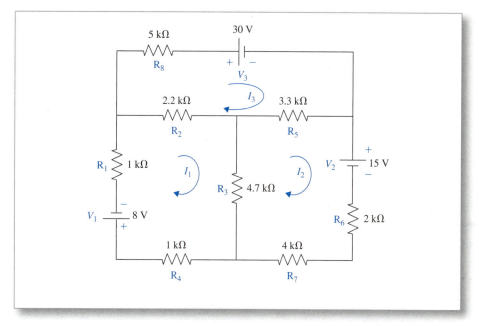

Loop 2: $-I_1 (4.7 \text{ k}\Omega) + I_2 (4.7 \text{ k}\Omega + 3.3 \text{ k}\Omega + 2 \text{ k}\Omega + 4 \text{ k}\Omega) - I_3 (3 \text{ k}\Omega) = -15$ V

Loop 3: $-I_1 (2.2 \text{ k}\Omega) - I_2 (3.3 \text{ k}\Omega) + I_3 (2.2 \text{ k}\Omega + 3.3 \text{ k}\Omega + 5 \text{ k}\Omega) = -30$ V

Collecting like terms, we obtain:

Loop 1: $I_1 (8.9 \text{ k}\Omega) - I_2 (4.7 \text{ k}\Omega) - I_3 (2.2 \text{ k}\Omega) = -8$ V

Loop 2: $-I_1 (4.7 \text{ k}\Omega) + I_2 (3.3 \text{ k}\Omega) - I_3 (3.3 \text{ k}\Omega) = -15$ V

Loop 3: $-I_1 (2.2 \text{ k}\Omega) - I_2 (3.3 \text{ k}\Omega) + I_3 (10.5 \text{ k}\Omega) = -30$ V

The numerical values for the loop currents can be obtained using the TI-89 calculator as:

$$I_1 = -3.95 \text{ mA}; \quad I_2 = -3.52 \text{ mA}; \quad I_3 = -4.79 \text{ mA}$$

The MultiSIM simulation is shown in Figure 8–9.

MULTISIM

FIGURE **8–9**

MultiSIM
Simulation
for Example 8–3
Showing
Loop Currents

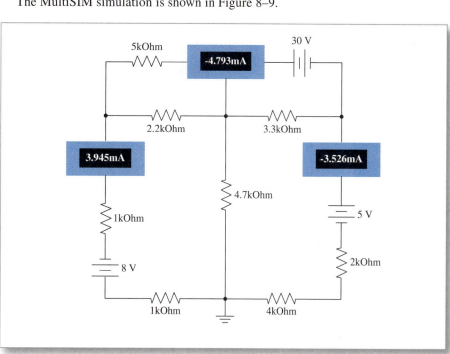

PRACTICE EXERCISES 8–1

Given the resistive network in Figure 8–10, write the format mesh equations and solve for the loop currents. Determine the current through each resistor in the network.

FIGURE **8–10**

Circuit for
Practice
Exercise 8–1

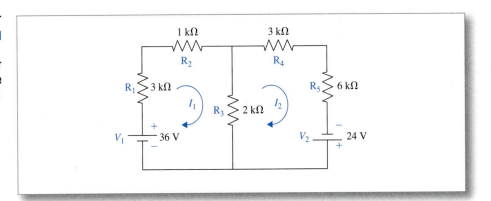

✓ ANSWERS

$I_1 = 7.16$ mA; $I_2 = 3.48$ mA; $I_{R1} = I_{R2} = 7.16$ mA;

$I_{R4} = I_{R5} = 3.48$ mA; $I_{R3} = 3.68$ mA

(Complete solutions to the Practice Exercises can be found in Appendix A.)

Format mesh analysis is valid for either DC or AC circuits. The coefficients of the loop currents for AC analysis are combinations of resistance, capacitive reactance, and inductive reactance to form the loop impedance. The AC network may contain multiple independent sinusoidal voltage signal sources of varying amplitude and phase. The procedure for writing the format mesh equations is identical to the previous examples involving DC resistive networks.

EXAMPLE 8–4   Format Mesh Analysis of an R-L-C Network

Given the R-L-C network in Figure 8–11, write the format mesh equations and solve for the loop currents. Determine the current for each component in the network.

FIGURE **8–11**

AC Circuit for
Example 8–4

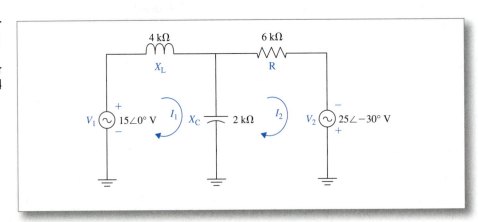

✔ **SOLUTION**

Loop 1: $I_1(4\angle 90° \text{ k}\Omega + 2\angle -90° \text{ k}\Omega) - I_2(2\angle -90° \text{ k}\Omega) = 15\angle 0° \text{ V}$

Loop 2: $-I_1(2\angle -90° \text{ k}\Omega) + I_2(2\angle -90° \text{ k}\Omega + 6\angle 0° \text{ k}\Omega) = 25\angle -30° \text{ V}$

The loop currents may be calculated using the TI-89 calculator as:

cSolve (x* ((4E3∠90)+(2E3∠−90)) − y* (2E3∠−90) = (15∠0) and −x* (2E3∠−90)+ y* ((6E3∠0) + (2E3∠−90)) = (25∠−30), {x,y})

CALCULATOR NOTE:

The calculator function to solve systems of equations involving phasor quantities is "cSolve".

The solutions to the format mesh equations are:

$I_1 = 6.79\angle -104.7° \text{ mA}$; $I_2 = 1.96\angle -283° \text{ mA}$;   and the capacitor current is calculated as $I_1 - I_2 = 6.58\angle -120.7° \text{ mA}$

Format mesh analysis may be applied to any electronic circuit or network. The method, however, is most useful for larger networks. The basic circuit analysis technique of series-parallel reduction presented in previous chapters is the most efficient and practical approach for less complex circuits.

8.2 FORMAT NODAL ANALYSIS

Nodal analysis is a systematic application of KCL at each node in a circuit. The resulting nodal equations can be solved to determine the voltage with respect to ground at each node in the circuit. Once the node voltages have been determined, the voltage across each circuit component may then be determined.

The procedure for writing format nodal equations is similar to the procedure for writing format mesh equations.

When writing format nodal equations, the equation variables are node voltages, and the signal sources should be represented as current sources.

Format nodal equations may be determined using the following procedure:

1. Identify the voltage at each node in the circuit with respect to ground as V_1, V_2, V_3, ... V_n, assuming all node voltages to be positive.

2. The left-hand side of the nodal equation contains a term for each node voltage, with the node voltage for the node being considered as the only positive term.

3. The coefficient of the node voltage for the node being considered is the sum of all conductances attached to that node.

4. The coefficient of the remaining node voltages is the mutual conductance connected between the node being considered and each respective remaining node.

5. The right-hand side of the equation is the algebraic sum of current sources attached to the node. If the source current is entering the node, it is a positive quantity in the nodal equation, and if the source current is leaving the node, it is a negative quantity in the nodal equation.

EXAMPLE 8–5 Nodal Analysis of a DC Resistive Network

Given the resistive network in Figure 8–12, write the format nodal equations. Solve for the node voltages. Determine the voltage across each resistor (conductance) in the network. Verify the solution using computer simulation.

FIGURE **8–12**

Circuit for Example 8–5 Indicating Node Voltages

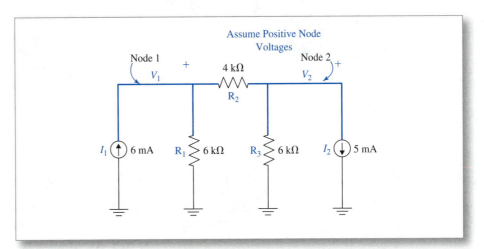

✓ **SOLUTION**

The node voltages in Figure 8–12 are designated as V_1 and V_2 as indicated. It is not necessary to write a nodal equation for the ground node because ground is the reference for all other voltage measurements and is considered to be zero.

Using the previously listed steps, the format nodal equations are written as:

Node 1: $V_1\left(\dfrac{1}{4\ \text{k}\Omega}+\dfrac{1}{2\ \text{k}\Omega}\right)-V_2\left(\dfrac{1}{4\ \text{k}\Omega}\right)=6\ \text{mA}$ (current entering node 1)

Node 2: $-V_1\left(\dfrac{1}{4\ \text{k}\Omega}\right)+V_2\left(\dfrac{1}{4\ \text{k}\Omega}+\dfrac{1}{6\ \text{k}\Omega}\right)=-5\ \text{mA}$ (current leaving node 2)

Note that the coefficient of node voltage V_1 in the first nodal equation is the sum of all conductances attached to node 1. The coefficient of node voltage V_2 is the mutual conductance between node 1 and node 2. A similar pattern is observed when writing the nodal equation at node 2.

Using the TI-89 calculator, we obtain the node voltages as:

Solve ((1/4E3) + (1/2E3) x−(1/4E3) y = 6E−3 and − (1/4E3) x + (1/4E3) + (1/6E3) y = −5E−3 ,{x,y}) ENTER

to obtain the respective node voltages as:

$$V_1 = 5\ \text{V}; \quad \text{and} \quad V_2 = -9\ \text{V}$$

Verification using MultiSIM is shown in Figure 8–13.

FIGURE **8-13**

MultiSIM
Simulation for
Example 8–5
Showing Node
Voltages

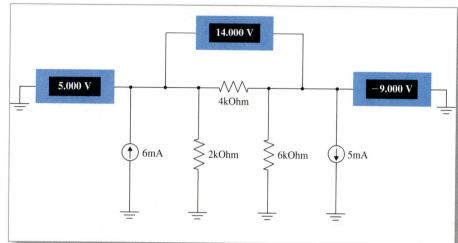

The analysis indicates that the voltage across R_1 is the voltage at node 1 or 5 V. The voltage across R_3 is the voltage at node 2 or -9 V. The voltage across R_2 is determined as the difference in node voltages 1 and 2 or 5 V $-$ (-9 V) = 14 V. The currents through each resistor can be calculated using Ohm's law if desired.

The results of the nodal analysis in Example 8–5 may be summarized as:

$$V_1 = 5 \text{ V}; \quad V_2 = -9 \text{ V};$$

$$V_{R1} = V_1 = 5 \text{ V}; \quad V_{R3} = V_2 = -9 \text{ V}; \quad V_{R2} = V_1 - V_2 = 5 \text{ V} - (-9 \text{ V}) = 14 \text{ V}$$

Format nodal analysis is valid for both DC and AC circuits. The coefficients of the node voltages for the AC analysis are combinations of conductance, capacitive susceptance, and inductive susceptance to form node admittances.

The AC network may contain multiple independent sinusoidal current sources of varying amplitude and phase. The procedure for writing the format nodal equations is identical to the procedure demonstrated in Example 8–5 for the DC resistive network.

EXAMPLE 8-6   Format Nodal Analysis for an R-L-C Network

Given the R-L-C network in Figure 8–14, write the format nodal equations.

FIGURE **8-14**

AC Circuit for
Example 8–6

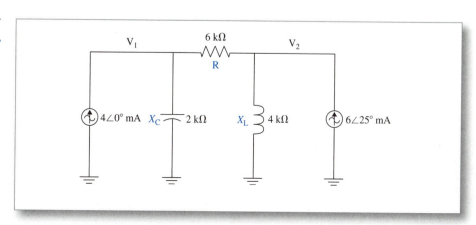

✓ SOLUTION

Node 1:
$$V_1\left(\frac{1}{6\angle 0°\text{ k}\Omega}+\frac{1}{2\angle -90°\text{ k}\Omega}\right)V_2-\left(\frac{1}{6\angle 0°\text{ k}\Omega}\right)=4\angle 0°\text{ mA}$$

Node 2:
$$-V_2\left(\frac{1}{6\angle 0°\text{ k}\Omega}\right)V_1+V_2\left(\frac{1}{6\angle 0°\text{ k}\Omega}+\frac{1}{4\angle 90°\text{ k}\Omega}\right)=6\angle 25°\text{ m}$$

Both KVL and KCL are often used to analyze circuits and networks without using the format procedures. The format procedures do, however, provide a systematic approach to circuit and network analysis that can be applied regardless of the complexity of the circuit or network to be solved. Most computer simulation programs rely upon either mesh or nodal analysis as the basis for solution.

PRACTICE EXERCISES 8−2

Given the resistive network in Figure 8–15, write the format nodal equations. Solve for the node voltages V_1 and V_2 as indicated. Use the node voltages to determine the voltage across each resistor (conductance) in the network.

FIGURE 8−15

Circuit for Practice Exercise 8–2

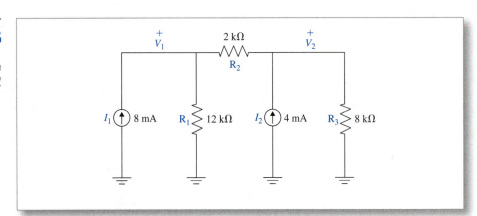

✓ ANSWERS

$V_1 = 61.1\text{ V};\quad V_2 = 55.2\text{ V};\quad V_{R1} = 61.1\text{ V};\quad V_{R2} = 13.9\text{ V};\quad V_{R3} = 55.2\text{ V}$

(Complete solutions to the Practice Exercises can be found in Appendix A.)

In the rare case where the input signal sources to an electronic circuit are a combination of voltage sources and current sources, it is common practice to convert the current sources to voltage sources and apply mesh analysis or to convert the voltage sources to current sources and apply nodal analysis. The procedure to convert signal sources is demonstrated in Appendix B.

8.3 THE SUPERPOSITION THEOREM

The **superposition theorem** is a method of analysis that is useful when solving circuits or networks that have more than one signal source. The method can be applied to circuits and networks that contain voltage sources, current sources, or both. The superposition theorem is particularly useful when analyzing circuits that contain both voltage and current sources.

When applying the superposition theorem, the value of the current or voltage for each component is determined, considering only one source at a time. The current and voltage obtained from the analysis considering each source may then be algebraically summed to determine the net current or voltage for each respective component.

The Superposition theorem is valid for circuits or networks that contain DC sources, sinusoidal sources, or a combination of both.

The superposition theorem requires the circuit to be analyzed considering only one source at time, which implies that the remaining sources must be temporarily removed from the circuit.

A voltage source is removed from a network by shorting its terminals. The voltage across the source is reduced to zero while a path through which current may flow is maintained.

A current source is removed from a network by opening its terminals. The current from the source is reduced to zero, since current cannot flow through the open terminals.

EXAMPLE 8–7   Superposition Analysis of a Resistive Network

Use the method of superposition to determine the current through each component for the network in Figure 8–16. Verify the solution using computer simulation.

FIGURE **8–16**

Circuit for
Example 8–7
Applying the
Superposition
Theorem

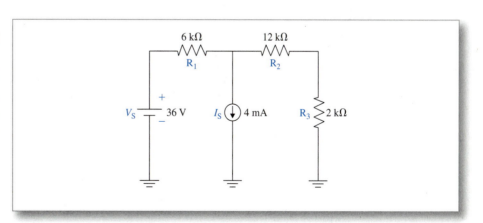

✓ SOLUTION

Considering the Voltage Source V_S:

The effects of the voltage source (V_S) may be determined by removing the current source (I_S) from the circuit. Recall that a current source is removed by opening its terminals as shown in Figure 8–17.

FIGURE **8-17**

Considering the Voltage Source (V_S)

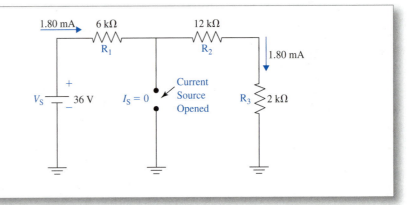

The network in Figure 8–17 may now be analyzed as a single source network. Since the current through the open circuit must be zero, the circuit in Figure 8–17 consists of V_S, R_1, R_2, and R_3 connected in series.

The total equivalent resistance for the circuit can be determined as:

$$R_T = R_1 + R_2 + R_3 = 6 \text{ k}\Omega + 12 \text{ k}\Omega + 2 \text{ k}\Omega = 20 \text{ k}\Omega$$

Ohm's law can be used to calculate the source current as:

$$I_S = \frac{V}{R_T} = \frac{36 \text{ V}}{20 \text{ k}\Omega} = 1.8 \text{ mA}$$

The current through each resistor in Figure 8–17 is shown, indicating the direction of current flow. It is important to indicate the direction of current flow because the superposition theorem requires the current from each respective source to be algebraically summed for each component.

The current directions will indicate whether currents will be added or subtracted to determine the net current.

$$I_{6 \text{ k}\Omega} = 1.8 \overrightarrow{\text{ mA}}; \quad I_{12 \text{ k}\Omega} = 1.8 \overrightarrow{\text{ mA}}; \quad I_{2 \text{ k}\Omega} = 1.8 \text{ mA} \downarrow$$

Considering the Current Source I_S:

The effects of the current source (I_S) may be determined by removing the voltage source (V_S) as shown in Figure 8–18. Recall that a voltage source is removed by shorting its terminals.

FIGURE **8-18**

Considering the Voltage Source (I_S)

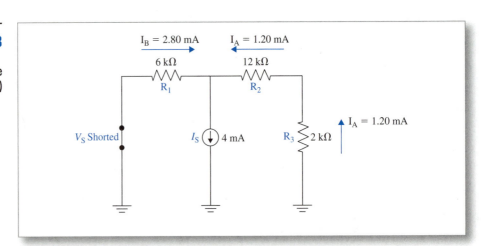

The 4 mA current from the current source divides between two branches. The resistance of branch A is the sum of 12 kΩ + 2 kΩ = 14 kΩ, and the resistance of branch B is 6 kΩ.

Branch A and branch B are connected in parallel, so that the total equivalent resistance can be calculated using the product-over-sum method to obtain:

$$R_A \, /\!/ \, R_B = \frac{(14 \text{ k}\Omega)(6 \text{ k}\Omega)}{(14 \text{ k}\Omega + 6 \text{ k}\Omega)} = 4.2 \text{ k}\Omega$$

The current divider rule can be used to calculate the current through each branch as:

$$I_A = 4 \text{ mA} \, \frac{4.2 \text{ k}\Omega}{14 \text{ k}\Omega} = 1.2 \text{ mA}$$

$$I_B = 4 \text{ mA} \, \frac{4.2 \text{ k}\Omega}{6 \text{ k}\Omega} = 2.8 \text{ mA}$$

The current through each resistor in Figure 8–18 is determined as:

$$I_{6 \text{ k}\Omega} = 2.8 \overrightarrow{\text{ mA}}; \quad I_{12 \text{ k}\Omega} = 1.2 \overleftarrow{\text{ mA}}; \quad \text{and } I_{2 \text{ k}\Omega} = 1.2 \text{ mA} \uparrow$$

The superposition theorem can be used to determine the net current through each resistor as the algebraic sum of the respective currents from both sources to obtain:

$$I_{6 \text{ k}\Omega} = 1.8 \text{ mA} + 2.8 \text{ mA} = 4.6 \overrightarrow{\text{ mA}}$$

$$I_{12 \text{ k}\Omega} = 1.8 \text{ mA} - 1.2 \text{ mA} = 0.6 \overrightarrow{\text{ mA}}$$

$$I_{2 \text{ k}\Omega} = 1.8 \text{ mA} - 1.2 \text{ mA} = 0.6 \text{ mA} \downarrow$$

The currents through each resistor are added when the arrows are in the same direction. The currents are subtracted when the arrows are in opposite directions.

The MultiSIM simulation is shown in Figure 8–19.

FIGURE **8–19**

MultiSIM
Simulation
for Example 8–7

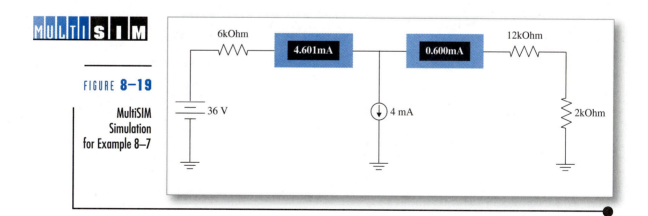

EXAMPLE 8–8   Superposition Analysis of a Resistive Network

Use the superposition theorem to determine the current through each component for the network in Figure 8–20. Verify the solution using computer simulation.

FIGURE **8–20**

Circuit for
Example 8–8

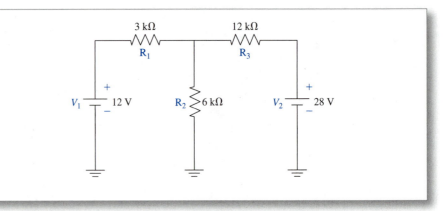

Considering the voltage source V_1 and removing (shorting) voltage source V_2, the network may be redrawn as shown in Figure 8–21.

FIGURE **8–21**

Considering the
Voltage Source V_1

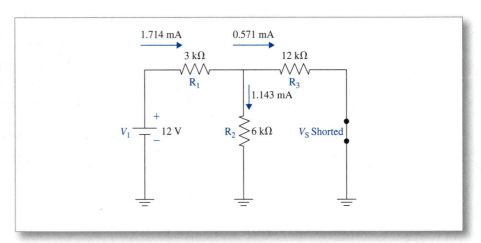

The total equivalent resistance as seen by V_1 in Figure 8–21 can be calculated as:

$$R_T = (R_2 /\!/ R_3) + R_1 = (6 \text{ k}\Omega /\!/ 12 \text{ k}\Omega) + 3 \text{ k}\Omega = 7 \text{ k}\Omega$$

Ohm's law may be used to determine the source current as:

$$I_S = \frac{V_1}{R_T} = \frac{12 \text{ V}}{7 \text{ k}\Omega} = 1.714 \text{ mA}$$

The source current I_S flows directly through R_1 and then divides between R_2 and R_3, so that:

$$I_{R1} = 1.714 \overrightarrow{\text{ mA}}$$

Using the current divider rule to determine the current through R_2 and R_3, we obtain:

$$I_{R2} = 1.714 \text{ mA} \; \frac{4 \text{ k}\Omega}{6 \text{ k}\Omega} = 1.143 \text{ mA} \downarrow$$

$$I_{R3} = 1.714 \text{ mA} \; \frac{4 \text{ k}\Omega}{12 \text{ k}\Omega} = 0.571 \overrightarrow{\text{ mA}}$$

Considering the voltage source V_2 and removing (shorting) voltage source V_1, the network may be redrawn as shown in Figure 8–22.

FIGURE 8-22

Considering the
Voltage Source V_2

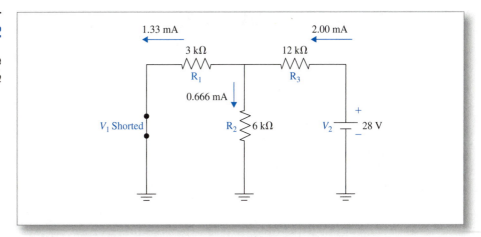

The total equivalent resistance as seen by V_2 in Figure 8–22 can be calculated as:

$$R_T = (R_2 // R_1) + R_3 = (3 \text{ k}\Omega // 6 \text{ k}\Omega) + 12 \text{ k}\Omega = 14 \text{ k}\Omega$$

Ohm's law can be used to determine the source current as:

$$I_{S2} = \frac{V_2}{R_T} = \frac{28 \text{ V}}{14 \text{ k}\Omega} = 2.00 \text{ mA}$$

The source current I_S flows directly through R_3 and then divides between R_1 and R_2, so that:

$$I_{R3} = 2.00 \overleftarrow{\text{ mA}}$$

Using the current divider rule to determine the current through R_1 and R_2, we obtain:

$$I_{R1} = 2 \text{ mA} \frac{2 \text{ k}\Omega}{3 \text{ k}\Omega} = 1.333 \overleftarrow{\text{ mA}}$$

$$I_{R2} = 2 \text{ mA} \frac{2 \text{ k}\Omega}{6 \text{ k}\Omega} = 0.666 \text{ mA} \downarrow$$

The total current through the resistors is determined as the algebraic sum of the currents from both sources to obtain:

$$I_{R1} = 1.714 \text{ mA} - 1.33 \text{ mA} = 0.384 \overrightarrow{\text{ mA}}$$
$$I_{R2} = 1.143 \text{ mA} + 0.666 \text{ mA} = 1.809 \text{ mA} \downarrow$$
$$I_{R3} = 2.00 \text{ mA} - 0.571 \text{ mA} = 1.429 \overleftarrow{\text{ mA}}$$

The MultiSIM simulation is shown in Figure 8–23.

FIGURE 8-23

MultiSIM
Simulation for
Example 8-8

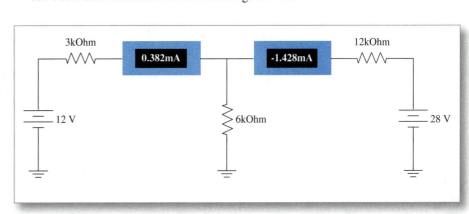

EXAMPLE 8-9   Superposition Analysis of an R-L-C Network

Use the superposition theorem to determine the current I_C for the network in Figure 8–24.

FIGURE **8-24**

AC Circuit for Example 8-9

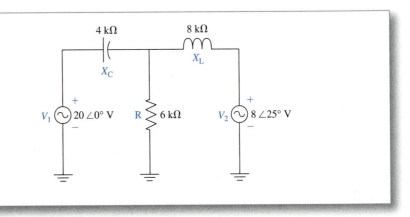

✓ **SOLUTION**

Considering the voltage source V_1 and removing (shorting) voltage source V_2, the network may be redrawn as shown in Figure 8–25.

FIGURE **8-25**

Considering the Sinusoidal Voltage Source V_1

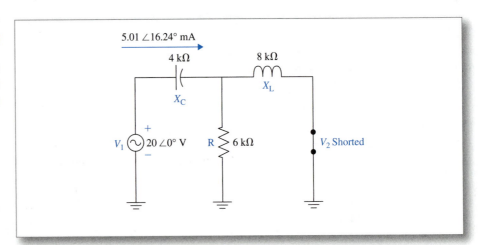

The total impedance of the circuit in Figure 8–25 may be calculated as:

$$\mathbf{Z} = (\mathbf{X_L} \,//\, \mathbf{R}) + \mathbf{X_C} = 3.99\angle{-16.24°}\ k\Omega$$

Ohm's law may be used to determine the source current as:

$$\mathbf{I_S} = \frac{\mathbf{V_1}}{\mathbf{Z}} = \frac{20\angle 0°\ V}{3.99\angle{-16.24°}\ k\Omega} 20\angle 0°/3.99\ k\Omega = 5.01\angle 16.24°\ mAz$$

The capacitor in Figure 8–25 is in series with the voltage source V_1, so that the capacitive current may be determined as:

$$\mathbf{I_{C1}} = \mathbf{I_S} = 5.01\angle 16.24°\ \overrightarrow{mA}$$

Considering the voltage source V_2 and removing (shorting) voltage source V_1, the network may be redrawn as shown in Figure 8–26.

FIGURE **8–26**

Considering the Sinusoidal Voltage Source V_2

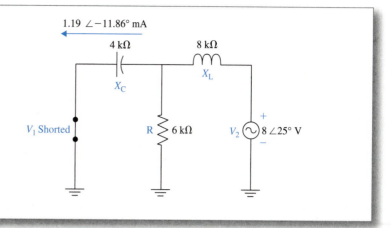

The total impedance of the circuit in Figure 8–26 can be calculated as:

$$\mathbf{Z} = (\mathbf{X_C} // \mathbf{R}) + \mathbf{X_L} = 5.55 \angle 70.56° \text{ k}\Omega$$

Ohm's law can be used to determine the source current as:

$$\mathbf{I_S} = \frac{\mathbf{V_2}}{\mathbf{Z}} = \frac{8 \angle 25° \text{ V}}{5.55 \angle 70.56° \text{ k}\Omega} = 1.44 \angle -45.56° \text{ mA}$$

Using the current divider rule to determine $\mathbf{I_C}$, we obtain:

$$\mathbf{I_{C2}} = \mathbf{I_S}\frac{\mathbf{R}//\mathbf{X_C}}{\mathbf{X_C}} = 1.44 \angle -45.56° \text{ mA} \frac{3.33 \angle -56.3° \text{ k}\Omega}{4 \angle -90° \text{ k}\Omega} = 1.19 \angle -11.86° \overleftarrow{\text{mA}}$$

The superposition theorem can be used to determine the net capacitive current $\mathbf{I_C}$ as:

$$\mathbf{I_C} = \mathbf{I_{C1}} - \mathbf{I_{C2}}$$

$$\mathbf{I_C} = 5.01 \angle 16.24° \text{ mA} - 1.19 \angle -11.86° \text{ mA} = 3.99 \angle 24.30° \text{ mA}$$

PRACTICE EXERCISES 8–3

Use the method of superposition to determine the current through each resistor for the network in Figure 8–27.

FIGURE **8–27**

Circuit for Practice Exercise 8–3

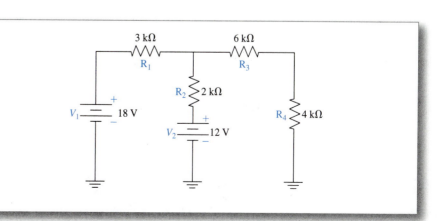

$I_{R1} = 1.71 \overrightarrow{\text{mA}}; \quad I_{R2} = 0.43 \text{ mA} \downarrow; \quad I_{R3} = 1.28 \text{ mA} \downarrow$

(Complete solutions to the Practice Exercises can be found in Appendix A.)

8.4 THEVENIN'S THEOREM

An electronic circuit is often designed to provide an output signal voltage or current to a specified load. A speaker connected to an amplifier is considered to be a load on the amplifier. A resistor connected to a DC voltage source is considered to be a load on the voltage source.

Thevenin's theorem is useful in cases where only the value of load current or load voltage is of interest. The theorem provides an efficient method of determining values of load current or load voltage for various values of load resistance or impedance.

Thevenin's theorem can reduce any complex circuit to an equivalent circuit consisting of a single voltage source and a series resistance or impedance.

The DC Thevenin equivalent circuit consists of a single voltage source (V_{Th}) in series with resistor R_{Th}. The AC Thevenin equivalent circuit consists of a single sinusoidal voltage source (V_{Th}) in series with impedance Z_{Th}.

The Thevenin equivalent DC and AC prototype circuits are shown in Figure 8–28.

FIGURE **8–28**

Prototype DC and AC Thevenin Equivalent Circuits

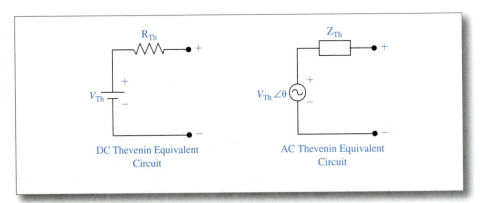

DC Thevenin Equivalent Circuit

AC Thevenin Equivalent Circuit

A DC Thevenin equivalent circuit may be determined using the following procedure:

1. Remove the designated load from the circuit. The open terminals created by removing the load are designated as the Thevenin terminals.

2. Remove all sources from the circuit. Voltage sources are shorted and current sources are opened.

3. Determine the total resistance between the Thevenin terminals. This resistance is the Thevenin resistance in the prototype Thevenin equivalent circuit shown in Figure 8–28.

4. Replace all sources in the circuit. The Thevenin voltage (V_{Th}) is determined as the open circuit voltage across the Thevenin terminals.

5. Connect the load resistor across the terminals of the Thevenin equivalent circuit to determine the voltage or current for specified load values.

EXAMPLE 8–10   Applying Thevenin's Theorem

Given the circuit in Figure 8–29, determine the Thevenin equivalent circuit for the designated load resistor (R_L). Use the Thevenin equivalent circuit to determine the voltage across R_L for values of 2 kΩ, 4 kΩ, 6 kΩ, and 10 kΩ. Use computer simulation to verify the solution for a 2 kΩ load and a 4 kΩ load.

FIGURE **8–29**

Circuit for
Example 8–10

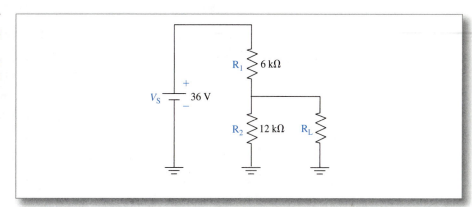

✓ SOLUTION

Steps 1 and 2 of the procedure can be combined to produce the circuit shown in Figure 8–30. The load resistor and source have been removed, and the Thevenin terminals are identified.

FIGURE **8–30**

Calculating
Thevenin
Resistance

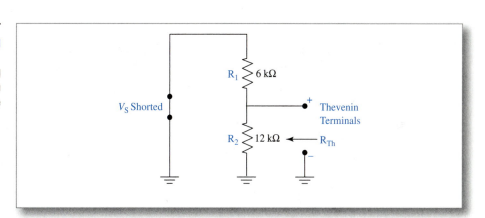

The Thevenin resistance can be calculated as:

$$R_{Th} = R_1 \; // \; R_2 = 6 \text{ k}\Omega \; // 12 \text{ k}\Omega = 4 \text{ k}\Omega$$

With V_S shorted, resistors R_1 and R_2 are in parallel when viewed from the Thevenin terminals.

Replacing the source (V_S), we obtain the circuit shown in Figure 8–31. Note that the load resistor is not replaced in this step.

FIGURE **8–31**

Calculating
Thevenin
Voltage

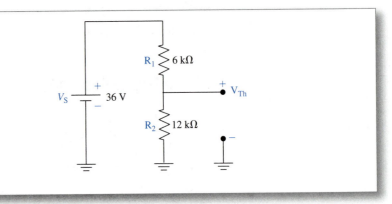

The Thevenin voltage is the voltage across the 12 kΩ resistor and may be calculated using the voltage divider rule to obtain:

$$V_{Th} = V \frac{R_2}{R_1 + R_2} = 36 \text{ V} \frac{12 \text{ k}\Omega}{6 \text{ k}\Omega} = 24 \text{ V}$$

The complete Thevenin equivalent circuit is shown in Figure 8–32.

FIGURE **8–32**

Thevenin
Equivalent
Circuit

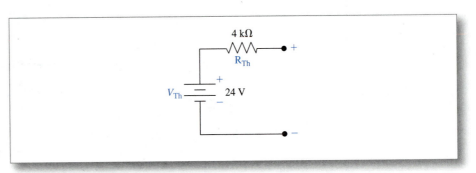

The Thevenin equivalent circuit with $R_L = 2$ kΩ connected is shown in Figure 8–33.

FIGURE **8–33**

Connecting a
2 kΩ
Load Resistor

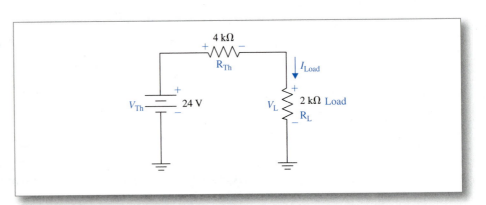

The voltage divider rule can be used to calculate the voltage across each respective given load value as:

For a 2 kΩ load:
$$V_L = V_{Th} \frac{R_L}{R_L + R_{Th}} = 24 \text{ V} \frac{2 \text{ k}\Omega}{6 \text{ k}\Omega} = 8 \text{ V}$$

For a 4 kΩ load:
$$V_L = V_{Th} \frac{R_L}{R_L + R_{Th}} = 24 \text{ V} \frac{4 \text{ k}\Omega}{8 \text{ k}\Omega} = 12 \text{ V}$$

For a 6 kΩ load:
$$V_{\mathrm{L}} = V_{\mathrm{Th}} \frac{R_{\mathrm{L}}}{R_{\mathrm{L}} + R_{\mathrm{Th}}} = 24 \text{ V} \frac{6 \text{ k}\Omega}{10 \text{ k}\Omega} = 14.4 \text{ V}$$

For a 10 kΩ load:
$$V_{\mathrm{L}} = V_{\mathrm{Th}} \frac{R_{\mathrm{L}}}{R_{\mathrm{L}} + R_{\mathrm{Th}}} = 24 \text{ V} \frac{10 \text{ k}\Omega}{14 \text{ k}\Omega} = 17.14 \text{ V}$$

Note that once the Thevenin equivalent circuit has been determined, the voltage across various load resistances can easily be calculated using the voltage divider rule. The respective load currents can be calculated using Ohm's law if desired.

The MultiSIM simulation in Figure 8–34 shows the original circuit and the Thevenin equivalent circuit with a 2 kΩ load, indicating a load voltage of 8 V in each circuit.

FIGURE 8–34

MultiSIM
Simulation
Showing Load
Voltage for
a 2 kΩ Load

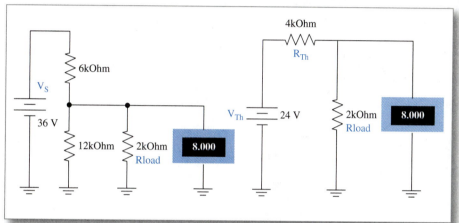

The MultiSIM simulation in Figure 8–35 shows the original circuit and the Thevenin equivalent circuit with a 4 kΩ load, indicating a load voltage of 12 V.

FIGURE 8–35

MultiSIM
Simulation
Showing Load
Voltage for
a 4 kΩ Load

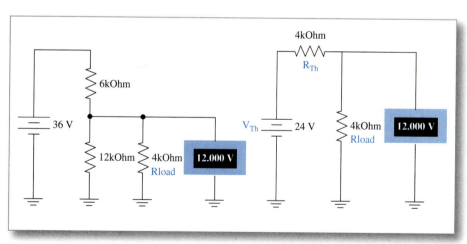

EXAMPLE 8–11   Applying Thevenin's Theorem

Given the circuit in Figure 8–36, determine the Thevenin equivalent circuit for the load resistor (R_{L}). Use the Thevenin equivalent circuit to determine the voltage across R_L for values of 8 kΩ, 12 kΩ, and 24 kΩ. Use computer simulation to verify the solution for an 8 kΩ load and a 24 kΩ load.

FIGURE **8–36**

Circuit for
Example 8–11

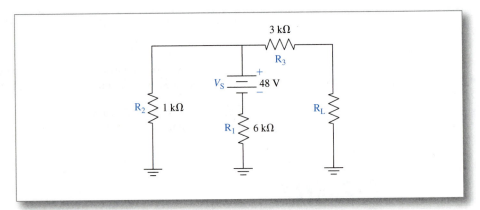

✓ **SOLUTION**

Removing the load resistor R_L, shorting the 48 V voltage supply, and identifying the Thevenin terminals, we obtain the circuit shown in Figure 8–37.

FIGURE **8–37**

Calculating
Thevenin
Resistance

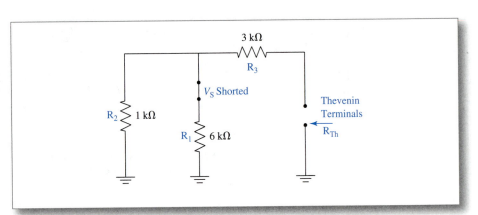

R_{Th} is calculated in Figure 8–37 as:

$$R_{Th} = (R_1 /\!/ R_2) + R_3 = 0.86 \text{ k}\Omega + 3 \text{ k}\Omega = 3.86 \text{ k}\Omega$$

Note that with the voltage source shorted, R_1 and R_2 are in parallel. Also note that the Thevenin resistance is determined as the resistance looking in from the Thevenin terminals.

Replacing the 48 V voltage source, we obtain the circuit in Figure 8–38.

FIGURE **8–38**

Calculating
Thevenin
Voltage

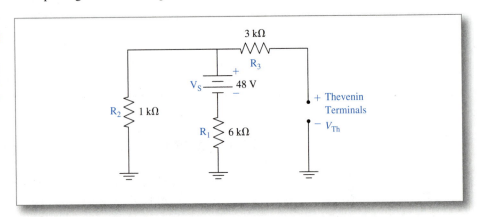

With the Thevenin terminals open as shown in Figure 8–38, no current can flow through R_3. Since the current through R_3 is zero, the voltage across R_3 must also be zero.

The total equivalent resistance as seen by the 48 V source in Figure 8–38 is $R_1 + R_2$. The resistor R_3 does not contribute to the total equivalent resistance because it is series with the open Thevenin terminals, so that:

$$R_T = R_1 + R_2 = 6\ k\Omega + 1\ k\Omega = 7\ k\Omega$$

Ohm's law can be used to calculate the source current in Figure 8–38 as:

$$I_S = \frac{48\ V}{7\ k\Omega} = 6.86\ mA$$

The source current flows through R_1, so that the voltage across R_1 can be calculated as:

$$V_{R1} = (6.86\ mA)(6\ k\Omega) = 41.16\ V$$

KVL can be expressed as:

$$-V_{R1} + 48\ V - V_{R3} - V_{Th} = 0\ V$$

so that: $\qquad V_{Th} = 48\ V - 41.16\ V - 0\ V = 6.84\ V$

The complete Thevenin equivalent circuit with an 8 kΩ load resistance connected is shown in Figure 8–39.

FIGURE **8–39**

Thevenin
Equivalent
Circuit with an
8 kΩ Load

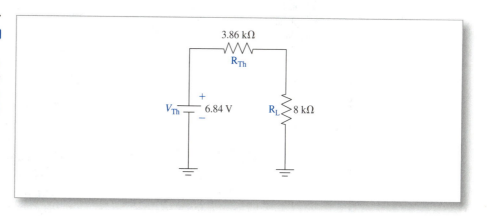

The Thevenin equivalent circuit and the voltage divider rule can be used to calculate the voltage across the specified load resistors as:

For an 8 kΩ load: $\quad V_L = V_{Th}\dfrac{R_L}{R_L + R_{Th}} = 6.84\ V\ \dfrac{8\ k\Omega}{(8\ k\Omega + 3.86\ k\Omega)} = 4.61\ V$

For a 12 kΩ load: $\quad V_L = V_{Th}\dfrac{R_L}{R_L + R_{Th}} = 6.84\ V\ \dfrac{12\ k\Omega}{(12\ k\Omega + 3.86\ k\Omega)} = 5.19\ V$

For a 24 kΩ load: $\quad V_L = V_{Th}\dfrac{R_L}{R_L + R_{Th}} = 6.84\ V\ \dfrac{24\ k\Omega}{(24\ k\Omega + 3.86\ k\Omega)} = 5.9\ V$

The MultiSIM simulation in Figure 8–40 shows the original circuit and the Thevenin equivalent circuit with an 8 kΩ load, indicating a load voltage of 4.26 V.

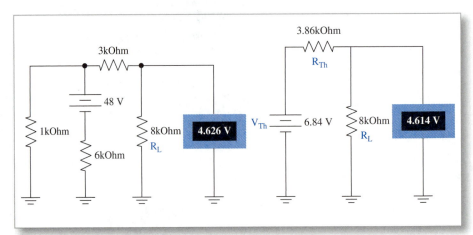

MULTISIM

FIGURE **8–40**

MultiSIM
Simulation
Showing Load
Voltage for an
8 kΩ Load

The MultiSIM simulation in Figure 8–41 shows the original circuit and the Thevenin equivalent circuit with a 24 kΩ load, indicating a load voltage of 5.9 V.

MULTISIM

FIGURE **8–41**

MultiSIM
Simulation
Showing Load
Voltage for a
24 kΩ Load

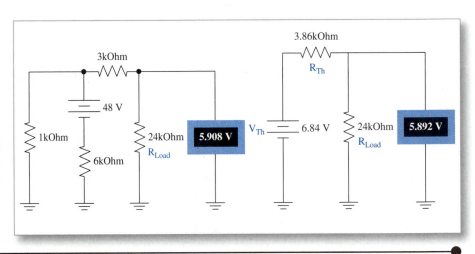

EXAMPLE 8–12   Applying Thevenin's Theorem to an AC Network

Given the circuit in Figure 8–42, determine the Thevenin equivalent circuit for the designated load capacitor (C). Use the Thevenin equivalent circuit to determine the voltage across the designated load capacitor C, given $C = 0.02$ μF. Use computer simulation to verify the solution.

FIGURE **8–42**

AC Circuit for
Example 8–12

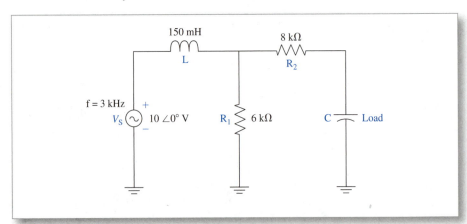

✓ SOLUTION

The Thevenin impedance is determined by shorting the voltage source (V_S) and removing the designated load capacitor as shown in Figure 8–43.

FIGURE **8–43**

Calculating
Thevenin
Impedance

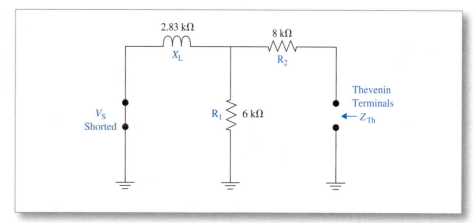

FIGURE **8–43**

Calculating
Thevenin
Impedance

With the voltage source shorted, R_1 and X_L are connected in parallel, so the Thevenin impedance is determined as:

$$\mathbf{Z}_{Th} = (\mathbf{R}_1 // \mathbf{X}_L) + \mathbf{R}_2$$

where: $$\mathbf{X}_L = 2\pi fL = 2\pi(3k\ Hz)(150\ mH) = 2.83\angle 90°\ k\Omega$$

so that: $$\mathbf{Z}_{Th} = (6\angle 0°\ k\Omega)//2.83\angle 90°\ k\Omega) + 8\angle 0°\ k\Omega = 9.38\angle 14.28°\ k\Omega$$

The circuit to determine the Thevenin voltage $\mathbf{V}_{Th}$ is shown in Figure 8–44.

FIGURE **8–44**

Calculating
Thevenin
Voltage

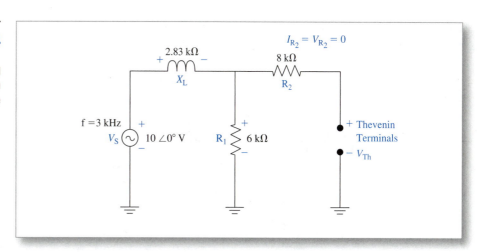

With the Thevenin terminals open as shown in Figure 8–44, the current through resistor R_2 will be zero and the voltage across R_2 is also zero, as indicated.

Expressing KVL around the closed loop beginning at the ground side of R_1, we obtain:

$$\mathbf{V}_{R1} + 0 - \mathbf{V}_{Th} = 0$$

so that: $$\mathbf{V}_{Th} = \mathbf{V}_{R1}$$

With the Thevenin terminals open, R_1 is in series with X_L, so that the voltage across R_1 may be calculated using the voltage divider rule as:

$$\mathbf{V_{Th}} = \mathbf{V_{R1}} = \mathbf{V_S}\frac{\mathbf{R_1}}{\mathbf{R_1} + \mathbf{X_L}} = 10\angle 0° \text{ V } \frac{6\angle 0° \text{ k}\Omega}{6\angle 0° \text{ k}\Omega + 2.983\angle 90° \text{ k}\Omega}$$

$$= 9.05\angle -25.23° \text{ V}$$

The complete Thevenin equivalent circuit with the load capacitor connected across the Thevenin terminals is shown in Figure 8–45.

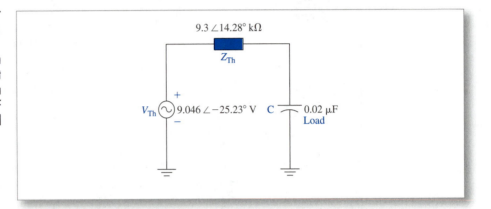

FIGURE 8–45

Thevenin
Equivalent
Circuit with a
0.02 μF
Capacitive Load

The capacitive reactance of the load capacitor is calculated as:

$$\mathbf{X_C} = \frac{1}{2\pi fC} = \frac{1}{2\pi (3 \text{ kHz})(0.02 \text{ μF})} = 2.65\angle -90° \text{ k}\Omega$$

The voltage divider rule may be used to determine $\mathbf{V_C}$ as:

$$\mathbf{V_C} = \mathbf{V_{Th}}\frac{\mathbf{X_C}}{\mathbf{X_C} + \mathbf{Z_{Th}}} = 9.05\angle -25.23° \text{ V } \frac{2.65\angle -90° \text{ k}\Omega}{2.65\angle -90° \text{ k}\Omega + 9.38\angle 14.28° \text{ k}\Omega}$$

$$\mathbf{V_C} = 2.61\angle -113.09° \text{ V}$$

The MultiSIM simulation is shown in Figure 8–46. Note that the peak voltage across the capacitor in Figure 8–46 is 2.61 V.

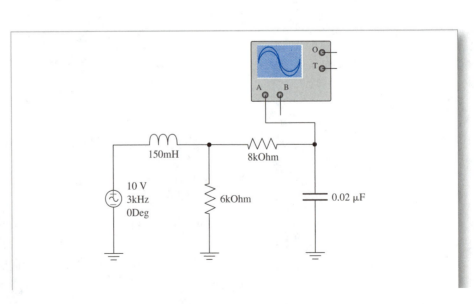

FIGURE 8–46

Example 8–12
Simulation
Showing
Voltage across a
0.02 μF
Capacitive
Load

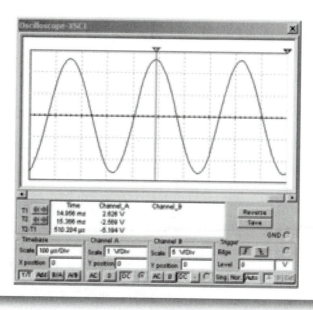

PRACTICE EXERCISES 8–4

a. Given the circuit in Figure 8–47, determine the Thevenin equivalent circuit for the designated load resistor (R_L). Use the Thevenin equivalent circuit to determine the voltage across R_L for values of 1 kΩ and 6 kΩ.

FIGURE **8–47**

Circuit for Practice
Exercise 8–4
Exercise a

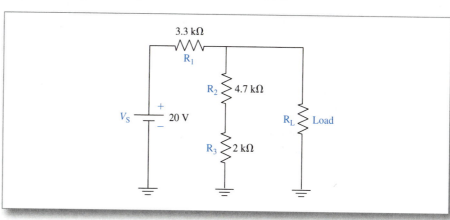

b. Given the circuit in Figure 8–48, determine the Thevenin equivalent circuit for the designated load resistor (R_L). Use the Thevenin equivalent circuit to determine the voltage across R_L for values of 4 kΩ and 12 kΩ.

FIGURE **8–48**

Circuit for Practice
Exercise 8–4
Exercise b

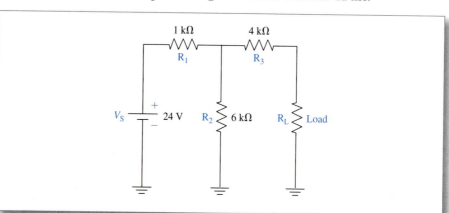

a. $R_{Th} = 2.21 \text{ k}\Omega$; $V_{Th} = 13.4 \text{ V}$;   For 1 kΩ: $R_L = 1 \text{ k}\Omega$; $V_{RL} = 4.17 \text{ V}$;

   For 6 kΩ : $R_L = 6 \text{ k}\Omega$; $V_{RL} = 9.79 \text{ V}$

b. $R_{Th} = 4.86 \text{ k}\Omega$; $V_{Th} = 20.57 \text{ V}$;   For 4 kΩ: $R_L = 4 \text{ k}\Omega$; $V_{RL} = 9.29 \text{ V}$;

   For 12 kΩ: $R_L = 12 \text{ k}\Omega$; $V_{RL} = 14.64 \text{ V}$

(Complete solutions to the Practice Exercises can be found in Appendix A.)

8.5 NORTON'S THEOREM

Norton's theorem is similar to Thevenin's theorem in that both are most useful when determining the voltage and current for one of the circuit components, typically a designated load.

The Norton equivalent circuit can also be obtained by converting the Thevenin equivalent circuit using the source conversion process illustrated in Appendix B. It is more appropriate, however, to determine the Norton's equivalent circuit directly using the procedure outlined in this section.

The Thevenin equivalent circuit consists of a single ideal voltage (V_{Th}, $\mathbf{V_{Th}}$) in series with a single resistance or impedance (R_{Th}, $\mathbf{Z_{Th}}$). The Norton equivalent circuit consists of a single current source (I_N, $\mathbf{I_N}$) in parallel with a single resistance or impedance (R_N, $\mathbf{Z_N}$). Since R_N and $\mathbf{Z_N}$ are connected in parallel with the current source, they may be referred to as G_N (Norton conductance) or $\mathbf{Y_N}$ (Norton admittance) respectively. This text will refer to the parallel component as R_N or $\mathbf{Z_N}$.

The Norton DC and AC equivalent prototype circuits are shown in Figure 8–49.

FIGURE **8–49**

Prototype DC and AC
Norton's Equivalent
Circuits

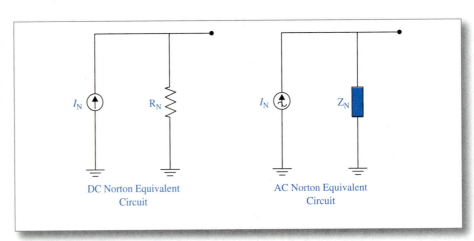

DC Norton Equivalent
Circuit

AC Norton Equivalent
Circuit

A DC Norton equivalent circuit can be determined using the following procedure:

1. Remove the component under consideration, typically designated as the load, from the circuit. The terminals created by removing the load are designated to be the Norton terminals.

2. Remove all sources from the circuit. Voltage sources are shorted and current sources are opened.

3. Determine the total resistance between the Norton terminals. This resistance is the parallel resistance in the prototype Norton equivalent circuits shown in Figure 8–49.

4. Replace all sources in the circuit. The value of the Norton current source I_N in Figure 8–49 is determined as the current that flows through the shorted Norton terminals.

5. Connect the designated load resistor across the Norton equivalent circuit terminals and use the current divider rule to determine the current through the specified load.

EXAMPLE 8–13   Applying Norton's Theorem

Given the circuit in Figure 8–50, determine the Norton equivalent circuit for the designated load resistor (R_L). Use the Norton equivalent circuit to determine the current through R_L for values of 8 kΩ, 12 kΩ, and 24 kΩ. Use computer simulation to verify the solution for the 8 kΩ load.

FIGURE **8–50**

Circuit for
Example 8–13

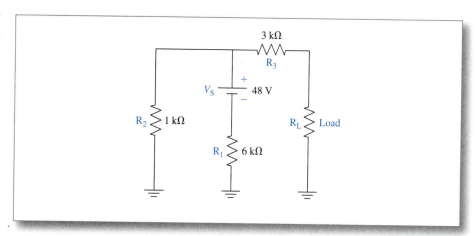

✓ SOLUTION

Steps 1 and 2 may be combined to produce the circuit shown in Figure 8–51.

FIGURE **8–51**

Calculating Norton
Resistance

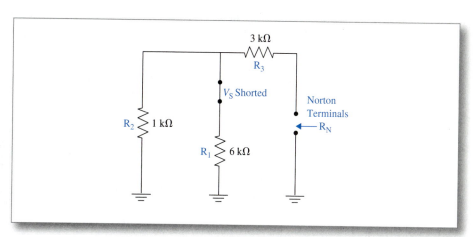

R_N is calculated in Figure 8–51 as:

$$R_N = (R_1 \,//\, R_2) + R_3 = 3.86 \text{ k}\Omega$$

Note that the procedure to determine the Norton resistance is identical to the procedure used to determine the Thevenin resistance, so that:

$$R_N = R_{Th}$$

The circuit to determine I_N is shown in Figure 8–52. Note that the Norton terminals are shorted when calculating I_N.

FIGURE **8–52**

Calculating Norton
Current

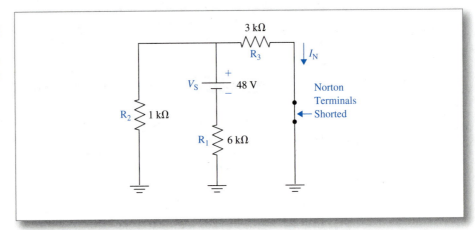

The total equivalent resistance as seen from the 48 V source in Figure 8–52 can be calculated as:

$$R_T = (R_2 \mathbin{/\mkern-3mu/} R_3) + R_1 = (1\ \text{k}\Omega \mathbin{/\mkern-3mu/} 3\ \text{k}\Omega) + 6\ \text{k}\Omega = 6.75\ \text{k}\Omega$$

Note that shorting the Norton terminals places R_2 in parallel with R_3.

Ohm's law can be used to determine the source current as:

$$I_S = \frac{48\ \text{V}}{6.75\ \text{k}\Omega} = 7.11\ \text{mA}$$

The current divider rule can be used to determine the Norton current, which is the current through R_3, to obtain:

$$I_N = I_S \frac{R_P}{R_3} = 7.11\ \text{mA} \; \frac{0.75\ \text{k}\Omega}{3\ \text{k}\Omega} = 1.78\ \text{mA}$$

where:

$$R_P = R_2 \mathbin{/\mkern-3mu/} R_3 = 0.75\ \text{k}\Omega$$

The Norton equivalent circuit with an 8 kΩ load resistor connected is shown in Figure 8–53.

FIGURE **8–53**

Norton Equivalent
Circuit with an
8 kΩ Load

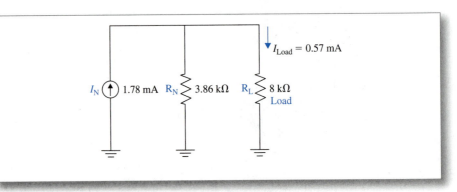

The current divider rule may be used to calculate the current through the 8 kΩ load resistor in Figure 8–53 as:

$$I_{R_L} = I_N \frac{R_P}{R_L} = 1.78\ \text{mA} \; \frac{2.52\ \text{k}\Omega}{8\ \text{k}\Omega} = 0.57\ \text{mA}$$

where R_P is the parallel combination of R_N and $R_L = 2.52\ \text{k}\Omega$.

The MultiSIM simulation in Figure 8–54 shows the original circuit and the Norton equivalent circuit with an 8 kΩ load connected, indicating a load current of 0.57 mA.

FIGURE 8–54

MultiSIM Simulation Showing Load Current

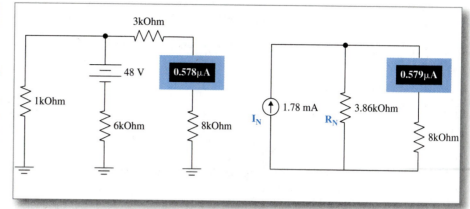

The currents through the 12 kΩ and 24 kΩ loads can be calculated using the current divider rule as:

For a 12 kΩ load: $I_{R_L} = I_N \dfrac{R_P}{R_L} = 1.78 \text{ mA} \dfrac{2.52 \text{ k}\Omega}{12 \text{ k}\Omega} = 0.374 \text{ mA}$

For a 24 kΩ load: $I_{R_L} = I_N \dfrac{R_P}{R_L} = 1.78 \text{ mA} \dfrac{2.52 \text{ k}\Omega}{24 \text{ k}\Omega} = 0.187 \text{ mA}$

where: $R_P = R_N // R_L$

The voltage across the 8 kΩ load can be calculated using Ohm's law as:

$$V_{R_L} = (I_{R_L})(R_L) = (0.57 \text{ mA})(8 \text{ k}\Omega) = 4.56 \text{ V}$$

The circuit in this example is the same circuit as Figure 8–36 in Example 8–11. Example 8–11 demonstrated the use of Thevenin's theorem to calculate the voltage across an 8 kΩ load to obtain 4.61 V, compared to the Norton's theorem solution above of 4.56 V. The slight differences between the solutions are due to rounding inaccuracies.

PRACTICE EXERCISES 8–5

Given the circuit in Figure 8–55, determine the Norton equivalent circuit for the designated load resistor (R_L). Use the Norton equivalent circuit to determine the current through R_L for values of 1 kΩ and 6 kΩ.

FIGURE 8–55

Circuit for Practice Exercise 8–5

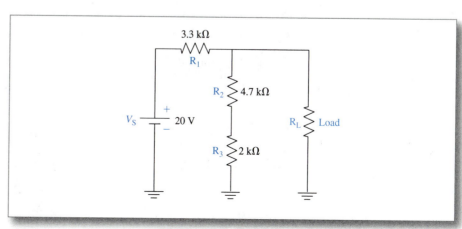

$R_N = 2.21 \text{ k}\Omega$; $I_N = 6.06 \text{ mA}$;   For 1 kΩ: $R_L = 1 \text{ k}\Omega$; $I_L = 4.17 \text{ mA}$;

For 6 kΩ: $R_L = 6 \text{ k}\Omega$; $I_L = 1.62 \text{ mA}$

(Complete solutions to the Practice Exercises can be found in Appendix A.)

AN APPLICATION OF THEVENIN'S THEOREM

Consider the circuit in Figure 8–56.

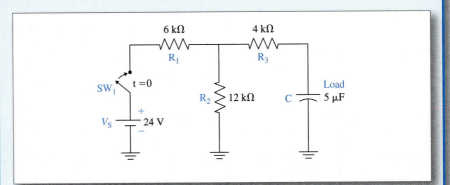

FIGURE **8–56**

An Application
of Thevenin's
Theorem

SPECIFICATIONS: Determine the voltage across the capacitor at time $t = 45$ ms.

✓ SOLUTION

The capacitor in Figure 8–56 will begin to charge when the switch is closed at time $t = 0$. The equation for the charging capacitor ($v_C = V_S(1 - e^{-t/\tau})$, where $\tau = RC$, was developed in the previous chapter for the prototype circuit consisting of the capacitor in series with a resistor and a voltage source.

The circuit in Figure 8–56 is not in the prototype configuration, so that the equation $v_C = V_S(1 - e^{-t/\tau})$ cannot be directly applied.

Thevenin's theorem may be used to reduce the circuit in Figure 8–56 to an equivalent circuit in the configuration of a capacitor in series with a single series resistance.

The equation defining the voltage across the charging capacitor can then be expressed using the Thevenin resistance and Thevenin voltage as:

$$v_C = V_{Th}(1 - e^{-t/\tau})$$

where: $$\tau = R_{Th}\,C$$

The Thevenin resistance can be determined from the circuit shown in Figure 8–57 as:

$$R_{Th} = R_1 \,//\, R_2 + R_3 = 8 \text{ k}\Omega$$

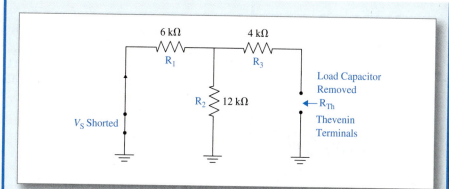

FIGURE **8-57**

Calculating Thevenin Resistance

The Thevenin voltage can be determined from Figure 8–58 as:

$$V_{Th} = V_S \frac{R_2}{R_1 + R_2} = 16 \text{ V}$$

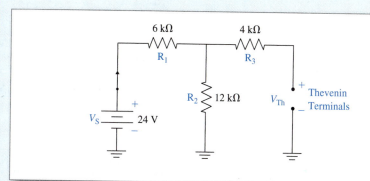

FIGURE **8-58**

Calculating Thevenin Voltage

The Thevenin equivalent circuit with the load capacitor connected is shown in Figure 8–59.

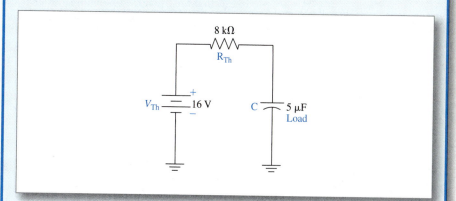

FIGURE **8-59**

Thevenin Equivalent Circuit with a 5 μF Capacitive Load Connected

Note that the voltage across the capacitor as a function of time in Figure 8–59 can be expressed as:

$$V_C = V_{Th}(1 - e^{-t/\tau})$$

where: $\tau = R_{Th}C = (8 \text{ k}\Omega)(5 \text{ μF}) = 40 \text{ ms}$

The voltage across the capacitor at the specified time $t = 45$ ms can be determined as:

$$V_C = 16(1 - e^{-45 \text{ ms}/40 \text{ ms}}) \text{ V} = 10.8 \text{ V}$$

The charging capacitor is shown in the MultiSIM transient analysis shown in Figure 8–60.

MULTISIM

FIGURE **8–60**

MultiSIM Simulation
Showing Voltage
across the Load
Capacitor

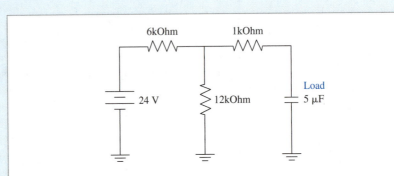

Note that the maximum voltage across the capacitor is 16 V, as indicated in the Thevenin equivalent circuit.

8.6 ALTERNATE METHODS OF SOLVING NETWORKS

Several network analysis methods and theorems have been presented in this chapter. The methods presented include mesh and nodal analysis, superposition, Thevenin's theorem, and Norton's theorem. Each of the methods has its own advantages and disadvantages, given the nature of the circuit or network to be analyzed. The examples that follow in this section are intended to demonstrate how each of the methods may be applied to a given circuit or network to achieve similar solutions.

It is common engineering practice to analyze or design a circuit using more than one method when possible to ensure the accuracy of the solution. Computer simulation is also an excellent resource to verify solutions.

EXAMPLE 8–14   An Alternate Solution to Example 8–9

Determine the current for the capacitor in Figure 8–24 of Example 8–9 using format mesh analysis.

✓ **SOLUTION**

The superposition solution for the circuit in Figure 8–24 can be verified using format mesh analysis as:

Loop 1: $I_1(4\angle -90° \text{ k}\Omega + 6\angle 0° \text{ k}\Omega) - I_2(6\angle 0° \text{ k}\Omega) = 20\angle 0° \text{ V}$

Loop 2: $-I_1(6\angle 0° \text{ k}\Omega) + I_2(8\angle 90° \text{ k}\Omega + 6\angle 0° \text{ k}\Omega) = -8\angle 25° \text{ V}$

Using the calculator to solve for I_1, we obtain:

cSolve(x*((4∠−90)+(6∠0))−y*(6∠0)=(20∠0) and −x*(6∠0)+y*((6∠0)+ (8∠90))=(−8∠25),x) ENTER

The calculator solution is x = I_1 = $\mathbf{I_C}$ = 3.98 ∠24.44 mA, which verifies the solution obtained using superposition in Example 8–9.

EXAMPLE 8–15   Alternate Methods of Solution

Given the network in Figure 8–61, determine the current and voltage for the 6 kΩ resistor using mesh analysis and the superposition theorem. Compare the solutions and verify the solution using computer simulation.

FIGURE 8–61

Circuit for
Example 8–15

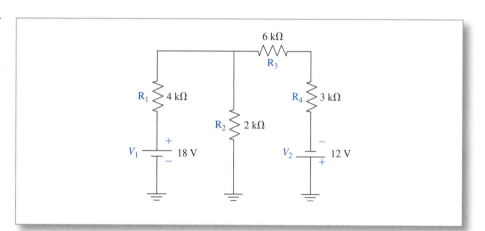

✓ SOLUTION

Solution using format mesh analysis:

The network in Figure 8–61 with loop currents indicated is shown in Figure 8–62.

FIGURE 8–62

Writing Format
Mesh Equations

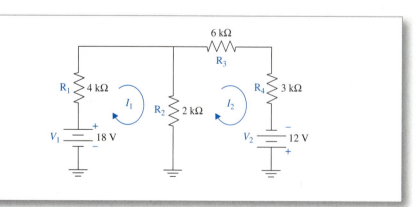

The format mesh equations are written as:

Loop 1: $I_1 (4 \text{ k}\Omega + 2 \text{ k}\Omega) - I_2 (2 \text{ k}\Omega) = 18 \text{ V}$

Loop 2: $-I_1 (2 \text{ k}\Omega) + I_2 (2 \text{ k}\Omega + 6 \text{ k}\Omega + 3 \text{ k}\Omega) = 12 \text{ V}$

Solving for loop currents I_1 and I_2, we obtain:

$$I_1 = 3.58 \text{ mA}$$

$$I_2 = 1.742 \text{ mA}$$

The current through the 6 kΩ resistor is loop current I_2, so that:

$$I_{6\text{ k}\Omega} = 1.742 \text{ mA} \rightarrow$$

Ohm's law may be used to determine the voltage across the 6 kΩ resistor as:

$$V_{6\text{ k}\Omega} = (1.742 \text{ mA})(6 \text{ k}\Omega) = 10.452 \text{ V}$$

✓ SOLUTION

Solution using superposition:

Considering the 18 V source and shorting the 12 V source, we obtain the circuit in Figure 8–63.

FIGURE **8–63**

Superposition
Considering the
Voltage Source V_1

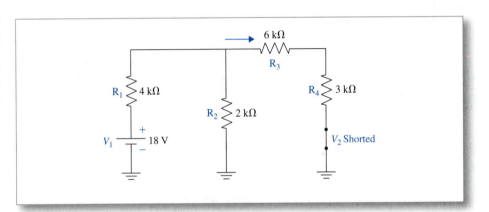

The total equivalent resistance as seen by the 18 V source in Figure 8–63 is:

$$R_T = (R_3 + R_4)//(R_2) + R_1 = 5.636 \text{ k}\Omega$$

Ohm's law can be used to calculate the source current as:

$$I_S = \frac{18 \text{ V}}{5.36 \text{ k}\Omega} = 3.194 \text{ mA}$$

The current through the 6 kΩ resistor can be calculated using the current divider rule as:

$$I_{6\text{ k}\Omega} = I_S \frac{R_P}{R_3 + R_4} = 3.194 \text{ mA} \frac{1.64 \text{ k}\Omega}{9 \text{ k}\Omega} = 0.581 \text{ mA} \rightarrow$$

Considering the 12 V source and shorting the 18 V source, we obtain the circuit in Figure 8–64.

The total equivalent resistance as seen by the 18 V source in Figure 8–64 is:

$$R_T = (R_1 // R_2) + R_3 + R_4$$

$$R_T = 10.33 \text{ k}\Omega$$

FIGURE **8–64**

Superposition
Considering the
Voltage Source V_2

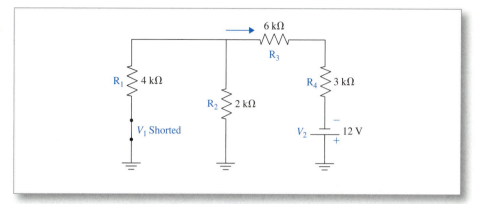

Ohm's law can be used to calculate the source current as:

$$I_S = \frac{12 \text{ V}}{10.33 \text{ k}\Omega} = 1.162 \text{ mA}$$

The source current flows through the 6 kΩ resistor, so that:

$$I_{6 \text{ k}\Omega} = 1.162 \text{ mA} \rightarrow$$

By using the superposition theorem, the total current through the 6 kΩ resistor is determined as:

$$I_{6 \text{ k}\Omega} = 0.581 \text{ mA} \rightarrow + 1.162 \text{ mA} \rightarrow = 1.743 \text{ mA}$$

The solution obtained using the superposition theorem agrees with the solutions obtained by mesh and nodal analysis.

The MultiSIM simulation for Example 8–15 is shown in Figure 8–65.

FIGURE **8–65**

MultiSIM
Simulation
Showing the
Current through
the 6 kΩ
Resistor

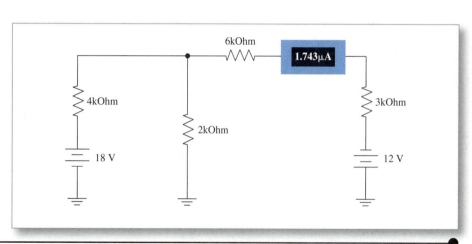

PRACTICE EXERCISES 8–6

a. Given the circuit in Figure 8–66, determine the Thevenin equivalent circuit for the designated load resistor R_L. Use the Thevenin equivalent circuit to determine the current through R_L for values of 4 kΩ and 12 kΩ. Determine the Norton equivalent circuit for the designated load resistor and use the Norton equivalent circuit to determine the current through R_L for values of 4 kΩ and 12 kΩ. Compare the solutions.

FIGURE **8–66**

Circuit for Practice
Exercise 8–6 Exercise 1

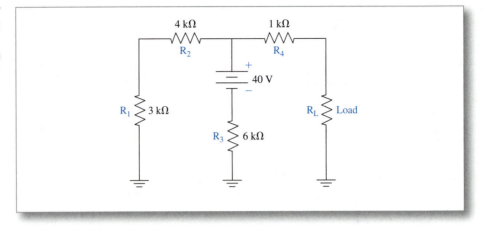

b. Given the circuit in Figure 8–67, determine the current through R_4 using format mesh analysis. Determine the current through R using superposition. Compare the solutions.

FIGURE **8–67**

Circuit for Practice
Exercise 8–6
Exercise 2

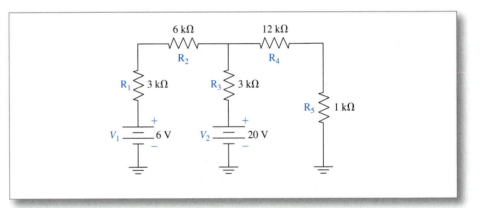

✔ **ANSWERS**

a. $R_{Th} = R_N = 4.23$ kΩ; $V_{Th} = 21.54$ V; $I_N = 5.11$ mA;

$I_{(RL=4\ k\Omega)} = 2.62$ mA; $I_{(RL=12\ k\Omega)} = 1.33$ mA

b. $I_{R4} = 1.08$ mA

(Complete solutions to the Practice Exercises can be found in Appendix A.)

SUMMARY

Format mesh analysis is a systematic procedure for writing a KVL equation for each of the closed loops in a given network. The resulting system of equations can be solved to obtain the magnitude and direction of each loop current. The current for each component in the network can be determined as one of the loop currents or as the sum or difference of two loop currents.

Format nodal analysis is a systematic procedure for writing a KCL equation for each of the nodes in a given network. The resulting system of equations can be solved to obtain the magnitude and polarity of each node voltage. The voltage across each component in the network may be determined as one of the node voltages or as the sum or difference of two node voltages.

Format mesh analysis and format nodal analysis are particularly useful when analyzing larger networks, especially those with multiple signal sources.

The **superposition theorem** is often used to determine current and voltage in a circuit or network that has more than one signal source. The superposition theorem states that a network with multiple input signals can be analyzed by considering one source at a time. The results of each individual analysis may then be algebraically summed to determine the current and voltage for each component in the network.

Thevenin's theorem and **Norton's theorem** are frequently used to determine load current and voltage under changing load conditions. A Thevenin equivalent circuit consists of an ideal voltage source (V_{Th}) in series with a resistance (R_{Th}) or impedance (Z_{Th}). A Norton equivalent circuit consists of an ideal current source (I_N) in parallel with a resistance (R_N) or impedance (Z_N).

Alternate methods of analysis can be used to solve electronic circuits. It is good practice to verify circuit solutions by employing at least one alternate method of analysis. Computer simulation is also an excellent tool to verify circuit analysis solutions.

ESSENTIAL LEARNING EXERCISES

(ANSWERS TO ODD-NUMBERED ESSENTIAL LEARNING EXERCISES CAN BE FOUND IN APPENDIX A.)

1. Review the source conversion procedures demonstrated in Appendix B. Convert a 20 V voltage source whose series internal resistance is 4 kΩ to an equivalent current source. Connect a 3 kΩ load resistor to each of the equivalent sources and verify that the load current from each source is equal.

2. Review the source conversion procedures demonstrated in Appendix B. Convert a 12 V voltage source whose series internal resistance is 5 kΩ to an equivalent current source. Connect a 3 kΩ load resistor to each of the equivalent sources and verify that the load current from each source is equal.

3. Review the source conversion procedures demonstrated in Appendix B. Convert a current source of 20 mA whose parallel internal resistance is 2.8 kΩ to an equivalent voltage source. Connect a 1.5 kΩ load resistor to each of the equivalent sources and verify that the load current from each source is equal.

4. Review the source conversion procedures demonstrated in Appendix B. Convert a current source of 32 mA whose parallel internal resistance is 3.5 kΩ to an equivalent voltage source. Connect a 4 kΩ load resistor to each of the equivalent sources and verify that the load current from each source is equal.

5. Review the source conversion procedures demonstrated in Appendix B. Convert a voltage source of $12\angle 0°$ V whose series internal impedance is $4\angle 26°$ kΩ to an equivalent current source.

6. Review the source conversion procedures demonstrated in Appendix B. Convert a voltage source of $21\angle 25°$ V whose series internal impedance is $5.6\angle 18°$ kΩ to an equivalent current source.

7. Review the source conversion procedures demonstrated in Appendix B. Convert a current source of $16.4\angle 40°$ mA whose parallel internal impedance is $3.9\angle -23°$ kΩ to an equivalent voltage source.

8. Review the source conversion procedures demonstrated in Appendix B. Convert a current source of $3.7\angle 15°$ mA whose parallel internal impedance is $3.9\angle 38°$ kΩ to an equivalent voltage source.

9. Given the network in Figure 8–68, determine the current through each resistor using format mesh analysis. Verify your solution using computer simulation.

FIGURE **8–68**

Circuit for Essential
Learning Exercise 9

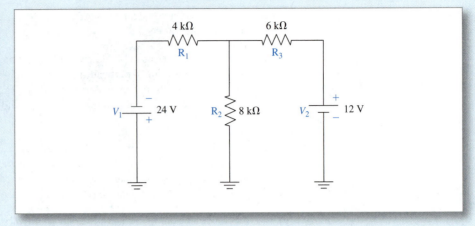

10. Given the network in Figure 8–69, determine the current through each resistor using format mesh analysis. Verify your solution using computer simulation.

FIGURE **8–69**

Circuit for Essential
Learning Exercise 10

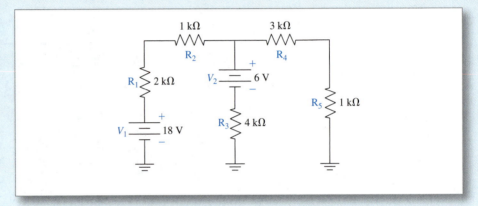

11. Given the network in Figure 8–70, determine the voltage across each resistor using format nodal analysis. Verify your solution using computer simulation.

FIGURE **8–70**

Circuit for Essential
Learning Exercise 11

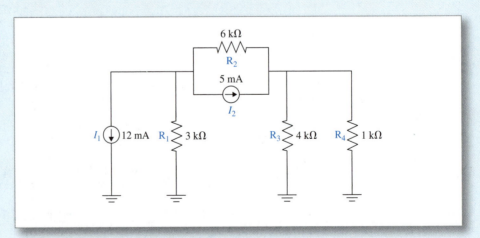

12. Given the network in Figure 8–71, determine the voltage across each resistor using format nodal analysis. Verify your solution using computer simulation.

FIGURE **8–71**

Circuit for Essential
Learning Exercise 12

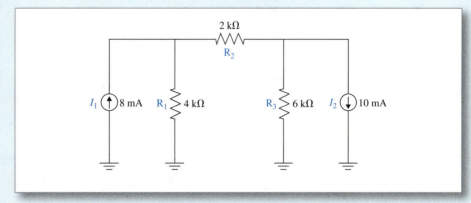

13. Determine the current through R_2 in Figure 8–72 using the superposition theorem. Verify your solution using computer simulation.

FIGURE **8–72**

Circuit for Essential
Learning Exercise 13

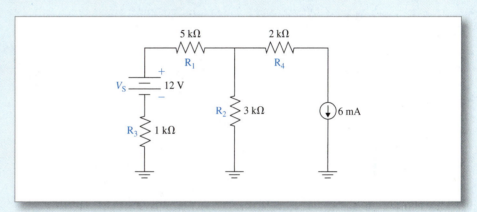

14. Determine the current through R_2 in Figure 8–73 using the superposition theorem. Verify your solution using computer simulation.

FIGURE **8–73**

Circuit for Essential
Learning Exercise 14

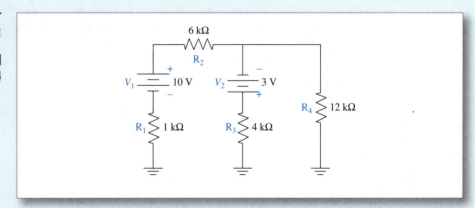

15. Determine the Thevenin equivalent circuit for the designated load resistor R_L in Figure 8–74. Use the Thevenin equivalent circuit to determine the current through R_L for $R_L = 4$ kΩ and $R_L = 12$ kΩ. Verify your solution using computer simulation to simulate the original circuit with the specified values of R_L.

FIGURE **8–74**

Circuit for Essential
Learning Exercise 15

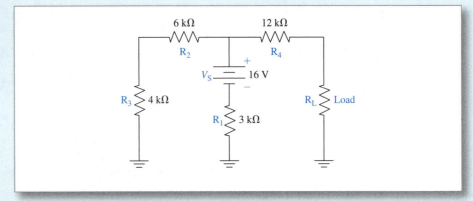

16. Determine the Thevenin equivalent circuit for the designated load resistor R_L in Figure 8–75. Use the Thevenin equivalent circuit to determine the current through R_L for $R_L = 2.5$ kΩ and $R_L = 8.5$ kΩ. Verify your solution using computer simulation to simulate the original circuit with the specified values of R_L.

FIGURE **8–75**

Circuit for Essential
Learning Exercise 16

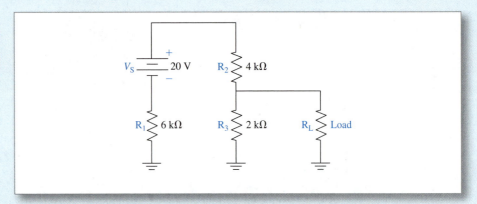

17. Determine the Norton equivalent circuit for the designated load resistor R_L in Figure 8–76. Use the Norton equivalent circuit to determine the current through R_L for $R_L = 1.5$ kΩ and $R_L = 4.5$ kΩ. Verify your solution using computer simulation to simulate the original circuit with the specified values of R_L.

FIGURE **8–76**

Circuit for Essential
Learning Exercise 17

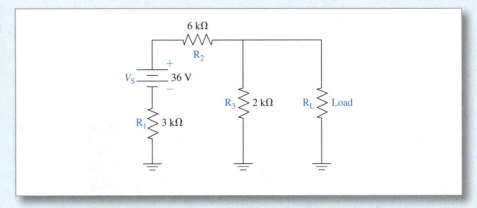

18. Determine the Norton equivalent circuit for the designated load resistor R_L in Figure 8–77. Use the Norton equivalent circuit to determine the current through R_L for $R_L = 12$ kΩ and $R_L = 18$ kΩ. Verify your solution using computer simulation to simulate the original circuit with the specified values of R_L.

FIGURE 8–77

Circuit for Essential Learning Exercise 18

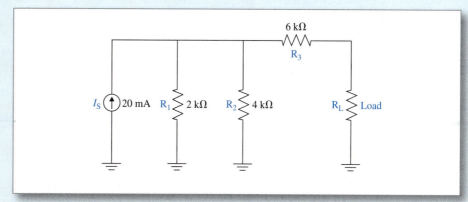

19. Determine the current through resistor R_2 in Figure 8–78 using format mesh analysis. Determine the current through R_2 in Figure 8–78 using the superposition theorem. Compare the solutions.

FIGURE 8–78

Circuit for Essential Learning Exercise 19

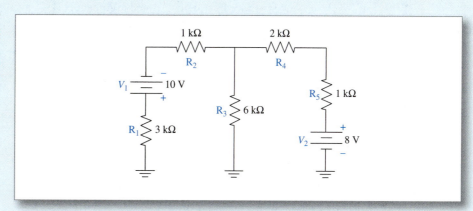

20. Determine the current through resistor R_2 in Figure 8–79 using format mesh analysis. Determine the current through R_2 in Figure 8–79 using the superposition theorem. Compare the solutions.

FIGURE 8–79

Circuit for Essential Learning Exercise 20

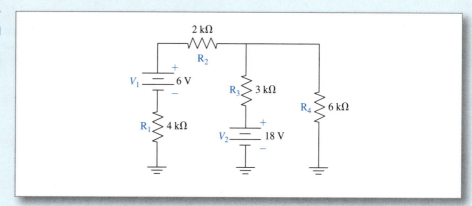

(ANSWERS TO SELECTED CHALLENGING LEARNING EXERCISES CAN BE FOUND IN APPENDIX A.)

1. Determine the current through the inductor in Figure 8–80 using format mesh analysis.

FIGURE **8–80**

Circuit for Challenging
Learning Exercise 1

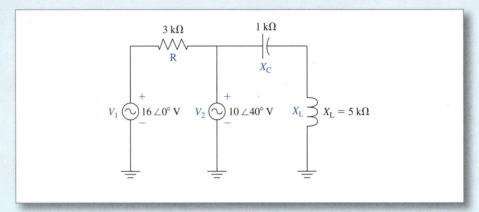

2. Determine the voltage across R_3 in Figure 8–81 using format nodal analysis. Review the source conversion procedures demonstrated in Appendix B. Convert the 16 V voltage source in series with R_1 to an equivalent current source before writing the nodal equations. Verify the voltage across R_3 using computer simulation.

FIGURE **8–81**

Circuit for Challenging
Learning Exercise 2, 3, 4

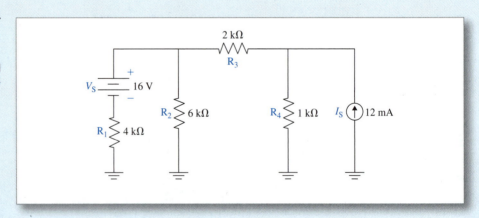

3. Determine the voltage across R_3 in Figure 8–81 using format mesh analysis. Review the source conversion procedures demonstrated in Appendix B. Convert the 12 mA current source in parallel with R_4 to an equivalent voltage source before writing the mesh equations. Compare the solution to Challenging Exercise 2.

4. Determine the voltage across R_3 in Figure 8–81 using the superposition theorem. Compare the solution to Challenging Exercises 2 and 3.

5. Determine the Thevenin equivalent circuit for the designated load resistor R_L in Figure 8–82. Use the Thevenin equivalent circuit to determine the current through the load resistor R_L if R_L is 2.5 kΩ.

FIGURE **8–82**

Circuit for
Challenging Learning
Exercise 5, 6

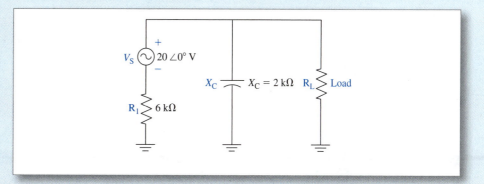

FIGURE **8–82**

Circuit for
Challenging Learning
Exercise 5, 6

6. Determine the Norton equivalent circuit for the designated load resistor R_L in Figure 8–82. Use the Norton equivalent circuit to determine the current through the load resistor R_L if R_L is 2.5 kΩ. Compare the solution to Challenging Exercise 5.

7. Determine the Thevenin equivalent circuit for the capacitor in Figure 8–83. Use the Thevenin equivalent circuit to determine the voltage across the capacitor 10 ms after the switch is closed in Figure 8–83.

FIGURE **8–83**

Circuit for Challenging
Learning Exercise 7

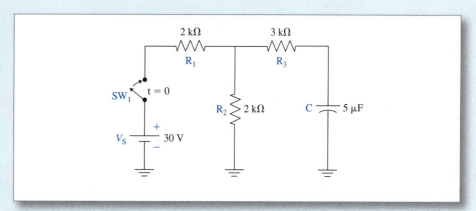

8. Determine the Thevenin equivalent circuit for the capacitor in Figure 8–84. Use the Thevenin equivalent circuit to determine the voltage across the 3 μF capacitor 3 ms after the switch is closed.

FIGURE **8–84**

Circuit for Challenging
Learning Exercise 8

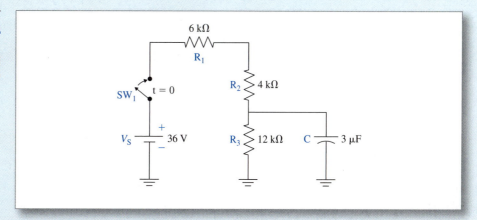

TEAM ACTIVITY

Construct the circuit in Figure 8–85 using computer simulation. Connect channel 1 and channel 2 of the oscilloscope as indicated.

FIGURE **8–85**

Team Activity Circuit 1

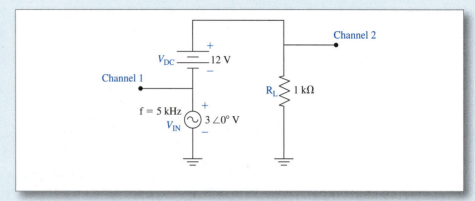

Observe that the output signal (channel 2) is the 3 V sinusoidal input signal V_{in} "riding" on top of the 12 V DC voltage. The sine wave is *superimposed* onto the DC voltage. Note that the peak value of the output waveform is $V_{DC} + V_P = 12\text{ V} + 3\text{ V} = 15\text{ V}$.

The circuit in Figure 8–86 is the same circuit as Figure 8–85 except that a capacitor has been placed in series with the load resistor.

FIGURE **8–86**

Team Activity Circuit 2

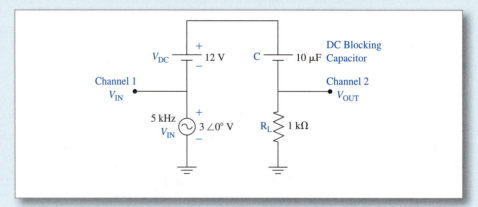

The capacitor is referred to as a "blocking capacitor" and its function is to remove the DC content of the output waveform. Construct the circuit in Figure 8–86 using computer simulation. Observe that the output signal no longer contains a DC component, so that the output signal is the same waveform as the sinusoidal input signal.

POWER CONSIDERATIONS IN ELECTRONIC CIRCUITS

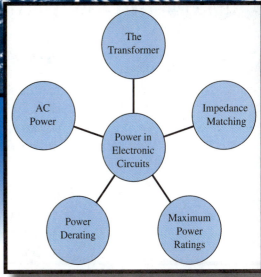

"DISAPPOINTMENT SHOULD ALWAYS BE
TAKEN AS A STIMULANT AND NEVER VIEWED
AS A DISCOURAGEMENT."

—C.B. Newcomb

CHAPTER OUTLINE

LEARNING OBJECTIVES

Upon successful completion of this chapter you will be able to:

- Calculate the real, reactive, and apparent power in an AC circuit.
- Determine the power factor for an AC circuit.
- Sketch the power triangle for an AC circuit.
- Determine the primary and secondary voltage and current for a step-up and step-down transformer.
- Determine the proper transformer turns ratio for impedance matching.
- Select circuit components based upon maximum power ratings.
- Derate maximum power ratings due to temperature effects.

CHAPTER NINE

INTRODUCTION

Power considerations for electronic circuits can be divided into three related but separate considerations. The first section of this chapter discusses power in an AC electronic circuit from the perspective of power source requirements. The source must supply power to the circuit, and so it is important to calculate how much power is required by the resistors, capacitors, and inductors that comprise the circuit.

The chapter then considers power from a signal processing perspective, which is to ensure that maximum power from a signal source is delivered to a specified load. The maximum power transfer theorem defines the conditions necessary for maximum power transfer, and techniques to achieve maximum power transfer are presented.

The final two sections of the chapter consider the power ratings required for the components that comprise the electronic circuit. Components must be selected so that their maximum power ratings are not exceeded. The power ratings of components vary with ambient temperature and must be adjusted or derated to compensate for temperature effects.

9.1 THE GENERAL POWER EQUATION

The general instantaneous power equation is expressed as the product of instantaneous voltage and current, so that:

$$p = vi \text{ (instantaneous power)} \tag{9.1}$$

The general expression of a sinusoidal voltage waveform is $v = V_P \sin(\omega t)$, assuming that the phase angle of the voltage is $0°$. The general expression of a sinusoidal current is $i = I_P \sin(\omega t + \theta)$, where θ is the phase difference between the current and voltage.

Substituting the general expressions for voltage and current into the instantaneous power Equation (9.1), we obtain:

$$p = vi = V_P \sin(\omega t) I_P \sin(\omega t + \theta) \tag{9.2}$$

where the phase angle of the voltage is assumed to be 0.

Given the trigonometric identity:

$$\sin(A) \sin(B) = \frac{1}{2}[\cos(A - B) - \cos(A + B)]$$

and letting:

$$A = (\omega t); \text{ and } B = (\omega t + \theta)$$

we obtain:

$$A - B = (-\theta); \quad \text{and} \quad A + B = (2\omega t + \theta)$$

so that the instantaneous power equation can be expressed as:

$$p = \frac{1}{2}V_P I_P \cos(-\theta) - \frac{1}{2}V_P I_P \cos(2\omega t + \theta) \tag{9.3}$$

The expression for instantaneous power in Equation (9.3) consists of two terms. The first term can be expressed as:

$$\frac{1}{2} V_P I_P \cos(-\theta)$$

Substituting the identity $\cos(-\theta) = \cos(\theta)$, the first term may be expressed as:

$$\frac{1}{2} V_P I_P \cos(\theta)$$

Note that the first term is a constant (not a function of time).

The second term can be expressed as:

$$-\frac{1}{2} V_P I_P \cos(2\omega t + \theta)$$

Note that the second term is a sinusoidal waveform at a frequency that is twice (2ω) the frequency of the current and voltage.

Figure 9–1 is a MATLAB plot of a sinusoidal current and voltage and instantaneous power, where the phase difference between the current and voltage (θ) is 60°.

FIGURE 9–1

Instantaneous Voltage, Current, and Power

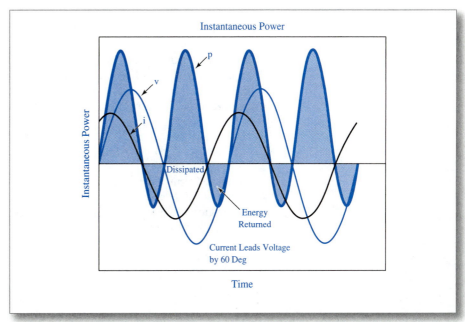

The instantaneous power waveform in Figure 9–1 is sinusoidal at twice the frequency of the current and voltage, as predicted by the second term of Equation (9.3). The instantaneous power waveform in Figure 9–1 is not symmetrical about the time axis. This offset is due to the DC component of the waveform, as predicted by the first term of Equation (9.3).

The positive portion of the waveform in Figure 9–1 represents the energy delivered to the circuit by the source, and the negative portion of the waveform represents the energy returned to the circuit that was temporarily stored in the reactive components. The difference between the energy delivered by the source and the energy returned to the circuit is the amount of energy that is consumed or dissipated by the resistive circuit components.

9.2 REAL POWER

The general instantaneous power equation can be used to determine an expression for AC power for a resistor.

The general instantaneous power equation expressed in Equation (9.3) is repeated here for convenience as:

$$p = \frac{1}{2}V_P I_P \cos(-\theta) - \frac{1}{2}V_P I_P \cos(2\omega t + \theta)$$

Recall that the current and voltage for a resistor are in phase, so that:

$$\theta = 0$$

The instantaneous power equation for a resistor can therefore be expressed as:

$$p = \frac{1}{2}V_P I_P \cos(0°) - \frac{1}{2}V_P I_P \cos(2\omega t + 0°) \tag{9.4}$$

Since $\cos(0°) = 1$, Equation (9.4) can be expressed as:

$$p = \frac{1}{2}V_P I_P - \frac{1}{2}V_P I_P \cos(2\omega t + 0°) \tag{9.5}$$

The instantaneous power equation is useful in determining the power dissipated by a circuit at any instant of time or in calculating the peak power for a given circuit. Since the effects of power in circuit components do not change instantaneously, a more useful measurement of power is the average value of the power over one complete cycle.

The average value of power over one cycle is referred to as **real power** and is measured in units of watts (W).

The second term of the instantaneous power for a resistor in Equation (9.5) is a sinusoidal function of time and therefore has an average value of zero over one complete cycle. The first term of Equation (9.5) is a constant (not a function of time) and represents real power measured in watts.

Real power for a resistor can be expressed as:

$$P_R = \frac{1}{2}V_P I_P \tag{9.6}$$

The RMS (root-mean-square) value of a sinusoidal current is defined as the equivalent value of DC current that would cause the same heating effect in a resistor. RMS value is often referred to as the effective value of the sine wave.

Root-mean-square is a mathematical procedure that is used to determine the effective value of a sinusoidal waveform. The relationship between the peak value of a sine wave and the RMS value of the sine wave is given as:

$$V_P = \sqrt{2}\ V_{RMS}; \quad I_P = \sqrt{2}\ I_{RMS}$$

or:

$$V_{RMS} = \frac{\sqrt{2}}{2}V_P; \quad I_{RMS} = \frac{\sqrt{2}}{2}I_P$$

The approximate relationships typically used by engineers are:

$$V_P = 1.414\ V_{RMS}; \quad I_P = 1.414\ I_{RMS}$$

or:

$$V_{RMS} = 0.707 V_P; \quad I_{RMS} = 0.707 I_P$$

It is customary to use RMS values for currents and voltages when making power computations, so that Equation (9.6) may be written as:

$$P_R = \frac{1}{2}(\sqrt{2}\ V_{RMS})(\sqrt{2}\ I_{RMS}) \tag{9.7}$$

Real power for a resistor in terms of RMS values for current and voltage can be expressed as:

$$P_R = V_{RMS}\ I_{RMS} \tag{9.8}$$

The real power (average power) dissipated by a resistor is calculated as the product of the RMS values of current and voltage and is measured in units of watts (W).

Substituting Ohm's law ($V_{RMS} = I_{RMS} R$) into Equation (9.8) indicates that real power may also be calculated as:

$$P_R = I_{RMS}{}^2\ R \tag{9.9}$$

Substituting ($I_{RMS} = V_{RMS}/R$) into Equation (9.8) indicates that real power may also be calculated as:

$$P_R = \frac{V_{RMS}{}^2}{R} \tag{9.10}$$

The relationship between instantaneous current, voltage, and power for a resistor are shown in the MATLAB plot in Figure 9–2. Note that the voltage and current are in phase for a resistor, as indicated in Figure 9–2.

FIGURE **9–2**

Instantaneous Voltage,
Current, and Power
for a Resistor

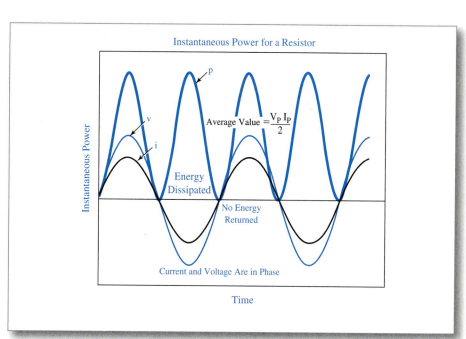

Note that the instantaneous power curve (p) in Figure 9–2 lies entirely above the zero axis and has an average value equal to $\frac{1}{2}V_P I_P = V_{RMS}\ I_{RMS}$, as indicated in Equations (9.6) and (9.8).

All of the energy supplied to a resistor is converted from electrical energy into heat, which is absorbed by the surrounding environment.

9.3 REACTIVE POWER

Capacitive Reactive Power

The general instantaneous power equation expressed in Equation (9.3) is repeated here for convenience as:

$$p = \frac{1}{2}V_P I_P \cos(-\theta) - \frac{1}{2}V_P I_P \cos(2\omega t + \theta)$$

Recall that capacitor current leads the capacitor voltage by 90°, and assuming that the voltage is at 0°, the phase difference can be expressed as:

$$\theta = 90° \tag{9.11}$$

Substituting Equation (9.11) into the general instantaneous power equation, the first term becomes:

$$\frac{1}{2}V_P I_P \cos(-90°)$$

Since $\cos(-90°) = 0$, the first term is equal to zero.

The second term of the instantaneous power equation for a capacitor becomes:

$$p = -\frac{1}{2}V_P I_P \cos(2\omega t + 90°)$$

Substituting the trigonometric identity $\cos(A) = \sin(A + 90°)$, we obtain:

$$p = -\frac{1}{2}V_P I_P \sin(2\omega t + 90° + 90°) \tag{9.12}$$

Substituting the trigonometric identity $\sin(A + 180°) = -\sin(A)$ into Equation (9.12), we obtain the expression for instantaneous power for a capacitor as:

$$p = \frac{1}{2}V_P I_P \sin(2\omega t) \tag{9.13}$$

The expression for instantaneous power for a capacitor in Equation (9.13) is sinusoidal and therefore has an average value of zero.

The average value of the instantaneous power for a capacitor is zero, indicating that a capacitor does not dissipate power. The energy delivered to a capacitor is temporarily stored in an electric field that develops between the capacitor plates as the capacitor charges. The energy is returned to the circuit as the capacitor discharges.

The current, voltage, and instantaneous power for a capacitor are shown in the MATLAB plot in Figure 9–3.

FIGURE **9–3**

Instantaneous Voltage,
Current, and Power
for a Capacitor

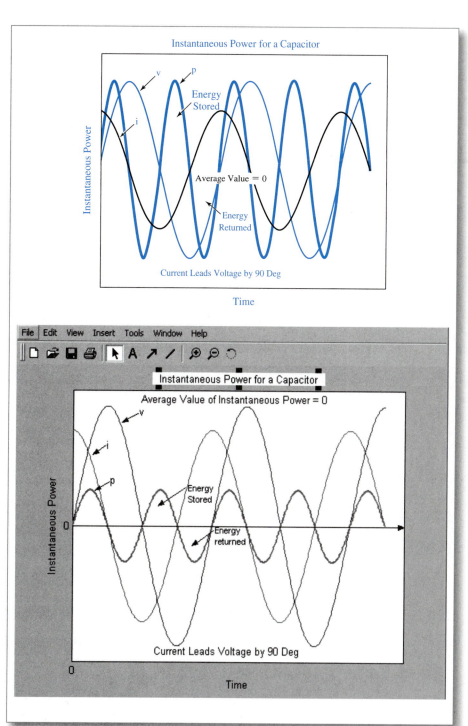

Although the average power consumed by a capacitor is zero, the source must still supply energy to the capacitor even though the energy is eventually returned to the circuit.

The magnitude of the quantity expressed in Equation (9.13) is referred to as **capacitive reactive power** and is measured in units of volt-amperes-reactive (VAR).

Capacitive reactive power may be expressed as:

$$Q_C = \frac{1}{2}V_P I_P \text{ VAR} \qquad (9.14)$$

where V_P and I_P are the peak values of capacitor voltage and current respectively. As indicated in Equation (9.14), reactive power is typically represented by the letter Q.

It is customary to use RMS values for currents and voltages for power calculations, so that Equation (9.14) can be written as:

$$Q_C = V_{RMS}\, I_{RMS} \text{ VAR} \qquad (9.15)$$

The following alternate forms of Equation (9.15) may also be used to calculate Q_C:

$$Q_C = I_{RMS}^2 X_C \text{ VAR} \qquad (9.16)$$

$$Q_C = \frac{V_{RMS}^2}{X_C} \text{ VAR} \qquad (9.17)$$

Inductive Reactive Power

The general instantaneous power equation is expressed in Equation (9.3) as:

$$p = \frac{1}{2}V_P I_P \cos(-\theta) - \frac{1}{2}V_P I_P \cos(2\omega t + \theta)$$

Recall that inductor current lags the inductor voltage by 90°, so that:

$$\theta = -90° \qquad (9.18)$$

Substituting Equation (9.18) into the general instantaneous power equation, the first term becomes:

$$\frac{1}{2}V_P I_P \cos(90°)$$

Since $\cos(90°) = 0$, the first term is equal to zero.

The second term of the instantaneous power equation for an inductor becomes:

$$p = -\frac{1}{2}V_P I_P \cos(2\omega t - 90°)$$

Substituting the trigonometric identity $\cos(A) = \sin(A + 90°)$, we obtain:

$$p = -\frac{1}{2}V_P I_P \sin(2\omega t + 90° - 90°) \qquad (9.19)$$

$$p = -\frac{1}{2}V_P I_P \sin(2\omega t) \qquad (9.20)$$

The expression for instantaneous power for an inductor in Equation (9.20) is sinusoidal and therefore has an average value of zero.

The average value of the instantaneous power for an inductor is zero, indicating that an inductor does not dissipate power. The energy delivered to an inductor is temporarily stored in an expanding magnetic field that develops around the inductor as current increases. The stored energy is returned to the circuit as the magnetic field collapses as the current decreases.

The current, voltage, and instantaneous power for an inductor are shown in the MATLAB plot in Figure 9–4.

FIGURE **9-4**

Instantaneous Voltage,
Current, and Power
for an Inductor

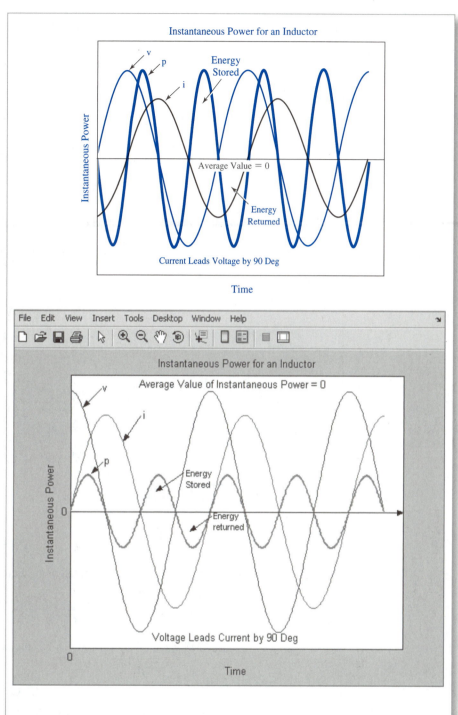

The quantity expressed in Equation (9.20) is referred to as **inductive reactive power** and is measured in units of volt-amperes-reactive (VAR).

Inductive reactive power may be expressed as:

$$Q_L = -\frac{1}{2} V_P I_P \text{ VAR} \tag{9.21}$$

As noted, it is customary to use RMS values for currents and voltages for power computations, so that Equation (9.21) can be written as:

$$Q_L = -V_{RMS} I_{RMS} \text{ VAR} \tag{9.22}$$

The following alternate forms of Equation (9.22) can also be used to calculate Q_L:

$$Q_L = -I_{RMS}^2 X_L \ \text{VAR} \tag{9.23}$$

$$Q_L = -\frac{V_{RMS}^2}{X_L} \ \text{VAR} \tag{9.24}$$

Note that the expression for inductive reactive power, Equation (9.22), is the negative of the expression for capacitive reactive power Equation (9.15), indicating that as a capacitor in an electronic circuit is storing energy, an inductor in the same circuit will be releasing or returning energy.

EXAMPLE 9–1   Calculating Real Power in a Resistive Circuit

Determine the real power dissipated by the resistor in Figure 9–5.

FIGURE **9–5**

Circuit for
Example 9–1

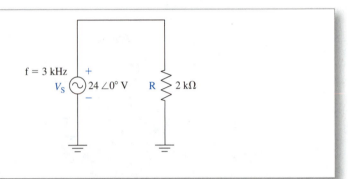

$f = 3\ \text{kHz}$

$V_S \quad 24 \angle 0° \ \text{V}$

$R \quad 2\ k\Omega$

✔ **SOLUTION**

The RMS value of the voltage across R can be calculated as:

$$V_{RMS} = (24 \ \text{V})(0.707) = 16.97 \ \text{V}$$

The real power can be calculated using Equation (9.10) as:

$$P_R = \frac{V_{RMS}^2}{R} = \frac{(16.97 \ \text{V})^2}{2 \ k\Omega} = 144 \ \text{mW}$$

The RMS value of the current can be calculated as:

$$I_{RMS} = \frac{V_{RMS}}{R} = \frac{16.97 \ \text{V}}{2 \ k\Omega} = 8.49 \ \text{mA}$$

Using Equation (9.9) to calculate real power, we obtain:

$$P_R = I_{RMS}^2 R = 144 \ \text{mW}$$

The real power can also be calculated using Equation (9.8) as:

$$P_R = V_{RMS} I_{RMS} = (16.97 \ \text{V})(8.49 \ \text{mA}) = 144 \ \text{mW}$$

EXAMPLE 9–2   Calculating Reactive Power in a Capacitive Circuit

Determine the reactive power for the capacitor in Figure 9–6.

FIGURE **9–6**

Circuit for
Example 9–2

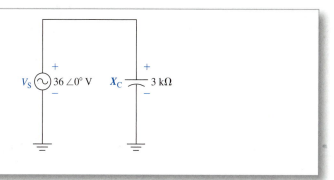

✓**SOLUTION**

The RMS values of the voltage and current can be calculated respectively as:

$$V_{RMS} = (36 \text{ V})(0.707) = 25.45 \text{ V}$$

$$I_{RMS} = \frac{V_{RMS}}{X_C} = \frac{25.45 \text{ V}}{3 \text{ k}\Omega} = 8.483 \text{ mA}$$

Using Equation (9.16), the capacitive reactive power can be calculated as:

$$Q_C = I_{RMS}^2 X_C = (8.483 \text{ mA})^2 (3 \text{ k}\Omega) = 215.9 \text{ mVAR}$$

The reactive power can also be calculated using Equation (9.17) as:

$$Q_C = \frac{V_{RMS}^2}{X_C} = \frac{(25.45 \text{ V})^2}{3 \text{ k}\Omega} = 215.9 \text{ mVAR}$$

The capacitive reactive power can also be calculated using Equation (9.15) as:

$$Q_C = V_{RMS} \, I_{RMS} = (25.45 \text{ V})(8.483 \text{ mA}) = 215.9 \text{ mVAR}$$

EXAMPLE 9-3 Calculating Reactive Power in an Inductive Circuit

Determine the reactive power for the inductor in Figure 9–7.

FIGURE **9–7**

Circuit for
Example 9–3

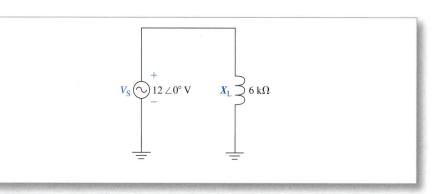

✓**SOLUTION**

The RMS values of the voltage and current can be calculated respectively as:

$$V_{RMS} = (12 \text{ V})(0.707) = 8.484 \text{ V}$$

$$I_{RMS} = \frac{V_{RMS}}{X_L} = \frac{8.484 \text{ V}}{6 \text{ k}\Omega} = 1.414 \text{ mA}$$

The inductive reactive power can be calculated using Equation (9.22) as:

$$Q_L = -V_{RMS} \, I_{RMS} = -(8.484 \text{ V})(1.414 \text{ mA}) = -12 \text{ mVAR}$$

The inductive reactive power can also be calculated using Equation (9.23) as:

$$Q_L = -I_{RMS}{}^2 X_L = -(1.414 \text{ mA})^2 (6 \text{ k}\Omega) = -12 \text{ mVAR}$$

The inductive reactive power can also be calculated using Equation (9.24) as:

$$Q_L = \frac{V_{RMS}{}^2}{X_L} = \frac{(8.484 \text{ V})^2}{6 \text{ k}\Omega} = -12 \text{ mVAR}$$

PRACTICE EXERCISES 9–1

a. The voltage across a 2.2 kΩ resistor is given as $\mathbf{V} = 24\angle 30°$ V. Determine the real (average) power dissipated by the resistor.

b. The current through a 3.3 kΩ resistor is given as $\mathbf{I} = 25\angle -60°$ mA. Determine the real (average) power dissipated by the resistor.

c. The voltage across a 0.1 μF capacitor is given as $\mathbf{V} = 48\angle -120°$ V at a frequency of 500 Hz. Determine the capacitive reactive power.

d. The current through a 2.5 mH inductor is given as $\mathbf{I} = 30\angle -30°$ mA at a frequency of 12 kHz. Determine the inductive reactive power.

✓ ANSWERS

a. $P_R = 131$ mW

b. $P_R = 1.03$ W

c. $Q_C = 362$ mVAR

d. $Q_L = -84.7$ mVAR

(Complete solutions to the Practice Exercises can be found in Appendix A.)

9.4 APPARENT POWER

The only power dissipated (lost) in an AC circuit is the real power due to resistive components. However, as noted, the power source must also supply energy to the reactive components measured in VARS even though the energy is ultimately returned to the circuit.

Apparent power can be defined as the total energy delivered to a circuit from a source and is measured in units of volt-amperes (VA).

Apparent power is calculated as the product of the RMS values of source voltage and source current. Representing apparent power by the symbol S, apparent power can be expressed as:

$$S = I_{S\,RMS} V_{S\,RMS} \text{ VA} \tag{9.25}$$

Apparent power may also be defined as the right-angle sum of real power and net reactive power. The net reactive power in an AC circuit is defined as the algebraic sum of capacitive reactive power and inductive reactive power, so that:

$$S = \sqrt{(\text{Real power})^2 + (\text{Net reactive power})^2} \ \text{VA}$$

$$S = \sqrt{P_R^{\ 2} + (Q_L + Q_C)^2} \ \text{VA} \tag{9.26}$$

Q_L is considered to be a negative quantity and Q_C is considered to be a positive quantity when evaluating Equation (9.26).

EXAMPLE 9-4 Calculating Apparent Power

Determine the apparent power delivered by the source for the AC power circuit in Figure 9–8.

FIGURE **9–8**

Circuit for Example 9–4

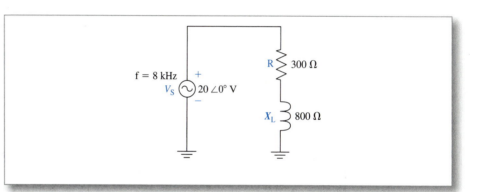

✓ SOLUTION

The total impedance of the circuit can be calculated as:

$$\mathbf{Z} = \mathbf{R} + \mathbf{X_L} = 300\angle 0°\ \Omega + 800\angle 90°\ \Omega = 854.4\angle 69.4°\ \Omega$$

The RMS value of the source voltage is calculated as:

$$V_{RMS} = (20\ \text{V})(0.707) = 14.14\ \text{V}$$

The RMS value of the source current is calculated as:

$$I_{RMS} = \frac{V_{RMS}}{Z} = \frac{14.14\ \text{V}}{854.4\ \Omega} = 16.55\ \text{mA}$$

Note that the impedance angle is not used when making calculations with RMS values.

Apparent power may be calculated as:

$$S = V_{RMS}\ I_{RMS} = (14.14\ \text{V})(16.55\ \text{mA}) = 234\ \text{mVA}$$

The real power for the circuit is calculated as:

$$P_R = I_{RMS}^{\ 2}R = (16.55\ \text{mA})^2(300\ \Omega) = 82.17\ \text{mW}$$

Reactive power is calculated as:

$$Q_L = I_{RMS}^{\ 2}X_L = (16.55\ \text{mA})^2(800\ \Omega) = -219.12\ \text{mVAR}$$

Apparent power can also be calculated as the right-angle sum of real power and net reactive power as:

$$S = \sqrt{(82.17 \text{ mW})^2 + (-219.12 \text{ mVAR})^2}$$

$$S = 234 \text{ mVA}$$

Slight differences in the calculations are due to rounding inaccuracies.

EXAMPLE 9-5 Calculating Apparent Power

Determine the apparent power delivered by the source for the AC power circuit in Figure 9–9.

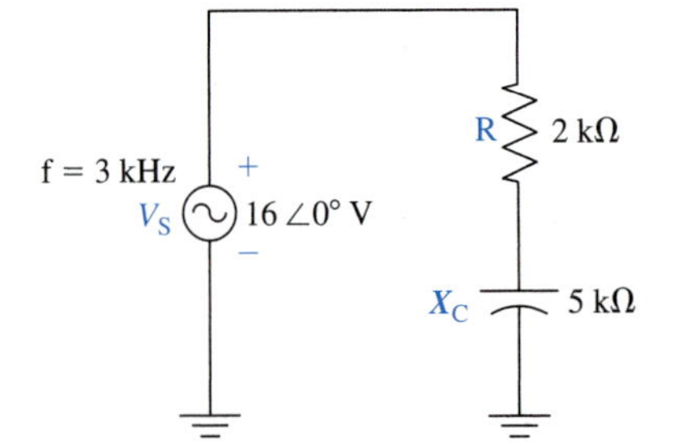

FIGURE **9-9**

Circuit for Example 9–5

✓ SOLUTION

The total impedance of the circuit can be calculated as:

$$\mathbf{Z} = \mathbf{R} + \mathbf{X_C} = 2\angle 0° \text{ k}\Omega + 5\angle -90° \text{ k}\Omega = 5.39\angle -68.2° \text{ k}\Omega$$

The RMS value of the source voltage is calculated as:

$$V_{RMS} = (16 \text{ V})(0.707) = 11.31 \text{ V}$$

The RMS value of the source current is calculated as:

$$I_{RMS} = \frac{V_{RMS}}{Z} = \frac{11.31 \text{ V}}{5.39 \text{ k}\Omega} = 2.1 \text{ mA}$$

Note that the impedance angle is not used when making calculations with RMS values.

Apparent power may be calculated as:

$$S = V_{RMS} I_{RMS} = (11.31 \text{ V})(2.1 \text{ mA}) = 23.75 \text{ mVA}$$

The real power for the circuit is calculated as:

$$P_R = I_{RMS}^2 R = (2.1 \text{ mA})^2 (2 \text{ k}\Omega) = 8.82 \text{ mW}$$

Inductive reactive power is calculated as:

$$Q_L = I_{RMS}^2 X_L = (2.1 \text{ mA})^2 (5 \text{ k}\Omega) = 22.05 \text{ mVAR}$$

Apparent power can also be calculated as the right-angle sum of real power and net reactive power.

$$S = \sqrt{(8.82 \text{ mW})^2 + (22.05 \text{ mVAR})^2}$$

$$S = 23.75 \text{ mVA}$$

EXAMPLE 9-6   **Calculating Apparent Power**

Determine the apparent power delivered by the source to the circuit in Figure 9–10.

FIGURE 9-10

Circuit for Example 9-6

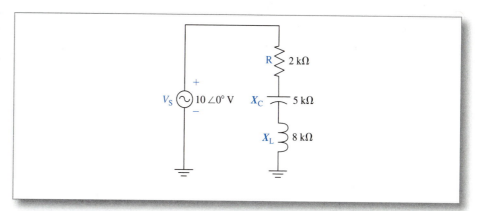

✓ SOLUTION

The total impedance of the circuit is calculated as:

$$\mathbf{Z} = \mathbf{R} + \mathbf{X_C} + \mathbf{X_L} = 2\angle 0° \text{ k}\Omega + 5\angle -90° \text{ k}\Omega + 8\angle 90° \text{ k}\Omega$$

$$\mathbf{Z} = 3.6\angle 56.31° \text{ k}\Omega$$

The RMS value of the source voltage is calculated as:

$$V_{S \text{ RMS}} = (10 \text{ V})(0.707) = 7.07 \text{ V}$$

Ohm's law can be used to calculate the RMS value of the source current as:

$$I_{S \text{ RMS}} = \frac{V_{S \text{ RMS}}}{Z} = \frac{7.07 \text{ V}}{3.6 \text{ k}\Omega} = 1.96 \text{ mA}$$

Apparent power can calculated using Equation (9.25) as:

$$S = I_{S \text{ RMS}} V_{S \text{ RMS}} = (7.07 \text{ V})(1.96 \text{ mA}) = 13.85 \text{ mVA}$$

The real power, capacitive reactive power, and inductive reactive power can be calculated respectively as:

$$P_R = I_{\text{RMS}}^2 R = (1.96 \text{ mA})^2 (2 \text{ k}\Omega) = 7.68 \text{ mW}$$

$$Q_C = I_{\text{RMS}}^2 X_C = (1.96 \text{ mA})^2 (5 \text{ k}\Omega) = 19.2 \text{ mVAR}$$

$$Q_L = I_{\text{RMS}}^2 X_L = -(1.96 \text{ mA})^2 (8 \text{ k}\Omega) = -30.7 \text{ mVAR}$$

The net reactive power is calculated as the algebraic sum of inductive reactive power and capacitive reactive power to obtain:

$$Q_L + Q_C = -30.7 \text{ mVAR} + 19.2 \text{ mVAR} = -11.5 \text{ mVAR (inductive)}$$

Apparent power can be calculated as the right-angle sum of real power and the net reactive power as:

$$S = \sqrt{P_{\mathrm{R}}^{2} + (Q_{\mathrm{L}} + Q_{\mathrm{C}})^{2}} = \sqrt{(7.68 \text{ mW})^{2} + (-11.5 \text{ mVAR})^{2}}$$

$$S = 13.83 \text{ mVA}$$

EXAMPLE 9-7   Calculating Apparent Power

Determine the real power, reactive power, and apparent power delivered by the commercial source for the AC power circuit in Figure 9–11.

FIGURE **9-11**

Circuit for
Example 9-7

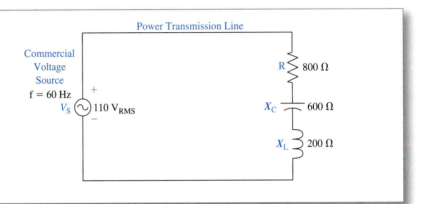

✔ SOLUTION

The circuit in Figure 9–11 illustrates a typical single phase commercial power distribution circuit. The RMS value of the voltage source is nominally 110 V, and the frequency of commercial power systems in the United States is 60 Hz, as indicated.

The same procedure is used to calculate real power, reactive power, and apparent power in a commercial power system as that used to calculate AC power in an electronic circuit.

The total impedance of the circuit can be calculated as:

$$\mathbf{Z} = \mathbf{R} + \mathbf{X_C} + \mathbf{X_L} = 800 \angle 0° \text{ } \Omega + 600 \angle -90° \text{ } \Omega + 200 \angle 90° \text{ } \Omega$$

$$\mathbf{Z} = 894.4 \angle -26.6° \text{ } \Omega$$

Ohm's law can be used to calculate the RMS value of the source current as:

$$I_{\mathrm{RMS}} = \frac{V_{\mathrm{RMS}}}{Z} = \frac{110 \text{ V}}{894.4 \text{ } \Omega} = 123 \text{ mA}$$

Apparent power can be calculated as:

$$S = V_{\mathrm{RMS}} I_{\mathrm{RMS}} = (110 \text{ V})(123 \text{ mA}) = 13.52 \text{ VA}$$

Apparent power can also be calculated as the right-angle sum of real power and net reactive power, where:

$$P_{\mathrm{R}} = I_{\mathrm{RMS}}^{2} R = (123 \text{ mA})^{2} (800 \text{ } \Omega) = 12.1 \text{ W}$$

$$Q_{\mathrm{C}} = I_{\mathrm{RMS}}^{2} X_{\mathrm{C}} = (123 \text{ mA})^{2} (600 \text{ } \Omega) = 9.08 \text{ VAR}$$

$$Q_L = -I_{RMS}^2 X_L = -(123 \text{ mA})^2 (200 \text{ }\Omega) = -3.03 \text{ VAR}$$

so that: $\qquad S = \sqrt{(12.1 \text{ W})^2 + (9.08 \text{ VAR} - 3.03 \text{ VAR})^2}$

$$S = 13.52 \text{ VA}$$

PRACTICE EXERCISES 9-2

a. Determine the real power and reactive power for each of the circuits in Figure 9–12.

FIGURE **9–12**

Circuit for Practice
Exercise 9–2
Exercise 1

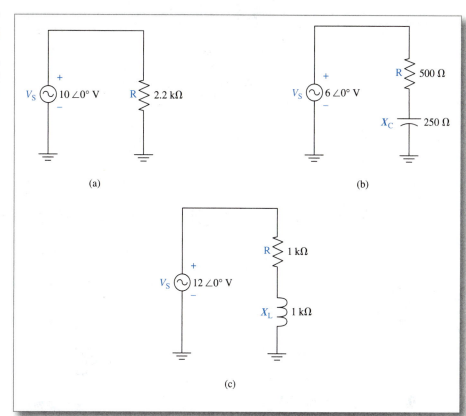

(a)   (b)

(c)

b. Determine the real power, reactive power, and apparent power for the commercial power distribution circuit in Figure 9–13.

FIGURE **9–13**

Circuit for Practice
Exercise 9–2
Exercise 2

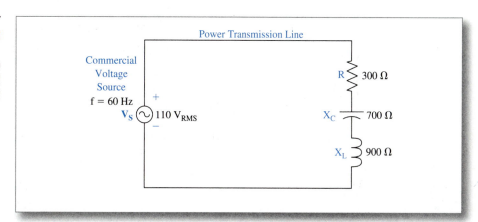

✓ **ANSWERS**

a. circuit a: $P_R = 22.72$ mW; $Q_C = Q_L = 0$;
    circuit b: $Q_C = 14.29$ mVAR; $P_R = 28.8$ mW; $S = 32.15$ mVA;
    circuit c: $Q_L = -36.24$ mVAR; $P_R = 36.24$ mW; $S = 51.25$ mVA

b. $P_R = 27.9$ W; $Q_C = 65.1$ VAR; $Q_L = -83.7$ VAR; $S = 33.66$ VA

(Complete solutions to the Practice Exercises can be found in Appendix A.)

9.5 POWER IN COMPLEX AC CIRCUITS

It is not necessary to consider whether circuit elements are connected in series, parallel, or series-parallel when making power calculations. The real power or reactive power for each element is calculated independently, and the results are summed to determine the total real power and total reactive power.

Apparent power is calculated as the right-angle sum of the total real power and net total reactive power. Apparent power may also be calculated as the product of the RMS values of the source voltage and current.

EXAMPLE 9-8   Calculating Apparent Power in a Complex AC Circuit

Determine the total real power, total net reactive power, and apparent power for the complex power circuit in Figure 9–14. Each box represents a circuit element for which the real power or reactive power has been determined as indicated.

FIGURE **9-14**

Circuit for
Example 9-8

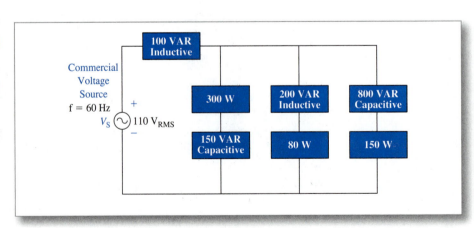

✓ **SOLUTION**

The total real power dissipated may be calculated as the sum of the real power dissipated by each respective resistive element as indicated in Figure 9–14 to obtain:

$$P_R \text{(total)} = 300 \text{ W} + 80 \text{ W} + 150 \text{ W} = 530 \text{ W}$$

Note that the configuration (series or parallel) of the individual elements is not considered when calculating the total real power for the circuit.

The total capacitive reactive power is determined as the sum of the individual capacitive reactive powers for each respective capacitive element as:

$$Q_C(\text{total}) = 150 \text{ VAR} + 800 \text{ VAR} = 950 \text{ VAR}$$

The total inductive reactive power is determined as the sum of the individual inductive reactive powers for each respective inductive element as:

$$Q_L(\text{total}) = -(100 \text{ VAR} + 200 \text{ VAR}) = -300 \text{ VAR}$$

Recall that Q_L is considered to be a negative quantity.

The apparent power may be calculated as the right-angle sum of the total real power and net reactive power as:

$$S = \sqrt{(530 \text{ W})^2 + (-300 \text{ VAR}) + (950 \text{ VAR})^2}$$

$$S = 838.7 \text{ VA}$$

Note that the values of real power and reactive power for each of the elements in the circuit of Example 9–8 are given directly in Figure 9–14. In cases where the circuit component values are given, the following procedure is required to determine apparent power.

The procedure to determine apparent power in a complex AC power circuit may be summarized as:

1. Calculate total circuit impedance.

2. Determine the current for each circuit component.

3. Calculate real power or reactive power for each respective component as $I_{RMS}^2 R$; $I_{RMS}^2 X_C$; or $I_{RMS}^2 X_L$ respectively.

4. Calculate the total real power as the sum of the real power for each resistive component.

5. Calculate the total capacitive reactive power as the sum of the reactive power for each capacitor.

6. Calculate the total inductive reactive power as the sum of the reactive power for each inductor.

7. Calculate the net total reactive power as $Q_L + Q_C$, recalling that Q_L is a negative quantity and Q_C is a positive quantity.

8. Calculate apparent power as the right-angle sum of total real power and total net reactive power as:

$$S = \sqrt{(\text{Real power})^2 + (\text{Net reactive power})^2} \text{ VA}$$

PRACTICE EXERCISES 9–3

Determine the total real power, total net reactive power, and apparent power for the complex power circuit in Figure 9–15.

FIGURE **9-15**

Circuit for Practice Exercise 9-3

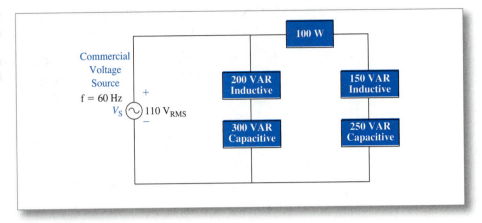

✓ ANSWERS

$P_R = 100$ W; $Q_C = 550$ VAR; $Q_L = -350$ VAR; $S = 223.6$ VA

(Complete solutions to the Practice Exercises can be found in Appendix A.)

9.6 POWER FACTOR AND THE POWER TRIANGLE

The **power triangle** is often used to show the right-angle relationship between real power, net reactive power, and apparent power in an AC power circuit. The general power triangle diagram is shown in Figure 9–16.

FIGURE **9-16**

The Power Triangle

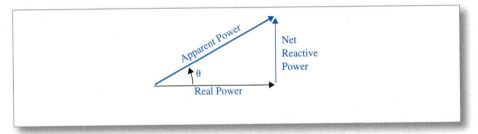

The **power factor** (*PF*) of an AC circuit is defined as the ratio of real power to apparent power, so that:

$$PF = \frac{\text{Real power}}{\text{Apparent power}} = \cos\theta$$

The current in an inductive circuit lags the voltage, so that an AC circuit with a net inductive reactive power is said to have a **lagging power factor.** The current in a capacitive circuit leads the voltage, so that an AC circuit with a net capacitive reactive power is said to have a **leading power factor.**

EXAMPLE 9-9   Calculating Power Factor

The real power and reactive power associated with the elements of a certain AC circuit are given respectively as:

$$P_{R1} = 120 \text{ W}$$

$$Q_{L1} = 225 \text{ VAR}$$

$$Q_{C1} = 100 \text{ VAR}$$

$$Q_{L2} = 345 \text{ VAR}$$

$$P_{R2} = 55 \text{ W}$$

$$Q_{C2} = 130 \text{ VAR}$$

Determine the power factor for the circuit and sketch the power triangle.

✓ **SOLUTION**

The total real power is determined as:

$$P_R \text{(total)} = 120 \text{ W} + 55 \text{ W} = 175 \text{ W}$$

The total inductive reactive power is determined as:

$$Q_L \text{(total)} = (-225 \text{ VAR}) + (-345 \text{ VAR}) = -570 \text{ VAR}$$

The total capacitive reactive power is determined as:

$$Q_C \text{(total)} = 100 \text{ VAR} + 130 \text{ VAR} = 230 \text{ VAR}$$

The net reactive power is determined as:

Net reactive power $= Q_C \text{ (total)} + Q_L \text{ (total)} = 230 \text{ VAR} - 570 \text{ VAR} = -340 \text{ VAR}$

Apparent power is calculated as the right-angle sum of real power and net reactive power to obtain:

$$S = \sqrt{(175 \text{ W})^2 + (-340 \text{ VAR})^2} = 382.4 \text{ VA}$$

The power factor (*PF*) may be calculated as the ratio of real power to apparent power to obtain:

$$PF = \frac{\text{Real power}}{\text{Apparent power}} = \frac{175 \text{ W}}{382.4 \text{ VA}} = 0.458 \text{ (lagging)}$$

The power factor is lagging because the net reactive power is inductive.

The power triangle is shown in Figure 9–17.

FIGURE **9–17**

The Power
Triangle for
Example 9–9

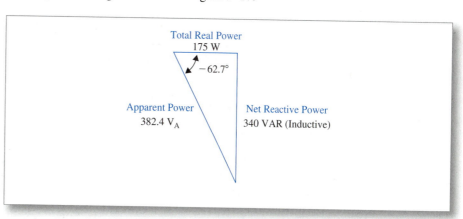

The angle θ may be calculated from the power triangle in Figure 9–17 as:

$$\theta = \tan^{-1} \frac{\text{Net reactive power}}{\text{Real power}}$$

$$\theta = \tan^{-1} \frac{340 \text{ VAR}}{175 \text{ W}} = -62.8°$$

The power factor may be calculated as cos θ = cos(−62.9°) = 0.458 (lagging).

The power associated with the elements of a certain AC circuit is given as:

$$P_{R1} = 285 \text{ W}$$

$$P_{R2} = 310 \text{ W}$$

$$Q_{C1} = 180 \text{ VAR}$$

$$Q_{C2} = 485 \text{ VAR}$$

$$Q_{L1} = 500 \text{ VAR}$$

$$Q_{L2} = 260 \text{ VAR}$$

Determine the apparent power and power factor for the circuit.

✓ ANSWERS

P_R (total) = 595 W; Q_L (total) = −760 VAR; Q_C (total) = 665 VAR;

S = 602.5 VA; PF = 0.988 (lagging)

(Complete solutions to the Practice Exercises can be found in Appendix A.)

Commercial utility power companies closely monitor the power factor of their generating and distribution systems. A commercial distribution system is optimized when the system load is purely resistive, which is indicated by a unity power factor ($PF=1$). Inductive system loads such as induction motors require large amounts of generation and transmission line capacity and cause the system power factor to be less than unity. Utility companies often insert capacitor loads to compensate for the large inductive loads required by many industrial customers. Most common household power requirements such as electrical heating elements, household lighting and appliances, and electronic devices are purely resistive loads.

9.7 THE IDEAL TRANSFORMER

Michael Faraday (1791–1867) discovered that a changing current through a coil will produce a changing magnetic field around the coil. He further demonstrated that the magnetic field produced could induce a changing current in a second coil in close proximity without any physical connection between the two coils. He also discovered that the voltage across a coil is directly proportional to the number of turns (N) and to the rate of change of the magnetic field, commonly referred to as change of flux represented by the Greek letter ϕ (phi), so that:

$$v = N \frac{d\phi}{dt} \text{ V (Faraday's law)}$$

where N is the number of turns and $\dfrac{d\phi}{dt}$ is the rate of change of flux.

A transformer consists of two coils or windings that exist in close proximity such that a magnetic field is shared between the two windings. In the case of an ideal transformer, the flux for both windings is considered to be equal, such that $\phi_1 = \phi_2$. To ensure that the magnetic field strength is equal for both windings, they are often coupled by a ferromagnetic iron core. The flux (magnetic field) flows from one winding to the other through the ferromagnetic core in much the same way as current flows through a conductor.

One of the windings of a transformer is called the *primary winding* and the other the *secondary winding*, as shown in Figure 9–18.

FIGURE **9–18**

Illustration
of an Ideal
Transformer

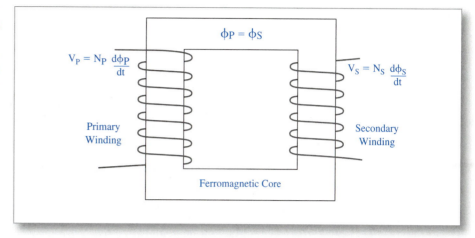

$\phi_P = \phi_S$

$V_P = N_P \dfrac{d\phi_P}{dt}$

$V_S = N_S \dfrac{d\phi_S}{dt}$

Primary
Winding

Secondary
Winding

Ferromagnetic Core

Using Faraday's law and letting $\phi_{\text{primary}} = \phi_{\text{secondary}}$, the primary and secondary voltage of a transformer may be expressed as:

$$V_{\text{primary}} = N_{\text{primary}} \frac{d\phi}{dt} \tag{9.27}$$

$$V_{\text{secondary}} = N_{\text{secondary}} \frac{d\phi}{dt} \tag{9.28}$$

Dividing Equation (9.27) by Equation (9.28), we obtain:

$$\frac{V_{\text{primary}}}{V_{\text{secondary}}} = \frac{N_{\text{primary}}}{N_{\text{secondary}}} \tag{9.29}$$

Considering the transformer to be ideal implies that there will be no power loss, so the primary power will equal the secondary power, such that:

Primary power = Secondary power

The respective power for each winding may be expressed in RMS values as:

$$P_{\text{primary}} = V_{\text{primary}} I_{\text{primary}}$$

$$P_{\text{secondary}} = V_{\text{secondary}} I_{\text{secondary}}$$

Setting the expressions for power equal to each other, we obtain:

$$V_{\text{primary}} I_{\text{primary}} = V_{\text{secondary}} I_{\text{secondary}}$$

so that:

$$\frac{I_{\text{primary}}}{I_{\text{secondary}}} = \frac{V_{\text{secondary}}}{V_{\text{primary}}} = \frac{N_{\text{secondary}}}{N_{\text{primary}}} \tag{9.30}$$

The relationships between the primary and secondary transformer voltage and current can be summarized as:

$$\frac{V_{\text{primary}}}{V_{\text{secondary}}} = \frac{N_{\text{primary}}}{N_{\text{secondary}}}$$

$$\frac{I_{\text{primary}}}{I_{\text{secondary}}} = \frac{N_{\text{secondary}}}{N_{\text{primary}}} \tag{9.31}$$

The number of turns on each transformer winding is not used for most calculations. Only the ratio of primary turns to secondary terms is typically given, so that:

$$TR = Turns\ ratio = \frac{N_{\text{primary}}}{N_{\text{secondary}}}$$

Substituting the turns ratio into Equation (9.31), we obtain:

$$\frac{V_{primary}}{V_{secondary}} = TR \qquad (9.32)$$

$$\frac{I_{primary}}{I_{secondary}} = \frac{1}{TR} \qquad (9.33)$$

9.8 VOLTAGE AND CURRENT TRANSFORMATION

EXAMPLE 9–10 The Step-Down Transformer

The RMS input voltage to a power transformer is given as 110 V, and the RMS primary current is given as 2.0 A. The primary winding has 50,000 turns and the secondary winding has 1000 turns. Determine the RMS value of the secondary voltage and current. Determine the primary and secondary power.

✔ SOLUTION

The turns ratio is determined as:

$$\frac{N_{primary}}{N_{secondary}} = \frac{50,000}{1000} = \frac{50}{1}$$

Equation (9.32) can be rearranged to determine the secondary voltage as:

$$V_{secondary} = V_{primary} \frac{1}{TR}$$

The RMS value of the secondary voltage is calculated as:

$$V_{secondary} = (100 \text{ V}) \left(\frac{1}{50} \right) = 2.2 \text{ V}$$

Equation (9.33) can be rearranged to determine the secondary current as:

$$I_{secondary} = I_{primary} \ TR = (2.0 \text{ A})(50) = 100 \text{ A}$$

The primary power may be calculated as:

$$P_{primary} = V_{primary} \ I_{primary} = (110 \text{ V})(2.0 \text{ A}) = 220 \text{ W}$$

The secondary power is calculated as:

$$P_{secondary} = V_{secondary} \ I_{secondary} = (2.2 \text{ V})(100 \text{ A}) = 220 \text{ W}$$

so that: $\qquad\qquad\qquad\qquad P_{primary} = P_{secondary}$

The secondary voltage for the transformer in Example 9–10 is less than the primary voltage, so that the transformer is referred to as a step-down transformer.

EXAMPLE 9–11 The Step-Up Transformer

The RMS input voltage to a power transformer is given as 110 V, and the RMS primary current is given as 300 A. The turns ratio is given as $TR = 1{:}250$. Determine the RMS value of the secondary voltage and current.

✓SOLUTION

The turns ratio is given as:

$$TR = 1{:}250 = \frac{1}{250}$$

Equation (9.32) can be rearranged to determine the secondary voltage as:

$$V_{secondary} = V_{primary}\frac{1}{TR}$$

The RMS value of the secondary voltage is calculated as:

$$V_{secondary} = (110\ V)(250) = 27.5\ kV$$

Equation (9.33) can be rearranged to determine the secondary current as:

$$I_{secondary} = I_{primary}\ TR = (300\ A)\left(\frac{1}{250}\right) = 1.2\ A$$

The primary power may be calculated as:

$$P_{primary} = V_{primary}\ I_{primary} = (110\ V)(300.0\ A) = 33\ kW$$

The secondary power is calculated as:

$$P_{secondary} = V_{secondary}\ I_{secondary} = (27.5\ kV)(1.2\ A) = 33\ kW$$

so that:
$$P_{primary} = P_{secondary}$$

The secondary voltage for the transformer in Example 9–11 is greater than the primary voltage, so that the transformer is referred to as a *step-up transformer*.

The voltages, currents, and power values in this example are very large compared to similar quantities that would be found in a practical electronic circuit. Such values as those calculated in this example, however, are common in large commercial power distribution systems. Power in a typical commercial distribution system is often measured in megawatts, current in kiloamps, and the voltage for many transmission lines may be near one million volts.

APPLICATION OF A STEP-UP TRANSFORMER

Example 9–11 illustrates that the secondary voltage of a step-up transformer is greatly increased at the secondary; however, the secondary current is greatly decreased. The voltage of typical commercial power companies is 110 V (RMS), and step-up transformers are used to increase the transmission line voltage while reducing the transmission line current. The reduction in current reduces the power loss (I^2R) due to the resistance in the long transmission lines.

SPECIFICATIONS: Determine the turns ratio required to step up the primary voltage of 110 V_{RMS} to a transmission voltage of 740 kV_{RMS}. Note that 740 kV_{RMS} is a common transmission voltage for commercial power companies.

✓SOLUTION

A step-up transformer is required to step-up the primary voltage.

The turns ratio is calculated using Equation (9.32) as:

$$TR = \frac{V_{primary}}{V_{secondary}} = \frac{110\ V}{740\ kV} = 1{:}6727\ (rounded)$$

APPLICATION OF A STEP-DOWN TRANSFORMER

DC power supplies for electronic systems are created by circuits designed to convert sinusoidal 110 V_{RMS} commercial voltage to low-voltage DC. The 110 V_{RMS} voltage must first be reduced to an appropriate voltage by a step-down transformer.

SPECIFICATIONS: Determine the transformer turns ratio required to reduce a 110 V_{RMS} commercial voltage to a 12 V_{RMS} voltage.

✓ SOLUTION

A step-down transformer is required to step down the primary voltage of 110 V_{RMS} to the required 12 V_{RMS}.

The turns ratio is calculated using Equation (9.32) as:

$$TR = \frac{V_{primary}}{V_{secondary}} = \frac{110 \text{ V}}{12 \text{ V}} = 9.166{:}1$$

PRACTICE EXERCISES 9–5

a. A transformer is designed with 60,000 primary turns and 1200 secondary turns. A sinusoidal voltage whose RMS value is 110 V is applied to the primary, and the RMS value of the primary current is given as 10 A. Determine the RMS value of the secondary voltage and current. Determine the primary and secondary power.

b. A commercial power company transformer has a turns ratio of 1:150. The RMS value of the primary voltage is 110 V, and the RMS value of the primary current is given as 2 kA. Determine the RMS value of secondary voltage and current. Determine the primary and secondary power.

✓ ANSWERS

a. $V_{secondary} = 2.2$ V; $I_{secondary} = 500$ mA;

$P_{primary} = P_{secondary} = 1.1$ W

b. $V_{secondary} = 16.5$ kV; $I_{secondary} = 13.3$ A;

$P_{primary} = P_{secondary} = 220$ kW

(Complete solutions to the Practice Exercises can be found in Appendix A.)

9.9 MAXIMUM POWER TRANSFER

The power dissipated by a resistor in a DC circuit can be calculated as:

$$P_{DC} = V_R I_R \text{ W} \tag{9.34}$$

or:

$$P_{DC} = I_R{}^2 R \text{ W} \tag{9.35}$$

or:

$$P_{DC} = \frac{V_R{}^2}{R} \text{ W} \tag{9.36}$$

Every practical voltage source can be modeled as an ideal voltage source in series with some internal impedance.

Consider the practical DC voltage source shown in Figure 9–19 with an arbitrary load resistor connected across the output terminals of the source.

FIGURE **9–19**

A Practical DC
Voltage Source with
Internal Resistance

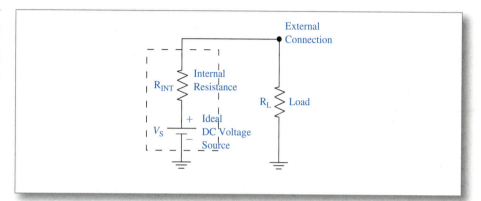

The total equivalent resistance can be calculated as:

$$R_T = R_{int} + R_L$$

Ohm's law can be used to determine the load current as:

$$I_L = \frac{V_S}{R_T} = \frac{V_S}{R_{int} + R_L}$$

The power dissipated by the load can be expressed as:

$$P_L = I_L^2 R_L = \left(\frac{V_S}{R_{int} + R_L}\right)^2 R_L \qquad (9.37)$$

Consider the circuit in Figure 9–20.

FIGURE **9–20**

A Practical
DC Voltage Source
with a 500 Ω
Load Resistor

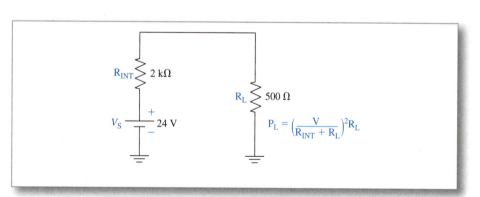

Equation (9.37) may be used to calculate the power dissipated by a 500 Ω (0.5 kΩ) load resistor as:

$$P_L = \left(\frac{V_S}{R_{int} + R_L}\right)^2 R_L = \left(\frac{24\ V}{2\ k\Omega + 0.5\ k\Omega}\right)^2 (0.5\ k\Omega) = 46.1\ mW$$

The computer-generated data in Figure 9–21 show the power dissipated by arbitrarily selected load resistors placed into the circuit in Figure 9–20.

FIGURE **9–21**

Power Dissipated by
Selected Load Resistors

RL	RT	IL	PL
5.00E+02	2.50E+03	9.60E−03	4.61E−02
8.00E+02	2.80E+03	8.57E−03	5.88E−02
1.00E+03	3.00E+03	8.00E−03	6.40E−02
1.20E+03	3.20E+03	7.50E−03	6.75E−02
1.30E+03	3.30E+03	7.27E−03	6.88E−02
1.60E+03	3.60E+03	6.67E−03	7.11E−02
1.80E+03	3.80E+03	6.32E−03	7.18E−02
1.90E+03	3.90E+03	6.15E−03	7.20E−02
2.00E+03	4.00E+03	6.00E−03	7.20E−02
2.20E+03	4.20E+03	5.71E−03	7.18E−02
2.40E+03	4.40E+03	5.45E−03	7.14E−02
2.80E+03	4.80E+03	5.00E−03	7.00E−02
3.00E+03	5.00E+03	4.80E−03	6.91E−02
3.50E+30	5.50E+03	4.36E−03	6.66E−02
4.00E+03	6.00E+03	4.00E−03	6.40E−02
5.00E+03	7.00E+03	3.43E−03	5.88E−02
6.00E+03	8.00E+03	3.00E−03	5.40E−02
7.00E+03	9.00E+03	2.67E−03	4.98E−02
8.00E+03	1.00E+03	2.40E−03	4.61E−02

The data in Figure 9–21 reveal that the power dissipated by the selected load resistors begins at 46.1 mW for the 500 Ω resistor as previously calculated, increases to a maximum value of 72 mW for $R_L = 2$ kΩ, and then decreases to a value of 46.1 mW for a load resistance of 8 kΩ.

Note that maximum power is dissipated by the load when the load resistance is 2 kΩ, which is the same value as the internal resistance of the source.

The **maximum power transfer theorem** states that maximum power is delivered to a load when the load resistance or impedance is equal to the series internal resistance or impedance of the voltage source.

A mathematical derivation of the maximum power transfer theorem is given in Appendix B.

9.10   MAXIMUM AC SIGNAL POWER TRANSFER

The theory of phasor mathematics states that the sum of a phasor and its complex conjugate is a real number whose phase angle is zero. Therefore, the sum of any impedance and its complex conjugate will result in pure resistance (angle 0°).

The conjugate of a complex number in rectangular form $\mathbf{N} = (a + jb)$ may be expressed as $\overline{\mathbf{N}} = (a - jb)$. The complex conjugate of a complex number in polar form $\mathbf{N} = M\angle\theta°$ may be expressed as $\overline{\mathbf{N}} = M\angle-\theta°$.

EXAMPLE 9–12   The Sum of Impedance and Its Conjugate

A load impedance is given as $\mathbf{Z} = 2.5 \angle 36°$ kΩ. Calculate the sum of the given load impedance and its conjugate, and show that the sum is a pure resistance.

✔ SOLUTION

The conjugate of the given impedance in polar form is determined as:

$$\overline{\mathbf{Z}} = 2.5 \angle - 36° \text{ k}\Omega$$

The sum of $\mathbf{Z}$ and $\overline{\mathbf{Z}}$ is calculated as:

$$\mathbf{Z} + \overline{\mathbf{Z}} = 2.5 \angle 36° \text{ k}\Omega + 2.5 \angle - 36° \text{ k}\Omega = 4.04 \angle 0° \text{ k}\Omega$$

The phase angle of 0° indicates that the sum is a pure resistance. The sum of any impedance and its conjugate is always a pure resistance.

EXAMPLE 9–13   Maximum AC Signal Power Transfer

Determine the load impedance for maximum signal power transfer in the circuit in Figure 9–22. Determine the maximum power delivered to the load.

FIGURE **9–22**

Circuit for
Example 9–13

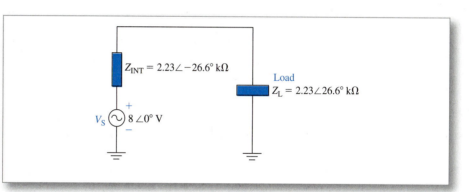

✔ SOLUTION

The maximum power transfer theorem developed for DC circuits in the previous section is also valid for AC circuits and may be stated as:

Maximum AC signal power is delivered to a load when the impedance of the load is equal to the complex conjugate of the internal impedance of the source.

Using the maximum AC signal power transfer theorem, the value of load impedance required for maximum signal power transfer is the conjugate of Z_{int}, so that:

$$\mathbf{Z}_{\text{L}} = \overline{\mathbf{Z}}_{\text{int}} = 2.23 \angle 26.6° \text{ k}\Omega$$

The total equivalent impedance of the circuit in Figure 9–22 may be calculated as:

$$\mathbf{Z}_{\text{T}} = \mathbf{Z}_{\text{int}} + \mathbf{Z}_{\text{L}} = \mathbf{Z}_{\text{int}} + \overline{\mathbf{Z}}_{\text{s}} = 2.23 \angle - 26.6° \text{ k}\Omega + 2.23 \angle 26.6° \text{ k}\Omega = 4 \angle 0° \text{ k}\Omega$$

The total equivalent impedance $\mathbf{Z}_{\text{T}}$ has a phase angle of 0°, indicating that $\mathbf{Z}_{\text{T}}$ is a pure resistance. Half of the resistance of $\mathbf{Z}_{\text{T}}$ is due to the source impedance and half is due to the load impedance. The circuit shown in Figure 9–23 shows the 8 V source and a load resistance of $\dfrac{4 \text{ k}\Omega}{2} = 2 \text{ k}\Omega$ at maximum power transfer.

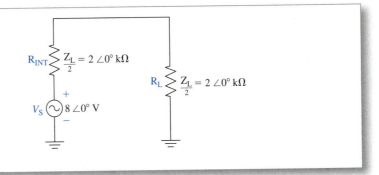

FIGURE **9–23**

Circuit for
Example 9–13 at
Maximum Power
Transfer

Ohm's law can be used to calculate the source current in Figure 9–23 as:

$$\mathbf{I_s} = \frac{\mathbf{V_s}}{\mathbf{Z_T}} = \frac{8\angle 0° \text{ V}}{4\angle 0° \text{ k}\Omega} = 2\angle 0° \text{ mA}$$

The power dissipated by the load can be calculated as:

$$P_{\text{load}} = (2 \text{ mA})^2 (2 \text{ k}\Omega) = 8 \text{ mW}$$

The maximum signal power transfer theorem is an important design criterion. Some signal sources exhibit high internal impedance and some exhibit low internal impedance.

Similarly, some loads exhibit high impedance and some low impedance. Interface circuits are often required to "match" the impedance of a source to the impedance of the load to ensure maximum signal power transfer.

9.11   TRANSFORMER IMPEDANCE MATCHING

The maximum power transfer theorem states that maximum power will be transferred from a source to a load when the resistance of the load is equal to the internal resistance of the source. If the internal resistance of the source and the load resistance are not equal, more of the signal power will be dissipated by the internal resistance of the source and less power will be delivered to the load. The transformer may be used to "match" impedances to ensure that maximum power is delivered to the load.

The transformer primary and secondary current and voltage relationships are expressed as:

$$\frac{V_{\text{primary}}}{V_{\text{secondary}}} = \frac{N_{\text{primary}}}{N_{\text{secondary}}} = TR \tag{9.38}$$

$$\frac{I_{\text{primary}}}{I_{\text{secondary}}} = \frac{N_{\text{secondary}}}{N_{\text{primary}}} = \frac{1}{TR} \tag{9.39}$$

Dividing Equation (9.38) by Equation (9.39), we obtain:

$$\frac{\dfrac{V_{\text{primary}}}{V_{\text{secondary}}}}{\dfrac{I_{\text{primary}}}{I_{\text{secondary}}}} = \frac{TR}{\dfrac{1}{TR}} = (TR)^2 \tag{9.40}$$

Recognizing that voltage divided by current is impedance, Equation (9.40) may be expressed as:

$$\frac{Z_{\text{primary}}}{Z_{\text{secondary}}} = (TR)^2 \tag{9.41}$$

Equation (9.41) indicates that the ratio of the transformer primary impedance to the secondary impedance is equal to the square of the turns ratio, $(TR)^2$.

Equation (9.41) can be rearranged as:

$$Z_{\text{primary}} = (TR)^2 Z_{\text{secondary}} \tag{9.42}$$

Equation (9.42) indicates that the secondary impedance will be *reflected* (appear) to the primary side of the transformer by a factor of $(TR)^2$. If TR is greater than 1 (step-down transformer), the secondary impedance will appear to be larger by a factor of $(TR)^2$ as viewed from the primary side. If TR is less than 1 (step-up transformer), the secondary impedance will appear to be smaller by a factor of $(TR)^2$ as viewed from the primary side.

The secondary impedance of a step-down transformer will be reflected (appear) larger by a factor of $(TR)^2$ as viewed from the primary side of the transformer.

EXAMPLE 9–14   **A Signal Source and an Unmatched Load**

A 500 mV signal source with an internal resistance of 375 Ω is directly connected to a resistive load of 15 Ω. Determine the power dissipated by the internal source resistance and the power dissipated by the load.

✓ SOLUTION

The given information is illustrated in the circuit in Figure 9–24.

FIGURE 9–24

Example 9–14
Source
Connected Directly
to Load

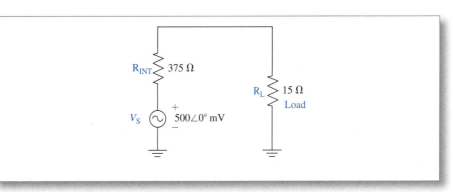

The total equivalent resistance can be determined as:

$$R_T = R_{\text{int}} + R_L = 375\ \Omega + 15\ \Omega = 390\ \Omega$$

Ohm's law can be used to calculate the source current as:

$$I = \frac{V_S}{R_T} = \frac{500\ \text{mV}}{390\ \Omega} = 1.28\ \text{mA}$$

The RMS value of current is calculated as:

$$I_{\text{RMS}} = (1.28\ \text{mA})(0.707) = 0.905\ \text{mA}$$

The power dissipated by the internal resistance of the source can be calculated as:

$$P_{R_{\text{int}}} = I^2 R_{R_{\text{int}}} = (0.905\ \text{mA})^2 (375\ \Omega) = 307.13\ \mu\text{W}$$

The power dissipated by the load can be calculated as:

$$P_{R_L} = I^2 R_L = (0.905\ \text{mA})^2 (15\ \Omega) = 12.28\ \mu\text{W}$$

Note that most of the signal source power is dissipated by the internal resistance of the source (307.13 μW) and relatively little power is delivered to the load (12.28 μW).

EXAMPLE 9-15   Impedance Matching with a Transformer

Determine the turns ratio of an impedance matching transformer to match the signal source of 375 Ω internal resistance to the 15 Ω load in Example 9–14. Connect the signal source to the primary side of the impedance matching transformer, and connect the load to the secondary side of the impedance matching transformer. Calculate the power dissipated by the internal resistance of the source and the power dissipated by the load.

✓ SOLUTION

The ratio of the internal resistance to the load resistance is calculated as:

$$\frac{375\ \Omega}{15\ \Omega} = 25$$

Using Equation (9.42), the ratio of the resistances is equal to the square of the turns ratio, such that:

$$\frac{Z_{primary}}{Z_{secondary}} = (TR)^2 = 25$$

so that: $TR = \sqrt{25} = 5$

The illustration in Figure 9–25 shows the 500 mV voltage source connected to the 15 Ω load resistor through the impedance matching transformer.

FIGURE **9-25**

Example 9–15
Using
an Impedance
Matching
Transformer

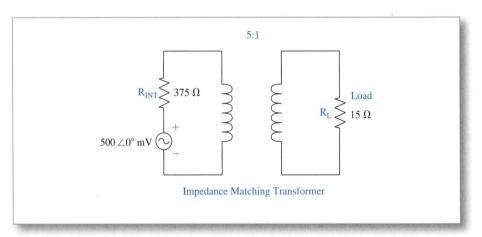

The magnitude of the primary current in Figure 9–25 can be calculated as:

$$I_{primary} = \frac{500\ mV}{375\ \Omega + 375\ \Omega} = 0.67\ mA$$

The first 375 Ω quantity in the denomination of the equation is the internal resistance of the source. The second 375 Ω quantity is the 15 Ω load "Reflected" to the primary as $(5)^2(15\ \Omega) = 375\ \Omega$.

The RMS value of the primary current is calculated as:

$$I_{primary} = (0.67\ mA)\,(0.707) = 0.474\ mA$$

The primary power is calculated as:

$$P_{primary} = (I_{RMS}{}^2)(R_{int}) = (0.474\ mA)^2\,(375\ \Omega) = 84.25\ \mu W$$

The RMS value of the secondary current can be calculated as:

$$I_{secondary} = I_{primary}(TR) = (0.474\ mA)(5) = 2.37\ mA$$

The power delivered to the load (secondary power) can be calculated as:

$$P_{\text{secondary}} = I_{\text{secondary}} \, R_L = (2.37 \text{ mA})^2 \, (15 \ \Omega) = 84.25 \ \mu\text{W}$$

Note that the power delivered to the load has increased from 12.25 μW to 84.25 μW when the load and internal resistance are matched. Also note that at maximum power transfer, the power dissipated by the source and the load is equal.

APPLICATION: MATCHING A SPEAKER TO AN AMPLIFIER

SPECIFICATIONS: The output impedance of an audio voltage amplifier is given as 3.2 kΩ. The amplifier is to be connected to a speaker whose impedance is given as 8 Ω. Determine the turns ratio of an impedance matching transformer for maximum power transfer from the amplifier to the speaker.

✓ SOLUTION

The prototype circuit indicating the given information is shown in Figure 9–26.

FIGURE **9–26**

Matching a Signal
Source to a Speaker

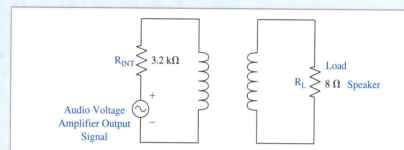

Impedance Matching Transformer

The secondary impedance must be made to appear larger from the primary side of the transformer, so a step-down transformer will be selected.

The ratio of impedances may be determined as:

$$\frac{Z_{\text{primary}}}{Z_{\text{secondary}}} = \frac{3.2 \text{ k}\Omega}{8 \ \Omega} = 400 = (TR)^2$$

The turns ratio may be calculated as:

$$TR = \sqrt{400} = 20 = \frac{N_{\text{primary}}}{N_{\text{secondary}}}$$

A step-down transformer with a turns ratio of 20:1 is required for maximum power transfer.

APPLICATION: A COAXIAL CABLE MATCHING TRANSFORMER

SPECIFICATIONS: The signal from a cable television company is connected to a television receiver through a coaxial cable designed to have a characteristic internal impedance of 50 Ω. The input impedance of the television receiver is given as 1.25 kΩ. Determine the turns ratio of an impedance matching transformer for maximum power transfer.

✓ **SOLUTION**

The secondary impedance (television) must be made to appear smaller to the primary side of the transformer, so a step-up transformer will be selected.

The impedance ratio is determined as:

$$\frac{Z_{secondary}}{Z_{primary}} = \frac{1.25 \text{ k}\Omega}{50 \text{ }\Omega} = 25 = (TR)^2$$

The turns ratio may be determined as:

$$TR = \sqrt{25} = 5$$

Since we are matching a low primary impedance to a higher secondary impedance, a step-up transformer with a turns ratio of 1:5 is required for maximum power transfer.

PRACTICE EXERCISES 9-6

a. A sinusoidal signal source of 300 mV (peak) with an internal resistance of 200 Ω is directly connected to a resistive load of 8 Ω. Determine the power dissipated by the internal source resistance and the power dissipated by the load.

b. Determine the turns ratio of an impedance matching transformer to match the signal source of 200 Ω internal resistance to the 8 Ω load in part a. Determine the power delivered to the matched load and the power dissipated by the internal resistance of the source.

✓ **ANSWERS**

a. $P_{int} = 208.08 \text{ }\mu\text{W}$; $P_L = 8.32 \text{ }\mu\text{W}$

b. $TR = 5:1$; $P_S = P_L = 56.18 \text{ }\mu\text{W}$

(Complete solutions to the Practice Exercises can be found in Appendix A.)

9.12 MAXIMUM POWER AND THEVENIN'S THEOREM

The prototype circuit for maximum power transfer is shown in Figure 9–27.

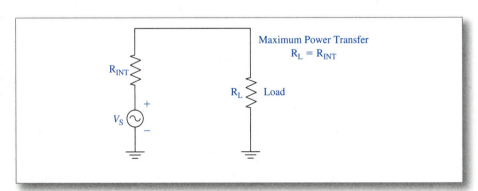

FIGURE 9–27

Prototype Circuit for Maximum Power Transfer

Maximum Power Transfer
$R_L = R_{INT}$

R_{INT}

R_L Load

V_S

Maximum power will be transferred to the load when the load resistance is equal to the internal resistance of the source ($R_L = R_{int}$). However, the load may not be directly connected to the source as shown in Figure 9–27. Note, however, that the prototype circuit in Figure 9–27 is in the same configuration as a Thevenin equivalent circuit.

Maximum power will be transferred to a load when the impedance of the source is equal to the impedance of the Thevenin equivalent circuit:

$$\mathbf{Z}_L = \overline{\mathbf{Z}}_{Th}$$

where $\overline{\mathbf{Z}}_{TH}$ is the complex conjugate of $\mathbf{Z}_{Th}$.

Consider the circuit in Figure 9–28.

FIGURE 9–28

Maximum Power Transfer and Thevenin's Equivalent Circuit

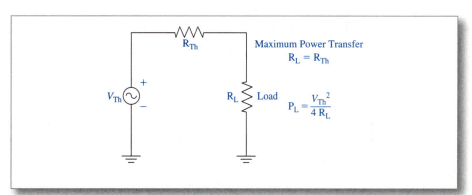

Since the load resistance and Thevenin resistance are equal at maximum power transfer, the voltage across each resistor may be determined using the voltage divider rule as:

$$V_L = \frac{V_{Th}}{2}$$

The power delivered to the load may be calculated as:

$$P_L = \frac{V_L{}^2}{R_L} = \left(\frac{V_{Th}}{2}\right)^2 \left(\frac{1}{R_L}\right)$$

$$P_L = \frac{V_{Th}{}^2}{4R_L} \tag{9.43}$$

EXAMPLE 9–16   **Maximum Power Transfer and Thevenin's Theorem**

Determine the value of load resistor R_L required for maximum power transfer for the circuit in Figure 9–29. Determine the power delivered to the load at maximum power transfer.

FIGURE 9–29

Circuit for Example 9–16

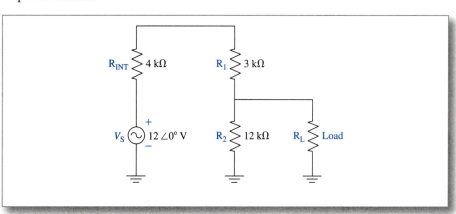

✓ **SOLUTION**

The circuit to determine the Thevenin equivalent impedance is shown in Figure 9–30.

FIGURE **9–30**

Determining
Thevenin's
Equivalent
Impedance

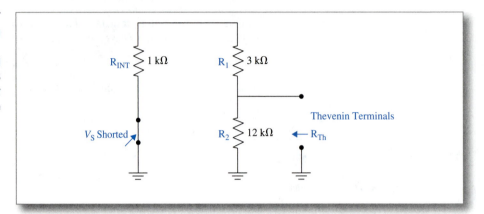

Note that the source has been shorted and the load has been removed to determine R_{Th}. The Thevenin resistance is calculated as:

$$R_{Th} = (R_1 + R_{int})//R_2 = (3\ k\Omega + 1\ k\Omega)//12\ k\Omega = 3\ k\Omega$$

Maximum power transfer will occur when the load resistance is equal to the Thevenin resistance, so that:

$$R_L = R_{Th} = 3\ k\Omega$$

The circuit to determine the Thevenin voltage is shown in Figure 9–31.

FIGURE **9–31**

Determining
Thevenin's
Equivalent
Voltage

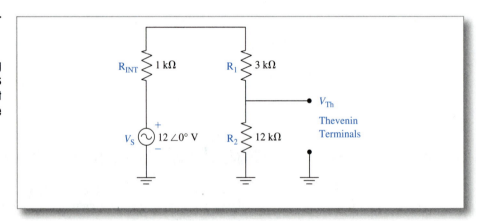

The Thevenin voltage can be calculated using the voltage divider rule as:

$$V_{Th} = 12\ V\ \frac{12\ k\Omega}{(12\ k\Omega + 3\ k\Omega + 1\ k\Omega)} = 9\ V$$

The RMS value of the Thevenin voltage is calculated as:

$$V_{Th(RMS)} = (9\ V)(0.707) = 6.36\ V$$

The power delivered to the load using Equation (9.43) is calculated as:

$$P_L = \frac{V_{Th(RMS)}^2}{4R_L} = \frac{6.36\ V^2}{(4)(3\ k\Omega)} = 3.37\ mW$$

Determine the value of load resistance for maximum power transfer to the load in Figure 9–32. Determine the maximum power dissipated by the load.

FIGURE **9–32**

Circuit for Practice
Exercise 9–32

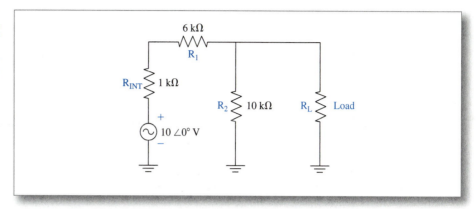

✓ANSWERS

$R_L = R_{Th} = 4.12$ kΩ; $P_{L(Max)} = 1.05$ mW

(Complete solutions to the Practice Exercises can be found in Appendix A.)

9.13 MAXIMUM POWER RATINGS

Electronic circuit component manufacturers typically indicate maximum ratings that must be considered when using their components in practical circuit design. Exceeding the maximum ratings may damage or destroy the component or deteriorate its performance in the circuit. In addition, commercial manufacturers of electronic components typically supply only selected values of maximum ratings, which are referred to as standard values. The standard maximum power ratings for common resistors include:

$$1/16 \text{ W} = 62.5 \text{ mW}$$

$$1/8 \text{ W} = 125 \text{ mW}$$

$$1/4 \text{ W} = 250 \text{ mW}$$

$$1/2 \text{ W} = 500 \text{ mW}$$

$$1 \text{ W} = 1000 \text{ mW}$$

$$2 \text{ W} = 2000 \text{ mW}$$

Resistors with higher power ratings are considered to be specialty power resistors and are manufactured specifically to dissipate excessive amounts of heat.

EXAMPLE 9–17   Calculating Resistor Power Rating Requirements in a DC Circuit

Consider the DC voltage divider circuit in Figure 9–33. Determine the minimum standard power rating for each resistor and the power delivered by the source.

FIGURE **9-33**

Circuit for
Example 9-17

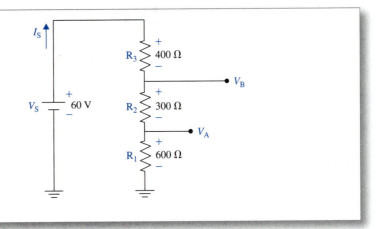

✓ **SOLUTION**

The circuit in Figure 9–33 is a series circuit, so the total equivalent resistance can be calculated as:

$$R_T = R_1 + R_2 + R_3 = 600\ \Omega + 300\ \Omega + 400\ \Omega = 1.3\ \text{k}\Omega$$

Ohm's law can be used to calculate the source current as:

$$I = \frac{V_S}{R_T} = \frac{60\ \text{V}}{1.3\ \text{k}\Omega} = 46.2\ \text{mA}$$

The power dissipated by each respective resistor can be calculated as:

$$P_{R_1} = I^2 R_1 = (46.2\ \text{mA})^2\,(600\ \Omega) = 1.28\ \text{W}$$

$$P_{R_2} = I^2 R_2 = (46.2\ \text{mA})^2\,(300\ \Omega) = 0.64\ \text{W}$$

$$P_{R_3} = I^2 R_3 = (46.2\ \text{mA})^2\,(400\ \Omega) = 0.854\ \text{W}$$

The minimum required standard wattage rating for each respective resistor is determined as:

$$R_1 = 2\ \text{W}; \quad R_2 = 1\ \text{W}; \quad R_3 = 1\ \text{W}$$

The power delivered by the source can be calculated as:

$$P_S = V_S\,I = (60\ \text{V})(46.2\ \text{mA}) = 2.77$$

Note that the sum of the power dissipated by the resistors is equal to the power delivered by the source:

$$P_{R_1} + P_{R_2} + P_{R_3} = 1.28\ \text{W} + 0.64\ \text{W} + 0.854\ \text{W} = 2.77\ \text{W} = P_S$$

EXAMPLE 9-18   Calculating Resistor Power Rating Requirements

Given the circuit in Figure 9–34, determine the minimum standard power rating for each resistor.

FIGURE **9–34**

Circuit for
Example 9–18

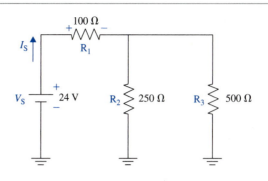

✔ **SOLUTION**

The total equivalent resistance of the circuit can be calculated as:

$$R_T = R_1 + (R_2 \mathbin{/\!/} R_3) = 100\ \Omega + (250\ \Omega \mathbin{/\!/} 500\ \Omega) = 100\ \Omega + 1.667\ \Omega = 266.7\ \Omega$$

Ohm's law can be used to calculate the source current as:

$$I_S = \frac{V_S}{R_T} = \frac{24\ \text{V}}{266.7\ \Omega} = 90\ \text{mA}$$

The current through R_1 is the source current, so that:

$$I_{R_1} = 90\ \text{mA}$$

The current through R_2 and R_3 may be determined using the current divider rule as:

$$I_{R_2} = I_T\left(\frac{R_P}{R_2}\right) = 89.9\ \text{mA}\ \frac{166.7\ \Omega}{250\ \Omega} = 60\ \text{mA}$$

$$I_{R_3} = I_T\left(\frac{R_P}{R_3}\right) = 89.9\ \text{mA}\ \frac{166.7\ \Omega}{500\ \Omega} = 30\ \text{mA}$$

where: $R_P = R_2 \mathbin{/\!/} R_3 = 166.7\ \Omega$

The power dissipated by each respective resistor can be calculated as:

$$P_{R_1} = I^2 R_1 = (90\ \text{mA})^2\,(100\ \Omega) = 810\ \text{mW}$$

$$P_{R_2} = I^2 R_2 = (60\ \text{mA})^2\,(250\ \Omega) = 900\ \text{mW}$$

$$P_{R_3} = I^2 R_3 = (30\ \text{mA})^2\,(500\ \Omega) = 450\ \text{mW}$$

The power delivered by the source can be calculated as:

$$P_S = V_S I = (24\ \text{V})(90\ \text{mA}) = 2.16\ \text{W}$$

Note that the sum of the power dissipated by the resistors is equal to the power delivered by the source:

$$P_{R_1} + P_{R_2} + P_{R_3} = 810\ \text{mW} + 900\ \text{mW} + 450\ \text{mW} = 2.16\ \text{W} = P_S$$

The minimum required standard wattage rating for the respective resistors is:

$$R_1 = 1\ \text{W}; \quad R_2 = 1\ \text{W}; \quad R_3 = 1/2\ \text{W}$$

EXAMPLE 9–19   Power Ratings for Resistors in an AC Circuit

Determine the minimum standard power rating for each resistor in the circuit in Figure 9–35.

FIGURE **9-35**

Circuit for
Example 9–19

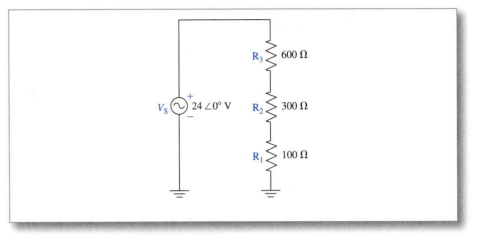

✓ SOLUTION

The maximum power rating for a resistor in an AC circuit is determined using RMS values for current and voltage.

The total equivalent resistance is calculated as:

$$R_T = 100 \ \Omega + 300 \ \Omega + 600 \ \Omega = 1 \ \text{k}\Omega$$

The RMS value of the source voltage is calculated as:

$$V_S(\text{RMS}) = (24 \ \text{V})(0.707) = 16.97 \ \text{V}$$

Ohm's law can be used to calculate the RMS value of the source current as:

$$I_S(\text{RMS}) = \frac{16.97 \ \text{V}}{1 \ \text{k}\Omega} = 16.97 \ \text{mA}$$

The power dissipated by each respective resistor is calculated as:

$$P_{R_1} = I^2 R_1 = (16.97 \ \text{mA})^2 \ (100 \ \Omega) = 28.79 \ \text{mW}$$

$$P_{R_2} = I^2 R_2 = (16.97 \ \text{mA})^2 \ (300 \ \Omega) = 86.39 \ \text{mW}$$

$$P_{R_3} = I^2 R_3 = (16.97 \ \text{mA})^2 \ (600 \ \Omega) = 172.78 \ \text{mW}$$

The power delivered by the source may be calculated as:

$$P_S = V_S \, I = (16.97 \ \text{V})(16.97 \ \text{mA}) = 287.96 \ \text{mW}$$

Note that the sum of the power dissipated by the resistors is equal to the power delivered by the source:

$$P_{R_1} + P_{R_2} + P_{R_3} = 28.79 \ \text{mW} + 86.39 \ \text{mW} + 172.78 \ \text{mW} = 287.96 \ \text{W} = P_S$$

The minimum required standard wattage ratings are determined as:

$$R_1 = 1/16 \ \text{W}; \quad R_2 = 1/8 \ \text{W}; \quad R_3 = 1/4 \ \text{W}$$

EXAMPLE 9-20   Maximum Power Ratings with a Safety Factor

Design engineers often include a "safety factor" when selecting components. The increase in power rating to include a safety factor must be weighed against the increase in cost and in the larger physical size of components with a higher power rating.

Determine the appropriate power rating for the resistors in the voltage divider circuit in Figure 9–36, including a safety factor of 25%.

FIGURE **9-36**

Circuit for
Example 9-20

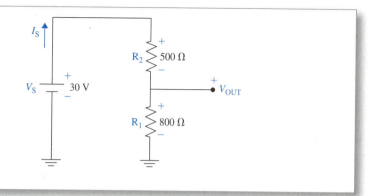

✓ **SOLUTION**

The total equivalent resistance is calculated as:

$$R_T = R_1 + R_2 = 1.3 \text{ k}\Omega$$

Ohm's law can be used to calculate the source current as:

$$\frac{V_S}{R_T} = \frac{30 \text{ V}}{1.3 \text{ }\Omega} = 23.08 \text{ mA}$$

The power dissipated by R_1 can be calculated as:

$$P_{R_1} = I^2 R_1 = (23.08 \text{ mA})^2 (800 \text{ }\Omega) = 426 \text{ mW}$$

The power dissipated by R_1 including a 25% safety factor is calculated as:

$$P_{R_1} = (426 \text{ mW})(1.25) = 532.5 \text{ mW}$$

The power dissipated by R_2 can be calculated as:

$$P_{R_2} = I^2 R_2 = (23.08 \text{ mA})^2 (500 \text{ }\Omega) = 266.34 \text{ mW}$$

The power dissipated by R_2 including a 25% safety factor is calculated as:

$$P_{R_2} = (266.34 \text{ mW})(1.25) = 332.93 \text{ mW}$$

The respective minimum power rating with a 25% safety factor for each resistor is:

$$R_1 = 1 \text{ W} \quad \text{and} \quad R_2 = 1/2 \text{ W}$$

Note that including a 25% safety factor increased the minimum power rating of R_1 from 1/2 W to 1 W.

PRACTICE EXERCISES 9-8

a. Given the circuit in Figure 9–37, determine the minimum standard power rating required for each resistor.

FIGURE **9-37**

Circuit for Practice
Exercise 9-8 Exercise 1a

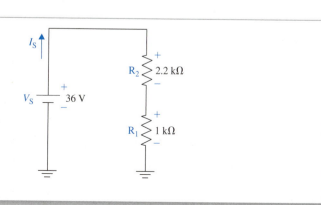

b. Given the circuit in Figure 9–38, determine the minimum standard power ratings for each resistor including a 25% safety factor.

FIGURE **9–38**

Circuit for Practice
Exercise 9–8
Exercise 2b

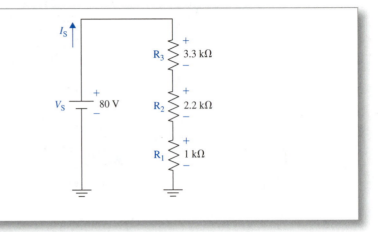

a. $P_{R1} = 126.56$ mW (standard value = 1/4 W);
$P_{R2} = 278.44$ mW (standard value = 1 W)

b. $P_{R1} = 189$ mW (standard value = 1/4 W);
$P_{R2} = 416$ mW (standard value = 1/2 W);
$P_{R3} = 625$ mW (standard value = 1 W)

(Complete solutions to the Practice Exercises can be found in Appendix A.)

9.14   POWER DERATING

The maximum power ratings for components are typically determined for a normal ambient temperature range of 25°C (normal room temperature) to 50° C. If a component is to be operated at temperatures higher than the specified range, the power rating must be *derated* (reduced). Derating information is typically included in the manufacturer's specification data and may be given as a derating coefficient or in graphical form as a derating curve.

EXAMPLE 9–21   Power Derating

Use the power derating curve shown in Figure 9–39 to determine the maximum power rating for a 1/2 W resistor at a temperature of 85°C.

✓ **SOLUTION**

The power derating curve in Figure 9–39 indicates that the maximum power rating for the resistor for lead temperatures up to 50°C is 1/2 W. Beyond 50°C,

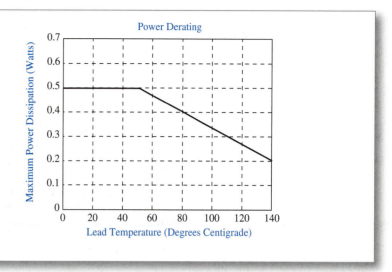

FIGURE **9–39**

Power Derating
Curve for
Example 9–21

the maximum power rating derates (decreases) as temperature increases. The derated wattage rating at 85° C can be determined from the graph as approximately 0.37 W (370 mW) at 85°C.

PRACTICE EXERCISES 9–9

Use the power derating curve in Figure 9–39 to determine the maximum power rating for a 1/2 W resistor at a lead temperature of 110°C.

✓ ANSWER

Maximum power = 0.30 W = 300 mW

(Complete solutions to the Practice Exercises can be found in Appendix A.)

SUMMARY

The electrical power is defined as the product of voltage and current ($p = vi$).

The power delivered to a resistor is converted into heat. The power converted into heat by a resistor is referred to as **real power** and is measured in units of watts. Real power may be calculated in the phasor domain as $P_R = (I_{RMS})^2 R$ W. DC power dissipated by a resistor is calculated as I^2R W.

The electrical energy delivered to a capacitor is temporarily stored by the capacitor and ultimately returned to the circuit. The energy delivered to a capacitor is referred to as **capacitive reactive power** and is measured in units of volt-ampere-reactive (VAR). Capacitive reactive power can be calculated in the phasor domain as $Q_C = (I_{RMS})^2 X_C$ VAR.

The electrical energy delivered to an inductor is temporarily stored by the inductor and ultimately returned to the circuit. The energy delivered to an inductor is referred to as **inductive reactive power** and is measured in units of volt-ampere-reactive (VAR). Inductive reactive power can be calculated in the phasor domain as $Q_L = -(I_{RMS})^2 X_L$ VAR.

The power delivered to a circuit by an AC source is referred to as **apparent power** and is measured in volt-amperes (VA). Apparent power is calculated as the product of the RMS value of the source voltage and the RMS value of the source current. Apparent power may also be calculated using the **power triangle** as the right-angle sum of real power and net reactive power.

The primary and secondary voltage and current for an ideal transformer are related by the ratio of turns on the respective windings. The primary and secondary impedance of an ideal transformer are related by the square of the turns ratio of the windings.

All practical voltage sources have internal impedance. Maximum power is delivered to a load when the load impedance is equal to the internal impedance of the source. In cases where the load is not directly connected to the source, the Thevenin impedance is equal to the load impedance required for maximum power transfer.

Electronic circuit designers must consider the maximum ratings for circuit components as specified by the manufacturer to avoid damaging or destroying the component. The maximum power ratings for electronic circuit components are typically specified at an ambient temperature of 25°C to 50°C. If the component is to be operated at higher temperatures, the maximum power rating must be derated using the manufacturer's derating information.

ESSENTIAL LEARNING EXERCISES

(ANSWERS TO ODD-NUMBERED ESSENTIAL LEARNING EXERCISES CAN BE FOUND IN APPENDIX A.)

1. The peak value of a sinusoidal current through a 3 kΩ resistor is given as 3 mA at a frequency of 2 kHz. Determine the real power dissipated by the resistor. Determine the reactive power dissipated by the resistor.

2. The peak value of a sinusoidal voltage across a 3 kΩ resistor is given as 12 V at a frequency of 6 kHz. Determine the real power dissipated by the resistor. Determine the reactive power dissipated by the resistor.

3. The peak value of a sinusoidal current for a 3 μF capacitor is given as 1.5 mA at a frequency of 500 Hz. Determine the reactive power for the capacitor. Determine the real power dissipated by the capacitor.

4. The peak value of a sinusoidal voltage across a 2 μF capacitor is given as 8 V at a frequency of 800 Hz. Determine the reactive power for the capacitor. Determine the real power dissipated by the capacitor.

5. The peak value of a sinusoidal current through a 2.2 mH inductor is given as 5.5 mA at a frequency of 4 kHz. Determine the reactive power for the inductor. Determine the real power dissipated by the inductor.

6. The peak value of a sinusoidal voltage across a 3.5 mH inductor is given as 24 V at a frequency of 2.5 kHz. Determine the reactive power for the inductor. Determine the real power dissipated by the inductor.

7. Determine the real power, reactive power, and apparent power for the circuit in Figure 9–40.

FIGURE **9–40**

Circuit for Essential
Learning Exercise 7

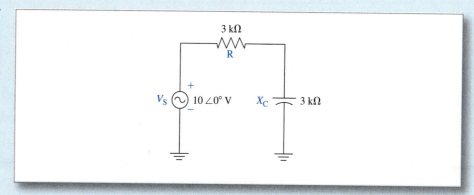

8. Determine the real power, reactive power, and apparent power for the circuit in Figure 9–41.

FIGURE **9–41**

Circuit for Essential
Learning Exercise 8

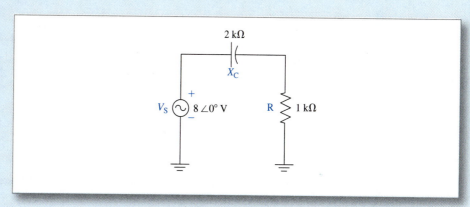

9. Determine the real power, reactive power, and apparent power for the circuit in Figure 9–42.

FIGURE **9–42**

Circuit for Essential
Learning Exercise 9

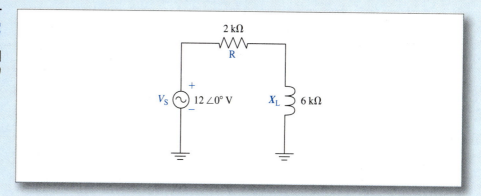

10. Determine the real power, reactive power, and apparent power for the circuit in Figure 9–43.

FIGURE **9–43**

Circuit for Essential
Learning Exercise 10

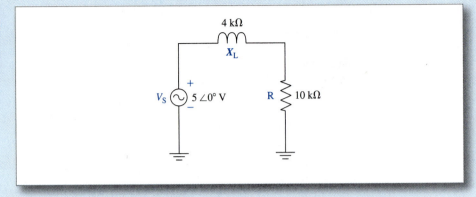

11. Determine the real power, reactive power, and apparent power for the circuit in Figure 9–44.

FIGURE **9–44**

Circuit for Essential
Learning Exercise 11

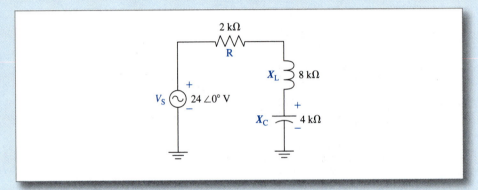

12. Determine the real power, reactive power, and apparent power for the circuit in Figure 9–45.

FIGURE **9–45**

Circuit for Essential
Learning Exercise 12

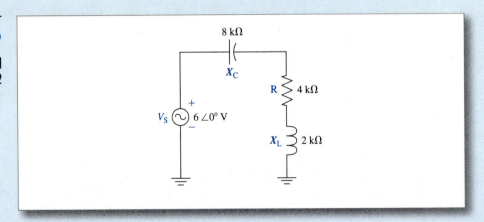

13. Sketch the power triangle for the circuit in Essential Learning Exercise 11.

14. Sketch the power triangle for the circuit in Essential Learning Exercise 12.

15. Determine the power factor for the circuit in Essential Learning Exercise 11.

16. Determine the power factor for the circuit in Essential Learning Exercise 12.

17. Determine the real power, reactive power, and apparent power for the circuit in Figure 9–46.

FIGURE **9–46**

Circuit for Essential
Learning Exercise 17

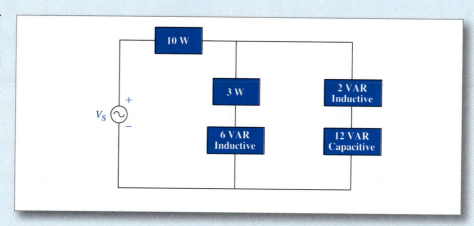

18. Determine the real power, reactive power, and apparent power for the circuit in Figure 9–47.

FIGURE **9–47**

Circuit for Essential
Learning Exercise 18

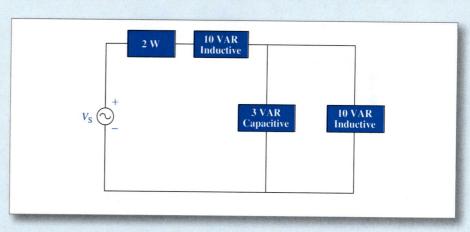

19. A certain transformer has a turns ratio of 25:1. Is the transformer a step-up or step-down transformer?

20. The RMS voltage across the primary winding of a transformer is measured as 100 V, and the RMS voltage across the secondary winding is measured as 4 kV. Is the transformer a step-up or step-down transformer?

21. The turns ratio of a transformer is given as 20:1. The RMS value of the primary voltage is given as 110 V, and the RMS value of the primary current is given as 12 mA. Determine the RMS value of the secondary voltage and current.

22. The turns ratio of a transformer is given as 1:36. The RMS value of the primary voltage is given as 12 V, and the RMS value of the primary current is given as 50 mA. Determine the RMS value of the secondary voltage and current.

23. Calculate the primary and secondary power for the transformer in Essential Learning Exercise 21.

24. Calculate the primary and secondary power for the transformer in Essential Learning Exercise 22.

25. Determine the value of the load resistor R_L for maximum power transfer for the circuit in Figure 9–48. Calculate the power dissipated by the load at maximum power transfer.

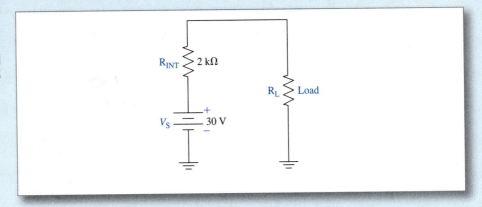

FIGURE **9–48**

Circuit for Essential Learning Exercise 25

26. Determine the value of the load resistor R_L for maximum power transfer for the circuit in Figure 9–49. Calculate the power dissipated by the load at maximum power transfer.

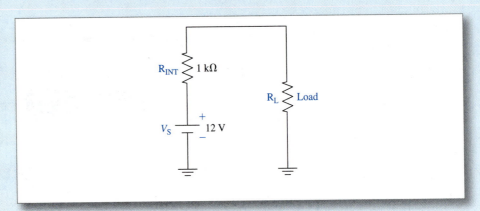

FIGURE **9–49**

Circuit for Essential Learning Exercise 26

27. Determine the value of the load resistor R_L for maximum power transfer for the circuit in Figure 9–50. Calculate the power dissipated by the load at maximum power transfer.

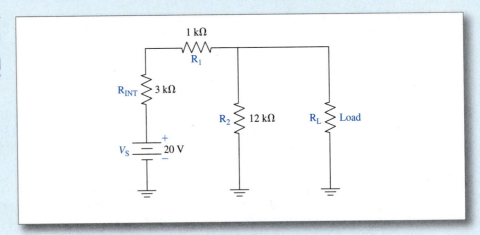

FIGURE **9–50**

Circuit for Essential Learning Exercise 27

28. Determine the value of the load resistor R_L for maximum power transfer for the circuit in Figure 9–51. Calculate the power dissipated by the load at maximum power transfer.

FIGURE **9–51**

Circuit for Essential Learning Exercise 28

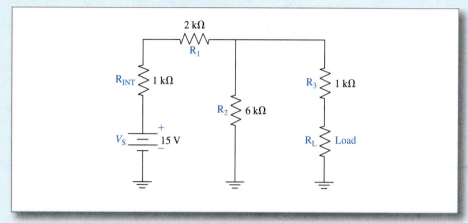

29. Determine the value of the load impedance $\mathbf{Z_L}$ for maximum power transfer for the circuit in Figure 9–52. Calculate the power dissipated by the load at maximum power transfer.

FIGURE **9–52**

Circuit for Essential Learning Exercise 29

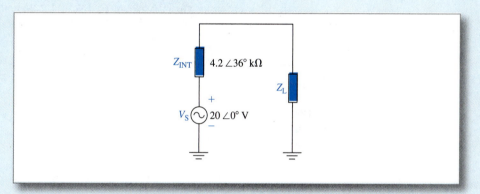

30. Determine the value of the load impedance $\mathbf{Z_L}$ for maximum power transfer for the circuit in Figure 9–53. Calculate the power dissipated by the load at maximum power transfer.

FIGURE **9–53**

Circuit for Essential Learning Exercise 30

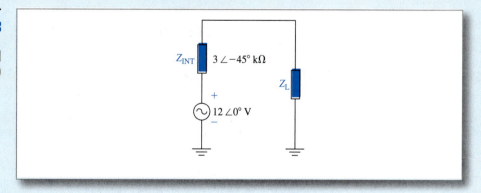

31. Determine the turns ratio of the transformer required to match a source with an internal source resistance of 400 Ω to a 4 Ω load.

32. Determine the turns ratio of the transformer required to match a source with an internal source resistance of 640 Ω to a 10 Ω load.

33. Determine the minimum standard power rating for each resistor in the circuit in Figure 9–54. (Standard power ratings are 1/16 W, 1/8 W, 1/4 W, 1/2 W, and 1 W.)

FIGURE **9–54**

Circuit for Essential
Learning Exercise 33

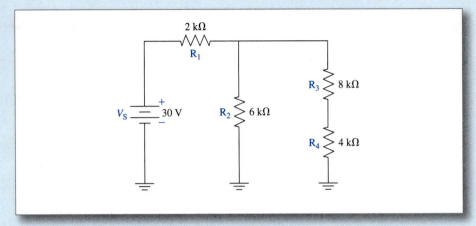

34. Determine the minimum standard power rating for each resistor in the circuit in Figure 9–55. (Standard power ratings are 1/16 W, 1/8 W, 1/4 W, 1/2 W, and 1 W.)

FIGURE **9–55**

Circuit for Essential
Learning Exercise 34

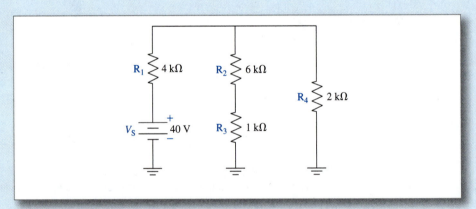

35. Repeat Essential Learning Exercise 33 including a 25% safety factor for the power rating of each resistor.

36. Repeat Essential Learning Exercise 34 including a 25% safety factor for the power rating of each resistor.

37. Figure 9–39 illustrates the power derating curve for a certain resistor. Determine the maximum power that can be dissipated by the resistor at a lead temperature of 110°C.

CHALLENGING LEARNING EXERCISES

(ANSWERS TO SELECTED CHALLENGING LEARNING EXERCISES CAN BE FOUND IN APPENDIX A.)

1. Given the circuit in Figure 9–56, determine the real power, reactive power, and apparent power.

FIGURE **9–56**

Circuit for Challenging
Learning Exercise 1

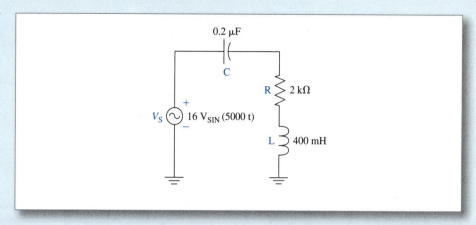

2. The total real power dissipated in an AC circuit is 146 W, and the net reactive power is 200 VAR (inductive). Determine the power factor for the circuit.

3. The power factor for an AC circuit is given as 0.86 (leading). The real power is measured as 80 W. Determine the net reactive power.

4. Prove that the product of impedance and its conjugate is always a pure resistance.

5. Prove that the maximum power delivered to a load resistor in a complex DC circuit is equal to $\dfrac{V_{\text{Th}}^{2}}{4R_{\text{L}}}$.

6. A DC source with an internal resistance of 2 kΩ is directly connected to a load resistance of 500 Ω. Determine the power dissipated by the 500 Ω load resistor. The load resistor value is changed from 500 Ω to 4 kΩ. Determine the power dissipated by the 4 kΩ load. The load resistor value is changed to 2 kΩ to equal the internal resistance of the source. Determine the power dissipated by the 2 kΩ load. What conclusion can you make from the calculations in this exercise?

7. The designer of the circuit in Figure 9–57 has used a 1/8 W resistor for each resistor in the circuit. Each resistor is to have a 35% safety factor for its power rating. Determine if the 1/8 W resistors meet the power requirements of the design specifications.

FIGURE **9–57**

Circuit for Challenging
Learning Exercise 7

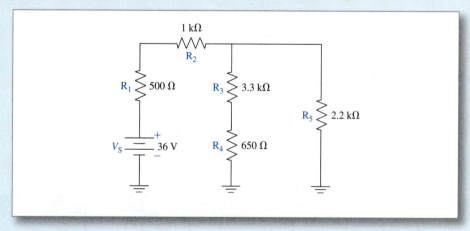

8. Show that an AC circuit for which the inductive reactive power and capacitive reactive power are equal has a power factor of 1.

9. A certain 1 kΩ, 1 W resistor is derated at 0.2%/C° for temperatures in excess of 100°C. Determine the maximum power rating for the resistor at a temperature of 125°C.

10. A transformer is designed with 60,000 primary turns and 15,000 secondary turns. The RMS values of the primary voltage and current are 110 V and 200 mA respectively. Determine the secondary voltage and current. Calculate the primary and secondary power.

11. A load resistance of 10 Ω is directly connected to an 8 V_{RMS} signal source whose internal resistance is 50 Ω. Determine the power dissipated by the internal resistance and the power dissipated by the load.

12. Determine the turns ratio of a transformer required to match the load and source impedance in Challenging Learning Exercise 2. Determine the power dissipated by the internal resistance of the source and the power dissipated by the matched load. Hint (the load is reflected to the primary side as $(TR)^2 RL$.)

13. Determine the minimum standard power rating for each resistor in Figure 9–58. Include a safety factor of 25%.

FIGURE **9–58**

Circuit for Challenging
Learning Exercise 13

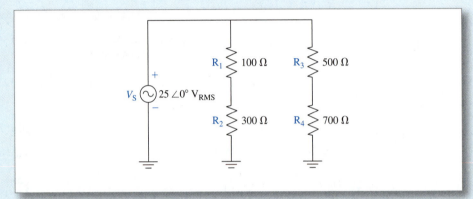

14. Determine the value of load resistor R_L for maximum power transfer in Figure 9–59. Determine the value of maximum power P_L.

FIGURE **9–59**

Circuit for Challenging
Learning Exercise 14

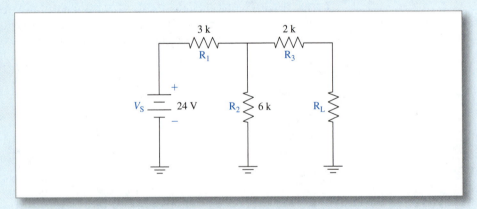

TEAM ACTIVITY

Contact the engineering department at the local electric company to see if you can obtain the following information:

    a. The RMS value of voltage produced at the generation site.

    b. The RMS value of the voltage used for long-distance distribution.

    c. The RMS value of the voltage used for local residential distribution.

    d. The power factor for the system (note that the power factor is continuously changing).

ANALYSIS OF NONLINEAR ANALOG CIRCUITS

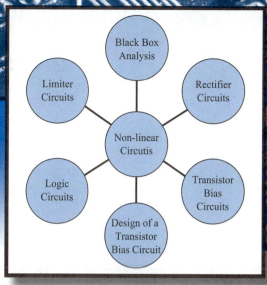

"HE WHO IS OVERCAUTIOUS WILL ACCOMPLISH LITTLE."

—Friedrich Schiller

CHAPTER OUTLINE

LEARNING OBJECTIVES

Upon successful completion of this chapter you will be able to:

- Describe the nature of the p-n junction.
- Determine the output of a diode half-wave rectifier circuit.
- Determine the output of a biased diode limiter circuit.
- Analyze a diode-resistor logic (DRL) circuit.
- Determine the DC operating point of a fixed bias transistor amplifier circuit.
- Determine the DC operating point of a voltage divider biased transistor amplifier circuit.
- Analyze a transistor logic circuit.
- Design a fixed bias transistor circuit.
- Analyze circuits that contain "black box" devices.

CHAPTER TEN

INTRODUCTION

The circuits presented in the previous chapters of this text consisted only of linear bilateral components including resistors, capacitors, and inductors. The current through a linear component is always proportional to the voltage across the device. A bilateral component permits current to flow in either direction through the device. Many other useful electronic circuits contain components such as the diode and transistor that are neither linear nor bilateral. The laws of circuit analysis, including Ohm's law and Kirchhoff's voltage and current laws, are valid for all electronic circuits including those that contain nonlinear circuit components.

This chapter introduces the characteristics of the p-n junction diode, junction transistor, and operational amplifier and demonstrates how to analyze several signal processing circuits that contain these devices.

10.1   THE P-N JUNCTION

Many contemporary electronic components and integrated circuits are manufactured from semiconductor materials such as silicon and germanium. Impurities may be added to these substances to produce either *p-material,* which has an excess of positive charge carriers referred to as holes, or *n-material,* which has an excess of negatively charged electrons. When p-material and n-material are joined, a *p-n junction* is formed. The discovery of the electrical characteristics of the p-n junction led to the invention of important circuit components such as the *junction diode* and *junction transistor*. Components manufactured from semiconductor materials are commonly referred to as *solid-state devices*.

Current can flow through a p-n junction only when the voltage applied to the p-material is more positive than the voltage applied to the n-material.

A p-n junction is **forward biased** when the voltage applied to the p-material is more positive than the voltage applied to the n-material. Current can flow through a silicon p-n junction that is forward biased with a minimum of 0.7 V. Current can flow through a germanium p-n junction that is forward biased with a minimum of 0.3 V.

A p-n junction is **reverse biased** when the voltage applied to the p-material is more negative than the voltage applied to the n-material. Current cannot flow through a p-n junction that is reverse biased.

10.2   THE P-N JUNCTION DIODE

The **p-n junction diode** is a two-terminal solid-state device. The anode of the diode consists of p-material and the cathode consists of n-material.

A p-n junction diode and its schematic symbol are illustrated in Figure 10–1.

The diode is said to be forward biased when the voltage at the anode of the device is more positive than the voltage at the cathode. The forward biased diode offers near zero resistance to current flow through the device.

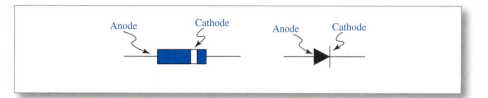

FIGURE **10-1**

A p-n Junction
Diode and
Schematic Symbol

The diode is said to be reverse biased when the voltage at the anode of the device is more negative than the voltage at the cathode. The reverse biased diode will not allow current to flow through the device, and the diode can be considered to be an open circuit.

The characteristics of electronic devices such as the p-n diode are often illustrated by an I-V characteristic curve that reveals how current through the device varies with voltage across the device. The I-V characteristic curves for a silicon diode and a germanium diode are shown in Figure 10–2. Note that the characteristic curve for a diode is not a straight line, indicating the device is nonlinear.

FIGURE **10-2**

I-V Characteristic
for a Silicon and
Germanium Diode

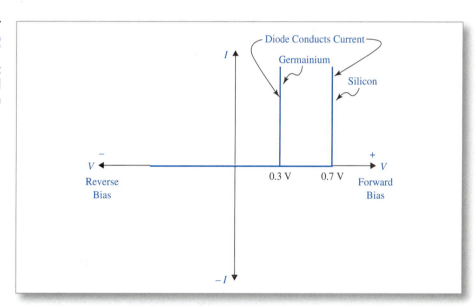

The I-V characteristic curves in Figure 10–2 indicate that a silicon diode will conduct current with near zero resistance when it is forward biased with a voltage greater than seven-tenths of a volt. A germanium diode will conduct current with near zero resistance when it is forward biased with a voltage greater than three-tenths of a volt. The I-V characteristic curves indicate that neither diode will conduct current when it is reverse biased.

The I-V characteristic curves shown in Figure 10–2 are linear approximations of ideal-ized diodes. More exact characteristics of a diode may be obtained if necessary; however, satisfactory results are normally obtained through analysis based on the approximate characteristics shown in Figure 10–2.

For most circuit applications, the diode is considered to be a solid-state electronic switch that is open when it is reverse biased and closed when it is forward biased. Due to the nature of a p-n junction, a voltage of approximately 0.7 V exists across the forward biased silicon diode.

Since the characteristics of a diode change dramatically when the polarity of voltage across the device is reversed, two models of the device are required, one model to represent the forward biased diode, and another model to represent the reverse biased diode. The approximate diode models for the silicon diode are shown in Figure 10–3 a and b.

FIGURE **10–3**

Circuit Model
for a Forward
and Reverse
Biased Silicon
Diode

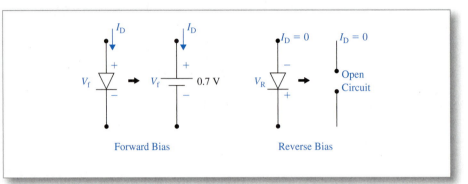

Forward Bias   Reverse Bias

The I-V curve illustrated in Figure 10–2 indicates that the silicon diode can be defined by the following set of equations:

Diode forward biased with a voltage greater than 0.7 V:

$$V_D = V_f \approx 0.7 \text{ V} \tag{10.1}$$

Diode reverse biased:

$$I_D = 0 \tag{10.2}$$

10.3   THE DIODE HALF-WAVE RECTIFIER

An important application of the diode is its use in circuits designed to convert sinusoidal commercial AC voltage (110 V_{RMS}; 60 Hz) into DC voltage. Recall that the average value (DC value) of a sine wave over one complete cycle is zero, so that a sine wave cannot be directly used to produce a DC voltage. A diode circuit can be used to convert a sinusoidal waveform into a waveform that has a nonzero DC value.

Rectification is defined as the process that converts a sinusoidal waveform to a waveform that has a nonzero average or DC value.

Consider the circuit in Figure 10–4.

FIGURE **10–4**

A Half-Wave
Rectifier Circuit

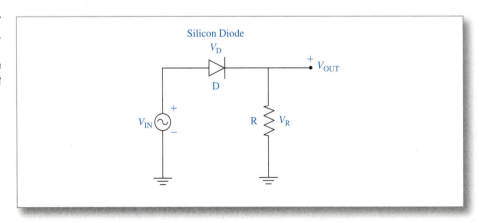

Since the model for the diode depends upon the polarity of the voltage across the device, the circuit in Figure 10–4 must be analyzed for both positive and negative values of input voltage.

Case 1: *Assume the input voltage is positive and less than 0.7 V.*

The silicon diode in Figure 10–5 will conduct current only when the input voltage is positive and greater than 0.7 V.

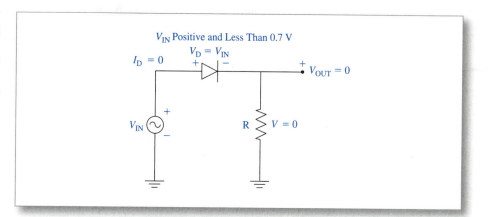

FIGURE 10–5

Half-Wave Rectifier Circuit with Positive $V_{in} < V_f$

If the input voltage is less than 0.7 V, the diode will not conduct current. Since current cannot flow in the circuit when the input bias voltage is less than 0.7 V. The current through the resistor and the voltage across the resistor are equal to zero.

Since the output signal is taken across the resistor, the output voltage is equal to zero, so that:

$$V_{out} = 0 \qquad (10.3)$$

KVL may be expressed as:

$$V_{in} - V_D - V_R = 0$$

Substituting $V_R = 0$, we obtain:

$$V_D = V_{in} \qquad (10.4)$$

Case 2: *Assume the input voltage is positive and greater than 0.7 V.*

When the input voltage is positive and greater than 0.7 V, as shown in Figure 10–6, the diode will conduct current.

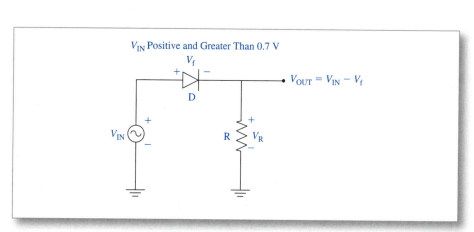

FIGURE 10–6

Half-Wave Rectifier Circuit with Positive $V_{in} > V_f$

The model for the forward biased diode shown in Figure 10–3 indicates that $V_f = 0.7$ V will appear across the diode and that current can flow through the diode in the direction that the arrowhead is pointing. The current flowing through the diode also flows through the resistor, creating a voltage across the resistor.

KVL may be expressed as:

$$V_{in} - V_f - V_R = 0$$

Since the output signal is taken across the resistor, the output voltage is equal to the voltage across the resistor, so that:

$$V_{out} = V_{in} - V_f \qquad (10.5)$$

Case 3: *Assume the input voltage is negative.*

When the input voltage is negative, as shown in Figure 10–7, the diode will be reverse biased.

FIGURE **10–7**

Half-Wave Rectifier
Circuit with
Negative V_{in}

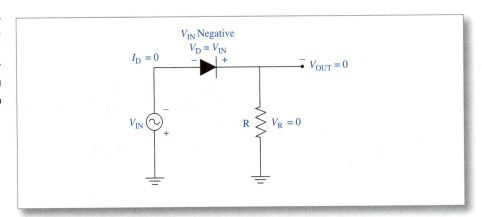

The model for the reverse biased diode shown in Figure 10–3 indicates that the diode is an open circuit, so that current cannot flow through the diode. Since the current through the circuit is zero, the voltage across the resistor will also be zero.

Since the output signal is taken across the resistor, the output voltage is equal to the voltage across the resistor, so that:

$$V_{out} = 0$$

KVL may be expressed as:

$$-V_{in} + V_D - V_R = 0$$

Since $V_R = 0$, the voltage across the diode is equal to the source voltage V_{in}, so that:

$$V_D = -V_{in} \qquad (10.6)$$

The illustration in Figure 10–8 shows the input source voltage, the output voltage, and the voltage across the diode over two complete cycles.

Note that the output voltage waveform consists only of positive values. The output waveform therefore has an average or DC value that is not zero.

FIGURE **10–8**

Waveforms for a
Half-Wave Rectifier

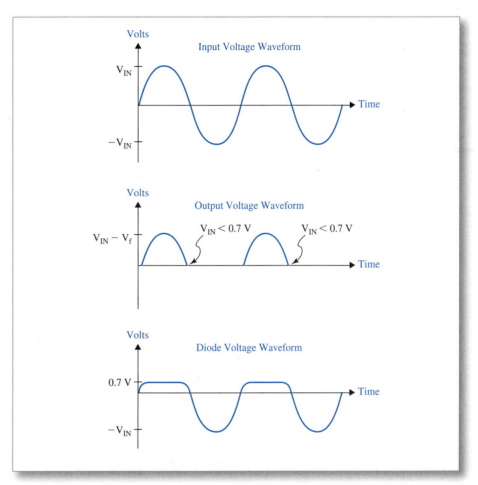

EXAMPLE 10–1   Analysis of a Half-Wave Rectifier Circuit

Determine the amplitude and shape of the output voltage waveform (V_{out}) for the circuit in Figure 10–9. Assume the forward voltage for the diode to be $V_{\text{f}} = 0.7$ V.

FIGURE **10–9**

Circuit for
Example 10–1

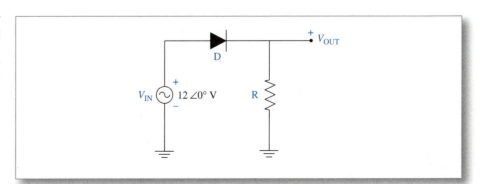

Case 1: *Assume the input signal is positive but less than 0.7 V.*

The circuit in Figure 10–9 may be redrawn, substituting the diode model for the case when the input bias voltage is positive and less than 0.7 V, as shown in Figure 10–10.

FIGURE **10–10**

Circuit with
Positive $V_{in} < V_f$

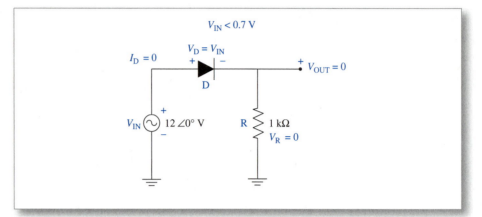

The circuit in Figure 10–10 is referred to as an equivalent circuit and can be considered to be a snapshot of the circuit in Figure 10–9 when the input signal voltage is equal to any positive voltage less than 0.7 V. With the input voltage less than 0.7 V, the diode will not conduct current, so the current through R and the voltage across R must be zero. Since the output signal is taken across the resistor R, the output voltage is equal to zero.

Case 2: *Assume the input signal is positive and greater than 0.7 V.*

The input signal voltage is now sufficient to forward bias the diode so that current can flow through the circuit. The voltage across the diode will be $V_f = 0.7$ V. The equivalent circuit is shown in Figure 10–11.

FIGURE **10–11**

Circuit with
Positive $V_{in} > V_f$

Note that the amplitude of the output signal is equal to $V_{in} - V_f$.

Case 3: *Assume the input signal is negative (reverse polarity).*

The diode is now reverse biased and the equivalent circuit may be drawn as shown in Figure 10–12.

FIGURE **10–12**

Circuit with
Negative V_{in}

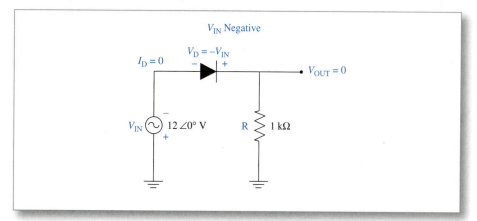

Current cannot flow through the reverse biased diode, so that the current through R is zero, and the output voltage will also be zero:

$$V_{out} = 0$$

The input and output voltage waveforms are shown in Figure 10–13.

FIGURE **10–13**

Waveforms for
Example 10–1

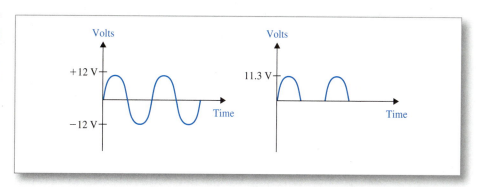

Note that the output waveform shown in Figure 10–13 contains only the positive values of the input sinusoidal waveform. The output waveform therefore has an average or DC value that is positive and nonzero. The MultiSIM simulation of the input and output waveforms is shown in Figure 10–14.

MULTISIM

FIGURE **10–14**

MultiSIM
Simulation for
Example 10–1

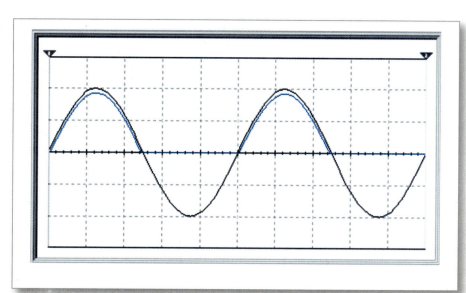

EXAMPLE 10-2 Analysis of a Half-Wave Rectifier Circuit

Given the circuit in Figure 10–15, sketch the output waveform, considering the forward bias voltage drop across the diode to be $V_f = 0.7$ V.

FIGURE **10–15**

Circuit for
Example 10–2

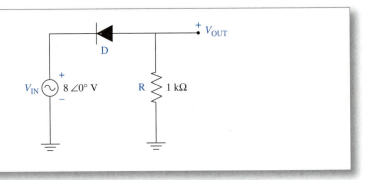

✓ **SOLUTION**

Case 1: *Assume that the input signal is positive.*

The first consideration when analyzing a circuit containing a diode is to determine whether the diode is forward or reverse biased. For all positive values of input signal voltage, as shown Figure 10–16, the diode is reverse biased.

FIGURE **10–16**

Circuit with
Positive V_{in}

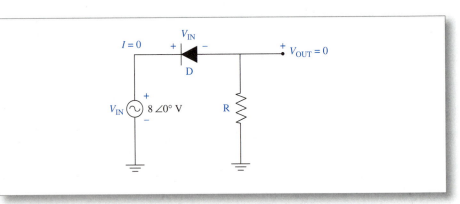

Current cannot flow through the circuit when the diode is reverse biased, so the voltage across the resistor will be zero, and because the output signal is taken across *R,* the output voltage will be zero:

$$V_{out} = 0$$

KVL can be used to determine the voltage across the diode as:

$$V_{in} - V_D - V_R = 0$$

With $V_R = 0$, we obtain:

$$V_D = V_{in}$$

Case 2: *Assume the input voltage is negative and less than 0.7 V.*

As the signal source reverses polarity, as shown in Figure 10–17, the diode will be forward biased.

However, the minimum forward bias voltage required for the diode to conduct current is 0.7 V. Therefore, with V_{in} less than 0.7 V, the current will be zero and

FIGURE **10–17**

Circuit with
Negative $V_{in} < V_f$

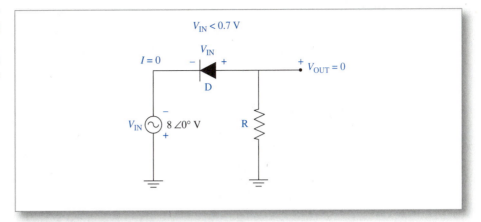

the output voltage will be zero. The voltage across the diode will equal the input voltage, so that:

$$V_{out} = 0$$

$$V_D = V_{in}$$

Case 3: *Assume the input voltage is negative and greater than 0.7 V.*

When the input voltage is negative and greater than 0.7 V, the diode will be forward biased with sufficient voltage to conduct current, as shown in Figure 10–18.

FIGURE **10–18**

Circuit with
Negative $V_{in} > V_f$

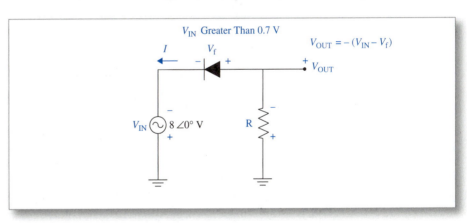

The current flowing through the resistor will develop a voltage across the resistor that is equal to V_{out}. The voltage across the forward biased diode will be equal to V_f.

We may write KVL as:

$$-V_{in} + V_f - V_{out} = 0$$

$$V_{out} = -V_{in} + V_f$$

The input and output waveforms are shown in Figure 10–19.

FIGURE **10–19**

Waveforms for
Example 10–2

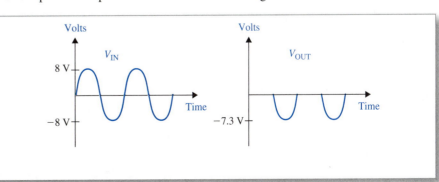

Note that the output waveform shown in Figure 10–19 contains only the negative values of the input sinusoidal waveform. The output waveform therefore has an average or DC value that is negative and nonzero.

The MultiSIM simulation of the input and output waveforms is shown in Figure 10–20.

FIGURE **10–20**

MultiSIM Simulation for Example 10–2

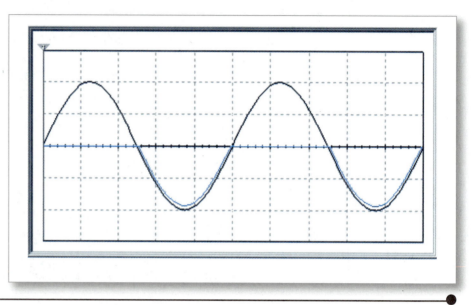

10.4 THE DIODE LIMITER

The diode limiter can be designed to limit the output signal voltage to any desired level. Limiter circuits are used to remove or limit the adverse effects of transient noise spikes that may be caused by voltage surges from electrostatic discharge, atmospheric disturbances, or other noise sources. Limiters are also used to protect electronic circuits from excessive voltage that may damage or destroy circuit components.

EXAMPLE 10–3 Analysis of a Diode Limiter

Consider the circuit in Figure 10–21 where V_B is referred to as the limiter bias voltage. Determine the shape and amplitude of the output signal (V_{out}). Consider the diode to be made of germanium, so that the forward bias voltage $V_f = 0.3$ V.

FIGURE **10–21**

Circuit for Example 10–3

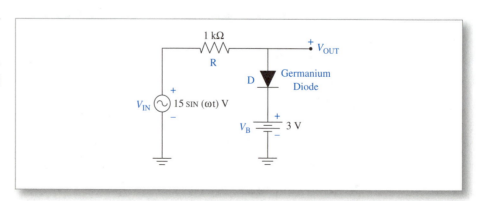

Case 1: *Assume the instantaneous value of the input voltage is positive and equal to 2 V.*

The cathode of the diode is at +3 V with respect to ground due to the bias voltage V_B. With the input signal less than 3 V, the cathode is more positive than the anode, so that the diode is reverse biased and cannot conduct current. The diode can be considered to be open, as indicated in Figure 10–22.

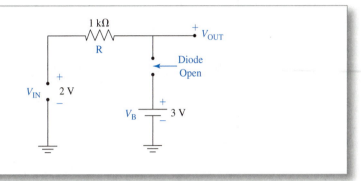

FIGURE **10-22**

Equivalent Circuit
with $V_{in} = +2\,V$

Since current cannot flow in the circuit, $I_R = 0$ and $V_R = 0$.

Applying KVL and moving clockwise from the source ground around the loop containing V_{out}, we obtain:

$$V_{in} - V_R - V_{out} = 0$$

Since $V_R = 0$, we obtain:

$$V_{out} = V_{in} = 3\ V$$

Case 2: *Assume the instantaneous value of the input voltage is positive and equal to 6 V.*

The cathode of the diode is at +3 V with respect to ground due to the bias voltage V_B. The input voltage of +6 V is sufficient to forward bias the diode, as shown in Figure 10–23.

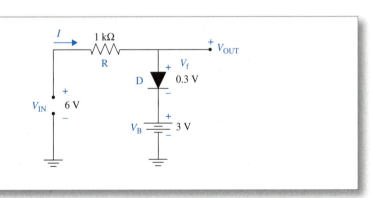

FIGURE **10-23**

Equivalent Circuit
with $V_{in} = +6\,V$

The diode is now forward biased, so that the voltage across the diode will be $V_f = 0.3\ V$ and current will flow through the circuit, causing voltage to be developed across the resistor.

Applying KVL and moving clockwise from the bias source (V_B) ground, we obtain:

$$V_B + V_f - V_{out} = 0$$

so that: $$V_{out} = V_B + V_f$$

With $V_f = 0.3$ V:

$$V_{out} = 3 \text{ V} + 0.3 \text{ V} = 3.3 \text{ V}$$

Any positive input voltage greater than 3.3 V will forward bias the diode and result in an output voltage of 3.3 V. For positive input voltage less than 3.3 V, the output voltage will be equal to the input voltage.

Case 3: *Assume the instantaneous value of the input voltage is negative and equal to 3 V.*

Negative values of V_{in} will aid the bias voltage V_B in reverse biasing the diode, as shown in Figure 10–24.

FIGURE **10–24**

Equivalent Circuit with $V_{in} = -3$ V

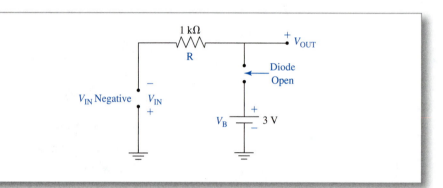

With the diode open no current will flow, so $I_R = V_R = 0$.

Applying KVL and moving clockwise from the source ground around the loop containing V_{out}, we obtain:

$$-V_{in} + V_R + V_{out} = 0$$

For all negative values of input voltage, the output voltage will be negative and equal to the input voltage.

The input and output waveforms for the limiter circuit are shown in Figure 10–25.

FIGURE **10–25**

Waveforms for Example 10–3

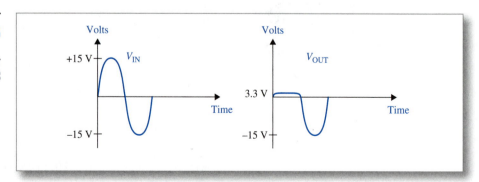

The biased diode limiter has limited the positive half cycle of the output voltage to 3.3 V.

▼

The MultiSIM simulation of the output signal of the limiter circuit is shown in Figure 10–26. Note that the positive half cycle voltage is limited to 3.3 V.

FIGURE **10–26**

MultiSIM
Simulation for
Example 10–3

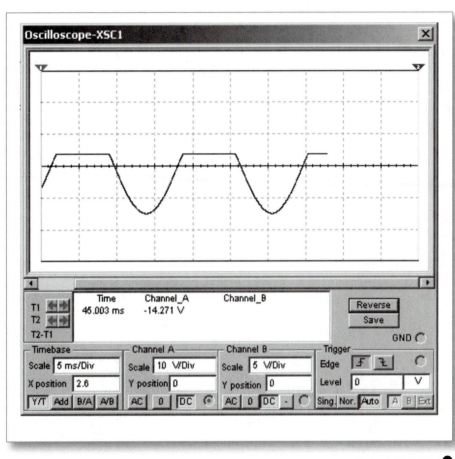

PRACTICE EXERCISES 10–1

a. Sketch the output signal for the diode rectifier circuit in Figure 10–27. Assume that the diode is silicon with $V_f = 0.7$ V. Use computer simulation to verify your solution.

FIGURE **10–27**

Circuit for Practice
Exercise 10–1
Exercise 1

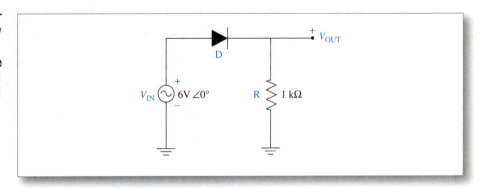

b. Sketch the output signal for the biased diode limiter circuit in Figure 10–28. Assume that the diode is germanium with $V_f = 0.3$ V. Use computer simulation to verify your solution.

FIGURE **10–28**

Circuit for Practice
Exercise 10–1
Exercise 2

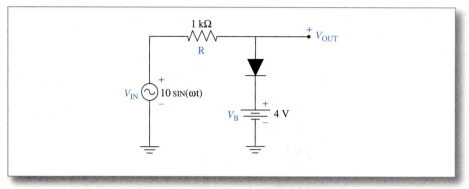

V_{IN} $\bigcirc$ 10 SIN(ωt)

1 kΩ
R

V_{OUT}

V_B 4 V

(Complete solutions to the Practice Exercises can be found in Appendix A.)

APPLICATION OF A DIODE RECTIFIER

Many electronic signal processing circuits require a DC voltage source. Batteries provide a DC voltage source for the signal processing circuits in portable electronic equipment. Electronic equipment designed to operate from a commercial voltage source (110 V_{RMS}; 60 Hz) requires circuits to convert the commercial sinusoidal voltage to DC voltage, as shown in Figure 10–29.

FIGURE **10–29**

A DC Voltage Supply Using a
Diode Rectifier Circuit

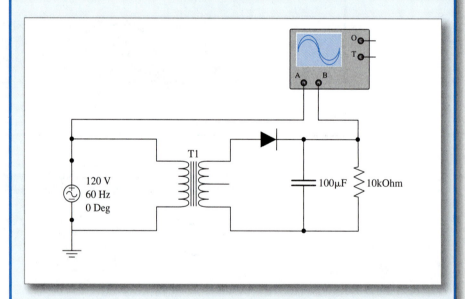

120 V
60 Hz
0 Deg

T1

100µF   10kOhm

SPECIFICATIONS: The transformer (T_1) in Figure 10–29 is used to step down the input voltage (110 V_{RMS}; 60 Hz) to the desired DC level. The diode rectifier converts the transformer secondary voltage to a half-wave rectified (positive values only) voltage waveform. The filter capacitor (C) will charge to the peak value of the rectified sine wave as the diode conducts. If the load resistor (R_L) is large, the capacitor will discharge only slightly through R_L as the rectified voltage decreases. The output voltage is therefore nearly constant in time.

✔ SOLUTION

The MultiSIM computer simulation of the circuit in Figure 10–29 is shown in Figure 10–30.

FIGURE **10–30**

MultiSIM Simulation
of a DC Voltage Supply

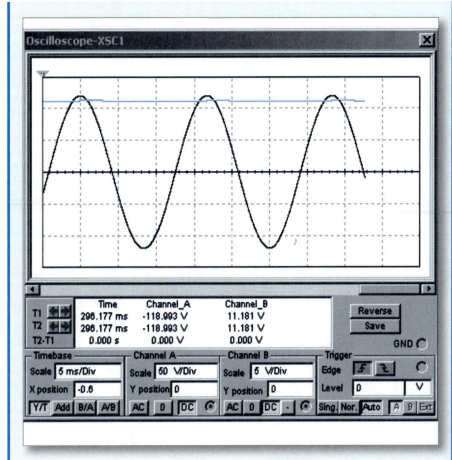

The sinusoidal waveform in Figure 10–30 is a 110 V, 60 Hz commercial voltage
source. The DC (blue waveform) is the voltage across the load.

10.5 A DIODE LOGIC CIRCUIT

The diode may be considered to be a solid-state switch whose state (on, off) is determined by the polarity of the voltage across the device. One of the many applications of
the diode is its use in diode-resistor logic (DRL) circuits. The binary nature of the device
(on, off) indicates its usefulness in digital circuit applications.

EXAMPLE 10–4   Analysis of a Diode Logic Circuit

The voltages V_1 and V_2 in Figure 10–31 are digital input signals whose values may
only be zero (ground) or 5 V. An input voltage of zero is said to be in its *low state*,
and an input voltage of 5 V is said to be in its *high state*. Determine the output voltage for all possible combinations of input voltage. Assume the forward bias voltage
drop for the diodes to be $V_f = 0.7$ V.

FIGURE **10-31**

A Diode-Resistor
Logic Circuit (DRL)
for Example 10-4

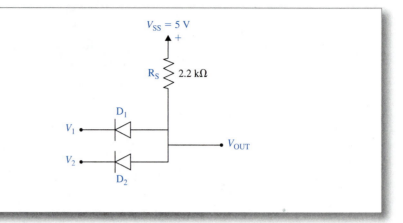

FIGURE **10-32**

DRL Circuit with
Both Input
Terminals at
Ground (0 V)

✓ **SOLUTION**

Case 1: *Assume $V_1 = V_2 = 0$ (ground).*

With V_1 and V_2 at ground, both diodes will be forward biased (on) by the voltage supply V_{SS}, so that current will flow from the 5 V source through the resistor R_S, through the forward biased diodes to ground, as indicated in Figure 10-32.

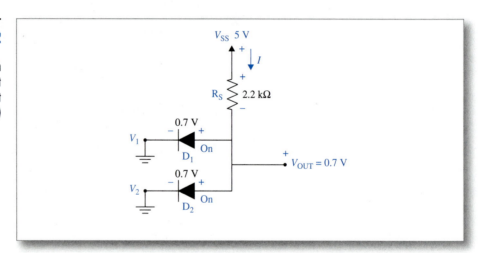

Beginning at the cathode of either diode, KVL can be expressed as:

$$0.7 \text{ V} - V_{out} = 0$$

so that:
$$V_{out} = 0.7 \text{ V}$$

Case 2: *Assume $V_1 = 5$ V and $V_2 = 0$ (ground).*

With $V_2 = 0$, D_2 is forward biased (on) and will conduct current. Current will flow from the voltage supply through R_S to ground through D_2, causing a voltage to be developed across R_S, as shown in Figure 10-33.

Beginning at the cathode of diode D_2, KVL can be expressed as:

$$0.7 \text{ V} - V_{out} = 0$$

so that:
$$V_{out} = 0.7 \text{ V}$$

The cathode of D_1 is at 5 V due to the input voltage V_1 and the anode of D_1 is at 0.7 V, so that D_1 is reverse biased (off), as shown in Figure 10–33.

FIGURE **10–33**

DRL Circuit with V_1 at +5 V and V_2 at Ground (0 V)

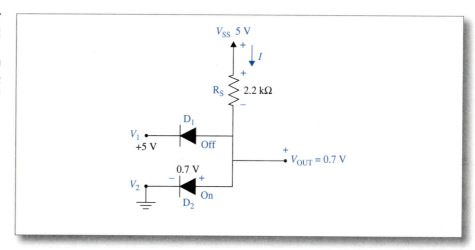

Case 3: *Assume that $V_1 = 0$ and $V_2 = 5$ V.*

Analysis of case 3 is similar to the analysis of case 2, so that D_1 is forward biased (on) and D_2 is reverse biased (off), as indicated in Figure 10–34.

FIGURE **10–34**

DRL Circuit with V_2 at +5 V and V_1 at Ground (0 V)

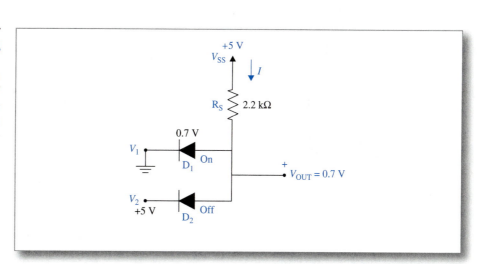

Beginning at the cathode of diode D_1, KVL can be expressed as:

$$0.7 \text{ V} - V_{out} = 0$$

so that:

$$V_{out} = 0.7 \text{ V}$$

Case 4: *Assume that $V_1 = 5$ V and $V_2 = 5$ V.*

With $V_1 = 5$ V and $V_2 = 5$ V, neither diode is forward biased, so that current cannot flow from the voltage source V_{SS}, as indicated in Figure 10–35.

Since current cannot flow, the voltage across R_S will be zero, and the output voltage will equal the source voltage V_{SS}, so that:

$$V_{out} = 5 \text{ V}$$

The circuit logic is summarized in the following table.

V_1	V_2	V_{out}
0 V	0 V	0 V
0 V	5 V	0 V
5 V	0 V	0 V
5 V	5 V	5 V

FIGURE **10–35**

DRL Circuit with V_1 at +5 V and V_2 at +5 V

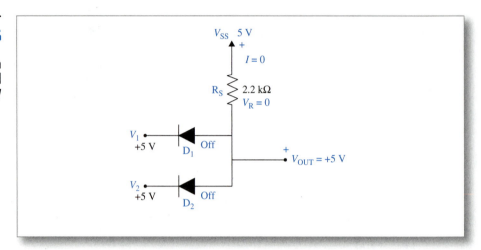

You may recognize that the circuit in this example executes the logic high AND function. The output voltage will acquire its high state (5 V) only when both input signals (V_1 **and** V_2) are in their high states (5 V).

The MultiSIM simulation for a low output is shown in Figure 10–36.

FIGURE **10–36**

MultiSIM Simulation of Low Output

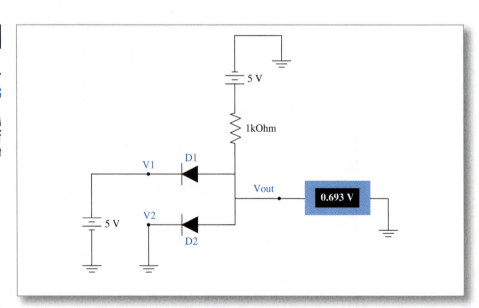

The simulation indicates the low state output as 0.693 V, which is the forward bias diode voltage V_f. The simulation in Figure 10–36 represents the output voltage whenever either input is zero (ground).

The MultiSIM simulation for the high output is shown in Figure 10–37.

FIGURE **10-37**

MultiSIM
Simulation of
High Output

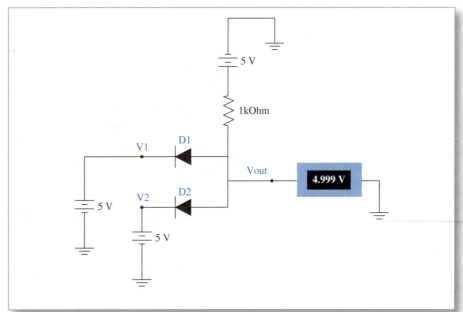

The output voltage will be high only when both input voltages V_1 and V_2 are high.

Diode-resistor logic (DRL) circuits were used extensively in the early development of modern digital computers. They were soon replaced by transistor-resistor logic (TRL) circuits, which gave way to the popular transistor-transistor logic (TTL) circuits found in many contemporary digital systems.

PRACTICE EXERCISES 10-2

Determine the logic function of the diode-resistor logic (DRL) circuit in Figure 10–38. The voltages V_1 and V_2 are digital input signals whose values may only be zero (ground) or 5 V. Assume the forward bias voltage drop for the diodes to be 0.7 V.

FIGURE **10-38**

Circuit for Practice
Exercise 10–2
Exercise 1

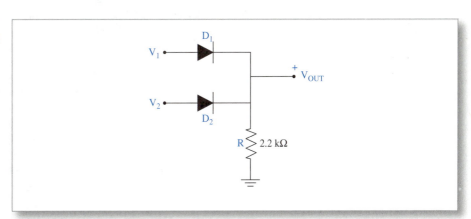

✓ ANSWER

OR gate

(Complete solutions to the Practice Exercises can be found in Appendix A.)

10.6 THE JUNCTION TRANSISTOR

The junction transistor consists of a region of p-material between two regions of n-material to form an n-p-n junction transistor. The regions are named base, collector, and emitter, as shown in Figure 10–39. For normal operation, the base-emitter junction of the transistor is forward biased by V_{BB} (positive voltage on p-material), the collector-base junction is reverse biased by V_{CC} (positive voltage on n-material), and the emitter is placed at ground (common emitter), as shown in Figure 10–39.

FIGURE **10–39**

Illustration of an n-p-n Junction Transistor

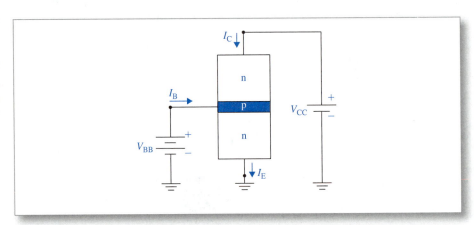

The current (I_C) that flows from the collector supply V_{CC} to ground through the emitter is directly proportional to the value of the base current (I_B) flowing into the base lead. The base current is considered to be the input current, and the collector current is considered to be the output current. An increase in base current will cause a proportional increase in collector current, and a decrease in base current will cause a proportional decrease in collector current.

The ratio of collector current to base current is referred to as the *forward current transfer ratio* and is given the symbol beta (β), so that:

$$\frac{I_C}{I_B} = \beta \tag{10.7}$$

The emitter current and collector current are approximately equal and for the purposes of this text will be assumed to be equal, so that:

$$I_E = I_C \tag{10.8}$$

Since the base-emitter junction is a forward biased p-n junction, the base-emitter voltage for a silicon junction transistor will be seven-tenths of a volt, so that:

$$V_{BE} = 0.7 \text{ V} \tag{10.9}$$

Since the base current controls the value of the collector current, the junction transistor is referred to as a *current-controlled device*. Equation (10.7) can be rearranged as $I_C = \beta I_B$, indicating that any change of I_B (input) will produce a change in I_C that is greater by a factor of beta (β). It may therefore be stated that the change of input current has been *amplified* by the change in output current.

One application of the junction transistor is to amplify an AC (sinusoidal) input signal (i_B). However, the output current can flow only from the collector to the emitter and cannot reverse direction. As a result, the positive values of a sinusoidal input current would be amplified, but the output current (I_C) would remain at zero for all negative (reverse direction) values of input current. Only the positive half of the input signal would be amplified. This problem is solved by creating circuits that set the collector current and collector-emitter voltage to a DC level before the AC input signal is applied. The output current can then increase and decrease around the DC level without reversing direction. Such circuits are referred to as **DC bias networks.** The DC value of collector current collector-emitter voltage is referred to as the **DC operating point** for the transistor.

10.7   ANALYSIS OF A FIXED BIAS NETWORK

The DC bias network in Figure 10–40 is referred to as a fixed bias network.

FIGURE **10–40**

A Fixed Bias
Transistor Circuit

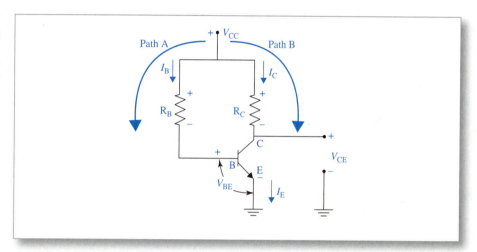

The junction transistor is represented by its schematic symbol with the base, collector, and emitter terminals labeled as indicated. Note that the positive voltage supply (V_{CC}) provides forward bias for the base-emitter junction (positive voltage on p-material) and reverse bias for the collector-base junction (positive voltage on n-material).

The schematic in Figure 10–40 may seem a bit overwhelming at first; however, a systematic application of Ohm's law and Kirchhoff's laws can be used to determine the DC current and voltage.

Two paths exist from V_{CC} to ground in Figure 10–40, so that KVL can be expressed for path A as:

$$V_{CC} - V_{RB} - V_{BE} = 0$$

so that:

$$V_{R_B} = V_{CC} - V_{BE}$$

Ohm's law can be used to solve for the current through R_B as:

$$I_{R_B} = \frac{V_{R_B}}{R_B}$$

Note that the current through R_B flows into the transistor base terminal as I_B, so that the base current can be expressed as:

$$I_{R_B} = \frac{V_{CC} - V_{BE}}{R_B} \qquad (10.10)$$

The forward transfer current ratio can be used to express the collector current as:

$$I_C = \beta I_B \qquad (10.11)$$

KVL may be expressed for path B as:

$$V_{CC} - V_{R_C} - V_{CE} = 0 \qquad (10.12)$$

The voltage across the collector resistor V_{RC} can be expressed using Ohm's law as:

$$V_{R_C} = I_C R_C$$

Equation (10.12) can be written as:

$$V_{CC} - I_C R_C - V_{CE} = 0$$

Solving for the collector-emitter voltage V_{CE}, we obtain:

$$V_{CE} = V_{CC} - I_C R_C \qquad (10.13)$$

The equations to determine the DC operating point for the fixed bias network may be summarized as:

$$I_B = \frac{(V_{CC} - V_{BE})}{R_B}; \quad I_C = \beta I_B; \quad V_{CE} = V_{CC} - I_C R_C$$

Note that the analysis of the fixed bias network was the result of applications of Kirchhoff's laws and Ohm's law.

EXAMPLE 10–5   Analysis of a Fixed Bias Transistor Circuit

Given the fixed bias transistor amplifier circuit in Figure 10–41, determine the values of base current I_B and collector current I_C. Also determine the collector-to-emitter voltage (V_{CE}). The collector DC supply voltage V_{CC} is given as 12 V DC. The value of β is given as 75, and the base-to-emitter voltage (V_{BE}) is given as 0.7 V.

FIGURE **10–41**

Circuit for
Example 10–5

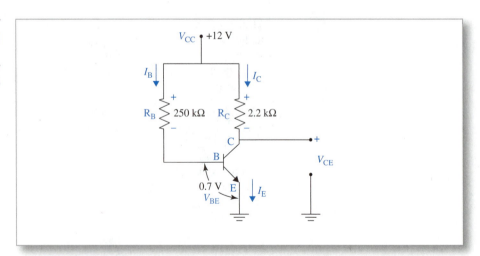

✓ SOLUTION

Equation (10.10) can be used to calculate the base current as:

$$I_B = \frac{V_{CC} - V_{BE}}{R_B} = \frac{12\text{ V} - 0.7\text{ V}}{250\text{ k}\Omega} = 45.2\ \mu\text{A}$$

The value of the collector current can be calculated using Equation (10.11) as:

$$I_C = \beta I_B = (75)(45.2\ \mu\text{A}) = 3.39\text{ mA}$$

Equation (10.13) can be used to determine the collector-emitter voltage as:

$$V_{CE} = V_{CC} - I_C R_C = 12\ V - (3.39\ mA)(2.2\ k\Omega) = 12\ V - 7.46\ V = 4.54\ V$$

The DC operating point for the transistor has been determined as:

$$I_C = 3.45\ mA; \quad V_{CE} = 4.41\ V$$

The MultiSIM simulation is shown in Figure 10–42.

FIGURE **10–42**

MultiSIM
Simulation
of a Fix Bias
Transistor Circuit

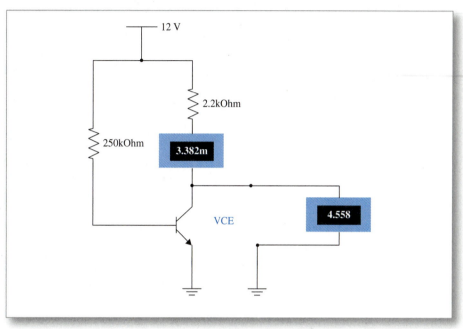

Note: The slight difference between calculated values and the values indicated by MultiSIM can be considered as insignificant.

PRACTICE EXERCISES 10–3

Consider a fixed bias circuit similar to the circuit in Figure 10–41. Given R_B = 300 kΩ, R_C = 1.5 kΩ, V_{CC} = 15 V, β = 90, and base-emitter voltage V_{BE} = 0.7 V, determine the value of collector current I_C and collector-to-emitter voltage V_{CE}.

✓ ANSWERS

I_C = 4.29 mA; V_{CE} = 8.56 V

(Complete solutions to the Practice Exercises can be found in Appendix A.)

10.8 ANALYSIS OF A VOLTAGE DIVIDER BIAS NETWORK

The fixed bias network is the minimum circuitry required to provide DC bias for a transistor. The voltage divider bias circuit shown in Figure 10–43 is somewhat more complex; however, the voltage divider bias network possesses some advantages over the fixed bias network.

FIGURE **10–43**

A Voltage Divider
Bias Transistor Circuit

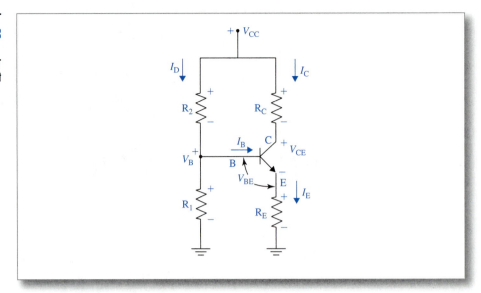

In practical applications, ambient temperature effects may cause the DC collector current to increase in the fixed bias network. The increase in collector current results in an undesirable shift of the DC operating point. Extreme effects of temperature may result in distortion in the output signal or even damage to the transistor. The voltage divider bias network provides operating point stability over a wide range of temperature variations.

A second advantage of the voltage divider bias circuit is that the operating point values do not depend upon the value of β for the transistor. Replacing a faulty transistor with a transistor with a different value of β will not affect the desired operating point values.

It is common in practical circuit analysis to simplify calculations when possible by introducing approximations into the analysis. As an example, the base current in Figure 10–43 is typically measured in microamperes. Given typical values for the resistors R_1 and R_2, the current I_D is measured in milliamperes. Since I_D is much larger than I_B, the current through R_2 is approximately equal to the current through R_1.

Assuming that R_1 and R_2 have the same current flowing through them, the two resistors can be considered to be in series. The voltage divider rule may be used to determine the approximate voltage at the base of the transistor as:

$$V_B = V_{CC} \frac{R_1}{R_1 + R_2} \tag{10.14}$$

KVL may be written starting at the base of the transistor as:

$$V_B - V_{BE} - V_{RE} = 0$$

V_{RE} may be expressed as:

$$V_{RE} = V_B - V_{BE}$$

Ohm's law can be used to determine the current through the emitter resistor (I_E) as:

$$I_E = \frac{V_{RE}}{R_E} = \frac{V_B - V_{BE}}{R_E} \tag{10.15}$$

Since the emitter current (I_E) and collector (I_C) are approximately equal, the collector current can be determined as:

$$I_C = I_E \tag{10.16}$$

KVL can be expressed beginning at V_{CC} to obtain:

$$V_{CC} - V_{RC} - V_{CE} - V_{RE} = 0 \tag{10.17}$$

Substituting $V_{RC} = I_C R_C$ and $V_{RE} = I_C R_E$, we obtain:

$$V_{CC} - I_C R_C - V_{CE} - I_C R_E = 0$$

so that:

$$V_{CE} = V_{CC} - I_C (R_C + R_E) \qquad (10.18)$$

The equations to determine the DC operating point for the voltage divider bias network can be summarized as:

$$V_B = V_{CC} \frac{R_1}{R_1 + R_2}; \quad V_{RE} = V_B - V_{BE}; \quad I_E = \frac{V_{RE}}{R_E}; \quad I_C = I_E, \ V_{CE} = V_{CC} - I_C(R_C + R_E)$$

Note that beta (β) does not appear in the equations to determine the operating point of the voltage divider bias circuit. The circuit is therefore referred to as beta-independent bias circuit.

EXAMPLE 10–6   Analysis of a Voltage Divider Bias Network

Determine the value of DC collector current and the collector-emitter voltage for the transistor in Figure 10–44. The base-emitter voltage V_{BE} is given as 0.7 V.

FIGURE **10–44**

Circuit for
Example 10–6

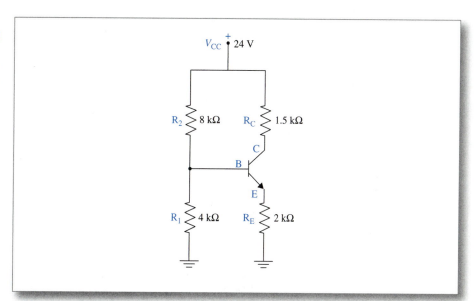

✓ **SOLUTION**

The base voltage (V_B) is calculated as:

$$V_B = V_{CC} \frac{R_1}{R_1 + R_2} = 25 \text{ V} \frac{4 \text{ k}\Omega}{(4 \text{ k}\Omega + 8 \text{ k}\Omega)} = 8 \text{ V}$$

The voltage across the emitter resistor is calculated as:

$$V_{RE} = V_B - V_{BE} = 8 \text{ V} - 0.7 \text{ V} = 7.3 \text{ V}$$

The emitter current is calculated as:

$$I_E = \frac{V_{RE}}{R_E} = \frac{7.3 \text{ V}}{2 \text{ k}\Omega} = 3.65 \text{ mA}$$

The emitter current and collector current can be considered to be equal, so that:

$$I_C = I_E = 3.65 \text{ mA}$$

The collector-to-emitter voltage (V_{CE}) is calculated as:

$$V_{CE} = V_{CC} - I_C(R_C + R_E) = 24 \text{ V} - 3.65 \text{ mA}(1.5 \text{ k}\Omega + 2 \text{ k}\Omega) = 11.3 \text{ V}$$

The DC operating point for the transistor has been determined as:

$$I_C = 3.65 \text{ mA}; \quad V_{CE} = 11.3 \text{ V}$$

The MultiSIM simulation is shown in Figure 10–45.

MuLTISIM

FIGURE **10–45**

MultiSIM
Simulation for
Example 10–6

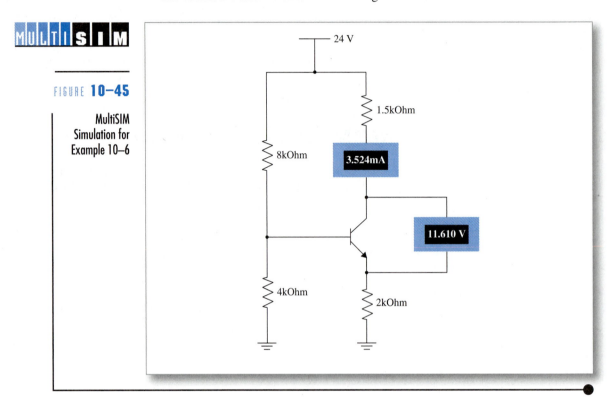

Note: The slight difference between calculated values and the values indicated by MultiSIM can be considered insignificant.

10.9   DESIGN OF A FIXED BIAS NETWORK

It may be necessary to design a bias circuit to produce a specified DC operating point. Manufacturers of transistors often recommend a DC operating point for a specific transistor, and circuits must be designed for the recommended values.

Assumptions are often made when designing electronic circuits. Many of the assumptions are commonly accepted engineering practices, while others may be based upon available components, cost, or other factors.

Several reasonable assumptions could be made that would lead to a suitable design of the fixed bias transistor circuit.

A common engineering practice used in the design of the fixed bias circuit is that the optimum collector-to-emitter voltage (V_{CE}) at the DC operating point is equal to one-half of the collector supply voltage (V_{CC}), so that:

$$V_{CE} = \frac{1}{2}V_{CC} \tag{10.19}$$

EXAMPLE 10-7   Design of a Fixed Bias Network

Using a collector supply voltage of 12 V and a silicon transistor with beta (β) equal to 95, design a fixed bias network so that the DC operating point current is 3.5 mA.

The known values (V_{CC} = 12 V, V_{BE} = 0.7 V, and I_C = 3.5 mA) are shown in the schematic in Figure 10–46. The collector-to-emitter voltage V_{CE} is assumed to be one half of the collector supply voltage such that: $V_{CE} = 1/2\ V_{CC} = 1/2\ (rV) = 6$ V.

FIGURE **10–46**

Prototype
Circuit with Design
Specifications for
Example 10–7

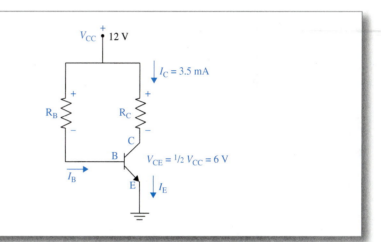

✓ **SOLUTION**

The objective of the circuit design is to determine the appropriate values for the base resistor R_B and collector resistor R_C to establish the desired operating point. The voltage across the collector resistor is determined using KVL as:

$$V_{RC} = V_{CC} - V_{CE} = 12\ V - 6\ V = 6\ V$$

The value of the collector resistor can be calculated using Ohm's law as:

$$R_C = \frac{V_{RC}}{I_{RC}} = \frac{6\ V}{3.5\ mA} = 1.71\ k\Omega$$

The voltage across the base resistor can be determined using KVL as:

$$V_{CC} - V_{RB} - V_{BE} = 0$$

so that: $$V_{RB} = V_{CC} - V_{BE} = 12\ V - 0.7\ V = 11.3\ V$$

The base current can be calculated as:

$$I_B = \frac{I_C}{\beta} = \frac{3.5\ mA}{95} = 36.84\ \mu A$$

The value of the base resistor can be calculated using Ohm's law as:

$$R_B = \frac{V_{RB}}{I_{RB}} = \frac{11.3\ V}{36.84\ \mu A} = 306.73\ k\Omega$$

The MultiSIM simulation of the design is shown in Figure 10–47.

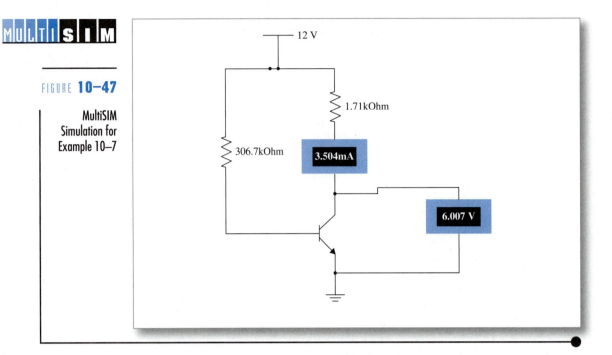

FIGURE **10–47**

MultiSIM
Simulation for
Example 10–7

Determine the DC operating point for the circuit in Figure 10–43, given $R_1 = 5$ kΩ, $R_2 = 15$ kΩ, $R_E = 500$ Ω, $R_C = 1.2$ kΩ, $V_{CC} = 12$ V, and $V_{BE} = 0.7$ V.

✓ ANSWERS

$I_C = 4.6$ mA; $V_{CE} = 4.18$ V

(Complete solutions to the Practice Exercises can be found in Appendix A.)

10.10 ANALYSIS OF A TRANSISTOR LOGIC CIRCUIT

Like the diode, the transistor can function as a binary device (on, off) and may be used in circuits designed to execute a variety of logic operations.

Consider the transistor-resistor logic (TRL) circuit in Figure 10–48.

FIGURE **10–48**

A Transistor-Resistor
Logic Circuit (TRL)

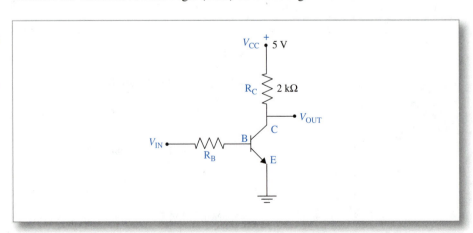

KVL can be expressed beginning at the collector supply V_{CC} as:

$$V_{CC} - V_{RC} - V_{CE} = 0$$

so that:

$$V_{CE} + V_{RC} = V_{CC} \qquad (10.20)$$

As the collector current increases, the voltage across the collector resistor (V_{RC}) will also increase, so that the collector-emitter voltage across the transistor (V_{CE}) must decrease to satisfy Equation (10.20). In the limit, the collector current becomes sufficiently large that the entire supply voltage is dropped across R_C, and the voltage across the transistor (V_{CE}) will be equal to zero. This condition is referred to as **saturation,** which is defined by the equations:

$$V_{CE(sat)} = 0 \qquad (10.21)$$

$$I_{C(sat)} = \frac{V_{CC}}{R_C} \qquad (10.22)$$

The collector current I_C is directly proportional to the base current, as indicated by the expression:

$$I_C = \beta I_B$$

If the base current I_B is equal to zero, then the collector current I_C must also be zero. If the collector current is zero, then the voltage across the collector resistor will also be zero. Equation (10.20) indicates that if V_{RC} is equal to zero, then the entire supply voltage is dropped across the transistor (V_{CE}). This condition is referred to as **cutoff,** which is defined by the equations:

$$I_C = 0 \qquad (10.23)$$

$$V_{CE} = V_{CC} \qquad (10.24)$$

A transistor logic circuit is designed to operate only in the saturation or cutoff condition. The output voltage is the collector-to-emitter voltage V_{CE}, which has only two permissible values, either zero (saturation) or V_{CC} (cutoff).

EXAMPLE 10-8   Analysis of a Transistor Logic Circuit

Given the transistor logic circuit in Figure 10–49:

a. Determine the value of V_{CE} when the input voltage V_{in} is equal to zero (ground).

b. Given $\beta = 175$, determine the minimum value of base current I_B for the saturation condition.

c. Given that $V_{BE} = 0.5$ V, determine the maximum permissible value of R_B so that the circuit is in saturation when the input signal $V_{in} = 5$ V.

FIGURE **10–49**

Circuit for
Example 10–8

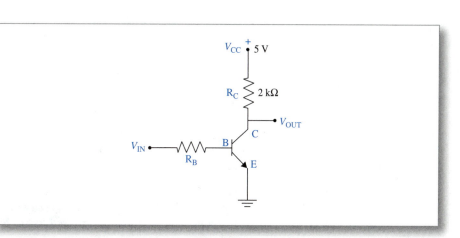

✓ **SOLUTION**

a. When the input signal voltage is zero, the base current will also be zero and the circuit is in the cutoff condition, so that Equations (10.22) and (10.23) indicate that:

$$I_C = 0$$

$$V_{CE} = V_{CC} = 5 \text{ V}$$

b. The saturation current may be calculated using Equation (10.22) as:

$$I_{C(sat)} = \frac{V_{CC}}{R_C} = \frac{5 \text{ V}}{2 \text{ k}\Omega} = 2.5 \text{ mA}$$

The minimum value of base current to cause saturation may be calculated as:

$$I_B = \frac{I_{C(sat)}}{\beta} = \frac{2.5 \text{ mA}}{175} = 14.3 \text{ }\mu\text{A}$$

c. KVL can be expressed moving clockwise from V_{in} to obtain:

$$V_{in} - V_{RB} - V_{BE} = 0$$

so that:
$$V_{RB} = V_{in} - V_{BE} = 5 \text{ V} - 0.5 \text{ V} = 4.5 \text{ V}$$

R_B may be calculated as:

$$R_B = \frac{V_{RB}}{I_B} = \frac{4.5 \text{ V}}{14.3 \text{ }\mu\text{A}} = 314.7 \text{ k}\Omega$$

Note that a high (5 V) input signal corresponds to a low (0 V) output signal, and that a low input signal corresponds to a high output signal, so that the circuit is referred to as a **signal inverter.**

PRACTICE EXERCISES 10–5

Given a transistor logic circuit similar to the circuit shown in Figure 10–48 with $V_{CC} = 5$ V, $\beta = 250$, $R_C = 1$ kΩ, and $V_{BE} = 0.7$ V, determine the value of saturation current $I_{C(sat)}$. Determine the minimum base current that will cause saturation. Determine the maximum permissible value of R_B when the input signal voltage is 5 V. Determine the value of V_{CE} when the input signal voltage is zero (ground).

✓ **ANSWERS**

$I_{C(sat)} = 5$ mA; $I_B = 20$ μA; $R_B = 215$ kΩ;

When $V_{in} = 0$: $V_{CE} = 5$ V

(Complete solutions to the Practice Exercises can be found in Appendix A.)

10.11   ANALYSIS OF CIRCUITS THAT CONTAIN "BLACK BOX" DEVICES

Electronic systems are often designed with small-, medium-, and large-scale integrated circuits. Electronic systems are typically viewed as a collection of electronic functional units commonly referred to as "black boxes" rather than a collection of discrete components. Consideration is not given to the internal composition of the boxes, but rather to the external connections and interface networks.

A "black box" is typically defined by parameters specified by the manufacturer. Values of input resistance (impedance) and output resistance (impedance) are normally specified. Depending upon the function of a particular "black box," other information such as gain, frequency response curves, external pin connections, as well as maximum ratings for current, voltage, and power may be given.

EXAMPLE 10-9 Analysis of a "Black Box" Circuit

Given the "black box" circuit in Figure 10–50, determine an expression for the voltage gain (A_v) of the network. Determine the numerical value for the circuit voltage gain given $R_f = 25$ kΩ and $R_1 = 2.5$ kΩ.

FIGURE 10–50

Circuit for Example 10–9

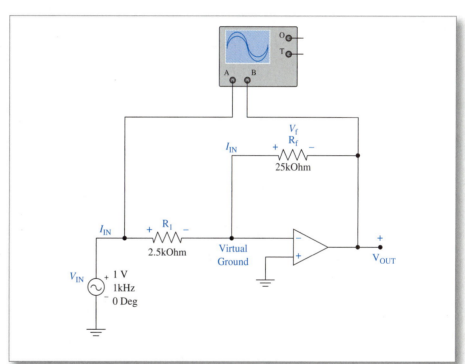

In this example, the "black box" has two input terminals labeled $(+)$ and $(-)$. It is also specified that the "black box" has infinite input impedance and the voltage between the input terminals is approximately zero The input signal V_{in} is applied to the $(-)$ input terminal through resistor R_1. Resistor R_2 connects the output terminal back to the input terminal and is referred to as a *feedback resistor*. The output signal, V_{out}, of the circuit is taken from the indicated output terminal to ground.

✓ SOLUTION

The given specifications and description indicate that the voltage between the $(+)$ and $(-)$ input terminals of the "black box" is approximately zero. We may therefore conclude that because the positive $(+)$ input terminal is at ground, then the negative $(-)$ input terminal must also be at ground or 0 V. The voltage at the negative input terminal is often referred to as *virtual ground*. Since one end of resistor R_1 is connected to the positive terminal of V_{in} and the other end to the negative ($-$) input terminal (virtual ground), the voltage across R_1 can be determined using KVL as:

$$V_{in} - V_{R1} = 0 \text{ V}$$

so that:

$$V_{R1} = V_{in}$$

Ohm's law can be used to calculate the current through R_1 as:

$$I_{in} = \frac{V_{in}}{R_1} \qquad (10.25)$$

Since the specifications indicate that the input impedance to the "black box" is infinite, current cannot flow into the "black box." Applying KCL at the negative ($-$) input terminal, the current flowing through the feedback resistor R_f must also be I_{in}, as indicated in Figure 10–50, so that:

$$I_{Rf} = I_{in} \qquad (10.26)$$

Since one end of the feedback resistor (R_f) is connected to V_{out} and the other end to the negative input terminal (virtual ground), the voltage across R_f can be determine as:

$$V_f - V_{out} = 0$$

so that:

$$V_f = V_{out} \qquad (10.27)$$

Since the current through R_f is I_{in}, as indicated by Equation (10.26), Ohm's law can be used to express the input current as:

$$I_{in} = \frac{-V_{out}}{R_f} \qquad (10.28)$$

The negative sign in Equation (10.28) is necessary because the direction of the current I_{in} as indicated is possible only if V_{out} is negative with respect to ground. Therefore, the output voltage V_{out} must be negative when the input voltage V_{in} is positive.

Two independent expressions for I_{in}, Equations (10.28) and (10.25), have been determined that can be set equal to each other to obtain:

$$I_{in} = \frac{-V_{out}}{R_f} = \frac{V_{in}}{R_1}$$

so that:

$$\frac{V_{out}}{V_{in}} = -\frac{R_f}{R_1} \qquad (10.29)$$

The expression $\dfrac{V_{out}}{V_{in}}$; is the voltage gain of the network A_v and may be expressed as:

$$A_v = -\frac{R_f}{R_1} \qquad (10.30)$$

Given $R_f = 100 \text{ k}\Omega$ and $R_1 = 2.5 \text{ k}\Omega$, the numerical value of gain may be calculated as:

$$A_v = -\frac{R_f}{R_1} = -\frac{25 \text{ k}\Omega}{2.5 \text{ k}\Omega} = -10$$

The "black box" in Example 10–9 is a popular integrated circuit called an **operational amplifier.** The circuit in the example is called an **inverting amplifier,** a circuit that is utilized in the design of many electronic systems.

The MultiSIM simulation is shown in Figure 10–51.

FIGURE **10-51**

MultiSIM
Simulation
for Example 10–9

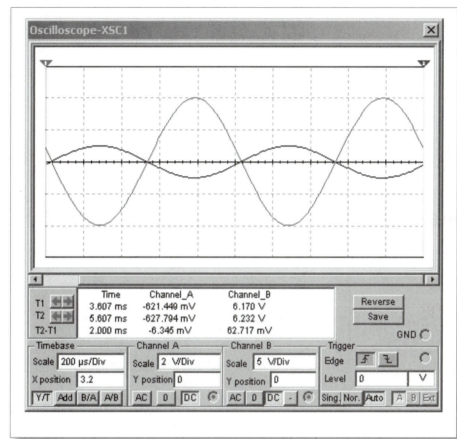

Note that the magnitude of the input signal is 1 V and the amplitude of the output signal is 10 V. Also note that the input and output signals are 180° out of phase, indicating that the input signal is inverted at the output.

The important observation in this example is that the analysis of a circuit containing a "black box" is accomplished through a systematic application of Kirchhoff's laws and Ohm's law. Example 10–9 also demonstrates that successful circuit analysis and design is often a process of reasoning as much as it is mathematical computation.

EXAMPLE 10-10   Analysis of a "Black Box" Circuit

Given the "black box" circuit in Figure 10–52, determine an expression for the voltage gain (A_v) of the network. Determine the numerical value for the voltage gain given $R_f = 36$ kΩ and $R_1 = 4$ kΩ.

The specifications and description of the "black box" device in this example are the same as the device described in Example 10–9.

✓ SOLUTION

The voltage divider rule may be used to obtain an expression for the feedback voltage V_f as:

$$V_f = V_{out} \frac{R_1}{R_f + R_1}$$

(10.31)

FIGURE **10-52**

Circuit for
Example 10–10

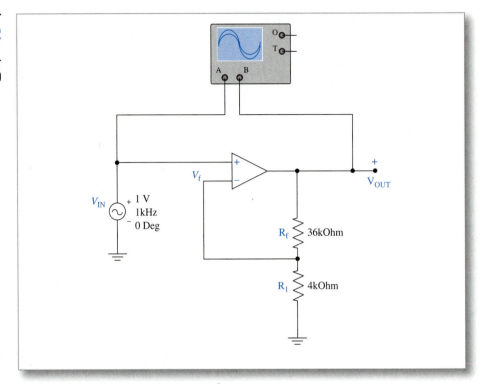

The specifications for the "black box" indicate that the voltage between the input terminals can be considered to be zero, so that:

$$V_f = V_{in} \qquad (10.32)$$

The two expressions for V_f, Equations (10.31) and (10.32), can be set equal to each other to obtain:

$$V_{in} = V_{out} \frac{R_1}{R_f + R_1}$$

The voltage gain may be expressed as:

$$A_V = \frac{V_{out}}{V_{in}} = \frac{R_f + R_1}{R_1}$$

$$A_V = \frac{R_f}{R_1} + 1 \qquad (10.33)$$

The voltage gain, after substituting the given resistor values, is calculated as:

$$A_V = \frac{R_f}{R_1} + 1 = \frac{36 \text{ k}\Omega}{4 \text{ k}\Omega} + 1 = 9 + 1 = 10$$

The circuit in this example is referred to as a **noninverting amplifier.**

The MultiSIM simulation is shown in Figure 10–53.

Note that the magnitude of the input signal is 1 V and the magnitude of the output voltage is 10 V. Also note that the input and output signals are in phase, indicating that the input signal has been amplified but not inverted.

FIGURE **10–53**

MultiSIM
Simulation for
Example 10–10

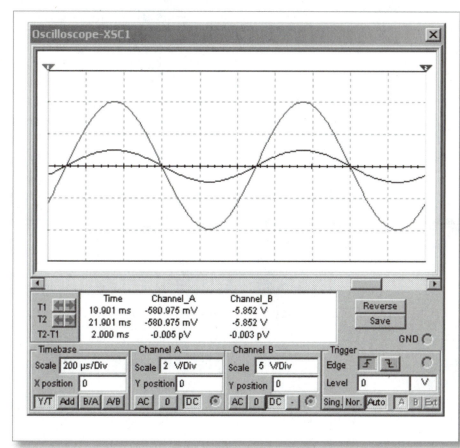

S U M M A R Y

Impurities can be added to semiconductor materials such as silicon and germanium to create n-material and p-material. A p-n junction is formed by joining p-material and n-material together to create a **p-n junction diode** and a junction transistor. Devices that are created using p-n junctions are referred to as solid-state devices.

A diode is a solid-state p-n junction that permits current to flow only in one direction through the device. When the anode (p-material) of the diode is more positive than the cathode (n-material), the diode is said to be forward biased and will conduct current. When the anode (p-material) of the diode is more negative than the cathode (n-material), the diode is said to be reverse biased and will not conduct current.

Diodes are used extensively in the design of rectifier circuits. Rectifier circuits are designed to convert an alternating sinusoidal waveform into a wave whose polarity does not change. Rectifier circuits can be described mathematically as circuits that convert a waveform whose average value is zero (sine wave) into a waveform whose average value is not zero. Rectifier circuits are an important part of DC power supply circuits that convert commercial utility voltage (120 V_{RMS}, 60 Hz sine wave) into DC voltage (constant voltage).

Diode limiters are designed to ensure that the amplitude of the output signal voltage does not exceed a specified value. Limiters are used to protect circuits from excessive voltage that may damage or destroy the component. Limiters can also be used to remove noise spike voltage from electronic signals. Diodes may also be used as binary switching devices (on, off) to create diode-resistor logic (DRL) circuits.

Transistor amplifier circuits are designed to increase the amplitude of the input signal without altering the shape of the input signal. In order for the amplifier to reproduce both positive and negative values, the output voltage V_{CE} is set around a DC operating point. Bias circuits are used to create a desired DC operating point for a transistor amplifier. The **DC operating point** is specified by the value of collector-to-emitter voltage V_{CE}, and collector current I_C.

A transistor may also function as a binary switching device (on, off) and may be used in circuits designed to execute a variety of logic operations. The transistor is on during the **saturation** condition and off during the **cutoff** condition. Transistor-transistor logic (TTL) circuits are often manufactured as integrated circuits that may contain thousands of transistors.

Electronic systems often consist of integrated circuits that are considered as "black boxes." Typically, the input and output specifications are given so that circuits containing "black boxes" can be analyzed using standard circuit analysis techniques. The **operational amplifier** is a "black box" device that finds many useful applications in a wide variety of circuit designs.

ESSENTIAL LEARNING EXERCISES

(ANSWERS TO ODD-NUMBERED ESSENTIAL LEARNING EXERCISES CAN BE FOUND IN APPENDIX A.)

1. Describe the polarity relationship between the anode voltage and cathode voltage for a forward biased diode.

2. Describe the polarity relationship between the anode voltage and cathode voltage for a reverse biased diode

3. Determine whether the diode in the circuits in Figure 10–54 is forward or reverse biased.

FIGURE **10–54**

Circuit for Essential Learning Exercise 3

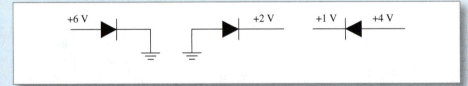

4. Determine whether the diode in the circuits in Figure 10–55 is forward or reverse biased.

FIGURE **10–55**

Circuit for Essential Learning Exercise 4

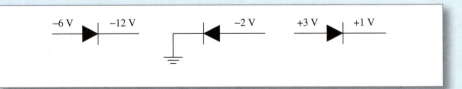

5. Sketch the output waveform for the circuit in Figure 10–56. Determine the peak value of V_{out}. Assume $V_f = 0.7$ V. Verify your solution using computer simulation.

FIGURE **10–56**

Circuit for Essential
Learning Exercise 5

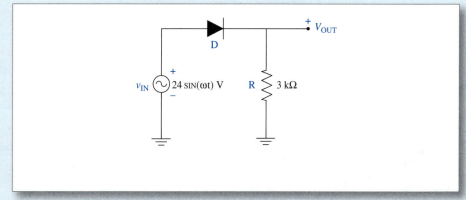

6. Sketch the output waveform for the circuit in Figure 10–57. Determine the peak value of V_{out}. Assume $V_f = 0.7$ V. Verify your solution using computer simulation.

FIGURE **10–57**

Circuit for Essential
Learning Exercise 6

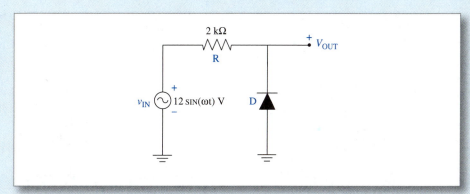

7. Sketch the output waveform for the circuit in Figure 10–58. Assume $V_f = 0.7$ V. Verify your solution using computer simulation.

FIGURE **10–58**

Circuit for Essential
Learning Exercise 7

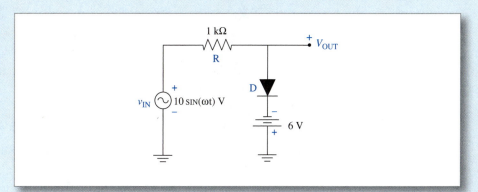

8. Sketch the output waveform for the circuit in Figure 10–59. Assume $V_f = 0.7$ V. Verify your solution using computer simulation.

FIGURE **10–59**

Circuit for Essential
Learning Exercise 8

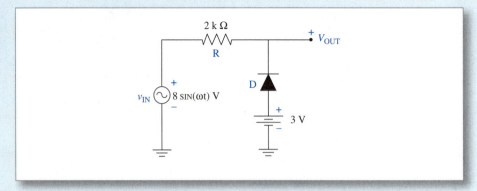

9. Determine the output of the logic circuit in Figure 10–60, given that V_1 is high (5 V) and V_2 is low (ground). Assume $V_f = 0.7$ V. Verify your solution using computer simulation.

FIGURE **10–60**

Circuit for Essential
Learning Exercise 9

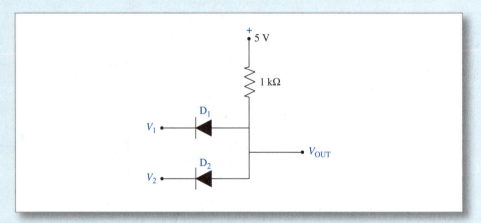

10. Determine the output of the logic circuit in Figure 10–60, given that V_1 is high (5 V) and V_2 is high (5 V). Assume $V_f = 0.7$ V. Verify your solution using computer simulation.

11. Determine the DC operating point (V_{CE}, I_C) for the fixed bias transistor circuit in Figure 10–61, given $R_C = 3.3$ kΩ, $R_B = 375$ kΩ, and beta (β) = 75. Assume $V_f = 0.7$ V.

FIGURE **10–61**

Circuit for Essential
Learning Exercise 11

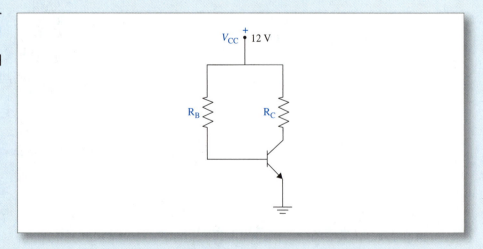

12. Determine the DC operating point (V_{CE}, I_C) for the fixed bias transistor circuit in Figure 10–61, given $R_C = 2.2$ kΩ, $R_B = 450$ kΩ, and beta (β) = 90. Assume $V_f = 0.7$ V.

13. Determine the DC operating point (V_{CE}, I_C) for the voltage divider biased transistor circuit in Figure 10–62, given $R_C = 1.5$ kΩ, $R_E = 1$ kΩ, $R_1 = 3$ kΩ, $R_2 = 6$ kΩ. Assume $V_f = 0.7$ V.

FIGURE **10–62**

Circuit for Essential
Learning Exercise 13

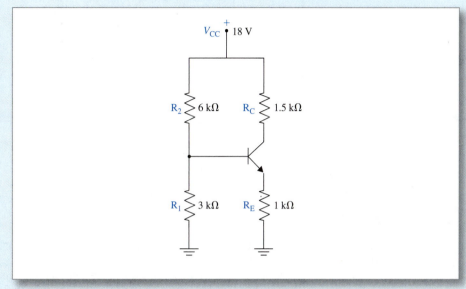

14. Determine the DC operating point (V_{CE}, I_C) for the voltage divider biased transistor circuit in Figure 10–62, given $R_C = 2.2$ kΩ, $R_E = 1.5$ kΩ, $R_1 = 8$ kΩ, $R_2 = 28$ kΩ. Assume $V_f = 0.7$ V.

15. Design a fixed bias transistor circuit similar to Figure 10–61 so that the DC operating point is $V_{CE} = 7$ V and $I_C = 2.5$ mA, given $V_{CC} = 14$ V. Assume $V_f = 0.7$ V and β = 100.

16. Design a fixed bias transistor circuit similar to Figure 10–61 so that the DC operating point is $V_{CE} = 10$ V and $I_C = 3$ mA, given $V_{CC} = 20$ V. Assume $V_f = 0.7$ V and β = 85.

17. Consider the transistor-resistor logic (TRL) circuit in Figure 10–63.

FIGURE **10–63**

Circuit for Essential
Learning Exercise 17

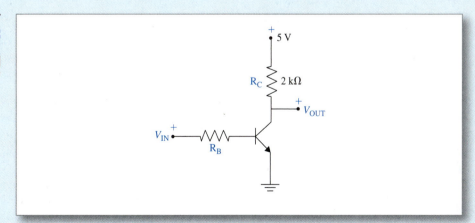

a. Determine the value of V_{out} when the input voltage V_{in} is equal to zero (ground).

b. Given β = 200, determine the minimum value of base current I_B for the saturation condition.

c. Determine the output voltage for the saturation condition.

18. Explain why the transistor logic circuit shown in Figure 10–63 is referred to as an inverter circuit.

19. Given the op-amp circuit in Figure 10–64, determine the numerical value of the voltage gain for the circuit. The input resistance (impedance) to the device can be considered to be infinite, and the voltage between the input terminals may be considered to be zero. Determine the output voltage for an input voltage of 1 V DC. Verify your solution using computer simulation.

FIGURE **10–64**

Circuit for Essential
Learning Exercise 19

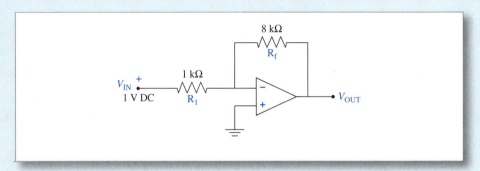

20. Given the op-amp circuit in Figure 10–64, determine the numerical value of the voltage gain for the circuit, given $R_1 = 3 \text{ k}\Omega$ and $R_f = 24 \text{ k}\Omega$. The input resistance (impedance) to the device can be considered to be infinite, and the voltage between the input terminals may be considered to be zero. Determine the output voltage for an input voltage of 0.5 V DC. Verify your solution using computer simulation.

21. Given the op-amp circuit in Figure 10–65, determine the numerical value of the voltage gain for the circuit. The input resistance (impedance) to the device can be considered to be infinite, and the voltage between the input terminals may be considered to be zero. Determine the output voltage for an input voltage of 0.8 V DC. Verify your solution using computer simulation.

FIGURE **10–65**

Circuit for Essential
Learning Exercise 21

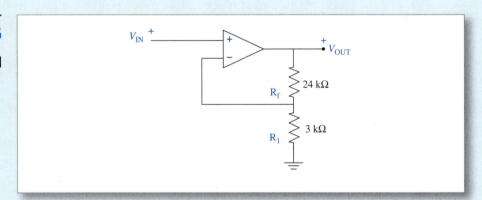

22. Given the op-amp circuit in Figure 10–65, determine the numerical value of the voltage gain for the circuit, given $R_1 = 6 \text{ k}\Omega$ and $R_f = 30 \text{ k}\Omega$. The input resistance (impedance) to the device can be considered to be infinite, and the voltage between the input terminals may be considered to be zero. Determine the output voltage for an input voltage of 0.4 V DC. Verify your solution using computer simulation.

CHALLENGING LEARNING EXERCISES

(ANSWERS TO SELECTED CHALLENGING LEARNING EXERCISES CAN BE FOUND IN APPENDIX A.)

1. Determine the output waveform for the diode limiter circuit in Figure 10–66. The input signal is $20 \sin(\omega t)$. Verify your solution using computer simulation.

FIGURE **10–66**

Circuit for Challenging
Learning Exercise 1

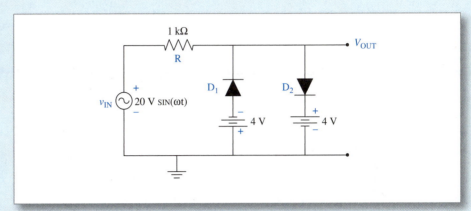

2. Determine the value of the collector current I_C and the collector-to-emitter voltage V_{CE} for the emitter resistor bias circuit in Figure 10–67, given $\beta = 125$. Assume $V_f = 0.7$ V. Verify your solution using computer simulation.

FIGURE **10–67**

Circuit for Challenging
Learning Exercise 2

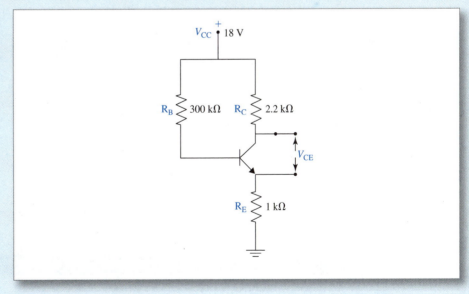

3. Determine the gain of the op-amp circuit in Figure 10–68. Also determine the magnitude, frequency, and phase angle of the output signal. Consider the input resistance (impedance) to be infinite and the voltage between the input terminals to be zero. Verify your solution using computer simulation.

FIGURE **10—68**

Circuit for Challenging
Learning Exercise 3

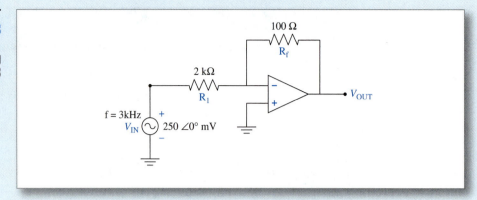

4. Determine the magnitude, phase, and frequency of the output signal V_{out} in Figure 10–69. The voltage gain of each "black box" is -20 as indicated.

FIGURE **10—69**

Circuit for Challenging
Learning Exercise 4

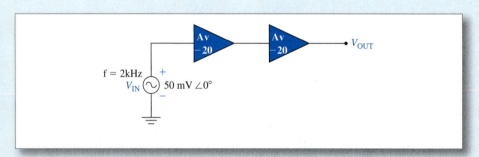

5. Determine the magnitude of the output signal V_{out} for the op-amp circuit shown in Figure 10–70, given $V_1 = 5$ V DC and $V_2 = 4$ V DC. Verify your solution using computer simulation.

FIGURE **10—70**

Circuit for Challenging
Learning Exercise 5

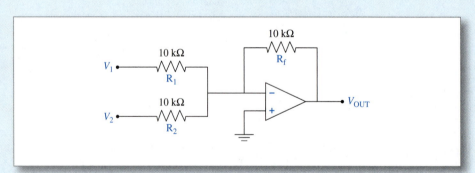

TEAM ACTIVITY

Design a fixed bias transistor circuit to have a DC operating point of $V_{CE} = 6$ V and $I_C = 1.5$ mA. Assume that beta (β) for the transistor is 125 and that the voltage supply $V_{CC} = 15$ V. Verify your design using computer simulation.

You may find a transistor in the circuit simulator software library that has a beta equal to 125. As an alternative, you can use the default n-p-n transistor for your design; however, you must edit the transistor model to set the forward current transfer ratio to 125 as indicated in the design specifications.

AN INTRODUCTION TO NONSINUSOIDAL SIGNALS

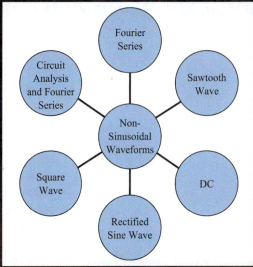

"THERE IS NOTHING LIKE A DREAM TO CREATE THE FUTURE."

—Victor Hugo

CHAPTER OUTLINE

LEARNING OBJECTIVES

Upon successful completion of this chapter you will be able to:

- Describe the purpose of the Fourier series.
- Determine the fundamental frequency of a periodic waveform.
- Determine the harmonic frequencies of a periodic waveform.
- Calculate the Fourier coefficients of a square wave.
- Calculate the Fourier coefficients of a sawtooth wave.
- Calculate the Fourier coefficients of a rectified sine wave.
- Determine the output waveform of an R-C circuit with a complex input waveform.
- Sketch a frequency spectrum diagram.
- Use MATLAB to create a plot of a Fourier series.

CHAPTER ELEVEN

INTRODUCTION

The previous chapters in this text have been dedicated to the analysis and design of discrete analog circuits whose voltage and current waveforms are either DC or sinusoidal. Other waveforms are common in electronic circuits although the sine wave remains the most important time-varying signal.

This chapter introduces circuit analysis techniques that may be used to analyze circuits whose input signals are nonsinusoidal. The material presented in this chapter is not intended to provide a rigorous mathematical analysis of Fourier series, nor is it intended to investigate overly complex nonsinusoidal waveforms. The chapter takes advantage of computer-assisted analysis to illustrate how circuits respond to nonsinusoidal signal input waveforms where appropriate.

11.1 AN INTRODUCTION TO FOURIER SERIES

Electronic systems often involve signals that are more complex than the pure sinusoidal waveform. Although electronic signals are often complex, they are typically repetitive with cycles occurring at regular intervals and so are referred to as periodic waveforms.

The French mathematician Jean Baptiste Fourier (1768–1830) discovered that any complex periodic waveform may be represented as a sum of sinusoidal waveforms referred to as a **Fourier series**. The Fourier series of some complex waveforms contains a DC component in addition to the sinusoidal waveforms.

The general form of the Fourier series representation of a complex periodic voltage waveform can be written as:

$$v(t) = A_0 + \sum_{n=1}^{\infty}(A_n \cos n\omega_0 t + B_n \sin n\omega_0 t) \tag{11.1}$$

where: A_0 = average or DC value of the waveform

A_n and B_n are called the Fourier coefficients

n = integers 1, 2, 3, ...

ω_0 = fundamental radian frequency of the waveform

Although the Fourier representation in Equation (11.1) contains an infinite number of sinusoidal terms ($n = 1 \ldots \infty$), most complex waveforms can be closely approximated by relatively few Fourier terms. The **fundamental frequency** of a complex periodic waveform is the reciprocal of the period. Integer multiples of the fundamental frequency are referred to as **harmonic frequencies.**

11.2 THE SQUARE WAVE

The square wave is an important nonsinusoidal waveform. The square wave may be defined as a special case of the rectangular pulse waveform for which the pulse width (t_w) is equal to one-half of the period (T). Consider the square wave in Figure 11–1.

The Fourier series representation of the square wave in Figure 11–1 is given without proof as:

$$v(t) = \frac{4A}{\pi} \sum_{n=1}^{\infty} \frac{1}{2n-1} \sin(2n-1)\omega_0 t \tag{11.2}$$

FIGURE **11-1**

A Square Wave

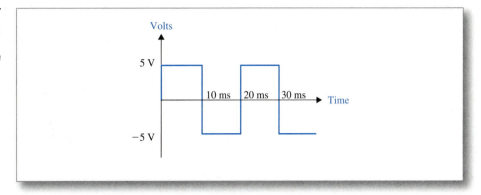

The expression $(2n - 1)$ represents all *odd* integers.

EXAMPLE 11-1   The Fourier Series Representation of a Square Wave

Determine the first three terms of the Fourier representation of the square wave in Figure 11–1.

✔ **SOLUTION**

Note that the square wave in Figure 11–1 has a period of 20 ms and an amplitude of 5 V. The fundamental frequency of the waveform can be calculated as the reciprocal of the period, such that:

$$f_0 = \frac{1}{T} = \frac{1}{20 \text{ ms}} = 50 \text{ Hz}$$

$$\omega_0 = 2\pi f_0 = 314.16 \text{ rad/s}$$

Equation (11.2) can be used to determine the first sinusoidal term $(n = 1)$ of the Fourier series as:

$$\frac{4A}{\pi} \sum_{n=1}^{\infty} \frac{1}{2n-1} \sin(2n-1)\omega_0 t = \frac{(4)(5)}{\pi} \sin(314.16t) = 6.366 \sin(314.16t)$$

Equation (11.2) can be used to determine the second term $(n = 2)$ of the Fourier series as:

$$\frac{4A}{\pi} \sum_{n=1}^{\infty} \frac{1}{2n-1} \sin(2n-1)\omega_0 t = \frac{(4)(5)}{3\pi} \sin(3\omega_0 t) = 2.122 \sin(942.48t)$$

Equation (11.2) can be used to determine the third term $(n = 3)$ of the Fourier series as:

$$\frac{4A}{\pi} \sum_{n=1}^{\infty} \frac{1}{2n-1} \sin(2n-1)\omega_0 t = \frac{(4)(5)}{5\pi} \sin(5\omega_0 t) = 1.273 \sin(1570.80t)$$

The square wave in Figure 11–2 can be approximated as the sum of the first three terms of the Fourier series as:

$$v(t) = 6.366 \sin(314.16t) + 2.122 \sin(942.48t) + 1.273 \sin(1570.80t)$$

The MATLAB plot in Figure 11–2 shows the first two sinusoidal terms of the Fourier series of a 5 V square wave whose fundamental frequency is $f_0 = 50$ Hz, so that $\omega_0 = 2\pi f_0 = 314.16$ rad/s. The sum of the first two sinusoidal terms is also shown.

FIGURE **11-2**

Sum of the First
Two Fourier Series
Terms of a
Square Wave

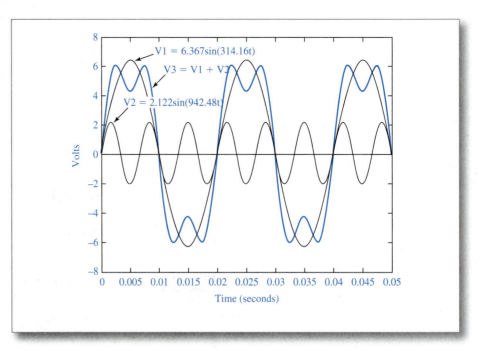

Note that the sum of the first two terms of the Fourier series represents a rough approximation of the square wave shown in Figure 11–1.

The MATLAB plot in Figure 11–3 shows the sum of the first three terms of the Fourier series of the 5 V square wave.

FIGURE **11-3**

Sum of the First
Three Fourier Series
Terms of a
Square Wave

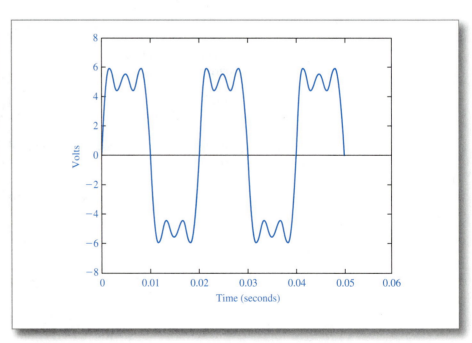

Note that the edges of the square wave in Figure 11–3 are steeper and the top of the wave is more flat. As the number of sinusoidal terms of the Fourier series is increased, the waveform becomes a closer approximation of the ideal 5 V square wave shown in Figure 11–1.

The sum of the first five Fourier terms, as shown in the MATLAB plot in Figure 11–4, is a very good approximation of a 5 V square wave with a period of 20 ms as shown in Figure 11–1.

FIGURE **11–4**

Sum of the First Five Fourier Series Terms of a Square Wave

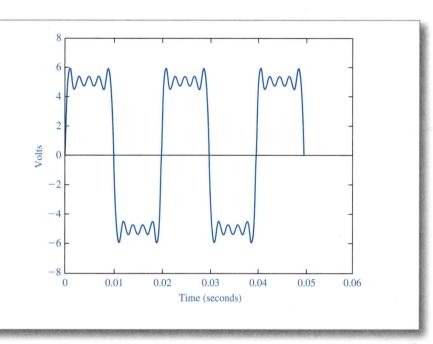

EXAMPLE 11–2   The Fourier Series of a Square Wave

Determine the first four sinusoidal Fourier series terms of a 10 V square wave with a period of 8 ms.

✓SOLUTION

The fundamental frequency of the square wave is calculated as the reciprocal of the period as:

$$f_0 = \frac{1}{T} = \frac{1}{8 \text{ ms}} = 125 \text{ Hz}$$

$$\omega_0 = 2\pi f_0 = 785.4 \text{ rad/s}$$

Equation (11.2) can be used to determine the first sinusoidal term ($n = 1$) of the Fourier series as:

$$\frac{4A}{\pi} \sum_{n=1}^{\infty} \frac{1}{2n-1} \sin(2n-1)\omega_0 t = \frac{(4)(10)}{\pi} \sin(785.4t) = 12.73 \sin(785.4t)$$

The sinusoidal terms for $n = 2$, 3, and 4 are:

$n = 2$: $4.244 \sin(2356t)$

$n = 3$: $2.55 \sin(3927t)$

$n = 4$: $1.82 \sin(5498t)$

The sum of the first four terms of the Fourier series representation of the square wave for Example 11–2 is shown in the MATLAB plot in Figure 11–5.

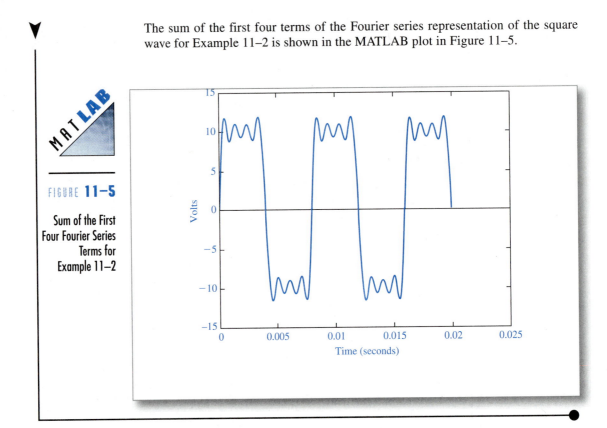

FIGURE **11–5**

Sum of the First Four Fourier Series Terms for Example 11–2

11.3 THE FREQUENCY SPECTRUM DIAGRAM

An alternate method of displaying the Fourier series representation of a complex periodic waveform is the frequency spectrum diagram. This diagram plots the amplitude of each of the sinusoidal Fourier terms indicating its respective frequency. The frequency spectrum diagram of the first five terms of the Fourier series representation of the 5 V square wave in Figure 11–1 is shown in Figure 11–6.

FIGURE **11–6**

Frequency Spectrum of a 5 V Square Wave with a Period of 10 ms.

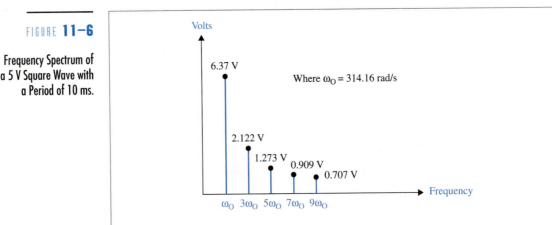

Frequency spectrum diagrams of complex signals are typically obtained through the use of an electronic instrument called a *frequency spectrum analyzer*. The frequency spectrum of a complex electronic signal is one of the most important considerations when designing and analyzing modern electronic communications circuits and systems. A typical spectrum analyzer display is shown in Figure 11–7.

FIGURE **11–7**

Display from a Commercial Spectrum Analyzer

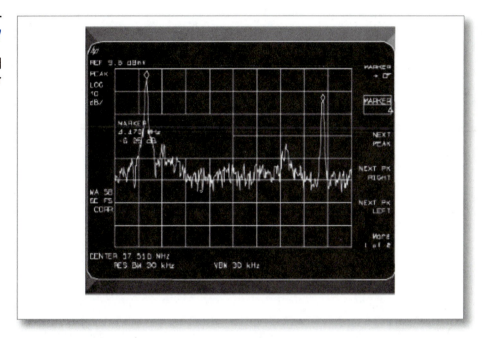

PRACTICE EXERCISES 11–1

Determine the first two sinusoidal terms of a square wave with an amplitude of 8 V and a period of 3 ms.

✓ANSWERS

DC component = 0;

$n = 1$: $v_1 = 10.2 \sin(2094t)$;

$n = 2$: $v_2 = 3.4 \sin(6282t)$

(Complete solutions to the Practice Exercises can be found in Appendix A.)

11.4   THE SAWTOOTH WAVE

The sawtooth waveform is used to create the time base for the video display in the cathode ray tube (CRT) oscilloscope, computer monitor, and television receiver. The Fourier series representation of a sawtooth wave is given without proof as:

$$v(t) = \frac{A}{2} - \frac{A}{\pi} \sum_{n=1}^{\infty} \frac{\sin n\omega_0 t}{n} \tag{11.3}$$

Consider the sawtooth wave in Figure 11–8.

FIGURE **11-8**

A Sawtooth (Sweep)
Waveform

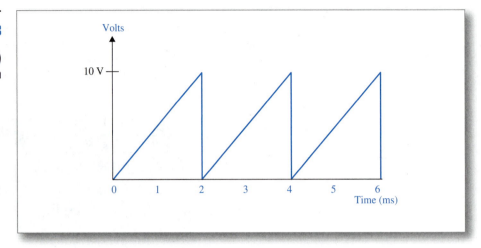

The waveform in Figure 11–8 is a sawtooth wave with amplitude of 10 V and a period of 2 ms.

The DC component of the Fourier series representation of a sawtooth waveform can be calculated using Equation (11.3) as:

$$\frac{A}{2} = \frac{10}{2} = 5 \text{ V}$$

The fundamental frequency of the sawtooth waveform is calculated as the reciprocal of the period so that:

$$f_0 = \frac{1}{T} = \frac{1}{2 \text{ ms}} = 500 \text{ Hz}$$

$$\omega_0 = 2\pi f_0 = 3142 \text{ rad/s}$$

The first sinusoidal term is calculated using Equation (11.3) with $n = 1$ as:

$$-\frac{A}{\pi} \sum_1^\infty \frac{\sin(n\omega_0 t)}{n} = \frac{-10}{\pi} \sin(\omega_0 t) = -3.18 \sin(3142t) = 3.18 \sin(3142t = 180°)$$

Note that adding 180° to the argument of the sine function replaces the negative sign in the expression.

The second sinusoidal term is calculated using Equation (11.3) with $n = 2$ as:

$$-\frac{A}{\pi} \sum_1^\infty \frac{\sin(n\omega_0 t)}{n} = -\frac{10}{2\pi} \sin(2\omega_0 t) = -1.59 \sin(6284t) = 1.59 \sin(6284t + 180°)$$

Again, adding 180° to the argument of the sine function replaces the negative sign in the expression.

The sinusoidal terms for $n = 3$, 4, and 5 are calculated in a similar manner and are:

$$v(n = 3) = 1.06 \sin(9426t + 180°)$$

$$v(n = 4) = 0.796 \sin(12{,}568t + 180°)$$

$$v(n = 5) = 0.637 \sin(15{,}710t + 180°)$$

The MATLAB plot showing the sum of the DC component and the first five sinusoidal terms of the Fourier series is shown in Figure 11–9.

FIGURE **11-9**

Sum of the First Five
Fourier Series Terms of a
Sawtooth Wave

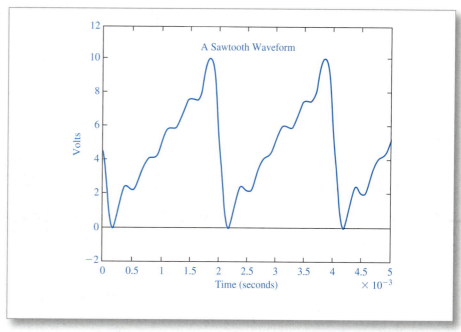

The plot in Figure 11–9 is a reasonable approximation of the given sawtooth waveform
with amplitude $A = 10$ V and period $T = 2$ ms.

EXAMPLE 11-3   Fourier Series of a Sawtooth Waveform

Determine the DC component and the first three sinusoidal terms of the Fourier
series representation of the sawtooth wave shown in Figure 11–10. Plot the sum of
the DC and the first three sinusoidal terms using MATLAB.

FIGURE **11-10**

Sawtooth
Waveform
for Example 11-3

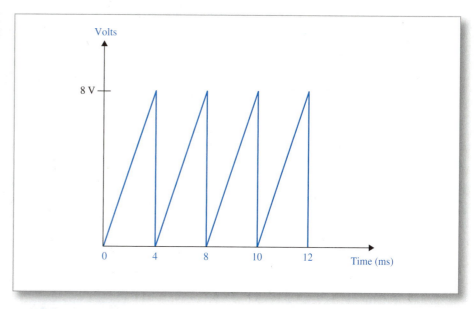

✔ **SOLUTION**

The amplitude of the sawtooth waveform shown in Figure 11–10 is 8 V and the
period is 4 ms.

The DC component of the Fourier representation of the sawtooth waveform is calculated as:

$$\frac{A}{2} = \frac{8}{2} = 4 \text{ V}$$

The fundamental frequency is calculated as:

$$f_0 = \frac{1}{T} = \frac{1}{4 \text{ ms}} = 250 \text{ Hz}$$

$$\omega_0 = 2\pi f_0 = 1570.8 \text{ rad/s}$$

The first sinusoidal term is determined for $n = 1$ as:

$$-\frac{A}{\pi} \sum_1^\infty \frac{\sin(n\omega_0 t)}{n} = -\frac{8}{\pi} \sin(\omega_0 t) = -2.546 \sin(1570.8 t) = 2.546 \sin(1570.8 t + 180°)$$

Note that adding 180° to the argument of the sine function replaces the negative sign in the expression.

The second sinusoidal term is calculated with $n = 2$ as:

$$-\frac{8}{2\pi} \sin(2\omega_0 t) = -1.732 \sin(3141.6 t) = 1.732 \sin(3141.6 t + 180°)$$

Again, adding 180° to the argument of the sine function replaces the negative sign in the expression.

The sinusoidal terms for $n = 3$ and 4 are calculated as:

$$v(n = 3) = 0.849 \sin(4712.4 t + 180°)$$

$$v(n = 4) = 0.637 \sin(6283.2 t + 180°)$$

The MATLAB plot for Example 11–3 is shown in Figure 11–11.

FIGURE 11–11

Fourier Representation of the Sawtooth Waveform for Example 11–3

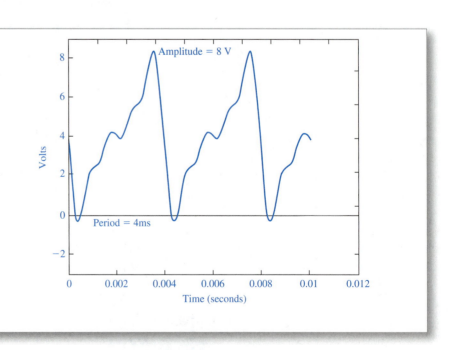

The frequency spectrum of the sawtooth waveform in Example 11–3 is shown in Figure 11–12.

FIGURE **11-12**

Frequency Spectrum of the Sawtooth Waveform for Example 11-3

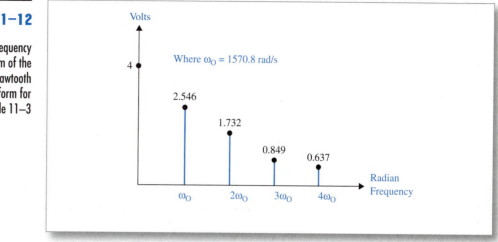

PRACTICE EXERCISES 11-2

Determine the DC value and first sinusoidal term of a 5 V sawtooth waveform with a period of 5 ms.

✓ ANSWERS

DC = 2.5 V; $v = 1.59 \sin(1257t + 180°)$

(Complete solutions to the Practice Exercises can be found in Appendix A.)

11.5   THE RECTIFIED SINE WAVE

The rectified sine wave is created by reversing the sign of all negative instantaneous values of the sine wave. A sine wave and its rectified waveform are shown in Figure 11-13.

FIGURE **11-13**

A Sine Wave and Its Rectified Waveform

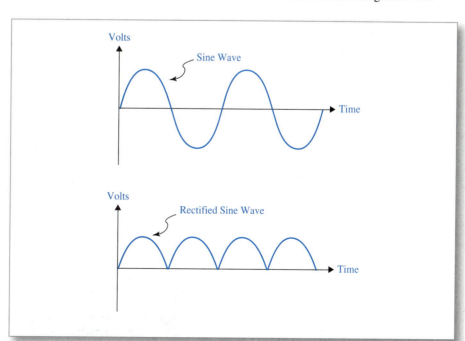

The sine wave has an average value of zero over a complete cycle. The rectified sine wave has only positive values and therefore has a nonzero average value. Rectifier circuits are used in DC power supplies to convert commercial AC (sinusoidal) voltage to a waveform that has a DC or average value.

The Fourier series representation of a rectified sine wave is given without proof as:

$$v(t) = \frac{2A}{\pi} + \frac{4A/\pi}{1-n^2} \cos(n\omega_0) \text{ for all even values of } n \qquad (11.4)$$

EXAMPLE 11-4   Fourier Series of a Rectified Sine Wave

A sine wave has an amplitude of 10 V and a period of 10 ms. Determine the first five terms of the Fourier series representation of the rectified sine wave.

✓ **SOLUTION**

Using the expression in Equation (11.4), the first five terms of the rectified sine wave are calculated as:

$$f_0 = \frac{1}{T} = \frac{1}{10 \text{ ms}} = 100 \text{ Hz}$$

$$\omega_0 = 2\pi f_0 = 628.3 \text{ rad/s}$$

$$\text{DC} = \frac{2A}{\pi} = \frac{20}{\pi} = 6.37 \text{ V}$$

$$v(n=2) = -4.24 \cos(1256.6t) \text{ V}$$

$$v(n=4) = -0.849 \cos(2513.2t) \text{ V}$$

$$v(n=6) = -0.364 \cos(3769.8t) \text{ V}$$

$$v(n=8) = -0.202 \cos(5026.4t) \text{ V}$$

The MATLAB plot of the sum of the first five terms of the Fourier series representation of the rectified sine wave is shown in Figure 11–14.

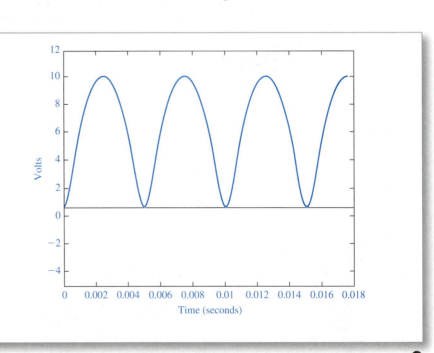

FIGURE 11–14

Fourier Representation of the Rectified Sine Wave for Example 11–4

PRACTICE EXERCISES 11-3

The amplitude of a rectified sine wave is given as 4 V and the period is given as 1 ms. Determine the DC and first sivnusoidal term of the rectified sine wave.

✓ **ANSWERS**

DC = 2.55 V; $v = -1.697\cos(12{,}566t)$ V

(Complete solutions to the Practice Exercises can be found in Appendix A.)

11.6 CIRCUIT ANALYSIS USING FOURIER SERIES

Now that we have learned that complex electronic signals can be represented as a sum of sinusoidal waveforms, we can use Fourier series to analyze circuits with complex input signals.

Recall that the **principle of superposition** states that a bilateral network with multiple input signals may be analyzed by considering each signal independently. The results of each analysis can then be summed (superimposed) to obtain the total circuit response. The Fourier representation of a complex waveform together with the application of the principle of superposition allows us to analyze many electronic circuits with complex input signals by using straightforward AC circuit analysis techniques.

EXAMPLE 11-5   Circuit Analysis Using Fourier Series

The input signal to the circuit shown in Figure 11–15 is a 5 V square wave with a period of 20 ms. Use Fourier series and the principle of superposition to determine the amplitude and phase of the first five components of the output signal.

FIGURE **11–15**

Circuit for
Example 11–5

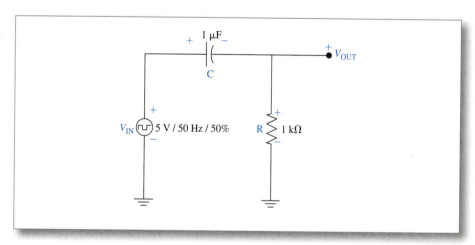

✓ **SOLUTION**

The frequency spectrum of the given 5 V square wave input signal is shown in Figure 11–6 and is shown again for convenience in Figure 11–16.

FIGURE **11–16**

Frequency
Spectrum
of a 5 V Square
Wave with a
Period of 10 ms.

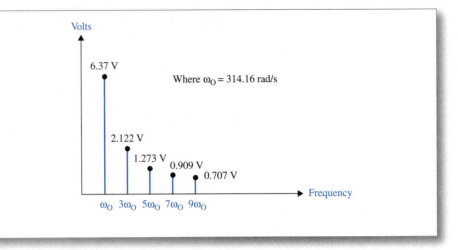

Each of the values in the frequency spectrum in Figure 11–16 represents the amplitude of a sinusoidal component of the input square wave. The DC component of the input square wave is zero.

The fundamental frequency of the input square wave is given in Figure 11–15 as $f_0 =$ 50 Hz, so ω_0 is calculated as:

$$\omega_0 = 2\pi f_0 = 314.16 \text{ rad/s}$$

The analysis of the circuit at the fundamental frequency ω_0 proceeds as follows:

$$\mathbf{X}_C = \frac{1}{\omega_0 C} = \frac{1}{(314 \text{ rad/s})(1 \text{ } \mu\text{F})} \angle -90° = 3183 \angle -90° \text{ } \Omega$$

The voltage divider rule can be used to obtain:

$$\mathbf{V}_{out} = \mathbf{V}_{in} \frac{\mathbf{R}}{\mathbf{R} + \mathbf{X}_C} = 6.367 \text{ V} \frac{1 \angle 0° \text{ k}\Omega}{1 \angle 0° \text{ k}\Omega + 3.18 \angle -90° \text{ k}\Omega} \qquad (11.5)$$

$$\mathbf{V}_{out} = 6.367 \text{ V} \frac{1 \angle 0° \text{ k}\Omega}{3.34 \angle -72.5° \text{ k}\Omega} = 1.9 \angle 72.5° \text{ V}$$

The first sinusoidal Fourier term of the output signal can be expressed in the time domain as:

$$v(t) = 1.9 \text{ V} \sin(314.16t + 72.5°)$$

The computer-generated table in Figure 11–17 shows the results of a similar analysis for each of the first five sinusoidal Fourier components of the given square wave input signal. (The values in Figure 11–17 have been rounded where appropriate.)

FIGURE **11–17**

Fourier Series
Calculations for
Example 11–5

ω(rad/s)	V_{in}	$X_C(\Omega)$	V_{out}
0	0	∞	0
314	6.37	3180	$1.90 \angle 72°$
942	2.12	1006	$1.46 \angle 47°$
1570	1.22	637	$1.07 \angle 32°$
2198	0.909	455	$0.83 \angle 24°$
2826	0.707	352	$0.67 \angle 19°$

The V_{out} column of Figure 11–17 contains the amplitude and phase angle of each of the first five sinusoidal Fourier components of the *output signal*.

The MATLAB plot shown in Figure 11–18 is a plot of the input square wave and the sum of the first four Fourier terms of the output waveform.

MATLAB

FIGURE **11–18**

Fourier
Representation
of the Input and
Output Signal for
Example 11–5

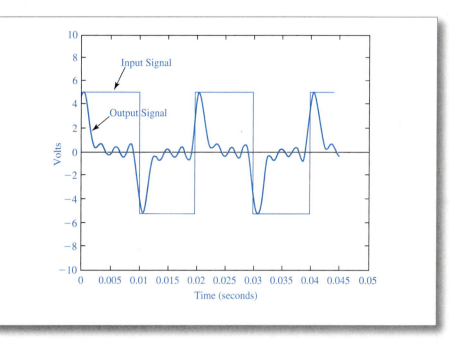

Note that the circuit in Example 11–5 is a *differentiator circuit*. The output signal obtained in Figure 11–18 is a close approximation of the expected output signal for a differentiator circuit. The output waveform indicates a positive spike at the leading edge of the input waveform and a negative spike at the trailing edge.

It is also interesting to note that the circuit in Figure 11–15 is the configuration for the R-C high-pass filter. The low-frequency component of the input square wave signal will appear across the capacitor, and only the high-frequency component will appear across the resistor at the output. We may further reason that the high-frequency component of the input signal would occur at the edges of the signal where the rapid transitions from high to low level or low to high level occur. This reasoning explains the shape of the output signal of the differentiator circuit. It is always enlightening to reason out circuit behavior rather than to rely solely upon mathematical analysis.

EXAMPLE 11–6   Circuit Analysis Using Fourier Series

Determine the first three Fourier terms of the output waveform for the circuit in Figure 11–19. Create a MATLAB plot of the first four Fourier terms.

✓ SOLUTION

The fundamental frequency of the input square wave shown in Figure 11–19 is given as 400 Hz, so ω_0 is calculated as:

$$\omega_0 = 2\pi f_0 = 2513 \text{ rad/s}$$

FIGURE **11-19**

Circuit for
Example 11–6

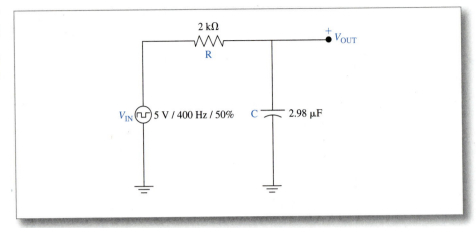

Referring again to the frequency spectrum of a 5 V square wave in Figure 11–16, and using Equation (11.2), the Fourier terms of the input signal may be determined as:

$$DC = 0$$

$$v(n=1) = \frac{(4)(5)}{\pi} \sin(\omega_0) = 6.37 \sin(2513t)$$

$$v(n=2) = \frac{(4)(5)}{3\pi} \sin(3\omega_0) = 2.122 \sin(7539t)$$

$$v(n=3) = \frac{(4)(5)}{5\pi} \sin(5\omega_0) = 1.273 \sin(12{,}565t)$$

The first Fourier term ($n = 1$) of the output signal can be determined as:

$$X_C = \frac{1}{\omega_0 C} = \frac{1}{(2531)(2.98\ \mu F)} = 132.6\angle -90°\ \Omega$$

The voltage divider rule can be used to calculate the output voltage as:

$$\mathbf{V}_{out} = \mathbf{V}_{in}\frac{\mathbf{X}_C}{(\mathbf{X}_C + \mathbf{R})} = 6.37\angle 0°\ V\frac{132.6\angle -90°\ \Omega}{132.6\angle -90°\ \Omega + 2000\angle 0°\ \Omega}$$

$$= 0.422\angle -86.2°\ V \qquad\qquad (11.6)$$

The first sinusoidal Fourier term of the output signal can be expressed in the time domain as:

$$v(n=1) = 0.425 \sin(2513t - 86.2°)$$

The remaining Fourier terms of the output signal are calculated in a similar manner and the results are:

$$v(n=2) = 0.046 \sin(7539t - 88.7°)$$

$$v(n=3) = 0.017 \sin(12{,}565t - 89.2°)$$

The MATLAB plot shown in Figure 11–20 is a plot of the input square wave and the sum of the first four Fourier terms of the output waveform.

FIGURE **11–20**

Fourier
Representation
of the Input and
Output Signal for
Example 11–6

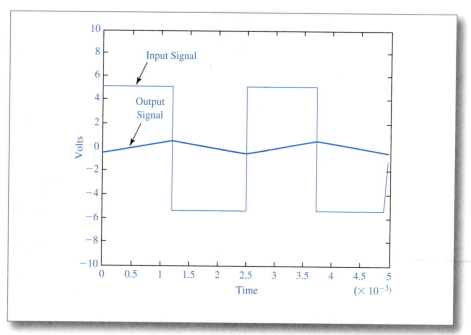

Note that the circuit in Figure 11–19 is an *integrator circuit*. The effect of the integrator circuit is to produce a waveform that is nearly constant around the average value (DC value) of the input waveform. The average value of the input sine wave is 0 V. The output waveform in Figure 11–20 approximates a constant value of 0 V.

EXAMPLE 11–7 Circuit Analysis Using Fourier Series

The input signal to the circuit shown in Figure 11–21 is a 10 V sawtooth wave with a period of 2 ms. Use the Fourier series representation obtained in Section 11.3 and the principle of superposition to determine the amplitude and phase of the first four Fourier terms of the output signal. Sum the output Fourier terms using MATLAB.

FIGURE **11–21**

Circuit for
Example 11–7

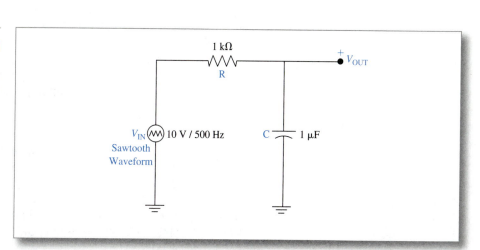

✓ SOLUTION

The Fourier terms for a 10 V sawtooth waveform with a period of 2 ms are calculated in Section 11.3 as:

$$DC = 5 \text{ V}$$

$$v(n = 1) = 3.18 \sin(3142t + 180°)$$

$$v(n = 2) = 1.59 \sin(6284t + 180°)$$

$$v(n = 3) = 1.06 \sin(9426t + 180°)$$

$$v(n = 4) = 0.796 \sin(12,568t + 180°)$$

DC component of the output waveform is calculated as:

$$\frac{A}{2} = \frac{10}{2} = 5 \text{ V}$$

The first sinusoidal term of the output waveform may be determined using the voltage divider rule as:

$$\mathbf{V}_{out} = \mathbf{V}_{in} \frac{\mathbf{X}_C}{\mathbf{Z}}$$

where: $\qquad v_{in} = 3.18 \sin(3142t + 180°)$

In the phasor domain:

$$\mathbf{V}_{in} = 3.18 \angle 180° \text{ V}$$

The capacitive reactance may be calculated at the fundamental frequency as:

$$\mathbf{X}_C = \frac{1}{\omega C} = 318.3 \angle -90° \text{ }\Omega$$

$$\mathbf{Z} = \mathbf{X}_C + \mathbf{R} = 318.3 \angle -90° \text{ }\Omega + 1 \angle 0° \text{ k}\Omega = 1.05 \angle -17.7° \text{ k}\Omega$$

The first sinusoidal term of the output signal can be determined as:

$$\mathbf{V}_{out} = 3.18 \angle 180° \text{ V} \frac{318.3 \angle -90° \text{ }\Omega}{1.05 \angle -17.7° \text{ k}\Omega} = 0.964 \angle 108° \text{ V} = 0.964 \sin(3142t + 108°)$$

The remaining sinusoidal terms of the output signal can be determined using a similar procedure as:

$$v(n = 2) = 0.251 \angle 99° = 0.251 \sin(6284t + 99°)$$

$$v(n = 3) = 0.112 \angle 96° = 0.112 \sin(9426t + 96°)$$

$$v(n = 4) = 0.063 \angle 94° = 0.063 \sin(12,568t + 94°)$$

The input sawtooth waveform and the sum of the DC and the first three sinusoidal terms of the Fourier series of the output waveform are shown in the MATLAB plot in Figure 11–22. The shape of the output signal in Figure 11–22 indicates that the rapid transitions in the input sawtooth waveform have been "smoothed" by the R-C integrator circuit. The amplitude of the output waveform approximates the average value of the input sawtooth waveform.

FIGURE **11–22**

Fourier
Representation
of the Input and
Output Signal for
Example 11–7

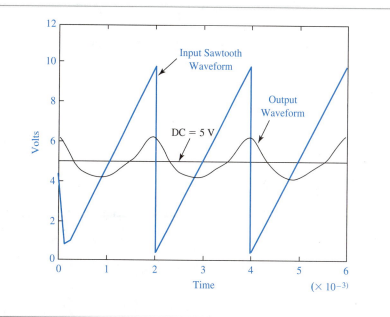

SUMMARY

The **Fourier series** is a mathematical procedure that represents a periodic complex waveform as a sum of sine or cosine terms of varying amplitude and frequency. Fourier series representation applies only to waveforms that are periodic.

The **fundamental frequency** of a periodic waveform is determined as the reciprocal of the period of the waveform: $f_0 = 1/T$. The remaining frequencies of the Fourier series sinusoidal terms are called harmonics and are integer multiples of the fundamental frequency.

The analysis of circuits that have complex periodic input signals may be accomplished by representing the input signal as a Fourier series. The Fourier series terms of the output signal may be obtained through standard AC circuit analysis techniques. The complete output signal is determined as the sum of the individual Fourier terms using the **principle of superposition.**

The frequency spectrum of a complex periodic waveform indicates the amplitude of the DC and sinusoidal Fourier terms. The amplitude of each sinusoidal term is shown at their respective frequency.

The frequency spectrum of complex signals such as those found in modern analog and digital communications systems is an important indication of signal quality. Unwanted harmonics that alter the wave shape of a signal can be identified through spectrum analysis. Electronic filters may then be designed to remove the unwanted frequencies and thereby restore signal quality.

The spectrum analyzer is an important instrument that reveals the frequency content of complex electronic signals. The analyzer displays the amplitude and phase of the fundamental and **harmonic frequencies** of the sampled signal.

ESSENTIAL LEARNING EXERCISES

(ANSWERS TO ODD-NUMBERED ESSENTIAL LEARNING EXERCISES CAN BE FOUND IN APPENDIX A.)

1. Explain why a periodic waveform that is symmetrical about the time axis has an average or DC value of zero.

2. Define what is meant by the fundamental frequency of a periodic waveform.

3. Define what is meant by a harmonic frequency of a periodic waveform.

4. The general expression of a Fourier series includes what two trigonometric functions?

5. Define what is meant by a frequency spectrum diagram.

6. Describe the characteristics of a square wave.

7. Sketch a 6 V square wave with a period of 5 ms. Determine the fundamental frequency of the square wave.

8. Sketch a 3 V sawtooth wave with a period of 8 μs. Determine the fundamental frequency of the sawtooth wave.

9. Describe the characteristics of a rectified sine wave.

10. A 12 V sine wave has a period of 10 ms. Sketch the waveform of the rectified sine wave.

11. Determine the DC and first two sinusoidal terms of a 4 V square wave with a period of 2 ms.

12. Determine the DC and first two sinusoidal terms of a 10 V square wave with a period of 3 ms.

13. Determine the DC and first two sinusoidal terms of a 12 V sawtooth wave with a period of 50 μs.

14. Determine the DC and first two sinusoidal terms of a 10 V sawtooth wave with a period of 75 μs.

15. Determine the DC and first two sinusoidal terms of a 3 V rectified sine wave with a period of 5 ms.

16. Determine the DC and first two sinusoidal terms of a 6 V rectified sine wave with a period of 1 ms.

17. The fundamental frequency of a periodic waveform is given as 500 Hz. Determine the first two even harmonic frequencies.

18. The fundamental frequency of a periodic waveform is given as 300 Hz. Determine the first two odd harmonic frequencies.

19. The DC value of a periodic waveform is given as 8 V, and the fundamental radian frequency is given as $\omega_0 = 650$ rad/s. The amplitude at ω_0, $3\omega_0$, $5\omega_0$, and $7\omega_0$ is given as 6.3 V, 4.2 V, 3.1 V, and 2.4 V respectively. Sketch the frequency spectrum of the waveform.

20. The DC value of a periodic waveform is given as 10 V, and the fundamental frequency is given as 600 Hz. The amplitude of the first, second, and third harmonic is given as 12 V, 9.2 V, and 7.1 V respectively. Sketch the frequency spectrum of the waveform.

CHALLENGING LEARNING EXERCISES

(ANSWERS TO SELECTED CHALLENGING LEARNING EXERCISES CAN BE FOUND IN APPENDIX A.)

1. Determine the DC and first four sinusoidal components of the Fourier representation of the square wave shown in Figure 11–23.

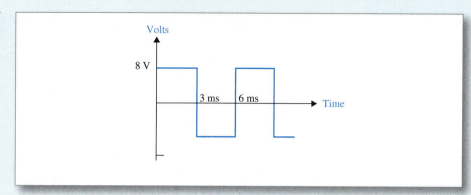

FIGURE **11–23**

Square Wave for
Challenging Exercise 1

2. The square wave shown in Figure 11–23 is the input signal to the circuit shown in Figure 11–24. Determine the DC and first two sinusoidal terms of the Fourier series of the output waveform.

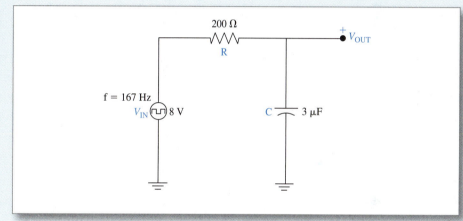

FIGURE **11–24**

Circuit for Challenging
Exercise 3

3. Determine the DC and first four sinusoidal Fourier terms of a 3 V square wave with a period of 150 μs.

4. Create a MATLAB plot of the DC and first four sinusoidal terms of the square wave given in Challenging Learning Exercise 3.

5. The waveform shown in Figure 11–25 is the input signal to a differentiator circuit. Sketch the output waveform.

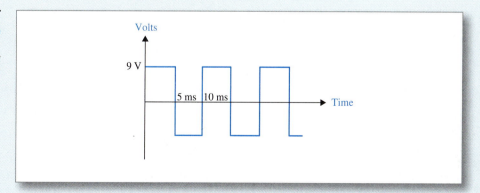

FIGURE **11–25**

Square Wave for
Challenging Exercise 7

TEAM ACTIVITY

If a spectrum analyzer is available at your institute, use the instrument to determine the spectrum of a 5 V square wave with a period of 100 μs. Compare the amplitudes of the DC and first four harmonics obtained from the frequency spectrum analyzer to those obtained through manual computation of the Fourier series.

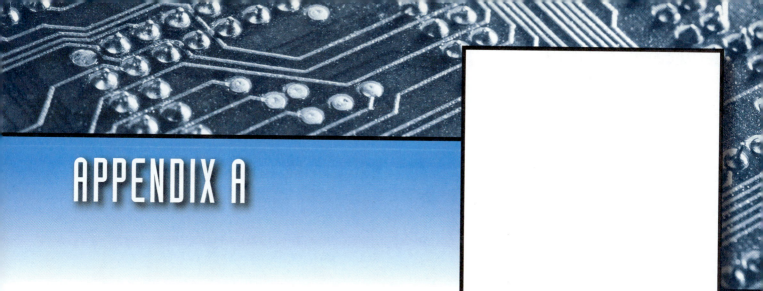

APPENDIX A

COMPLETE SOLUTIONS TO PRACTICE EXERCISES

NOTE: Solutions may vary slightly due to rounding difference.

CHAPTER 1

Practice Exercise 1.1: a. $I = 2.1\ \mu A$; b. $I = 43\ \mu A$; c. $V = 129\ mV$

CHAPTER 2

Practice Exercise 2.1: a. $I = V/R = 18\ V/5\ k\Omega = 3.6\ mA$;

b. $V = IR = (2.8\ mA)(4.7\ k\Omega) = 13.16\ V$

Practice Exercise 2.2: Starting at circuit ground and traveling clockwise:
$-3\ V + 6\ V - V_X(+-) - 3\ V + 8\ V + 4\ V$
$-2\ V + 12\ V = 0$;
Solving for V_X to obtain $V_X = 22\ V$ (assumed polarity is correct)

Practice Exercise 2.3: $R_T = 2\ k\Omega + 4\ k\Omega + 3\ k\Omega = 9\ k\Omega$;

$I_S = V_S/R_T = 12\ V/9\ k\Omega = 1.33\ mA$;
$V_{R1} = I_S R_1 = (1.33\ mA)(2\ k\Omega) = 2.67\ V$;
$V_{R2} = I_S R_2 = (1.33\ mA)(4\ k\Omega) = 5.33\ V$;
$V_{R3} = I_S R_3 = (1.33\ mA)(3\ k\Omega) = 4\ V$;

KVL: $2.66\ V + 5.32\ V + 4\ V = 12\ V$

Practice Exercise 2.4: $V_{R1} = V_S(R_1/R_T) = 12\ V(2\ k\Omega/9\ k\Omega) = 2.66\ V$;
$V_{R2} = V_S(R_2/R_T) = 12\ V(4\ k\Omega/9\ k\Omega) = 5.33\ V$;
$V_{R3} = V_S(R_3/R_T) = 12\ V(3\ k\Omega/9\ k\Omega) = 4\ V$

Practice Exercise 2.5: Starting at circuit ground: $-6\ V + 3\ V = -3\ V = V_A$;
Starting at circuit ground: $+1\ V + 4\ V = +5\ V = V_B$

Practice Exercise 2.6: $G_T = 1/2\ k\Omega + 1/8\ k\Omega + 1/16\ k\Omega + 1/24\ k\Omega$
$= 729.127\ \mu S$;
$R_T = 1/G_T = 1/729.127\ \mu S = 1.37\ k\Omega$;

$$R_1 \,/\!/\, R_2 = (2 \text{ k}\Omega)(8 \text{ k}\Omega)/(2 \text{ k}\Omega + 8 \text{ k}\Omega) = 1.6 \text{ k}\Omega;$$
$$R_1 \,/\!/\, R_2 \,/\!/\, R_3 = (1.6 \text{ k}\Omega)(16 \text{ k}\Omega)/(1.6 \text{ k}\Omega + 16 \text{ k}\Omega)$$
$$= 1.454 \text{ k}\Omega;$$
$$R_1 \,/\!/\, R_2 \,/\!/\, R_3 \,/\!/\, R_4 = (1.454 \text{ k}\Omega)(24 \text{ k}\Omega)/$$
$$(1.6 \backslash 1.454 \text{ k}\Omega + 24 \text{ k}\Omega) = R_\text{T} = 1.37 \text{ k}\Omega$$

Practice Exercise 2.7:

Using Ohm's law: $I_{\text{R1}} = V_\text{S}/R_1 = 36 \text{ V}/3 \text{ k}\Omega = 12 \text{ mA};$
$I_{\text{R2}} = V_\text{S}/R_2 = 36 \text{ V}/6 \text{ k}\Omega = 6 \text{ mA};$
$I_{\text{R3}} = V_\text{S}/R_3 = 36 \text{ V}/12 \text{ k}\Omega = 3 \text{ mA};$
KCL: $I_\text{S} = 12 \text{ mA} + 6 \text{ mA} + 3 \text{ mA} = 21 \text{ mA};$
$R_\text{T} = V_\text{S}/I_\text{S} = 36 \text{ V}/21 \text{ mA} = 1.714 \text{ k}\Omega;$
$G_\text{T} = 1/R_\text{T} = 583.33 \text{ mS}$

Practice Exercise 2.8:

$G_\text{T} = 1/R_\text{T} = 1/8 \text{ k}\Omega + 1/6 \text{ k}\Omega + 1/4 \text{ k}\Omega = 541.67 \text{ μS};$
$R_\text{T} = 1/G_\text{T} = 1/541.67 \text{ μS} = 1.85 \text{ k}\Omega;$
$I_\text{S} = 36 \text{ V}/1.85 \text{ k}\Omega = 19.46 \text{ mA};$

Using the current divider rule:
$I_{\text{R1}} = I_\text{S}(R_\text{T}/R_1) = 19.46 \text{ mA}(1.85 \text{ k}\Omega/8 \text{ k}\Omega) = 4.5 \text{ mA};$
$I_{\text{R2}} = I_\text{S}(R_\text{T}/R_2) = 19.46 \text{ mA}(1.85 \text{ k}\Omega/6 \text{ k}\Omega) = 6.0 \text{ mA};$
$I_{\text{R3}} = I_\text{S}(R_\text{T}/R_3) = 19.46 \text{ mA}(1.85 \text{ k}\Omega/4 \text{ k}\Omega) = 9.0 \text{ mA};$

KCl: $I_\text{S} = I_{\text{R1}} + I_{\text{R2}} + I_{\text{R3}} = 4.5 \text{ mA} + 6.0 \text{ mA} + 9.0 \text{ mA}$
$= 19.5 \text{ mA}$

CHAPTER 3

Practice Exercise 3.1:

$R_\text{T} = 2.2 \text{ k}\Omega + (3.3 \text{ k}\Omega \,/\!/\, 4.7 \text{ k}\Omega) = 4.14 \text{ k}\Omega;$
$I_\text{S} = 20 \text{ V}/4.14 \text{ k}\Omega = 4.83 \text{ mA}; I_{\text{R1}} = I_\text{S} = 4.83 \text{ mA};$
(current divider rule) $I_{\text{R2}} = 4.84 \text{ mA}(1.94 \text{ k}\Omega/3.3 \text{ k}\Omega)$
$= 2.84 \text{ mA};$
$I_{\text{R3}} = 4.84 \text{ mA}(1.94 \text{ k}\Omega/4.7 \text{ k}\Omega) = 1.99 \text{ mA};$

$V_{\text{R1}} = IR = (4.83 \text{ mA})(2.2 \text{ k}\Omega) = 10.63 \text{ V};$
$V_{\text{R2}} = IR = (2.84 \text{ mA})(3.3 \text{ k}\Omega) = 9.37 \text{ V};$
$V_{\text{R3}} = IR = (1.99 \text{ mA})(4.7 \text{ k}\Omega) = 9.35 \text{ V};$

KCL: $2.84 \text{ mA} + 1.99 \text{ mA} = 4.84 \text{ mA};$
KVL: $20 \text{ V} - 10.63 \text{ V} - 9.37 \text{ V} = 0;$
$V_\text{A} = V_{\text{R2}} = 9.37 \text{ V}$

Practice Exercise 3.2:

$I_{\text{R1}} = V_{\text{R1}}/R_1 = (36 \text{ V} - 0 \text{ V})/6 \text{ k}\Omega = 6.0 \text{ mA};$
$I_{\text{R2}} = V_{\text{R2}}/R_2 = (36 \text{ V} - 14 \text{ V})/4 \text{ k}\Omega = 5.5 \text{ mA};$
$I_{\text{R3}} = V_{\text{R3}}/R_3 = (14 \text{ V} - 0 \text{ V})/3 \text{ k}\Omega = 4.67 \text{ mA};$
$V_\text{A} = V_\text{S} = 36 \text{ V}$

Practice Exercise 3.3:

a. Given the open circuit shown in Figure 3–27, R_1 and R_2 are connected in series with V_S. Using the voltage divider rule:
$V_{\text{R2}} = 15 \text{ V}(9 \text{ k}\Omega/11 \text{ k}\Omega) = 12.27 \text{ V};$
$V_{\text{open}} = V_{\text{R2}} = 12.27 \text{ V};$

b. Given R_1 shorted as shown in Figure 3–28, R_2 and R_3 are connected in parallel with V_S. $R_\text{T} = R_2 \,/\!/\, R_3 = 4.8 \text{ k}\Omega;$
$I_\text{S} = 24 \text{ V}/4.8 \text{ k}\Omega = 5 \text{ mA}; I_{\text{short}} = I_\text{S} = 5 \text{ mA}$

CHAPTER 4

Practice Exercise 4.1:

a. $v = 25 \sin(1000t)$; $V_p = A = 25$ V; $\omega = 1000$ rad/s;
$f = \omega/2\pi = 159.15$ Hz; $T = 1/f = 6.28$ ms; $\theta = 0°$;

b. $v = 16 \sin(3000t + 81°)$; $V_p = A = 16$ V;
$\omega = 3000$ rad/s;
$f = \omega/2\pi = 477.46$ Hz; $T = 1/f = 2.09$ ms; $\theta = 81°$

Practice Exercise 4.2:

a. $v = 6 \sin(2500t)$ V at 1.3 ms $= 6 \sin(2500 \times 1.3$ ms$)$;
$v = 6(-0.108) = -0.65$ V;

b. $i = 9 \sin(6000t)$ mA at 250 μs $= 9 \sin(6000 \times 250 \; \mu$s$)$;
$i = 9(0.997) = 8.98$ mA;

c. $v = 24 \sin(4000t + 30°)$ V at 0.5 ms
$= 24 \sin(4000 \times 0.5$ ms $+ 0.5236$ rad$)$;
$v = 24(0.579) = 13.89$ V

Practice Exercise 4.3:

a. $\mathbf{V_1} + \mathbf{V_2} = 6\angle 75°$ V $+ 11.5\angle -42°$
$V = 10.27\angle -10.65°$ V;

b. $(\mathbf{I_1})(\mathbf{I_2}) = (5.3\angle 53°$ mA$)(2.6\angle 15°$ mA$)$
$= 13.78\angle 68°$ mA^2

c. $\mathbf{V_2} + \mathbf{V_1} = 6.4\angle 20°$ V $- 7.8\angle -39°$ V
$= 7.09\angle 90.38°$ V

Practice Exercise 4.4:

a. Set calculator complex format to RECTANGULAR. calculator keystrokes: $(8\angle 58°)$ ENTER to obtain 4.24 + *j*6.78 V;

b. Set calculator complex format to RECTANGULAR. calculator keystrokes: $(36\angle -35°)$ ENTER to obtain 29.49 − *j*20.64 mA;

c. Set calculator complex format to POLAR. calculator keystrokes: $(14 + 24i)$ ENTER to obtain $27.78 \angle 59.74°$ V;

d. Set calculator complex format to POLAR. calculator keystrokes: $(16 - 12.5i)$ ENTER to obtain $20.30 \angle -37.99°$ mA

Practice Exercise 4.5:

a. $v = iR = [5.3 \sin(300t + 18°)$ mA$](4$ k$\Omega)$
$= 21.2 \sin(300t + 18°)$ V (i and v are in phase);

b. $i = v/R = 24 \sin(2000t + 120°)/10$ kΩ
$= 2.4 \sin(2000t + 120°)$ mA (i and v are in phase);

c. $R = v/i = 15 \sin(300t)$ V $/ 20 \sin(300t)$ mA $= 750 \; \Omega$

Practice Exercise 4.6:

a. $\mathbf{I} = \mathbf{V/R} = 13\angle 29°$ V $/ 4.7\angle 0°$ k$\Omega = 2.77\angle 29°$ mA;

b. $\mathbf{V} = \mathbf{IR} = (41\angle 68°$ mA$)(500\angle 0°) = 20.5\angle 68°$ V

Practice Exercise 4.7:

a. $\mathbf{X_L} = \omega L = (4000)(30$ mH$) = 120\angle 90° \; \Omega$;
$\mathbf{I} = \mathbf{V/X_L} = 21\angle 34°$ V $/ 120\angle 90° \; \Omega = 175\angle 124°$ mA;

b. $\mathbf{X_L} = \omega L = (2500)(75$ mH$) = 187.5\angle 90° \; \Omega$;
$\mathbf{V} = \mathbf{I X_L} = (40\angle -41°$ mA$)(187.5\angle 90° \; \Omega)$
$= 7.5\angle 49°$ V;

c. $\mathbf{X_L} = V_L / I_p = 36.9$ V $/ 12.3$ mA $= 3\angle 90°$ kΩ

Practice Exercise 4.8:

a. $\mathbf{X_C} = 1/\omega C = 1/(5000)(0.47\ \mu F) = 425.53\angle -90°\ \Omega;$

b. $\mathbf{X_C} = 1/\omega C = 1/(4000)(0.22\ \mu F) = 1.14\angle -90°\ k\Omega;$
$\mathbf{I} = \mathbf{V}/\mathbf{X_C} = 20\angle 25°\ V/1.14\angle -90°\ k\Omega$
$= 17.54\angle 115°\ mA;$
$i = 17.54\sin(4000t + 115°)\ mA$

Practice Exercise 4.9:

a. $\mathbf{X_C} = 1/\omega C = 1/(2500)(0.33\ \mu F) = 1.21\angle -90°\ k\Omega;$
$\mathbf{I} = \mathbf{V}/\mathbf{X_C} = 6\angle 0°\ V/1.21\angle -90°\ k\Omega$
$= 4.96\angle 90°\ mA;$

b. $\mathbf{X_C} = 1/\omega C = 1/(1000)(0.22\ \mu F) = 4.54\angle -90°\ k\Omega;$
$\mathbf{I} = \mathbf{V}/\mathbf{X_C} = 14\angle 23°\ V/4.54\angle -90°\ k\Omega$
$= 3.08\angle 113°\ mA$

CHAPTER 5

Practice Exercise 5.1:

a. $\mathbf{R_T} = 6\angle 0°\ k\Omega + 4\angle 0°\ k\Omega = 10\ k\Omega;$
$\mathbf{I_S} = 20\angle 45°\ V/10\angle 0°\ k\Omega = 2\angle 45°\ mA;$
$\mathbf{V_{R1}} = (2\angle 45°\ mA)(6\angle 0°\ k\Omega) = 12\angle 45°\ V;$
$\mathbf{V_{R2}} = (2\angle 45°\ mA)(4\angle 0°\ k\Omega) = 8\angle 45°\ V;$

Using the voltage divider rule:
$\mathbf{V_{R1}} = 20\angle 45°\ V(6\angle 0°\ k\Omega/10\angle 0°\ k\Omega) = 12\angle 45°\ V;$
$\mathbf{V_{R2}} = 20\angle 45°\ V(4\angle 0°\ k\Omega/10\angle 0°\ k\Omega) = 8\angle 45°\ V;$

KVL: $2\angle 45°\ V - 8\angle 45°\ V - 12\angle 45°\ V = 0;$

b. $\mathbf{X_L} = 2\pi fL = 2.01\angle 90°\ k\Omega;$
$\mathbf{Z} = \mathbf{R} + \mathbf{X_L} = 2.2\angle 0°\ k\Omega + 2.01\angle 90°\ k\Omega$
$= 2.98\angle 42.42°\ k\Omega;$
$\mathbf{I_S} = 12\angle 0°\ V/2.98\angle 42.42°\ k\Omega$
$= 4.03\angle -42.42°\ mA;$

$\mathbf{V_R} = (4.03\angle -42.42°\ mA)(2.2\angle 0°\ k\Omega)$
$= 8.87\angle -42.42°\ V;$
$\mathbf{V_L} = (4.03\angle -42.42°\ mA)(2.01\angle 90°\ k\Omega)$
$= 8.10\angle 47.58°\ V;$

Using the voltage divider rule:
$\mathbf{V_R} = 12\angle 0°\ V(2.2\angle 0°\ k\Omega/2.98\angle 42.42°\ k\Omega)$
$= 8.86\angle -42.42°\ V;$
$\mathbf{V_L} = 12\angle 0°\ V(2.01\angle 90°\ k\Omega/2.98\angle 42.42°\ k\Omega)$
$= 8.1\angle 47.58°\ V;$

KVL: $8.87\angle -42.42°\ V + 8.10\angle 47.58°\ V = 12\angle 0°\ V;$

c. $\mathbf{X_C} = 1/2\pi fC = 3.54\angle -90°\ k\Omega;$
$\mathbf{Z} = \mathbf{R} + \mathbf{X_C} = 5\angle 0°\ k\Omega + 3.54\angle -90°\ k\Omega$
$= 6.12\angle -35.3°\ k\Omega;$
$\mathbf{I_S} = 9\angle 0°\ V/6.12\angle -35.3°\ k\Omega = 1.47\angle 35.03°\ mA;$

$\mathbf{V_R} = (1.47\angle 35.03°\ mA)(5\angle 0°\ k\Omega) = 7.35\angle 35.03°\ V;$
$\mathbf{V_L} = (1.47\angle 35.03°\ mA)(3.54\angle -90°\ k\Omega)$
$= 5.20\angle -54.97°\ V;$

Using the voltage divider rule:
$\mathbf{V_R} = 9\angle 0°\ V(5\angle 0°\ k\Omega/6.12\angle -35.3°\ k\Omega)$
$= 7.35\angle 35.03°\ V;$

$$\mathbf{V_L} = 9\angle 0° \text{ V} (3.54 \angle -90° \text{ k}\Omega/6.12 \angle -35.3° \text{ k}\Omega)$$
$$= 5.20 \angle -54.97° \text{ V};$$

KVL: $7.35 \angle 35.03° \text{ V} + 5.20 \angle -54.97° \text{ V} = 9\angle 0° \text{ V}$

Practice Exercise 5.2:
$$\mathbf{Z} = \mathbf{R} + \mathbf{X_L} + \mathbf{X_C} = 6.04 \angle 38.96° \text{ k}\Omega;$$
$$\mathbf{I_S} = 36 \angle 0° \text{ V}/6.04 \angle 38.96° \text{ k}\Omega = 5.96 \angle -38.96° \text{ mA};$$

$$\mathbf{V_R} = (5.96 \angle -38.96° \text{ mA})(4.7 \angle 0° \text{ k}\Omega)$$
$$= 28.01 \angle -38.96° \text{ V};$$
$$\mathbf{V_L} = (5.96 \angle -38.96° \text{ mA})(10 \angle 90° \text{ k}\Omega)$$
$$= 59.6 \angle 51.04° \text{ V};$$
$$\mathbf{V_C} = (5.96 \angle -38.96° \text{ mA})(6.2 \angle -90° \text{ k}\Omega)$$
$$= 36.96 \angle -128.96° \text{ V};$$

KVL: $28.01 \angle -38.96° \text{ V} + 59.6 \angle 51.04° \text{ V} +$
$36.95 \angle -128.96° \text{ V} = 36 \angle 0° \text{ V}$

Practice Exercise 5.3:
$$\mathbf{Y} = 1/\mathbf{R} + 1/\mathbf{X_L} + 1/\mathbf{X_C} = 250 \angle 36.87° \text{ }\mu\text{S};$$
$$\mathbf{Z} = 1/\mathbf{Y} = 4 \angle -36.87° \text{ k}\Omega; \mathbf{I_S} = \mathbf{V_S}/\mathbf{Z} = 5 \angle 36.87° \text{ mA};$$
$$\mathbf{I_R} = \mathbf{V_S}/\mathbf{R} = 4 \angle 0° \text{ mA}; \mathbf{I_L} = \mathbf{V_S}/\mathbf{X_L} = 2 \angle -90° \text{ mA};$$
$$\mathbf{I_C} = \mathbf{V_S}/\mathbf{X_C} = 5 \angle 90° \text{ mA};$$

KCL: $4 \angle 0° \text{ mA} + 2 \angle -90° \text{ mA} + 5 \angle 90° \text{ mA}$
$= 5 \angle 36.87° \text{ mA}$

Practice Exercise 5.4:
a. $\mathbf{Z} = \mathbf{X_C} + (\mathbf{R}/\!/\mathbf{X_L}) = 3 \angle 16.26° \text{ k}\Omega;$
$\mathbf{I_S} = 20 \angle 0° \text{ V}/3 \angle 16.26° \text{ k}\Omega;$
$\mathbf{I_S} = 6.67 \angle -16.26° \text{ mA} = \mathbf{I_C};$

Using the current divider rule:
$\mathbf{I_R} = \mathbf{I_S}(\mathbf{Z_P}/\mathbf{R}) = 6.67 \angle -16.26° \text{ mA}$
$(4.8 \angle 53.13° \text{ k}\Omega/8 \angle 0° \text{ k}\Omega)$
$\mathbf{I_R} = 4.0 \angle 36.87° \text{ mA};$
$\mathbf{I_L} = \mathbf{I_S}(\mathbf{Z_P}/\mathbf{X_L}) = 6.67 \angle -16.26° \text{ mA}$
$(4.8 \angle 53.13° \text{ k}\Omega/6 \angle 90° \text{ k}\Omega)$
$\mathbf{I_L} = 5.34 \angle -53.13° \text{ mA};$

$\mathbf{V_C} = (6.67 \angle -16.26° \text{ mA})(3 \angle -90° \text{ k}\Omega)$
$= 20.01 \angle -106.26° \text{ V};$
$\mathbf{V_R} = (4.0 \angle 36.87° \text{ mA})(8 \angle 0° \text{ k}\Omega) = 32.0 \angle 36.87° \text{ V};$
$\mathbf{V_L} = (5.34 \angle -53.13° \text{ mA})(6 \angle 90° \text{ k}\Omega) = 32.0 \angle 36.87° \text{ V};$

b. $\mathbf{R} + \mathbf{X_C} = 6.24 \angle -41.1° \text{ k}\Omega = \text{k}\Omega$
$\mathbf{I_R} = \mathbf{I_C} = \mathbf{V_S}/\mathbf{R} + \mathbf{X_C} = 15.0 \angle 0° \text{ V}/6.24 \angle -41.1° \text{ k}\Omega$
$= 2.4 \angle 41.1° \text{ mA};$
$\mathbf{I_L} = \mathbf{V_S}/\mathbf{X_L} = 15.0 \angle 0° \text{ V}/6.2 \angle 90° \text{ k}\Omega = 2.42 \angle -90° \text{ mA};$
$\mathbf{I_S} = 2.40 \angle 41.1° \text{ mA} + 2.42 \angle -90° \text{ mA}$
$= 1.99 \angle -24.98° \text{ mA};$

$\mathbf{V_R} = (2.4 \angle 41.1° \text{ mA})(4.7 \angle 0° \text{ k}\Omega) = 11.28 \angle 41.1° \text{ V};$
$\mathbf{V_C} = (2.4 \angle 41.1° \text{ mA})(4.1 \angle -90° \text{ k}\Omega) = 9.84 \angle -48.9° \text{ V};$
$\mathbf{V_L} = (2.42 \angle -90° \text{ mA})(6.2 \angle 90° \text{ k}\Omega) = 15 \angle 0° \text{ V}$

CHAPTER 6

Practice Exercise 6.1:
$\mathbf{T.F.} = \mathbf{R_2}/(\mathbf{R_1} + \mathbf{R_2}) = 9 \angle 0° \text{ k}\Omega/(6 \angle 0° \text{ k}\Omega + 9 \angle 0° \text{ k}\Omega);$
$\mathbf{T.F.} = 0.6 \angle 0°;$

$A_V = 0.6; A_{V(dB)} = 20 \log (0.6) = -4.44 \text{ dB}; \theta = 0°;$
$V_{out} = (V_{in})(A_V) = (5 \text{ V})(0.6) = 3 \text{ V}; f = 10 \text{ kHz}$

Practice Exercise 6.2:

$f_C = 1/2\pi RC = 468.1 \text{ Hz};$
At cutoff by definition: $A_V = 0.707; A_{V(dB)} = -3.0 \text{ dB};$
$\theta = -45°;$

At 1 kHz: $\mathbf{X_C} = 318.31\angle -90° \; \Omega; \mathbf{T.F.} = \mathbf{X_C}/(\mathbf{R} + \mathbf{X_C});$
$\mathbf{T.F.} = 318.31\angle -90° \; \Omega/(680\angle 0° \; \Omega + 318.31\angle -90° \; \Omega);$
$\mathbf{T.F.} = 0.424\angle -64.9°; A_V = 0.424; \theta = -64.9°;$
$A_{V(dB)} = 20 \log (0.424) = -7.45 \text{ dB}; \mathbf{V_{out}} = \mathbf{V_{in}} \; \mathbf{T.F.};$
$\mathbf{V_{out}} = (5\angle 0° \text{ V})(0.424\angle -64.9°) = 2.12\angle -64.9° \text{ V}$

Practice Exercise 6.3:

$f_C = 1/2\pi RC = 1.21 \text{ kHz};$
At cutoff by definition: $A_V = 0.707; \quad A_{v(dB)} = -3.0 \text{ dB};$
$\theta = 45°;$

At 3 kHz: $\mathbf{X_C} = 132.63\angle -90° \; \Omega; \quad \mathbf{T.F.} = \mathbf{R}/(\mathbf{R} + \mathbf{X_C});$
$\mathbf{T.F.} = 330\angle 0° \; \Omega/(330\angle 0° \; \Omega + 132.63\angle -90° \; \Omega);$
$\mathbf{T.F.} = 0.928\angle 21.9°; \quad A_V = 0.928; \quad \theta = 21.9°;$
$A_{v(dB)} = 20 \log (0.928) = -0.65 \text{ dB}; \quad \mathbf{V_{out}} = \mathbf{V_{in}} \; \mathbf{T.F.};$
$\mathbf{V_{out}} = 11.14\angle 21.9° \text{ V}$

Practice Exercise 6.4:

$f_C = R/2\pi L = 1.33 \text{ kHz};$
At cutoff by definition: $A_V = 0.707; \quad A_{v(dB)} = -3.0 \text{ dB};$
$\theta = -45°;$

At 500 Hz: $\mathbf{X_L} = 56.55\angle 90° \; \Omega; \quad \mathbf{T.F.} = \mathbf{R}/(\mathbf{R} + \mathbf{X_L});$
$\mathbf{T.F.} = 150\angle 0° \; \Omega/(150\angle 0° \; \Omega + 56.55\angle 90° \; \Omega);$
$\mathbf{T.F.} = 0.936\angle -20.7°; \quad A_V = 0.936; \quad \theta = -20.7°;$
$A_{v(dB)} = 20 \log (0.936) = -0.57 \text{ dB}; \quad \mathbf{V_{out}} = \mathbf{V_{in}} \; \mathbf{T.F.};$
$\mathbf{V_{out}} = 7.49\angle -20.7° \text{ V}$

Practice Exercise 6.5:

$f_C = R/2\pi L = 4.78 \text{ kHz};$
At cutoff by definition: $A_V = 0.707; \quad A_{v(dB)} = -3.0 \text{ dB};$
$\theta = 45°;$

At 500 Hz: $\mathbf{X_L} = 31.42\angle 90° \; \Omega; \quad \mathbf{T.F.} = \mathbf{X_L}/(\mathbf{R} + \mathbf{X_L});$
$\mathbf{T.F.} = 31.42\angle 90° \; \Omega/(300\angle 0° \; \Omega + 31.42\angle 90° \; \Omega);$
$\mathbf{T.F.} = 0.104\angle 84°; \quad A_V = 0.104; \quad \theta = 84°;$
$A_{v(dB)} = 20 \log (0.104) = -19.7 \text{ dB}; \quad \mathbf{V_{out}} = \mathbf{V_{in}} \; \mathbf{T.F.};$
$\mathbf{V_{out}} = 2.49\angle 84° \text{ V}$

Practice Exercise 6.6:

$f_r = \dfrac{1}{2\pi\sqrt{LC}} = 5.81 \text{ kHz};$

$Q = X_L/R \text{ (at resonant frequency)} = 912.6 \; \Omega/150 \; \Omega = 6.1;$
$BW = f_r/Q = 5.81 \text{ kHz}/6.1 = 953 \text{ Hz};$
$f_{HC} = f_r + \frac{1}{2} BW = 6.29 \text{ kHz};$
$f_{LC} = f_r - \frac{1}{2} BW = 5.33 \text{ kHz}$

CHAPTER 7

Practice Exercise 7.1:

a. $\tau = RC = 20 \text{ ms};$
$V_C (3.5 \text{ ms}) = 30 \text{ V}(1 - e^{-3.5 \text{ ms}/20 \text{ ms}}) = 4.8 \text{ V};$
$V_C (6 \text{ ms}) = 30 \text{ V}(1 - e^{-6 \text{ ms}/20 \text{ ms}}) = 7.8 \text{ V};$

$V_C(10 \text{ ms}) = 30 \text{ V}(1 - e^{(-10 \text{ ms}/20 \text{ ms})}) = 11.8 \text{ V};$
$t_{SS} = 5\tau = 100 \text{ ms};$

b. $\tau = RC = 20 \text{ ms};$
$V_R(3.5 \text{ ms}) = 30 \text{ V}(e^{-3.5 \text{ ms}/20 \text{ ms}}) = 25.2 \text{ V};$
$V_R(6 \text{ ms}) = 30 \text{ V}(e^{-6 \text{ ms}/20 \text{ ms}}) = 22.2 \text{ V};$
$V_R(10 \text{ ms}) = 30 \text{ V}(e^{-10 \text{ ms}/20 \text{ ms}}) = 18.2 \text{ V}$

Practice Exercise 7.2: $\tau = RC = 10 \text{ ms};$ $V_C(4 \text{ ms}) = 45 \text{ V}(e^{-4 \text{ ms}/10 \text{ ms}}) = 30.2 \text{ V};$
$V_C(8 \text{ ms}) = 45 \text{ V}(e^{-8 \text{ ms}/10 \text{ ms}}) = 20.2 \text{ V};$
$V_C(35 \text{ ms}) = 45 \text{ V}(e^{-35 \text{ ms}/10 \text{ ms}}) = 1.4 \text{ V};$
$t_{SS} = 5\tau = 50 \text{ ms}$

Practice Exercise 7.3:   Substituting given data: $8 \text{ V} = 12 \text{ V}(1 - e^{-t/\tau})$
where $\tau = RC = 0.495 \text{ s};$
Dividing both sides by 12 V, we obtain: $0.67 = (1 - e^{-t/\tau});$
Subtracting 1 from each side, we obtain: $-0.33 = -e^{-t/\tau};$
Multiplying both sides by -1, we obtain: $0.33 = e^{-t/\tau};$
Taking the natural log of each side, we obtain:
$-1.1 = -t/0.495 \text{ s};$
Solving for t, we obtain: $t = 0.545 \text{ s}$

Practice Exercise 7.4: $\tau = L/R = 2.67 \text{ ms};$
$V_L(0.5 \text{ ms}) = 45 \text{ V}(e^{-0.5 \text{ ms}/2.67 \text{ ms}}) = 37.3 \text{ V};$
$V_L(3 \text{ ms}) = 45 \text{ V}(e^{-3 \text{ ms}/2.67 \text{ ms}}) = 14.6 \text{ V};$

$I_L(0.5 \text{ ms}) = 45 \text{ V}/30 \text{ } \Omega \text{ }(1 - e^{-0.5 \text{ ms}/2.67 \text{ ms}}) = 0.26 \text{ A};$
$I_L(3 \text{ ms}) = 45 \text{ V}/30 \text{ } \Omega \text{ }(1 - e^{-3 \text{ ms}/2.67 \text{ ms}}) = 1 \text{ A};$
$t_{SS} = 5\tau = 13.4 \text{ ms}$

Practice Exercise 7.5:   Amplitude $= 10 \text{ V} - 0 \text{ V} = 10 \text{ V};$ $t_w = 4 \text{ ms} - 0 \text{ ms} = 4 \text{ ms};$
$T = 6 \text{ ms} - 0 \text{ ms} = 6 \text{ ms};$ $f_P = 1/T = 166.7 \text{ kHz};$
$D = t_w / T = 66.7\%$

CHAPTER 8

Practice Exercise 8.1: $(1 \text{ k}\Omega + 2 \text{ k}\Omega + 3 \text{ k}\Omega) I_1 - (2 \text{ k}\Omega) I_2 = 36 \text{ V};$
$-(2 \text{ k}\Omega) I_1 + (3 \text{ k}\Omega + 6 \text{ k}\Omega + 2 \text{ k}\Omega) I_2 = 24 \text{ V}$

Calculator solution: $I_1 = 7.16 \text{ mA}$ and $I_2 = 3.48 \text{ mA};$
$I_{R1} = I_{R2} = I_1 = 7.16 \text{ mA};$ $I_{R4} = I_{R5} = I_2 = 3.48 \text{ mA};$
$I_{R3} = I_1 - I_2 = 3.68 \text{ mA}$

Practice Exercise 8.2: $(1/2 \text{ k}\Omega + 1/2 \text{ k}\Omega) V_1 - (1/2 \text{ k}\Omega) V_2 = 8 \text{ mA};$
$-(1/2 \text{ k}\Omega) V_1 + (1/2 \text{ k}\Omega + 1/8 \text{ k}\Omega) V_2 = 4 \text{ mA};$

Calculator solution: $V_1 = 61.1 \text{ V};$ $V_2 = 55.2 \text{ V};$
$V_{R1} = 61.1 \text{ V}; V_{R3} = 55.2 \text{ V}; V_{R2} = V_1 - V_2 = 13.9 \text{ V}$

Practice Exercise 8.3:   Consider source V_1 (V_2 shorted):
$R_T = [(R_3 + R_4 /\!/ R_2) + R_1] = 4.67 \text{ k}\Omega;$
$I_S = 18 \text{ V}/4.67 \text{ k}\Omega = 3.85 \text{ mA};$
$I_{R1} = I_S = 3.85 \text{ mA} \rightarrow;$

Using the current divider rule:
$I_{R2} = 3.21 \text{ mA} \downarrow$ and $I_{R3} = I_{R4} = 0.64 \text{ mA} \downarrow;$

Consider source V_2 (V_1 shorted):
$R_T = [(R_3 + R_4 // R_1) + R_2] = 4.31 \text{ k}\Omega$;
$I_S = 12 \text{ V}/4.31 \text{ k}\Omega = 2.78 \text{ mA}$;
$I_{R2} = I_S = 2.78 \text{ mA} \uparrow$;

Using the current divider rule:
$I_{R1} = 2.14 \text{ mA} \leftarrow$ and $I_{R3} = I_{R4} = 0.64 \text{ mA} \downarrow$;

Summing respective currents:
$I_{R1} = 3.85 \text{ mA} - 2.14 \text{ mA} = 1.71 \text{ mA} \rightarrow$;
$I_{R2} = 3.21 \text{ mA} + 2.78 \text{ mA} = 0.43 \text{ mA} \downarrow$;
$I_{R3} = I_{R3} = 0.64 \text{ mA} + 0.64 \text{ mA} = 1.28 \text{ mA} \downarrow$

Practice Exercise 8.4:   a. Open R_L and short V_S;
$R_{Th} = (4.7 \text{ k}\Omega + 2 \text{ k}\Omega) // 3.3 \text{ k}\Omega = 2.21 \text{ k}\Omega$;
Replace V_S and using the voltage divider rule:
$V_{Th} = 20 \text{ V}[(4.7 \text{ k}\Omega + 2 \text{ k}\Omega)/(4.7 \text{ k}\Omega + 2 \text{ k}\Omega + 3.3 \text{ k}\Omega)]$
$\quad = 13.4 \text{ V}$;

For 1 kΩ load:
$V_L = 13.4 \text{ V} (1 \text{ k}\Omega/1 \text{ k}\Omega + 2.21 \text{ k}\Omega) = 4.17 \text{ V}$;
For 6 kΩ load:
$V_L = 13.4 \text{ V} (6 \text{ k}\Omega/6 \text{ k}\Omega + 2.21 \text{ k}\Omega) = 9.79 \text{ V}$;

b. Open R_L and short V_S;
$R_{Th} = (6 \text{ k}\Omega // 1 \text{ k}\Omega) + 4 \text{ k}\Omega = 4.86 \text{ k}\Omega$;
Replace V_S and using the voltage divider rule:
$V_{Th} = 24 \text{ V}[6 \text{ k}\Omega/(6 \text{ k}\Omega + 1 \text{ k}\Omega)] = 20.57 \text{ V}$;

For 4 kΩ load:
$V_L = 20.57 \text{ V} [4 \text{ k}\Omega/(4 \text{ k}\Omega + 4.86 \text{ k}\Omega)] = 9.29 \text{ V}$
For 12 kΩ load:
$V_L = 20.57 \text{ V} [12 \text{ k}\Omega/(12 \text{ k}\Omega + 4.86 \text{ k}\Omega)] = 14.64 \text{ V}$

Practice Exercise 8.5:   Open R_L and short V_S;
$R_N = (4.7 \text{ k}\Omega + 2 \text{ k}\Omega) // 3.3 \text{ k}\Omega = 2.21 \text{ k}\Omega$;
Replace V_S and short load terminals:
$I_N = I_S = 20 \text{ V}/3.3 \text{ k}\Omega = 6.06 \text{ mA}$;
Note: Shorting load terminals shorts R_2 and R_3;

For 1 kΩ load using the current divider rule:
$I_L = 6.06 \text{ mA} (0.68 \text{ k}\Omega/1 \text{ k}\Omega) = 4.12 \text{ mA}$;
For 6 kΩ load using the current divider rule:
$I_L = 6.06 \text{ mA} (0.68 \text{ k}\Omega/6 \text{ k}\Omega) = 1.62 \text{ mA}$

Practive Exercise 8.6:   a. Open R_L and short V_S;
$R_{Th} = [(4 \text{ k}\Omega + 3 \text{ k}\Omega) // 6 \text{ k}\Omega] + 1 \text{ k}\Omega$;
$R_{Th} = 4.23 \text{ k}\Omega$;
Replace V_S and using the voltage divider rule:
$V_{R3} = 40 \text{ V}[6 \text{ k}\Omega // (6 \text{ k}\Omega + 4 \text{ k}\Omega + 3 \text{ k}\Omega)] = 18.46 \text{ V}$;
$V_{Th} = 40 \text{ V} - 18.46 \text{ V} = 21.54 \text{ V}$;

For 4 kΩ load:
$I_L = 21.54 \text{ V}/(4 \text{ k}\Omega + 4.23 \text{ k}\Omega) = 2.62 \text{ mA}$;
For 12 kΩ load:
$I_L = 21.54 \text{ V}/(12 \text{ k}\Omega + 4.23 \text{ k}\Omega) = 1.33 \text{ mA}$;

$R_N = R_{Th} = 4.23 \text{ k}\Omega$;

With load terminals shorted, total equivalent resistance:

$R_T = [(R_1 + R_2) // R_4] + R_3 = 6.88 \text{ k}\Omega$;

$I_S = 40 \text{ V}/6.88 \text{ k}\Omega = 5.81 \text{ mA}$;

Using the current divider rule:

$I_N = I_{R4} = 5.81 \text{ mA} (0.88 \text{ k}\Omega/1 \text{ k}\Omega) = 5.11 \text{ mA}$;

For 4 kΩ load: $I_L = 5.11 \text{ mA} (2.01 \text{ k}\Omega/4 \text{ k}\Omega) = 2.57 \text{ mA}$;
For 12 kΩ load:

$I_L = 5.11 \text{ mA} (3.13 \text{ k}\Omega/12 \text{ k}\Omega) = 1.33 \text{ mA}$;

Note that there may be slight differences in answers due to rounding.

b. $(3 \text{ k}\Omega + 6 \text{ k}\Omega + 3 \text{ k}\Omega) I_1 - (3 \text{ k}\Omega) I_2$
$= 6 \text{ V} - 20 \text{ V} = 14 \text{ V}$;

$-(3 \text{ k}\Omega) I_1 + (3 \text{ k}\Omega + 12 \text{ k}\Omega + 1 \text{ k}\Omega) I_2 = 20 \text{ V}$;

Calculator solution: $I_1 = -0.89 \text{ mA}$ and $I_2 = 1.08 \text{ mA}$;
$I_{R4} = I_2 = I_1 = 1.08 \text{ mA}$;

Using superposition:
Consider source V_1 (V_2 shorted):
$R_T = [(R_3 // R_4 + R_5) + R_1 + R_2] = 11.44 \text{ k}\Omega$;
$I_S = 6 \text{ V}/11.44 \text{ k}\Omega = 0.52 \text{ mA}$;
Using the current divider rule:
$I_{R4} = 0.52 \text{ mA} (2.44 \text{ k}\Omega/13 \text{ k}\Omega) = 0.098 \text{ mA} \rightarrow$;

Consider source V_2 (V_1 shorted):
$R_T = [(R_1 + R_2 // R_4 + R_5) + R_3] = 8.32 \text{ k}\Omega$;
$I_S = 20 \text{ V}/8.32 \text{ k}\Omega = 2.40 \text{ mA}$;
Using the current divider rule:
$I_{R4} = 2.40 \text{ mA} (5.32 \text{ k}\Omega/13 \text{ k}\Omega) = 0.98 \rightarrow$;

Summing respective currents:
$I_{R4} = 0.098 \text{ mA} + 0.98 \text{ mA} = 1.08 \text{ mA} \rightarrow$

CHAPTER 9

Practice Exercise 9.1:

a. $P_R = (V_{RMS})^2 / R$; $V_{RMS} = (24 \text{ V})(.0707) = 16.97 \text{ V}$;
$P_R = (16.97 \text{ V})^2 / 2.2 \text{ k}\Omega = 131 \text{ mW}$;

b. $P_R = (I_{RMS})^2 R$; $I_{RMS} = (25 \text{ mA})(0.707) = 17.67 \text{ mA}$;
$P_R = (17.67 \text{ mA})^2 (3.3 \text{ k}\Omega) = 1.03 \text{ W}$;

c. $X_C = 1/(2\pi)(500 \text{ Hz})(0.1 \text{ μF}) = 3.18 \text{ k}\Omega$;
$Q_C = (V_{RMS})^2 / X_C$; $V_{RMS} = (48 \text{ V})(0.707) = 33.94 \text{ V}$;
$Q_C = (33.94 \text{ V})^2 / 3.18 \text{ k}\Omega = 362 \text{ mVAR}$;

d. $X_L = (2\pi)(12 \text{ kHz})(2.5 \text{ mH}) = 188.5 \text{ }\Omega$;
$Q_L = -(I_{RMS})^2 X_L$; $I_{RMS} = (30 \text{ mA})(0.707) = 21.2 \text{ mA}$;
$Q_L = (22.1 \text{ mA})^2 (188.5 \text{ }\Omega) = -84.7 \text{ mVAR}$

Practice Exercise 9.2:

a. $P_R = (V_{RMS})^2 / R$; $V_{RMS} = (10 \text{ V})(0.707) = 7.07 \text{ V}$;
$P_R = (7.07 \text{ V})^2 / 2.2 \text{ k}\Omega = 22.72 \text{ mW}$;
$Q_C = Q_L = S = 0$

b. $\mathbf{Z} = \mathbf{R} + \mathbf{X}_C = 559 \angle -26.6° \text{ }\Omega$;
$\mathbf{I_S} = 6 \angle 0° \text{ V}/559 \angle -26.6° \text{ }\Omega$;

$$\mathbf{I}_s = 10.73 \angle 26.6° \text{ mA};$$
$$I_{RMS} = (10.73 \text{ mA})(0.707) = 7.59 \text{ mA};$$
$$P_R = (I_{RMS})^2 R = (7.59 \text{ mA})^2 (500 \ \Omega) = 28.8 \text{ mW};$$
$$Q_C = (I_{RMS})^2 X_C = (7.59 \text{ mA})^2 (250 \ \Omega) = 14.29 \text{ mVAR};$$
$$S = \sqrt{(28.8 \text{ mW}^2) + (14.29 \text{ mVAR})^2} = 32.15 \text{ mVA};$$

c. $$\mathbf{Z} = \mathbf{R} + \mathbf{X}_L = 1.41 \angle 45° \text{ k}\Omega;$$
$$\mathbf{I}_s = 12 \angle 0° \text{ V} / 1.41 \angle 45° \text{ k}\Omega;$$
$$\mathbf{I}_s = 8.51 \angle -45° \text{ mA};$$
$$I_{RMS} = (8.51 \text{ mA})(0.707) = 6.02 \text{ mA};$$
$$P_R = (I_{RMS})^2 R = (6.02 \text{ mA})^2 (1 \text{ k}\Omega) = 36.24 \text{ mW};$$
$$Q_L = -(I_{RMS})^2 X_C = -(6.02 \text{ mA})^2 (250 \ \Omega)$$
$$= -36.24 \text{ mVAR};$$
$$S = \sqrt{(36.24 \text{ mW}^2) + (-36.24 \text{ mVAR})^2} = 51.25 \text{ mVA};$$

d. $$\mathbf{Z} = \mathbf{R} + \mathbf{X}_C + \mathbf{X}_L = 360.55 \angle 33.7° \ \Omega;$$
$$I_{RMS} = V_{RMS} / Z = 110 \text{ V} / 360.55 \ \Omega = 305.08 \text{ mA};$$
$$P_R = (I_{RMS})^2 R = (305.08 \text{ mA})^2 (300 \ \Omega) = 27.9 \text{ mW};$$
$$Q_C = (I_{RMS})^2 X_C = (305.08 \text{ mA})^2 (700 \ \Omega) = 65.1 \text{ mVAR};$$
$$Q_L = -(I_{RMS})^2 X_L = -(305.08 \text{ mA})^2 (900 \ \Omega)$$
$$= -83.7 \text{ mVAR};$$
$$S = \sqrt{(27.9 \text{ mW})^2 + (-83.7 \text{ mVAR} + 65.1 \text{ mVAR})^2}$$
$$= 33.66 \text{ mVA}$$

Practice Exercise 9.3: Total real power $= 100 \text{ W};$
Total inductive reactive power:
$$Q_L = (-200 \text{ VAR}) + (-150 \text{ VAR}) = -350 \text{ VAR};$$
Total capacitive reactive power:
$$Q_C = 300 \text{ VAR} + 250 \text{ VAR} = 550 \text{ VAR};$$
$$S = \sqrt{(100 \text{ W})^2 + (550 \text{ VAR} - 350 \text{ VAR})^2} = 223.6 \text{ VA}$$

Practice Exercise 9.4: Total real power $= 285 \text{ W} + 310 \text{ W} = 595 \text{ W};$
Total inductive reactive power:
$$Q_L = -(500 \text{ VAR} + 260 \text{ VAR}) = -760 \text{ VAR};$$
Total capacitive reactive power:
$$Q_C = 180 \text{ VAR} + 485 \text{ VAR} = 665 \text{ VAR};$$
$$S = \sqrt{(595 \text{ W})^2 + (-760 \text{ VAR} + 665 \text{ VAR})^2} = 602.5 \text{ VA};$$

PF $=$ Total real power/Apparent power $= 0.988$ lagging

Practice Exercise 9.5: a. $V_{secondary} = 110 \text{ V} (1200/60{,}000) = 2.2 \text{ V};$
$$I_{secondary} = 10 \text{ mA} (60{,}000/1200) = 500 \text{ mA};$$
$$P_{primary} = V_{primary} I_{primary} = (110 \text{ V})(10 \text{ mA}) = 1.1 \text{ W};$$
$$P_{secondary} = V_{secondary} I_{secondary} = (2.2 \text{ V})(500 \text{ A}) = 1.1 \text{ W};$$

b. $V_{secondary} = 110 \text{ V} (150/1) = 16.5 \text{ kV};$
$$I_{secondary} = 2 \text{ kA} (1/150) = 13.33 \text{ A};$$
$$P_{primary} = V_{primary} I_{primary} = (110 \text{ V})(2 \text{ kA}) = 220 \text{ kW};$$
$$P_{secondary} = V_{secondary} I_{secondary} = (16.5 \text{ kV})(13.33 \text{ A}) = 220 \text{ kW}$$

Practice Exercise 9.6: a. $V_S = 300 \text{ mV (peak)} = 212.10 \text{ mV (RMS)};$
$$I_S \text{ (RMS)} = 212.10/208 \ \Omega = 1.02 \text{ mA};$$
$$P_{RL} = I_S \text{ (RMS)}^2 (R_L) = (1.02 \text{ mA})^2 (8 \ \Omega) = 8.32 \ \mu\text{W};$$
$$P_{Rint} = I_S \text{ (RMS)}^2 (R_{int}) = (1.02 \text{ mA})^2 (200 \ \Omega) = 208.8 \ \mu\text{W};$$

b. $Z_P / Z_S = 200\ \Omega / 8\ \Omega = 25 = $ Turns ratio squared;

$TR = \sqrt{25} = 5$;   the transformer is step down 5:1;

The 8 Ω load now "appears" in the primary as 200 Ω;

$V_{\text{primary}} = 300$ mV (peak) $= 212.10$ mV (RMS);

$I_{\text{primary}} = 212.10$ mV (RMS)$/(200\ \Omega + 200\ \Omega)$
$= 0.53$ mA (RMS);

$P_{\text{primary}} = P_{\text{secondary}} = (0.53\ \text{mA})^2\ (200\ \Omega) = 56.18\ \mu\text{W}$;

The power dissipated by the source equals the power delivered to the load at maximum power transfer. Note the increase in power to the load when the impedances are matched using a matching transformer.

Practice Exercise 9.7: Determine the Thevenin equivalent circuit for R_L:

$R_{\text{Th}} = (1\ \text{k}\Omega + 6\ \text{k}\Omega)//10\ \text{k}\Omega = 4.12\ \text{k}\Omega$;

$V_{\text{Th}} = V_{R2} = (10 \angle 0°\ \text{V})[10\ \text{k}\Omega//(10\ \text{k}\Omega + 6\ \text{k}\Omega + 1\ \text{k}\Omega)]$;

$V_{\text{Th}} = 5.88 \angle 0°\ \text{V}$;

$V_{\text{Th(RMS)}} = (5.88\ \text{V})(0.707) = 4.16\ \text{V}$;

The required load for maximum power transfer is

$R_{\text{Th}} = 4.12\ \text{k}\Omega$;

$P_L = V_{\text{Th(RMS)}}^2 / 4\,R_L = (4.16\ \text{V})^2 [4\,(4.12\ \text{k}\Omega)] = 1.05\ \text{mW}$

Practice Exercise 9.8: a. $I_S = 36\ \text{V}/(2.2\ \text{k}\Omega + 1\ \text{k}\Omega) = 11.25\ \text{mA}$;

$P_{R1} = (11.25\ \text{mA})^2\ (1\ \text{k}\Omega) = 126.56\ \text{mW}$
(requires 1/4 W rating);

$P_{R2} = (11.25\ \text{mA})^2\ (2.2\ \text{k}\Omega) = 278.44\ \text{mW}$
(requires 1/2 W rating);

b. $I_S = 80\ \text{V}/(1\ \text{k}\Omega + 2.2\ \text{k}\Omega + 3.3\ \text{k}\Omega) = 12.31\ \text{mA}$;

$P_{R1} = (12.31\ \text{mA})^2\ (1\ \text{k}\Omega) = 189.42\ \text{mW}$
(requires 1/4 W rating);

$P_{R2} = (12.31\ \text{mA})^2\ (2.2\ \text{k}\Omega) = 416.7\ \text{mW}$
(requires 1/2 W rating);

$P_{R3} = (12.31\ \text{mA})^2\ (3.3\ \text{k}\Omega) = 625\ \text{mW}$
(requires a 1 W rating);

Including a 25% safety factor:

$P_{R1} = 151.54\ \text{mW}\,(1.25) = 189\ \text{mW}$ (requires 1/4 W rating);

$P_{R2} = 333.38\ \text{mW}\,(1.25) = 416\ \text{mW}$ (requires 1/2 W rating);

$P_{R3} = 500\ \text{mW}\,(1.25) = 625\ \text{mW}$ (requires 1 W rating)

Practice Exercise 9.9: Refer to Figure 9–39: The power rating at 110° C is 300 mW

CHAPTER 10

Practice Exercise 10.1:

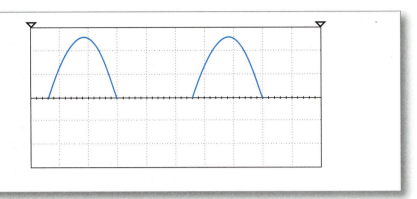

Practice Exercise 10.1a

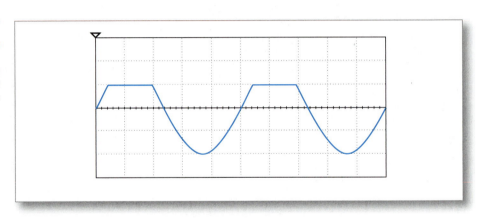

Practice Exercise 10.1b

Practice Exercise 10.2:   The output voltage will be 5 V − 0.5 V = 4.5 V when either input voltage is high (5 V) or when both input voltages are high. The circuit executes the logic OR function.

Practice Exercise 10.3:
$I_B = (V_{CC} - V_{BE}) / R_B = 14.3 \text{ V} / 300 \text{ k}\Omega = 47.67 \text{ }\mu\text{A};$
$I_C = \beta I_B = (90)(47.67 \text{ }\mu\text{A}) = 4.29 \text{ mA};$
$V_{RC} = I_C R_C = (4.29 \text{ mA})(1.5 \text{ k}\Omega) = 6.44 \text{ V};$
$V_{CE} = (V_{CC} - V_{RC}) = 15 \text{ V} - 6.44 \text{ V} = 8.56 \text{ V}$

Practice Exercise 10.4:   Using the voltage divider rule:
$V_B = V_{CC}[R_1 / (R_1 + R_2)] = 3 \text{ V};$
$V_{RE} = V_B - V_{BE} = 3 \text{ V} - 0.7 \text{ V} = 2.3 \text{ V};$
$I_{RE} = V_{RE} / R_E = 2.3 \text{ V} / 500 \text{ }\Omega = 4.6 \text{ mA};$
$I_C = I_E = 4.6 \text{ mA};$
$V_{RC} = I_C R_C = (4.6 \text{ mA})(1.2 \text{ k}\Omega) = 5.52 \text{ V};$
$V_{CE} = (V_{CC} - V_{RC} - V_{RE}) = 12 \text{ V} - 5.52 \text{ V} - 2.3 \text{ V} = 4.18 \text{ V}$

Practice Exercise 10.5:   Saturation occurs when $V_{CE} = 0$;
$I_{C(sat)} = V_{CC} / R_C = 5 \text{ V} / 1 \text{ k}\Omega = 5 \text{ mA};$
$I_{B(sat)} = I_{C(sat)} / \beta = 5 \text{ mA} / 250 = 20 \text{ }\mu\text{A};$
$R_{B(max)} = (V_{in} - V_{BE}) / I_{B(sat)} = 215 \text{ k}\Omega;$

With V_{in} at ground, $I_B = I_C = 0$; $V_{RC} = 0$;
$V_{CE} = V_{CC} = 5$ V

CHAPTER 11

Practice Exercise 11.1: $f_0 = 1/T = 1/3$ ms $= 333.33$ Hz; $\omega_0 = 2\pi f_0 = 2094$ rad/s;

$n = 1$: $\dfrac{4A}{\pi} \displaystyle\sum_{n=1}^{\infty} \dfrac{1}{2n-1} \sin(2n-1)\omega_0 t = \dfrac{(4)(8)}{\pi} \sin(2092t)$

$= 10.2 \sin(2094t)$;

$n = 2$: $\dfrac{4A}{\pi} \displaystyle\sum_{n=1}^{\infty} \dfrac{1}{2n-1} \sin(2n-1)\omega_0 t = \dfrac{(4)(8)}{3\pi} \sin(6282t)$

$= 3.4 \sin(6282t)$

Practice Exercise 11.2:   The DC component $= A/2 = 5$ V$/2 = 2.5$ V;

$f_0 = 1/T = 1/5$ ms $= 200$ Hz; $\omega_0 = 2\pi f_0 = 1257$ rad/s;

$n = 1$: $-\dfrac{A}{\pi} \displaystyle\sum_{1}^{\infty} \dfrac{\sin(n\omega_0 t)}{n} = -\dfrac{5}{\pi} \sin(\omega_0 t)$

$= -1.59 \sin(1257t) = 1.59 \sin(1257t + 180°)$

Practice Exercise 11.3:   The DC component $= 2A/\pi = 2.55$ V;

$f_0 = 1/T = 1/1$ ms $= 1$ kHz; $\omega_0 = 2\pi f_0 = 6283$ rad/s;

$v(t) = \dfrac{2A}{\pi} + \dfrac{4A/\pi}{1-n^2} \cos(n\omega_0)$ for all even values of n;

$n = 2$: $\dfrac{4A/\pi}{1-n^2} \cos(n\omega_0) = 5.1/(1-4)\cos(2\omega_0)$

$= -1.69 \cos(12{,}566t)$

CHAPTER 1

ANSWERS TO ODD-NUMBERED ESSENTIAL LEARNING EXERCISES

1. *Signal*: A signal may be defined as any scheme by which information is transferred from one location to another.
Electronic signal: Electronic signals exist in the form of either a voltage or a current.
Current: Current is defined as the controlled flow of electrical charge.
Voltage: Voltage is defined as the force that sustains the flow of current.

3. Conductors

5. Magnitude; Polarity

7. Current that flows from positive to negative

9. Microphone

11. Oscilloscope
13. A voltage or current that varies in time

15. V_{AB}

17. 10^{-3}; 10^{-6}; 10^{-9}; 10^{-12}

19. 8.5 mA; 26 mV

CHAPTER 1

ANSWERS TO SELECTED CHALLENGING LEARNING EXERCISES

3. Voltage is a measure of the imbalance of charge between two points, one of which is called the reference.

CHAPTER 2

ANSWERS TO ODD-NUMBERED ESSENTIAL LEARNING EXERCISES

1. $4.7 \text{ k}\Omega$

3. $R_T = 12 \text{ k}\Omega$; $I_S = 2 \text{ mA}$; $V_{R1} = 4 \text{ V}$;

   $V_{R2} = 12 \text{ V}$; $V_{R3} = 8 \text{ V}$;
   KVL: $4 \text{ V} + 12 \text{ V} + 8 \text{ V} = 24 \text{ V}$

5. $V_A = 14.4 \text{ V}$

7. $V_{R1} = 27 \text{ V}$; $V_{R2} = 12 \text{ V}$; $V_{R3} = 21 \text{ V}$;
   KVL: $27 \text{ V} + 12 \text{ V} + 21 \text{ V} = 60 \text{ V}$

9. $V_A = 9 \text{ V}$

11. a. $-6 \text{ V} + 10 \text{ V} + 1 \text{ V} - 5 \text{ V} = 0$
    b. $10 \text{ V} + 12 \text{ V} - 8 \text{ V} - 14 \text{ V} = 0$

13. $V_{R1} = 4.8 \text{ V}$; $V_{R2} = 9.6 \text{ V}$; $V_{R3} = 6.4 \text{ V}$; $V_{R4} = 3.2 \text{ V}$;
    KVL: $4.8 \text{ V} + 9.6 \text{ V} + 6.4 \text{ V} + 3.2 \text{ V} = 24 \text{ V}$

15. $R_1 = 3 \text{ k}\Omega$; $R_3 = 2 \text{ k}\Omega$

17. $V_{R1} = 5.17 \text{ V}$; $V_{R2} = 7.76 \text{ V}$;
    $V_{R3} = 11.05 \text{ V}$; $V_A = V_{R3} = -11.05 \text{ V}$

19. $R_T = 3 \text{ k}\Omega$; $G_T = 333.33 \text{ }\mu\text{S}$

21. $I_{R1} = 2 \text{ mA}$; $I_{R2} = 6 \text{ mA}$; $I_{R3} = 4 \text{ mA}$

23. $I = 5 \text{ mA}$ leaving the node

25. $I_{R1} = 12 \text{ mA}$; $I_{R2} = 3 \text{ mA}$;
    KCL: $12 \text{ mA} + 3 \text{ mA} = 15 \text{ mA}$

27. $I_{R1} = 8.87 \text{ mA}$; $I_{R2} = 13.30 \text{ mA}$;
    $I_{R3} = 4.43 \text{ mA}$; $I_{R4} = 13.30 \text{ mA}$;
    KCL: $8.87 \text{ mA} + 13.30 \text{ mA} + 4.43 \text{ mA}$
    $+ 13.30 \text{ mA} = 39.9 \text{ mA}$

29. $I_{R1} = 18 \text{ mA}$; $I_{R2} = 6 \text{ mA}$; $I_{R3} = 3 \text{ mA}$; $I_{R4} = 4.5 \text{ mA}$

CHAPTER 2

ANSWERS TO SELECTED CHALLENGING LEARNING EXERCISES

1. $V_{out} = 8 \text{ V}$

3. $V_{out} = 8 \text{ V}$

5. $R_4 = 2 \text{ k}\Omega$

7. $V_A = -25.1 \text{ V}$; $V_B = -12.32 \text{ V}$

12. $I_X = 7.51 \text{ mA}$

CHAPTER 3

ANSWERS TO ODD-NUMBERED ESSENTIAL LEARNING EXERCISES

1. $R_T = 8\text{ k}\Omega$; $I_S = 5\text{ mA}$;
   $I_{R1} = 5\text{ mA}$; $I_{R2} = 1.67\text{ mA}$; $I_{R3} = 3.33\text{ mA}$;
   KCL: $1.67\text{ mA} + 3.33\text{ mA} = 5\text{ mA}$;
   KVL: $V_S - V_{R1} - V_{R2} = 40\text{ V} - 20\text{ V} - 20\text{ V} = 0$

3. $R_T = 8\text{ k}\Omega$; $I_S = 2\text{ mA}$;
   $I_{R1} = 2\text{ mA}$; $I_{R2} = I_{R3} = 1\text{ mA}$; $I_{R4} = 1\text{ mA}$;
   KCL: $1\text{ mA} + 1\text{ mA} = 2\text{ mA}$;
   KVL: $V_S - V_{R1} - V_{R4} = 16\text{ V} - 6\text{ V} - 10\text{ V} = 0$

5. $R_T = 8\text{ k}\Omega$; $I_S = 6\text{ mA}$;
   $I_{R1} = I_{R6} = 6\text{ mA}$; $I_{R2} = I_{R3} = 3\text{ mA}$; $I_{R4} = I_{R5} = 3\text{ mA}$;
   KCL: $3\text{ mA} + 3\text{ mA} = 6\text{ mA}$;
   KVL: $V_S - V_{R1} - V_{R4} - V_{R5} - V_{R6}$
   $= 48\text{ V} - 18\text{ V} - 18\text{ V} - 6\text{ V} - 6\text{ V} = 0$

7. $R_T = 4.14\text{ k}\Omega$; $I_S = 7.25\text{ mA}$;
   $I_{R1} = 7.25\text{ mA}$; $I_{R2} = 4.26\text{ mA}$; $I_{R3} = 2.97\text{ mA}$;
   KCL: $4.26\text{ mA} + 2.97\text{ mA} = 7.23\text{ mA}$;
   KVL: $30\text{ V} - V_{R1} - V_{R2} = 30\text{ V} - 15.95\text{ V} - 14.06\text{ V} = 0$

9. $I_{R1} = 6\text{ mA}$; $I_{R2} = 0.95\text{ mA}$;
   $I_{R3} = 0.63\text{ mA}$; $I_{R4} = 1.58\text{ mA}$;
   KCL: $0.95\text{ mA} + 0.63\text{ mA} = 1.58\text{ mA}$

11. $V_A = 6\text{ V}$; $V_B = 18\text{ V}$; $V_C = 36\text{ V}$

13. $V_{open} = 7.2\text{ V}$

15. $V_{open} = V_S = 40\text{ V}$

17. $I_{short} = 4\text{ mA}$;   Note: Since R_2 is shorted, R_3 is also shorted.

19. $V_{open} = 0$; The bridge circuit is balanced.

21. $I_{R2} = 0.67\text{ mA}$; $I_{R3} = 1.33\text{ mA}$

23. $I_{12\text{ k}\Omega} = 2.33\text{ mA}$

25.

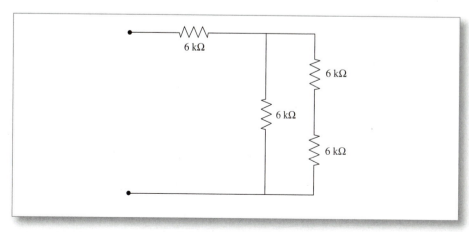

Solution to Exercise 25

CHAPTER 3

ANSWERS TO SELECTED CHALLENGING LEARNING EXERCISES

1. $I_{R1} = I_{R2} = 0.67$ mA; $I_{R3} = I_{R4} = I_{R5} = 0.5$ mA; $I_{R6} = 1.5$ mA

3. $V_{open} = 18$ V $- (-18$ V$) = 36$ V

8. $I_{R1} = 3$ mA; $I_{R2} = 2.25$ mA;
   $I_{R3} = I_{R4} = 0.75$ mA; $I_{R5} = 3$ mA

CHAPTER 4

ANSWERS TO ODD-NUMBERED ESSENTIAL LEARNING EXERCISES

1. V_P (volts); I_P (amperes); f (hertz) ; ω (radians/s); θ (degrees)

3. At t = 1 ms: $V = 3.7 \sin(0.377 \text{ rad}) = 1.36$ V
   At t = 2.4 ms: $V = 3.7 \sin(0.904 \text{ rad}) = 2.91$ V
   At t = 9.4 ms: $V = 3.7 \sin(3.54 \text{ rad}) = -1.44$ V
   At t = 12.62 ms: $V = 3.7 \sin(4.76 \text{ rad}) = -3.58$ V
   At t = 16.67 ms: $V = 3.7 \sin(6.28 \text{ rad}) = -11.79$ V

5. $\mathbf{V_1} = 3.19 \angle 35°$ V; $\mathbf{V_2} = 1.04 \angle 16°$ V

7. $\mathbf{I} = 3.6 \angle 25°$ mA $= 3.26 + j1.52$ mA

9. In time domain: $V_P = 6.5$ V, $T = 12.57$ ms; $\theta = 36°$;
   In phasor domain: $V = 6.5 \angle 36°$ V; Sketches are left as a student exercise.

11. a. $25.35 \angle 81°$;   b. $32.61 \angle 74°$;   c. $13.13 \angle -43°$

13. $v = 14.81 \sin(5500t + 38°)$ V

15. $X_C = 425.53$ Ω; $\mathbf{X_C} = 425.53 \angle -90°$ Ω

17. $i_L = 32 \sin(5000t - 90°)$ mA;   Current lags voltage by 90°.

19. $i_C = 4.85 \sin(2000t + 60°)$ mA; Current leads voltage by 90°.

21. $\mathbf{V_R} = 21.12 \angle -68°$ V: Phasor diagram sketch is left as a student exerise.

23. $\mathbf{I_L} = 11.33 \angle -110°$ mA: Phasor diagram sketch is left as a student exercise.

25. $\mathbf{I_C} = 70 \angle 170°$ μA: Phasor diagram sketch is left as a student exercise.

CHAPTER 4

ANSWERS TO SELECTED CHALLENGING LEARNING EXERCISES

2. $V = 1.43$ V

4. $V_{1P} = 6$ V; $V_{2P} = 4$ V; $T = 1.55$ ms; $f = 654.2$ Hz; $\theta = 18.6°$; All answers are approximate values obtained from the plot in Figure 4–40.

7. MATLAB commands:

```
> t=linspace (0,3*pi/1000,500);
>> v=20*sin(1000*t);
>> plot(t,v)
```

Time	Estimates from Plot	Calculated Values
1.5 ms	20 V	19.95 V
3.9 ms	−13 V	−13.75 V
5.2 ms	−17.5 V	−17.67 V

CHAPTER 5

ANSWERS TO ODD-NUMBERED ESSENTIAL LEARNING EXERCISES

1. $\mathbf{Z} = 7.21\angle -33.7°\ \text{k}\Omega$

3. $\mathbf{Z} = 5.66\angle -45°\ \text{k}\Omega$

5. $\mathbf{R_T} = 16\angle 0°$; $\mathbf{V_{R1}} = 18\angle 0°\ \text{V}$; $\mathbf{V_{R2}} = 6\angle 0°\ \text{V}$

7. $\mathbf{Z} = 3.79\angle 29.7°\ \text{k}\Omega$; $\mathbf{V_R} = 6.96\angle -29.7°\ \text{V}$;
   $\mathbf{V_L} = 3.97\angle 60.3°\ \text{V}$

9. $\mathbf{Z} = 1.28\angle -38.5°\ \text{k}\Omega$; $\mathbf{I_R} = \mathbf{I_L} = 14.06\angle 38.5°\ \text{mA}$;
   $\mathbf{V_R} = 14.06\angle 38.5°\ \text{V}$; $\mathbf{V_L} = 11.19\angle -51.5°\ \text{V}$

11. $\mathbf{Z} = 2.91\angle 75.9°\ \text{k}\Omega$; $\mathbf{Y} = 343.64\angle -75.9°\ \mu\text{S}$;
    $\mathbf{I_L} = 12\angle -90°\ \text{mA}$; $\mathbf{I_R} = 3\angle 0°\ \text{mA}$

13. $\mathbf{Z} = 2.85\angle -18.4°\ \text{k}\Omega$; $\mathbf{Y} = 351.36\angle 18.4°\ \mu\text{S}$;
    $\mathbf{I_R} = 6\angle 0°\ \text{mA}$; $\mathbf{I_C} = 2\angle 90°\ \text{mA}$

15. $\mathbf{Z} = 3.84\angle 1.8°\ \text{k}\Omega$; $\mathbf{I_R} = 5.04\angle -38.7°\ \text{mA}$;
    $\mathbf{I_C} = 3.79\angle 51.3°\ \text{mA}$; $\mathbf{V_R} = 30.24\angle -38.7°\ \text{V}$;
    $\mathbf{V_L} = 30.24\angle -38.7°\ \text{V}$

17. $\mathbf{I_R} = \mathbf{I_C} = 2.02\angle 45°\ \text{mA}$; $\mathbf{I_L} = 4\angle -90°\ \text{mA}$;
    $\mathbf{V_R} = 14.14\angle 45°\ \text{V}$; $\mathbf{V_C} = 14.14\angle -45°\ \text{V}$;
    $\mathbf{V_L} = 20\angle 0°\ \text{V}$

19. $\mathbf{V_{out}} = 5.68\angle 18.4°\ \text{V}$

21. $\mathbf{V_R} = 6.75\angle -32.5°\ \text{V}$; $\mathbf{V_C} = 4.29\angle -122.5°\text{V}$;
    $\mathbf{V_L} = 8.59\angle 57.5°\ \text{V}$; $\mathbf{V_{AB}} = 8\angle -64.9°\ \text{V}$

CHAPTER 5

ANSWERS TO SELECTED CHALLENGING LEARNING EXERCISES

1. $\mathbf{I_X} = 3.08\angle -90°\ \text{mA}$

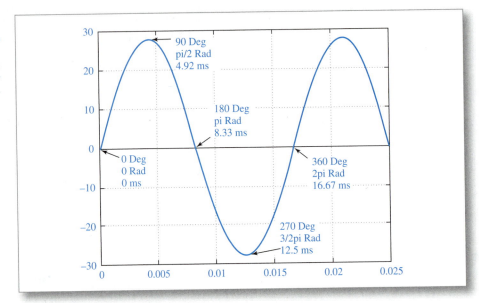

Sketch for Challenging Learning Exercise 6

3. $\mathbf{Z} = 11.1\angle 26.6° \ \Omega$; $\mathbf{V_R} = 10.81\angle -26.6°$ V;
$\mathbf{V_C} = 21.62\angle -116.6°$ V; $\mathbf{V_L} = 27.03\angle 67.8°$ V

6. MATLAB commands:

```
>> t=linspace(0,3*pi/377,500);
>> i= 30*sin(377*t);
>> plot(t,i)
```

CHAPTER 6

ANSWERS TO ODD-NUMBERED ESSENTIAL LEARNING EXERCISES

1. $\mathbf{TF} = \mathbf{V_{out}}/\mathbf{V_{in}} = \mathbf{R_1}/(\mathbf{R_1} + \mathbf{R_2})$

3. $V_{out} = 9.91$ V (peak); $f = 500$ Hz

5. a. $f_C = 1.59$ kHz; $A_V = 0.707$; $A_{V(dB)} = -3$ dB; $\theta = -45°$;
   b. $V_{out} = 4.24$ V (peak);
   c. $\mathbf{TF} = \mathbf{V_{out}}/\mathbf{V_{in}} = \mathbf{X_C}/(\mathbf{X_C} + \mathbf{R})$;
   d. $A_V = 0.468$; $A_{V(dB)} = -6.59$ dB; $\theta = -62.1°$
   At $f = 3$k Hz: $\mathbf{TF} = 0.468\angle -62.1°$;

7. a. $f_C = 2.41$ kHz; $A_V = 0.707$; $A_{V(dB)} = -3$ dB; $\theta = 45°$;
   b. $V_{out} = 2.83$ V (peak);

   c. $\mathbf{TF} = \mathbf{V_{out}}/\mathbf{V_{in}} = \mathbf{R}/(\mathbf{R} + \mathbf{X_C})$;
   d. At $f = 10$ kHz: $\mathbf{TF} = 0.972\angle 13.6°$;
   $A_V = 0.972$; $A_{V(dB)} = -0.25$ dB; $\theta = 13.6°$

9. a. $f_C = 530.52$ Hz; $A_V = 0.707$;
   $A_{V(dB)} = -3$ dB; $\theta = 45°$;
   b. $V_{out} = 4.24$ V (peak);

c. $\mathbf{TF} = \mathbf{V_{out}}/\mathbf{V_{in}} = \mathbf{X_L}/(\mathbf{X_L}+\mathbf{R})$;

d. At $f = 2$ kHz: $\mathbf{TF} = 0.256\angle -75.1°$;

$A_V = 0.256$; $A_{V(dB)} = -11.8$ dB; $\theta = -75.1°$

11. a. $f_C = 1$ Hz; $A_V = 0.707$; $A_{V(dB)} = -3$ dB; $\theta = -45°$;

b. $V_{out} = 10.61$ V (peak);

c. $\mathbf{TF} = \mathbf{V_{out}}/\mathbf{V_{in}} = \mathbf{R}/(\mathbf{R}+\mathbf{X_L})$;

d. At $f = 3$ kHz: $\mathbf{TF} = 0.949\angle 18.3°$;

$A_V = 0.949$; $A_{V(dB)} = -0.45$ dB; $\theta = 18.3°$

13. a. $f_r = 5.03$ kHz

b. $Q = 10.53$

c. $BW = 477.68$ Hz

d. $f_{HC} = 5.27$ kHz; $f_{LC} = 4.79$ kHz

e. The Bode diagram sketch is left as a student exercise.

f. Computer simulation is left as a student exercise.

15. a. $f_r = 4.59$ kHz;

b. $Q = 231.12$;

c. $BW = 19.86$ Hz;

d. $f_{HC} = 4.6$ kHz; $f_{LC} = 4.58$ kHz; Note the narrow bandwidth for the parallel resonant band-pass filter.

17. $f_r = 1.99$ kHz

19. $f_C = 1/(2\pi RC)$

21. $C = 0.22$ μF (given); $R = 1.2$ kΩ (calculated)

23. $C = 0.1$ μF (given); $R = 1.99$ kΩ (calculated)

25. $R = 65$ Ω (given); $L = 34$ mH (calculated)

27. $R = 25$ Ω (given); $L = 26.53$ mH (calculated)

CHAPTER 6

ANSWERS TO SELECTED CHALLENGING LEARNING EXERCISES

1. a. At point a: $TF = R_1/(R_1 + R_2 + R_3)$;

b. $A_V = 0.46$; $A_{V(dB)} = -6.4$ dB;

c. At point b: $TF = (R_1 + R_2)/(R_1 + R_2 + R_3)$;

d. $A_V = 0.78$; $A_{V(dB)} = -2.16$ dB

2. a. $\mathbf{TF} = \mathbf{V_{out}}/\mathbf{V_{in}} = \mathbf{X_C}/(\mathbf{X_C}+\mathbf{R})$;

b. At 3 kHz: $TF = 0.197\angle -78.6°$;

$$A_V = 0.197; \quad A_{V(dB)} = -14.1 \text{ dB};$$

c. $\theta = -78.6°$

4. Note: The design component values will vary depending on the value of the arbitrarily selected component values.

   Let $R = 1 \text{ k}\Omega$ (arbitrary selection); $C = 63.6$ nF (calculated value)

8. Note: The design component values will vary depending on the value of the arbitrarily selected component values.

   Let $C = 0.1$ µF (arbitrary selection); $L = 10.1$ mH (calculated); $R = 31.5 \, \Omega$ (calculated)

CHAPTER 7

ANSWERS TO ODD-NUMBERED ESSENTIAL LEARNING EXERCISES

1. $v_C = V_S (1 - e^{-t/\tau}); \quad i_C = \dfrac{V_S}{R}(e^{-t/\tau}); \quad$ where $\tau = RC$

3. $\tau = RC = 9.4$ ms;
   At $t = 3$ ms: $v_C = 6.56$ V; $i_C = 3.71$ mA;
   At $t = 10$ ms: $v_C = 15.72$ V; $i_C = 1.76$ mA;
   At $t = 15$ ms: $v_C = 19.13$ V; $i_C = 1.03$ mA;
   The sketch is left as a student exercise.

5. $t_{SS} = 5\tau = 5(9.4 \text{ ms}) = 47$ ms; $\quad i_C$ (steady state) $= 0$

7. $\tau = RC = 6.6$ ms;
   At $t = 3.5$ ms: $v_C = 4.94$ V; $\quad v_R = 7.06$ V;
   KVL: 4.94 V + 7.06 V = 12 V;
   At $t = 6$ ms: $v_C = 7.17$ V; $\quad v_R = 4.83$ V;
   KVL: 7.17 V + 4.83 V = 12 V;
   The sketch is left as a student exercise.

9. $\tau = RC = 8$ ms;
   At $t = 4$ ms: $v_C = 30.32$ V; $\quad i_C = 15.16$ mA;
   At $t = 12$ ms: $v_C = 11.16$ V; $\quad i_C = 5.58$ mA;
   The sketch is left as a student exercise.

11. At 4 ms: $v_R = 30.32$ V; KVL: 30.32 V − 30.32 V = 0 V
    At 12 ms: $v_R = 11.16$ V; KVL: 11.16 V − 11.16 V = 0 V

13. $v_L = V_S(e^{-t/\tau}); \quad i_L = \dfrac{V_S}{R}(1 - e^{-t/\tau}); \quad$ where $\tau = L/R$

15. $\tau = L/R = 5$ ms;
    At $t = 4$ ms: $v_L = 16.18$ V; $\quad i_L = 1.98$ A;
    At $t = 13$ ms: $v_L = 2.67$ V; $\quad i_L = 3.33$ A

17. $t_{SS} = 5\tau = 5(5 \text{ ms}) = 25$ ms; $\quad i_L$ (steady state) $= 3.6$ A

19. $\tau = L/R = 250$ µs;
    At $t = 100$ µs: $v_L = 6.70$ V; $\quad v_R = 3.3$ V;
    KVL: 6.70 V + 3.3 V = 10 V;

At $t = 500$ μs: $v_L = 1.35$ V; $v_R = 8.65$ V;
KVL: 1.35 V $+ 8.65$ V $= 10$ V

21. $\tau = L/R = 600$ μs;
At $t = 350$ μs: $v_L = 1.40$ V; $i_L = 27.90$ mA;
At $t = 900$ μs: $v_L = 0.56$ V; $i_L = 1.16$ mA

23. At $t = 350$ μs: $v_R = 1.40$ V;   KVL: 140 V $- 1.40$ V $= 0$;
At $t = 900$ μs: $v_R = 0.56$ V;   KVL: 0.56 V $- 0.56$ V $= 0$

25. Connected in series: $1/C_T = 1/C_1 + 1/C_2 + 1/C_3 = 3.33$ μF;
Connected in parallel: $C_T = C_1 + C_2 + C_3 = 30$ μF

27. $T = 5$ μs; $f = 200$ kHz; $t_w = 2$ μs; $D = 40\%$

29. The sketch is left as a student exercise. Refer to Figure 7–48.

31. The sketch is left as a student exercise. Refer to Figure 7–48.

CHAPTER 7

ANSWERS TO SELECTED CHALLENGING LEARNING EXERCISES

1. $\tau = RC = 1.5$ s;
12 V $= 24$ V $(1 - e^{t/1.5\ s})$;
$0.5 = (1 - e^{t/1.5\ s})$;
$-0.5 = -e^{t/1.5\ s}$;
$\ln(0.5) = t/1.5$ s (rule of logarithms);
$-0.693 = -(t/1.5\ s)$;
$t = 1.04$ s

3. Position 1:

At $t = 5$ ms: $v_C = 9.29$ V; At $t = 8$ ms: $v_C = 12.64$ V; At $t = 10$ ms: $v_C = 14.27$ V;

Position 2:

At $t = 15$ ms: $v_C = 9.41$ V; At $t = 25$ ms: $v_C = 4.09$ V;
The sketch is left as a student exercise.

CHAPTER 8

ANSWERS TO ODD-NUMBERED ESSENTIAL LEARNING EXERCISES

1. $I_S = 20$ V/4 kΩ $= 5$ mA; $R_{int} = 4$ kΩ (series);
For 3 kΩ load:

Considering the voltage source:
$I_L = 20$ V/(4 kΩ $+ 3$ kΩ) $= 2.86$ mA;
Considering the current source:
$I_L = 5$ mA $(1.71$ kΩ/3 kΩ) $= 2.85$ mA (current divider rule)

3. $V_S = 56$ V; $R_{int} = 2.8$ kΩ (series);
For 3 kΩ load:

Considering the voltage source:
$I_L = 56 \text{ V}/(2.8 \text{ k}\Omega + 1.5 \text{ k}\Omega) = 13.07 \text{ mA}$;
Considering the current source:
$I_L = 20 \text{ mA}(0.98 \text{ k}\Omega/1.5 \text{ k}\Omega)$
$= 13.02 \text{ mA}$ (current divider rule)

5. $\mathbf{I}_S = 3\angle - 26° \text{ mA}$; $\mathbf{Z}_{int} = 4\angle 26° \text{ k}\Omega$ (parallel)

7. $V_S = 63.96\angle - 23°$; $\mathbf{Z}_{int} = 3.9\angle - 23° \text{ k}\Omega$ (series)

9. $I_{R1} = 4.15 \text{ mA}$; $I_{R2} = 0.92 \text{ mA}$; $I_{R3} = 3.23 \text{ mA}$

11. $V_1 = -34.16 \text{ V}$; $V_2 = -0.49 \text{ V}$ (node voltages);
$V_{R1} = 34.16 \text{ V}$; $V_{R2} = 33.67 \text{ V}$; $V_{R3} = 0.49 \text{ V}$

13. $I_{R2} = 2.67 \text{ mA}$

15. $R_{Th} = 14.31 \text{ k}\Omega$; $V_{Th} = 12.31 \text{ V}$; $I_{Load} = 0.67 \text{ mA}$

17. $R_N = 1.64 \text{ k}\Omega$; $I_N = 4 \text{ mA}$; $I_{Load} = 1.07 \text{ mA}$

19. $I_{R2} = 2.56 \text{ mA}$

CHAPTER 8

ANSWERS TO SELECTED CHALLENGING LEARNING EXERCISES

1. $\mathbf{I}_L = 0.0187\angle - 54.6° \text{ mA}$

5. $\mathbf{Z}_{Th} = 1.90\angle - 71.6° \text{ k}\Omega$; $\mathbf{V}_{Th} = 6.32\angle - 71.6° \text{ V}$;
$\mathbf{I}_{Load} = 1.76\angle - 41.4° \text{ mA}$

CHAPTER 9

ANSWERS TO ODD-NUMBERED ESSENTIAL LEARNING EXERCISES

1. $P_R = 8.99 \text{ mW}$;   reactive power $= 0$

3. $Q_C = 677.76 \text{ mVAR}$;   real power $= 0$

5. $Q_L = -836.65 \text{ mVAR}$;   real power $= 0$

7. $P_R = 8.37 \text{ mW}$; $Q_C = 8.37 \text{ mVAR}$; $S = 11.84 \text{ mVA}$

9. $P_R = 3.59 \text{ mW}$; $Q_L = -10.77 \text{ mVAR}$; $S = 11.35 \text{ mVA}$

11. $P_R = 29.03 \text{ mW}$; $Q_C = 116.1 \text{ mVAR}$;
$Q_L = -58.06 \text{ mVAR}$; $S = 64.89 \text{ mVA}$

13. The power triangle sketch is left as a student exercise.

15. $PF = P_R/S = 29.03 \text{ mW}/64.89 \text{ mVA} = 0.447$ lagging

17. Total real power $= 13 \text{ W}$; Total inductive reactive power $= 8 \text{ VAR}$;
Total capacitive reactive power $= 12 \text{ VAR}$; $S = 13.60 \text{ VA}$

19. Step-down transformer

21. $V_{secondary} = 5.5 \text{ V}$; $I_{secondary} = 240 \text{ mA}$

23. $P_{primary} = (110 \text{ V})(12 \text{ mA}) = 1.32 \text{ W}$;

$P_{scondary} = (5.5 \text{ V})(240 \text{ mA}) = 1.32 \text{ W}$

25. For maximum power transfer:
$R_L = R_{int} = 2$ kΩ; $P_L = V^2/4R_L = 112.5$ mW

27. $R_{Th} = 3$ kΩ; $V_{Th} = 15$ V;
Maximum power $P_L = (V_{Th})^2/(4R_L) = 18.75$ mW

29. $P_L = 14.71$ mW, Hint: Set $\mathbf{Z_L}$
equal to the complex conjugate of $\mathbf{Z_{int}}$.

31. $TR = 10:1$

33. Standard power ratings:
$R_1 = 1/16$ W; $R_2 = 1/8$ W; $R_3 = 1/16$ W; $R_4 = 1/16$ W

35. With 25% safety factor:
$R_1 = 1/8$ W; $R_2 = 1/8$ W; $R_3 = 1/16$ W; $R_4 = 1/16$ W

37. Refer to the power derating curve in figure 9–39.
Maximum power dissipation at $T = 110°C = 0.3$ W $= 300$ mW

CHAPTER 9

ANSWERS TO SELECTED CHALLENGING LEARNING EXERCISES

2. $S = 247.62$ VA; $PF = 0.589$ lagging

4. Let $\mathbf{Z} = M \angle \theta$, so the complex conjugate is $\overline{Z} = M\angle - \theta$;
Converting from polar form to rectangular form:
$\mathbf{Z} = M \cos \theta + j\sin \theta$ and $\overline{Z} = M \cos (-\theta) + j\sin (-\theta)$

Using the identities: $\sin (-\theta) = -\sin \theta$ and $\cos (-\theta) = \cos \theta$,
the imaginary parts of Equation (1) cancel, so that:
$\mathbf{Z} + \overline{Z} = 2M \cos\theta$, which is a pure real number

8. With $Q_C = Q_L$: $S = \sqrt{P_R{}^2 + 0} = P_R$: $PF = P_R/S = P_R/P_R = 1$;
or: With $Q_C = Q_L$, then θ in the power triangle $= 0$, so that $PF = \cos 0° = 1$

11. $P_{Rint} = 884.45$ μW; $P_L = 176.89$ μW

14. $R_L = R_{Th} = 4$ kΩ; $P_L = 16$ mW

CHAPTER 10

ANSWERS TO ODD-NUMBERED ESSENTIAL LEARNING EXERCISES

1. The anode must be positive with respect to the cathode to forward bias the diode. The minimum forward bias voltage for the silicon diode is 0.7 V, and the minimum forward bias voltage for the germanium diode is 0.3 V.

3. a. Forward bias; b. Reverse bias; c. Forward bias

5. The peak value of the output waveform is 23.3 V.

7. The peak value of the output waveform is -10 V.

An Electronic Signal
Waveform Simulated using
MultiSIM

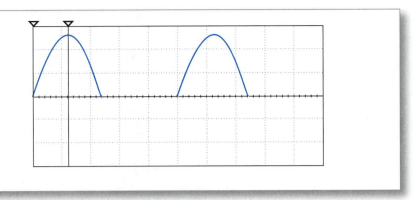

MultiSIM plot for Exercise 5

A Voltage Source Causing
Conventional Current Flow
through a Conductor

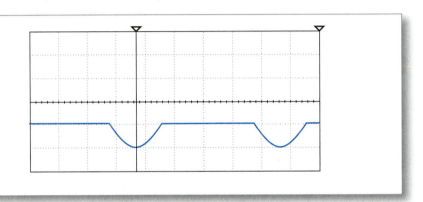

MultiSIM plot for Exercise 7

9. $V_{out} = 5$ V

11. $I_C = 2.26$ mA; $V_{CE} = 4.54$ V

13. $I_C = 5.3$ mA; $V_{CE} = 4.75$ V

15. $R_C = 2.8$ kΩ; $R_B = 2.8$ kΩ

17. a. $V_{out} = 5$ V; b. $I_{C(sat)} = 2.5$ mA; $I_{B(sat)} = 12.5$ μA

19. $A_V = -8$; $V_{out} = -8$ V

21. $A_V = 9$; $V_{out} = 7.2$ V

CHAPTER 10

ANSWERS TO SELECTED CHALLENGING LEARNING EXERCISES

2. $I_C = 5.09$ mA; $V_{CE} = 1.71$ V

4. $V_{out} = 20$ V; The input and output signals are in phase. The output signal has the same frequency as the input signal = 2 kHz.

5. $V_{out} = -9$ V; The circuit is a summing amplifier.

CHAPTER 11

ANSWERS TO ODD-NUMBERED ESSENTIAL LEARNING EXERCISES

1. The area above the time axis equals the area below the time axis over one cycle, so that the average value over one cycle is 0.

3. A harmonic frequency is an integer multiple of the fundamental frequency.

5. A frequency spectrum diagram indicates the amplitude of each Fourier series term at each respective frequency.

7. $f_0 = 200$ Hz; The sketch of the waveform is left as a student exercise.

9. A rectified sine wave consists of either only the positive or only the negative values of the input sine wave.

11. $DC = 0$; $n = 1$: $v = 5.09 \sin(3141.6t)$ V;
    $n = 2$: $v = 1.69 \sin(9424.8t)$ V

13. $DC = 6$ V; $n = 1$: $v = 3.82 \sin(125.7\,kt - 180°)$ V;
    $n = 2$: $v = 1.91 \sin(251.3\,kt - 180°)$ V

15. $DC = 1.91$ V; $n = 2$: $v = -1.27 \cos(25,313.2t)$ V;
    $n = 4$: $v = -0.25 \cos(5025.4t)$ V

17. First even harmonic = 1 kHz; second even harmonic = 2 kHz

19. The frequency spectrum sketch is left as a student exercise. Refer to Figure 11–6.

CHAPTER 11

ANSWERS TO SELECTED CHALLENGING LEARNING EXERCISES

1. $DC = 0$; $n = 1$: $v = 10.18 \sin(166.7t)$ V;
   $n = 2$: $v = 3.39 \sin(500t)$ V;
   $n = 3$: $v = 2.04 \sin(833.5t)$ V;
   $n = 4$: $v = 1.46 \sin(1166.9t)$ V

2. $DC = 0$; First term of the output signal:
   $v = 10.76 \sin(166.7t - 5.7°)$;
   Second term of output signal: $v = 3.25 \sin(500t - 16.7°)$

APPENDIX B

SOURCE CONVERSIONS

Electronic signals may be in the form of a voltage or current waveform. If the impedance of a particular circuit is high, the input signal source is typically represented as a voltage source. If the circuit impedance is low, the input signal source is typically represented as a current source. *Any signal source may be represented as either a voltage or current source.*

It is possible to convert a voltage source into an equivalent current source or to convert a current source into an equivalent voltage source. The current and voltage for a given load are identical for either equivalent source representation.

The prototype representations of a DC voltage and current source and a voltage and current source in the phasor domain are illustrated in Figure B–1.

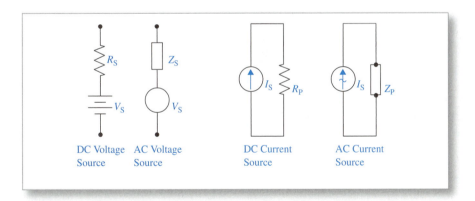

FIGURE **B-1**

A practical voltage source will always possess some series internal resistance or impedance, and a practical current source will always possess some parallel internal resistance or impedance. The internal series resistance R_S and the internal series impedance Z_S are associated with voltage sources, as indicated in Figure B–1. The internal parallel resistance R_P and the internal parallel impedance Z_P are associated with current sources in Figure B–1.

Connecting a given load resistor R_L to the DC voltage and DC current source in Figure B–1, we obtain the circuits shown in Figure B–2.

A P P E N D I X

FIGURE **B-2**

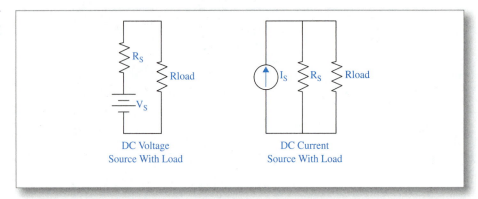

DC Voltage
Source With Load

DC Current
Source With Load

The current through the load from the **voltage source** may be determined using Ohm's law applied to the series circuit as:

$$I_{load} = \frac{V_S}{(R_S + R_L)}$$

The current through the load from the **current source** may be determined using the current divider rule as:

$$I_{load} = I_S \frac{R_P \; // \; R_L}{R_L}$$

Applying the product-over-sum method to determine the parallel resistance R_P, we obtain:

$$I_{load} = I_S \frac{R_P R_L}{(R_P + R_L) R_L} = I_S \frac{R_P}{(R_P + R_L)}$$

If we consider the sources to be equivalent, then the load current from the sources must be equal, so that:

$$I_{load} = \frac{V_S}{(R_S + R_L)} = \frac{I_S R_P}{(R_P + R_L)} \tag{B.1}$$

The expressions for I_{Load} in Equation (B.1) will be equal when the numerator and denominator of both expressions are equal. Setting the respective numerators and denominators equal to each other respectively, we obtain:

$$V_S = I_S R_P \tag{B.2}$$

and
$$(R_S + R_L) = (R_P + R_L) \tag{B.3}$$

Equations (B.2) and (B.3) indicate that the voltage source and current source will be equivalent when:

$$V_S = I_S R_P$$

and
$$R_S = R_P \tag{B.4}$$

Equations (B.2) and (B.3) also indicate that the voltage source and current source will be equivalent when:

$$I_S = \frac{V_S}{R_S}$$

and
$$R_S = R_P \tag{B.5}$$

The equivalent sources are illustrated in Figure B–3.

FIGURE **B-3**

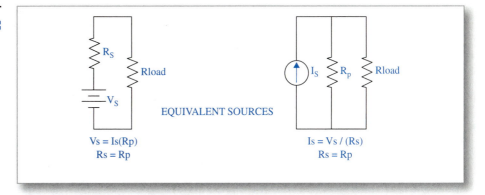

FIGURE B-3 diagram:

Vs = Is(Rp)
Rs = Rp

Is = Vs / (Rs)
Rs = Rp

EQUIVALENT SOURCES

EXAMPLE B-1   Converting a DC Voltage Source to an Equivalent DC Current Source

Convert a 12 V voltage source with an internal resistance of 2 kΩ to an equivalent current source. Verify that the two sources are equivalent given a load resistance of 3 kΩ.

✔ **SOLUTION**

Equation (B.5) may be used to determine I_S and R_P for the equivalent current source as:

$$I_S = \frac{V_S}{R_S} = \frac{12\ \text{V}}{2\ \text{k}\Omega} = 6\ \text{mA}$$

and
$$R_P = R_S = 2\ \text{k}\Omega$$

The specified voltage source and equivalent current source are shown in Figure B–4.

FIGURE **B-4**

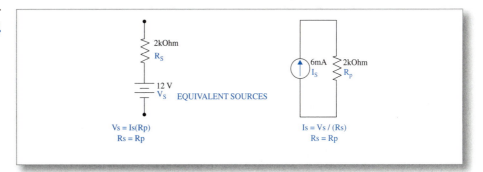

Vs = Is(Rp)
Rs = Rp

Is = Vs / (Rs)
Rs = Rp

EQUIVALENT SOURCES

Figure B–5 shows the equivalent sources shown in Figure B–4 with a 3 kΩ load resistor R_L connected to each source.

FIGURE **B-5**

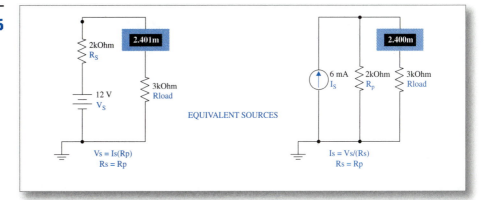

Vs = Is(Rp)
Rs = Rp

Is = Vs/(Rs)
Rs = Rp

EQUIVALENT SOURCES

The current through the 3 kΩ load resistor from the voltage source may be calculated using Ohm's law as:

$$I_{\text{load}} = \frac{V_S}{R_S + R_L} = \frac{12\ \text{V}}{2\ \text{k}\Omega + 3\ \text{k}\Omega} = 2.4\ \text{mA}$$

The current through the 3 kΩ load resistor from the current source may be calculated using the current divider rule as:

$$I_{\text{load}} = I_S \frac{R_P \mathbin{/\!/} R_L}{R_L}$$

where:

$$R_P \mathbin{/\!/} R_L = \frac{R_P R_L}{(R_P + R_L)} = 1.2\ \text{k}\Omega$$

so that:

$$I_{\text{load}} = 6\ \text{mA} = \frac{1.2\ \text{k}\Omega}{3\ \text{k}\Omega} = 2.4\ \text{mA}$$

Note that the load current through RL from both sources has the same magnitude and direction.

EXAMPLE B-2   Converting a Current Source to an Equivalent Voltage Source

Given a current source of 3 mA with an internal resistance of 6 kΩ, determine the equivalent voltage source and verify that the load current to a 4 kΩ load is equal for both sources.

✓ SOLUTION

The voltage of the equivalent voltage source may be determined using Equation (B.4) as:

$$V_S = I_S R_P = (3\ \text{mA})(6\ \text{k}\Omega) = 18\ \text{V}$$

The internal resistance of the voltage source can be determined as:

$$R_S = R_P = 6\ \text{k}\Omega$$

The equivalent sources with the specified 4 kΩ load are shown in Figure B–6.

FIGURE **B-6**

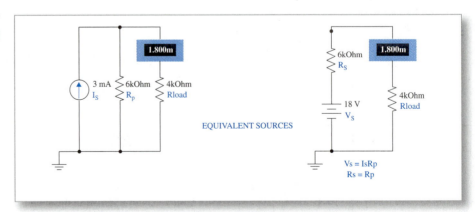

EQUIVALENT SOURCES

The load current from the *voltage source* may be calculated using Ohm's law as:

$$I_L = \frac{V_S}{R_S + R_L} = \frac{18\ \text{V}}{6\ \text{k}\Omega + 4\ \text{k}\Omega} = 1.8\ \text{mA}$$

And the load current from the *current source* can be calculated using the current divider rule as:

$$I_{\text{load}} = I_S \frac{R_P /\!/ R_L}{R_L}$$

where:

$$R_P /\!/ R_L = \frac{R_P R_L}{(R_P + R_L)} = 2.4 \text{ k}\Omega$$

so that:

$$I_{\text{load}} = 3 \text{ mA} \frac{2.4 \text{ k}\Omega}{4 \text{ k}\Omega} = 1.8 \text{ mA}$$

The load current has the same magnitude and direction for either of the equivalent sources.

Sinusoidal source conversions are accomplished in a similar manner in the phasor domain by substituting Z_P for R_P and Z_S for R_S. The voltage and current must be represented in phasor notation.

EXAMPLE B-3   Source Conversion in the Phasor Domain

1. Convert a sinusoidal voltage source $\mathbf{V_S} = 12 \angle 0°$ V with an internal series impedance $\mathbf{Z_S} = 2.5 \angle 45°$ kΩ to an equivalent current source.

✔ **SOLUTION**

$$\mathbf{I_S} = \frac{\mathbf{V_S}}{\mathbf{Z_S}} = \frac{12 \angle 0° \text{ V}}{2.5 \angle 45° \text{ k}\Omega} = 4.8 \angle -45° \text{ mA}$$

and

$$\mathbf{Z_P} = \mathbf{Z_S} = 2.5 \angle 45° \text{ k}\Omega$$

2. Convert a current source $\mathbf{I_S} = 3.2 \angle 30°$ mA with a parallel internal impedance of $\mathbf{Z_P} = 2.1 \angle 75°$ kΩ to an equivalent voltage source.

✔ **SOLUTION**

$$\mathbf{V_S} = \mathbf{I_S} \mathbf{Z_P} = (3.2 \angle 30° \text{ mA})(2.1 \angle 75° \text{ k}\Omega) = 6.72 \angle 105° \text{ V}$$

and

$$\mathbf{Z_S} = \mathbf{Z_P} = 2.1 \angle 75° \text{ k}\Omega$$

MATHEMATICAL DERIVATION OF THE MAXIMUM POWER TRANSFER THEOREM

Every practical voltage source can be modeled as an ideal voltage source in series with some internal impedance. Signal generators, DC power supplies, and all other voltage sources contain some internal impedance. Consider the practical DC voltage source shown in Figure B–7 with an arbitrary load R_L connected.

FIGURE **B–7**

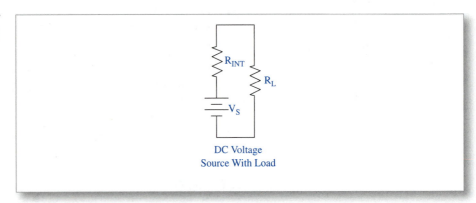

DC Voltage
Source With Load

The total equivalent resistance of the circuit in Figure B–7 may be calculated as:

$$R_T = R_{int} + R_L$$

Ohm's law may be used to determine the load current as:

$$I_L = \frac{V}{R_T} = \frac{V}{R_{int} + R_L}$$

The power dissipated by the load may be expressed as:

$$P_L = I_L^2 R_L = \left(\frac{V}{R_{int} + R_L}\right)^2 R_L$$

A fundamental notion of calculus defines the derivative of a function as the slope of a line tangent to the curve that describes the function. A bit of reasoning would convince us that the slope of a curve at its maximum point would be zero. That is to say, the tangent to the curve at its maximum point would be a horizontal line, which has a slope of zero. We may use this reasoning to determine the condition for maximum power transfer.

We have determined that the power dissipated by a load resistor in Figure B–7 may be expressed as:

$$P_L = \left(\frac{V}{R_S + R_L}\right)^2 R_L$$

Using MATLAB, we may determine the derivative of P_L with respect to R_L to obtain:

```
EDU≫ syms RL V Rs PL;
EDU≫ PL=[V/(RL+Rs)]^2*RL;
EDU≫; diff (PL,RL)
```

$-2*V^2/(RL+Rs)^3*RL+V^2/(RL+Rs)^2$

Setting the expression equal to zero to determine the condition for maximum power, we obtain:

$$\frac{-2V^2 R_L}{(R_L + R_S)^3} + \frac{V^2}{(R_L + R_S)^2} = 0$$

Dividing both numerators by V^2 and dividing both denominators by $(R_L + R_S)^2$, we obtain:

$$\frac{-2R_L}{R_L + R_S} + 1 = 0$$

so that

$$-2R_L + R_L + R_S = 0$$

and

$$R_L = R_S$$

We have now shown mathematically that maximum power is delivered to a load when the load resistance is equal to the internal resistance of the source. The maximum power transfer theorem may be applied in the phasor domain so that maximum power is transferred when the load impedance is equal to the impedance of the voltage source.

INDEX